Statistical Theory and Inference

David J. Olive

Statistical Theory and Inference

Springer

David J. Olive
Department of Mathematics
Southern Illinois University
Carbondale, IL, USA

ISBN 978-3-319-37589-2 ISBN 978-3-319-04972-4 (eBook)
DOI 10.1007/978-3-319-04972-4
Springer Cham Heidelberg New York Dordrecht London

Mathematics Subject Classification (2010): 62 01, 62B05, 62F03, 62F10, 62F12

Softcover reprint of the hardcover 1st edition 2014

Printed on acid-free paper

Springer is part of Springer Science+Business Media (www.springer.com)

Preface

Many math and some statistics departments offer a one-semester graduate course in statistical theory using texts such as Casella and Berger (2002), Bickel and Doksum (2007) or Mukhopadhyay (2000, 2006). The course typically covers minimal and complete sufficient statistics, maximum likelihood estimators (MLEs), method of moments, bias and mean square error, uniform minimum variance unbiased estimators (UMVUEs) and the Fréchet–Cramér–Rao lower bound (FCRLB), an introduction to large sample theory, likelihood ratio tests, and uniformly most powerful (UMP) tests and the Neyman–Pearson Lemma. A major goal of this text is to make these topics much more accessible to students by using the theory of exponential families.

One of the most important uses of exponential families is that the theory often provides two methods for doing inference. For example, minimal sufficient statistics can be found with either the Lehmann Scheffé theorem or by finding $\boldsymbol{T}$ from the exponential family parameterization. Similarly, if $Y_1, \ldots, Y_n$ are iid from a one-parameter regular exponential family with complete sufficient statistic $T(\boldsymbol{Y})$, then one-sided UMP tests can be found by using the Neyman–Pearson lemma or by using exponential family theory.

The prerequisite for this text is a calculus-based course in statistics at the level of Hogg and Tanis (2005), Larsen and Marx (2011), Wackerly et al. (2008), or Walpole et al. (2006). Also see Arnold (1990), Gathwaite et al. (2002), Spanos (1999), Wasserman (2004), and Welsh (1996).

The following intermediate texts are especially recommended: DeGroot and Schervish (2012), Hogg et al. (2012), Rice (2006), and Rohatgi (1984).

A less satisfactory alternative prerequisite is a calculus-based course in probability at the level of Hoel et al. (1971), Parzen (1960), or Ross (2009).

A course in Real Analysis at the level of Bartle (1964), Gaughan (2009), Rosenlicht (1985), Ross (1980), or Rudin (1964) would be useful for the large sample theory chapter.

The following texts are at a similar to higher level than this text: Azzalini (1996), Bain and Engelhardt (1992), Berry and Lindgren (1995), Cox and Hinkley (1974), Ferguson (1967), Knight (2000), Liero and Zwanzig (2012), Lindgren (1993),

Lindsey (1996), Mood et al. (1974), Roussas (1997), and Silvey (1970). Also see online lecture notes Marden (2012).

The texts Bickel and Doksum (2007), Lehmann and Casella (1998), and Rohatgi and Ehsanes Saleh (2001) are at a higher level as are Poor (1994) and Zacks (1971). The texts Bierens (2004), Cramér (1946), Keener (2010), Lehmann and Romano (2005), Rao (1973), Schervish (1995), and Shao (2003) are at a much higher level.

Some other useful references include a good low-level probability text Ash (1993) and a good introduction to probability and statistics Dekking et al. (2005). Also see Ash (2011), Spiegel (1975), Romano and Siegel (1986), and online lecture notes Ash (2013).

Many of the most important ideas in statistics are due to Fisher, Neyman, E.S. Pearson and K. Pearson. For example, David (2006–7) says that the following terms were due to Fisher: consistency, covariance, degrees of freedom, efficiency, information, information matrix, level of significance, likelihood, location, maximum likelihood, multinomial distribution, null hypothesis, pivotal quantity, probability integral transformation, sampling distribution, scale, statistic, Student's t, studentization, sufficiency, sufficient statistic, test of significance, uniformly most powerful test and variance.

David (2006–7) says that terms due to Neyman and E.S. Pearson include alternative hypothesis, composite hypothesis, likelihood ratio, power, power function, simple hypothesis, size of critical region, test criterion, test of hypotheses, and type I and type II errors. Neyman also coined the term confidence interval.

David (2006–7)says that terms due to K. Pearson include binomial distribution, bivariate normal, method of moments, moment, random sampling, skewness, and standard deviation.

Also see, for example, David (1995), Fisher (1922), Savage (1976), and Stigler (2007). The influence of Gosset (Student) on Fisher is described in Zabell (2008) and Hanley et al. (2008). The influence of Karl Pearson on Fisher is described in Stigler (2008).

This book covers some of these ideas and begins by reviewing probability, counting, conditional probability, independence of events, the expected value, and the variance. Chapter 1 also covers mixture distributions and shows how to use the kernel method to find $E(g(Y))$. Chapter 2 reviews joint, marginal, and conditional distributions; expectation; independence of random variables and covariance; conditional expectation and variance; location–scale families; univariate and multivariate transformations; sums of random variables; random vectors; the multinomial, multivariate normal, and elliptically contoured distributions. Chapter 3 introduces exponential families, while Chap. 4 covers sufficient statistics. Chapter 5 covers maximum likelihood estimators and method of moments estimators. Chapter 6 examines the mean square error and bias as well as uniformly minimum variance unbiased estimators, Fisher information, and the Fréchet–Cramér–Rao lower bound. Chapter 7 covers uniformly most powerful and likelihood ratio tests. Chapter 8 gives an introduction to large sample theory, while Chap. 9 covers confidence intervals. Chapter 10 gives some of the properties of 54 univariate distributions, many of which are exponential families. Chapter 10 also gives over 30 exponential family parameterizations, over 28 MLE examples, and over 27 UMVUE examples. Chapter 11 gives a brief introduction to Bayesian methods, and Chap. 12 gives some hints for the problems.

Some highlights of this text are as follows:

- Exponential families, indicator functions, and the support of the distribution are used throughout the text to simplify the theory.
- Section 1.5 describes the kernel method, a useful technique for computing the expected value of a function of a random variable, $E[g(Y)]$.
- Theorem 2.2 shows the essential relationship between the independence of random variables $Y_1, \ldots, Y_n$ and the support in that the random variables are dependent if the support is not a cross product. If the support is a cross product and if the joint pdf or pmf factors on the support, then $Y_1, \ldots, Y_n$ are independent.
- Theorems 2.17 and 2.18 give the distribution of $\sum Y_i$ when $Y_1, \ldots, Y_n$ are iid for a wide variety of distributions.
- Chapter 3 presents exponential families. The theory of these distributions greatly simplifies many of the most important results in mathematical statistics.
- Corollary 4.6 presents a simple method for finding sufficient, minimal sufficient and complete statistics for k-parameter exponential families.
- Section 5.4.1 compares the "proofs" of the MLE invariance principle due to Zehna (1966) and Berk (1967). Although Zehna (1966) is cited by most texts, Berk (1967) gives a correct elementary proof.
- Theorem 6.5 compares the UMVUE and the estimator that minimizes the MSE for a large class of one-parameter exponential families.
- Theorem 7.3 provides a simple method for finding uniformly most powerful tests for a large class of one-parameter exponential families. Power is examined in Example 7.12.
- Theorem 8.4 gives a simple proof of the asymptotic efficiency of the complete sufficient statistic as an estimator of its expected value for one-parameter regular exponential families.
- Theorem 8.21 provides a simple limit theorem for the complete sufficient statistic of a k-parameter regular exponential family.
- Chapter 10 gives information for 54 "brand name" distributions.
- Chapter 11 shows how to use the shorth to estimate the highest posterior density region.

In a semester, I cover Sects. 1.1–1.6, 2.1–2.9, 3.1, 3.2, 4.1, 4.2, 5.1, 5.2, 6.1, 6.2, 7.1, 7.2, 7.3, 8.1–8.4, and 9.1.

Much of the text material is on parametric frequentist methods, but the most used methods in statistics tend to be semi-parametric. Many of the most used methods originally based on the univariate or multivariate normal distribution are also semi-parametric methods. For example, the t-interval works for a large class of distributions if σ^2 is finite and n is large. Similarly, least squares regression is a semi-parametric method. Multivariate analysis procedures originally based on the multivariate normal distribution tend to also work well for a large class of elliptically contoured distributions.

Warning: For parametric methods that are not based on the normal distribution, often the methods work well if the parametric distribution is a good approximation to the data, but perform very poorly otherwise.

Acknowledgments

Teaching the material to Math 580 students at Southern Illinois University in 2001, 2004, 2008 (when the text was first used), 2012, and 2014 was very useful. Some of the Chap. 8 material came from a reading course in Large Sample Theory taught to SIU students on two occasions. Many of the problems were taken from SIU qualifying exams, and some of these problems were written by Sakthivel Jeyaratnam, Bhaskar Bhattacharya, and Abdel Mugdadi. The latter two professors also contributed several solutions. A small part of the research in Chaps. 9 and 10 was supported by NSF grant DMS 0600933.

Thanks also goes to Springer, Springer's Associate Editor Donna Chernyk, and to several reviewers.

Contents

Chapter 1
Probability and Expectations

The first two chapters of this book review some of the tools from probability that are useful for statistics. These two chapters are no substitute for the prerequisite of a calculus-based course in probability and statistics at the level of Hogg and Tanis (2005), Larsen and Marx (2011), Wackerly et al. (2008), Walpole et al. (2006), DeGroot and Schervish (2012), Hogg et al. (2012), Rice (2006), or Rohatgi (1984).

Most of the material in Sects. 1.1–1.4 should be familiar to the reader and might be reviewed in one 50-min lecture. Section 1.5 covers the kernel method, a useful technique for computing the expectation of a function of a random variable, while Sect. 1.6 introduces mixture distributions.

Definition 1.1. *Statistics* is the science of extracting useful information from data.

1.1 Probability

The following terms from set theory should be familiar. A *set* consists of distinct elements enclosed by *braces*, e.g., $\{1,5,7\}$. The *universal set* S is the set of all elements under consideration while the *empty set* Ø is the set that contains no elements. The set A is a *subset* of B, written $A \subseteq B$, if every element in A is in B. The *union* $A \cup B$ of A with B is the set of all elements in A or B or in both. The *intersection* $A \cap B$ of A with B is the set of all elements in A and B. The *complement* of A, written $\overline{A}$ or A^c, is the set of all elements in S but not in A. The following theorem is used to find complements of unions and intersections of two sets.

Theorem 1.1. DeMorgan's Laws:

a) $\overline{A \cup B} = \overline{A} \cap \overline{B}$.
b) $\overline{A \cap B} = \overline{A} \cup \overline{B}$.
c) $(\cup_{i=1}^{\infty} A_i)^c = \cap_{i=1}^{\infty} A_i^c$.
d) $(\cap_{i=1}^{\infty} A_i)^c = \cup_{i=1}^{\infty} A_i^c$.

D.J. Olive, *Statistical Theory and Inference*, DOI 10.1007/978-3-319-04972-4_1,

Proof. The proofs of a) and b) are similar to those of c) and d), and "iff" means "if and only if."

c) $(\cup_{i=1}^{\infty} A_i)^c$ occurred iff $\cup_{i=1}^{\infty} A_i$ did not occur, iff none of the A_i occurred, iff all of the A_i^c occurred, iff $\cap_{i=1}^{\infty} A_i^c$ occurred.

d) $(\cap_{i=1}^{\infty} A_i)^c$ occurred iff not all of the A_i occurred, iff at least one of the A_i^c occurred, iff $\cup_{i=1}^{\infty} A_i^c$ occurred. □

If $S = \{1,2,3,4,5\}, A = \{1,2\}$ and $B = \{4,5\}$, then $\overline{A \cup B} = \{1,2,4,5\}^c = \{3\} = \{3,4,5\} \cap \{1,2,3\} = \overline{A} \cap \overline{B}$. Similarly, $\overline{A \cap B} = \emptyset^c = S = \{3,4,5\} \cup \{1,2,3\} = \overline{A} \cup \overline{B}$.

Sets are used in probability, but often different notation is used. For example, the universal set is called the sample space S.

Definition 1.2. The *sample space* S is the set of all possible outcomes of an experiment.

Definition 1.3. Let $\mathscr{B}$ be a special field of subsets of the sample space S forming the class of events. Then A is an *event* if $A \in \mathscr{B}$.

In the definition of an event above, the special field of subsets $\mathscr{B}$ of the sample space S forming the class of events will not be formally given. However, $\mathscr{B}$ contains all "interesting" subsets of S and every subset that is easy to imagine. The point is that not necessarily all subsets of S are events, but every event A is a subset of S.

Definition 1.4. If $A \cap B = \emptyset$, then A and B are *mutually exclusive* or *disjoint events.* Events $A_1, A_2, \ldots$ are *pairwise disjoint* or *mutually exclusive* if $A_i \cap A_j = \emptyset$ for $i \neq j$.

If A_i and A_j are disjoint, then $P(A_i \cap A_j) = P(\emptyset) = 0$. A *simple event* is a set that contains exactly one element s_i of S, e.g., $A = \{s_3\}$. A *sample point* s_i is a possible outcome. Simple events $\{s_i\}$ and $\{s_j\}$ are disjoint if $s_i \neq s_j$.

Definition 1.5. A **discrete sample space** consists of a finite or countable number of outcomes.

Notation. Generally we will assume that all events under consideration belong to the same sample space S.

The *relative frequency interpretation of probability* says that the probability of an event A is the proportion of times that event A would occur if the experiment was repeated again and again infinitely often.

Definition 1.6: Kolmogorov's Definition of a Probability Function. Let $\mathscr{B}$ be the class of events of the sample space S. A **probability function** $P : \mathscr{B} \rightarrow [0,1]$ is a set function satisfying the following three properties:

P1) $P(A) \geq 0$ for all events A,
P2) $P(S) = 1$, and

P3) if $A_1, A_2, \ldots$ are pairwise disjoint events, then $P(\cup_{i=1}^{\infty} A_i) = \sum_{i=1}^{\infty} P(A_i)$.

Example 1.1. Flip a coin and observe the outcome. Then the sample space $S = \{H, T\}$. Note that $\{H\}$ and $\{T\}$ are disjoint simple events. Suppose the coin is biased with $P(\{H\}) = 1/3$. Then $P(\{T\}) = 2/3$. Often the notation $P(H) = 1/3$ will be used.

Theorem 1.2. Let A and B be any two events of S. Then

i) $0 \leq P(A) \leq 1$.
ii) $P(\emptyset) = 0$ where $\emptyset$ is the empty set.
iii) **Complement Rule:** $P(A) = 1 - P(\overline{A})$.
iv) **General Addition Rule:** $P(A \cup B) = P(A) + P(B) - P(A \cap B)$.
v) If $A \subseteq B$, then $P(A) \leq P(B)$.
vi) **Boole's Inequality:** $P(\cup_{i=1}^{\infty} A_i) \leq \sum_{i=1}^{\infty} P(A_i)$ for any events $A_1, A_2, \ldots$. vi) **Bonferonni's Inequality:** $P(\cap_{i=1}^{n} A_i) \geq \sum_{i=1}^{n} P(A_i) - (n-1)$ for any events $A_1, A_2, \ldots, A_n$.

Note that A and $\overline{A}$ are disjoint and $A \cup \overline{A} = S$. Hence $1 = P(S) = P(A \cup \overline{A}) = P(A) + P(\overline{A})$, proving the complement rule. Note that S and $\emptyset$ are disjoint, so $1 = P(S) = P(S \cup \emptyset) = P(S) + P(\emptyset)$. Hence $P(\emptyset) = 0$. If $A \subseteq B$, let $C = \overline{A} \cap B$. Then A and C are disjoint with $A \cup C = B$. Hence $P(A) + P(C) = P(B)$, and $P(A) \leq P(B)$ by i). The general addition rule for two events is very useful. Given three of the 4 probabilities in iv), the 4th can be found. $P(A \cup B)$ can be found given $P(A)$, $P(B)$ and that A and B are disjoint or independent. So if $P(A) = 0.2$, $P(B) = 0.3$, and A and B are disjoint, then $P(A \cup B) = 0.5$. The addition rule can also be used to determine whether A and B are independent (see Sect. 1.3) or disjoint.

Following Casella and Berger (2002, p. 13), $P(\cup_{i=1}^{n} A_i^c) = P[(\cap_{i=1}^{n} A_i)^c] = 1 - P(\cap_{i=1}^{n} A_i) \leq \sum_{i=1}^{n} P(A_i^c) = \sum_{i=1}^{n} [1 - P(A_i)] = n - \sum_{i=1}^{n} P(A_i)$, where the first equality follows from DeMorgan's Laws, the second equality holds by the complement rule, and the inequality holds by Boole's inequality $P(\cup_{i=1}^{n} A_i^c) \leq \sum_{i=1}^{n} P(A_i^c)$. Hence $P(\cap_{i=1}^{n} A_i) \geq \sum_{i=1}^{n} P(A_i) - (n-1)$, and Bonferonni's inequality holds.

1.2 Counting

The *sample point method* for finding the probability for event A says that if $S = \{s_1, \ldots, s_k\}$, then $0 \leq P(s_i) \leq 1$, $\sum_{i=1}^{k} P(s_i) = 1$, and $P(A) = \sum_{i: s_i \in A} P(s_i)$. That is, $P(A)$ is the sum of the probabilities of the sample points in A. If all of the outcomes s_i are *equally likely*, then $P(s_i) = 1/k$ and $P(A) =$ (number of outcomes in $A)/k$ if S contains k outcomes.

Counting or combinatorics is useful for determining the number of elements in S. The *multiplication rule* says that if there are n_1 ways to do a first task, n_2 ways to do a 2nd task, …, and n_k ways to do a kth task, then the number of ways to perform the total act of performing the 1st task, then the 2nd task, …, then the kth task is $\prod_{i=1}^{k} n_i = n_1 \cdot n_2 \cdot n_3 \cdots n_k$.

Techniques for the multiplication principle:

a) Use a slot for each task and write n_i above the ith task. There will be k slots, one for each task.
b) Use a tree diagram.

Definition 1.7. A *permutation* is an ordered arrangement using r of n distinct objects and the *number of permutations* $= P_r^n$. A special case of permutation formula is

$$P_n^n = n! = n \cdot (n-1) \cdot (n-2) \cdot (n-3) \cdots 4 \cdot 3 \cdot 2 \cdot 1 =$$

$n \cdot (n-1)! = n \cdot (n-1) \cdot (n-2)! = n \cdot (n-1) \cdot (n-2) \cdot (n-3)! = \cdots$. Generally n is a positive integer, but define $0! = 1$. An application of the multiplication rule can be used to show that $P_r^n = n \cdot (n-1) \cdot (n-2) \cdots (n-r+1) = \dfrac{n!}{(n-r)!}$.

The quantity $n!$ is read "n factorial." Typical problems using $n!$ include the number of ways to arrange n books, to arrange the letters in the word CLIPS (5!), etc.

Example 1.2. The formula for $n!$ can be derived using n slots and the multiplication rule. Suppose there are n distinct books to be arranged in order. The first slot corresponds to the first book, which can be chosen in n ways. Hence an n goes in the first slot. The second slot corresponds to the second book, which can be chosen in $n-1$ ways since there are $n-1$ books remaining after the first book is selected. Similarly, the ith slot is for the ith book after the first $i-1$ books have been selected. Since there are $n-(i-1) = n-i+1$ books left, an $n-i+1$ goes in ith slot for $i = 1, 2, \ldots, n$. The formula for the number of permutations P_r^n can be derived in a similar manner, with r slots. Hence if r people are chosen from n and made to stand in a line, the ith slot corresponds to the ith person in line, $i-1$ people have already been chosen, so $n-i+1$ people remain for the ith slot for $i = 1, 2, \ldots, r$.

A story problem is asking for the permutation formula if the story problem has r slots and *order is important*. No object is allowed to be repeated in the arrangement. Typical questions include *how many ways* are there to "to choose r people from n and arrange in a line," "to make r letter words with no letter repeated," or "to make 7 digit phone numbers with no digit repeated." Key words include *order, no repeated* and *different.*

Notation. The symbol $\equiv$ below means the first three symbols are equivalent and equal, but the fourth term is the formula used to compute the symbol. This notation will often be used when there are several equivalent symbols that mean the same thing. The notation will also be used for functions with subscripts if the subscript is usually omitted, e.g., $g_X(x) \equiv g(x)$. The symbol $\binom{n}{r}$ is read "n choose r" and is called a binomial coefficient.

Definition 1.8. A *combination* is an unordered selection using r of n distinct objects. The *number of combinations* is

$$C(n,r) \equiv C_r^n \equiv \binom{n}{r} = \frac{n!}{r!(n-r)!}.$$

Combinations are used in story problems where *order is not important.* Key words include *committees, selecting* (e.g., four people from ten), *choose*, *random sample* and *unordered.*

1.3 Conditional Probability and Independence

Definition 1.9. The **conditional probability** of A **given** B is

$$P(A|B) = \frac{P(A \cap B)}{P(B)}$$

if $P(B) > 0$.

It is often useful to think of this probability as an experiment with sample space B instead of S.

Definition 1.10. Two events A and B are **independent**, written $A \perp\!\!\!\perp B$, if

$$P(A \cap B) = P(A)P(B).$$

If A and B are not independent, then A and B are *dependent.*

Definition 1.11. A collection of events $A_1, \ldots, A_n$ are *mutually independent* if for *any* subcollection $A_{i_1}, \ldots, A_{i_k}$,

$$P(\cap_{j=1}^k A_{i_j}) = \prod_{j=1}^k P(A_{i_j}).$$

Otherwise the n events are *dependent.*

Theorem 1.3. Assume that $P(A) > 0$ and $P(B) > 0$. Then the two events A and B are *independent* if any of the following three conditions hold:

i) $P(A \cap B) = P(A)P(B)$,
ii) $P(A|B) = P(A)$, or
iii) $P(B|A) = P(B)$.

If *any of these conditions fails to hold*, then A and B are dependent.

The above theorem is useful because only one of the conditions needs to be checked, and often one of the conditions is easier to verify than the other two conditions. If $P(A) = 0.5$ and $P(B) = 0.8$, then A and B are independent iff $P(A \cap B) = 0.4$.

Theorem 1.4. a) *Multiplication rule:* If $A_1, \ldots, A_k$ are events with $P(A_1 \cap A_2 \cap \cdots \cap A_{k-1}) > 0$, then
$P(\cap_{i=1}^k A_i) = P(A_1)P(A_2|A_1)P(A_3|A_1 \cap A_2) \cdots P(A_k|A_1 \cap A_2 \cap \cdots \cap A_{k-1})$.
In particular, $P(A \cap B) = P(A)P(B|A) = P(B)P(A|B)$.

b) *Multiplication rule for independent events:* If $A_1, A_2, \ldots, A_k$ are independent, then $P(A_1 \cap A_2 \cap \cdots \cap A_k) = P(A_1) \cdots P(A_k)$. If A and B are independent ($k = 2$), then $P(A \cap B) = P(A)P(B)$.
c) *Addition rule for disjoint events:* If A and B are disjoint, then $P(A \cup B) = P(A) + P(B)$. If $A_1, \ldots, A_k$ are pairwise disjoint, then $P(\cup_{i=1}^k A_i) = P(A_1 \cup A_2 \cup \cdots \cup A_k) = P(A_1) + \cdots + P(A_k) = \sum_{i=1}^k P(A_i)$.

Example 1.3. The above rules can be used to find the probabilities of more complicated events. The following probabilities are closely related to Binomial experiments. Recall that for Bernoulli trials there are two outcomes called "success" and "failure" where a "success" is the outcome that is counted. Suppose that there are n independent identical trials, that Y counts the number of successes and that $\rho =$ probability of success for any given trial. Let D_i denote a success in the ith trial. Then

i) P(none of the n trials were successes) $= (1-\rho)^n = P(Y=0) = P(\overline{D}_1 \cap \overline{D}_2 \cap \cdots \cap \overline{D}_n)$.
ii) P(at least one of the trials was a success) $= 1-(1-\rho)^n = P(Y \geq 1) = 1 - P(Y=0) = 1 - P(none) = P(\overline{\overline{D}_1 \cap \overline{D}_2 \cap \cdots \cap \overline{D}_n})$.
iii) P(all n trials were successes) $= \rho^n = P(Y=n) = P(D_1 \cap D_2 \cap \cdots \cap D_n)$.
iv) P(not all n trials were successes) $= 1 - \rho^n = P(Y < n) = 1 - P(Y=n) = 1 -$ P(all).
v) P(Y was at least k) $= P(Y \geq k)$.
vi) P(Y was at most k) $= P(Y \leq k)$.

If $A_1, A_2, \ldots$ are pairwise disjoint and if $\cup_{i=1}^\infty A_i = S$, then the collection of sets $A_1, A_2, \ldots$ is a *partition* of S. By taking $A_j = \emptyset$ for $j > k$, the collection of pairwise disjoint sets $A_1, A_2, \ldots, A_k$ is a partition of S if $\cup_{i=1}^k A_i = S$.

Theorem 1.5: Law of Total Probability. If $A_1, A_2, \ldots, A_k$ form a partition of S such that $P(A_i) > 0$ for $i = 1, \ldots, k$, then

$$P(B) = \sum_{j=1}^k P(B \cap A_j) = \sum_{j=1}^k P(B|A_j)P(A_j).$$

Theorem 1.6: Bayes' Theorem. Let $A_1, A_2, \ldots, A_k$ be a partition of S such that $P(A_i) > 0$ for $i = 1, \ldots, k$, and let B be an event such that $P(B) > 0$. Then

$$P(A_i|B) = \frac{P(B|A_i)P(A_i)}{\sum_{j=1}^k P(B|A_j)P(A_j)}.$$

Proof. Notice that $P(A_i|B) = P(A_i \cap B)/P(B)$ and $P(A_i \cap B) = P(B|A_i)P(A_i)$. Since $B = (B \cap A_1) \cup \cdots \cup (B \cap A_k)$ and the A_i are disjoint, $P(B) = \sum_{j=1}^k P(B \cap A_j) = \sum_{j=1}^k P(B|A_j)P(A_j)$. □

Example 1.4. There are many medical tests for rare diseases and a positive result means that the test suggests (perhaps incorrectly) that the person has the disease.

Suppose that a test for disease is such that if the person has the disease, then a positive result occurs 99 % of the time. Suppose that a person without the disease tests positive 2 % of the time. Also assume that 1 in 1,000 people screened have the disease. If a randomly selected person tests positive, what is the probability that the person has the disease?

Solution: Let A_1 denote the event that the randomly selected person has the disease and A_2 denote the event that the randomly selected person does not have the disease. If B is the event that the test gives a positive result, then we want $P(A_1|B)$. By Bayes' theorem,

$$P(A_1|B) = \frac{P(B|A_1)P(A_1)}{P(B|A_1)P(A_1) + P(B|A_2)P(A_2)} = \frac{0.99(0.001)}{0.99(0.001) + 0.02(0.999)}$$

≈ 0.047. Hence instead of telling the patient that she has the rare disease, the doctor should inform the patient that she is in a high risk group and needs further testing.

Bayes' theorem is very useful for including prior information into a statistical method, resulting in Bayesian methods. See Chap. 11. Chapters 3–10 cover frequentist methods which are based on the relative frequency interpretation of probability discussed above Definition 1.6.

1.4 The Expected Value and Variance

Definition 1.12. A *random variable* Y is a real valued function with a sample space as a domain: $Y : S \rightarrow \mathbb{R}$ where the set of real numbers $\mathbb{R} = (-\infty, \infty)$.

Definition 1.13. Let S be the sample space and let Y be a random variable. Then the *(induced) probability function* for Y is $P_Y(Y = y_i) \equiv P(Y = y_i) =$ $P_S(\{s \in S : Y(s) = y_i\})$. The sample space of Y is $S_Y = \{y_i \in \mathbb{R} : \text{there exists an } s \in S \text{ with } Y(s) = y_i\}$.

Definition 1.14. The *population* is the entire group of objects from which we want information. The *sample* is the part of the population actually examined.

Example 1.5. Suppose that 5-year survival rates of 100 lung cancer patients are examined. Let a 1 denote the event that the ith patient died within 5 years of being diagnosed with lung cancer, and a 0 if the patient lived. The outcomes in the sample space S are 100-tuples (sequences of 100 digits) of the form $s = 1010111 \cdots 0111$. Let the random variable $X(s) =$ the number of 1's in the 100-tuple = the sum of the 0's and 1's = the number of the 100 lung cancer patients who died within 5 years of being diagnosed with lung cancer. Then the sample space of X is $S_X = \{0, 1, \ldots, 100\}$. Notice that $X(s) = 82$ is easier to understand than a 100-tuple with 82 ones and 18 zeroes. Note that there are 2^{100} outcomes in S and 101 outcomes in S_X.

For the following definition, F is a right continuous function if for every real number x, $\lim_{y \downarrow x} F(y) = F(x)$. Also, $F(\infty) = \lim_{y \to \infty} F(y)$ and $F(-\infty) = \lim_{y \to -\infty} F(y)$.

Definition 1.15. The **cumulative distribution function** (cdf) of any random variable Y is $F(y) = P(Y \leq y)$ for all $y \in \mathbb{R}$. If $F(y)$ is a cumulative distribution function, then $F(-\infty) = 0$, $F(\infty) = 1$, F is a nondecreasing function and F is right continuous.

Definition 1.16. A random variable is **discrete** if it can assume only a finite or countable number of distinct values. The collection of these probabilities is the *probability distribution* of the discrete random variable. The **probability mass function** (pmf) of a discrete random variable Y is $f(y) = P(Y = y)$ for all $y \in \mathbb{R}$ where $0 \leq f(y) \leq 1$ and $\sum_{y: f(y)>0} f(y) = 1$.

Remark 1.1. The cdf F of a discrete random variable is a step function with a jump of height $f(y)$ at values of y for which $f(y) > 0$.

Definition 1.17. A random variable Y is **continuous** if its distribution function $F(y)$ is absolutely continuous.

The notation $\forall y$ means "for all y."

Definition 1.18. If Y is a continuous random variable, then a **probability density function** (pdf) $f(y)$ of Y is an integrable function such that

$$F(y) = \int_{-\infty}^{y} f(t)dt \tag{1.1}$$

for all $y \in \mathbb{R}$. If $f(y)$ is a pdf, then $f(y)$ is continuous except at most a countable number of points, $f(y) \geq 0 \ \forall y$, and $\int_{-\infty}^{\infty} f(t)dt = 1$.

Theorem 1.7. If Y has pdf $f(y)$, then $f(y) = \frac{d}{dy}F(y) \equiv F'(y)$ wherever the derivative exists (in this text the derivative will exist and be continuous except for at most a finite number of points in any finite interval).

Theorem 1.8. i) $P(a < Y \leq b) = F(b) - F(a)$.
ii) If Y has pdf $f(y)$, then $P(a < Y < b) = P(a < Y \leq b) = P(a \leq Y < b) = P(a \leq Y \leq b) = \int_a^b f(y)dy = F(b) - F(a)$.
iii) If Y has a probability mass function $f(y)$, then Y is discrete and $P(a < Y \leq b) = F(b) - F(a)$, but $P(a \leq Y \leq b) \neq F(b) - F(a)$ if $f(a) > 0$.

Definition 1.19. Let Y be a discrete random variable with probability mass function $f(y)$. Then the *mean* or **expected value** of Y is

$$EY \equiv E(Y) = \sum_{y: f(y)>0} y\, f(y) \tag{1.2}$$

if the sum exists when y is replaced by $|y|$. If $g(Y)$ is a real valued function of Y, then $g(Y)$ is a random variable and

$$E[g(Y)] = \sum_{y:f(y)>0} g(y)\, f(y) \tag{1.3}$$

if the sum exists when $g(y)$ is replaced by $|g(y)|$. If the sums are not absolutely convergent, then $E(Y)$ and $E[g(Y)]$ do not exist.

Example 1.6. Common low level problem. The sample space of Y is $S_Y = \{y_1, y_2, \ldots, y_k\}$ and a table of y_j and $f(y_j)$ is given with one $f(y_j)$ omitted. Find the omitted $f(y_j)$ by using the fact that $\sum_{i=1}^k f(y_i) = f(y_1) + f(y_2) + \cdots + f(y_k) = 1$. Hence if $S_Y = \{1,2,3\}$ with $f(1) = 0.01$ and $f(2) = 0.1$, then $f(3) = 0.89$. Thus $E(Y) = 0.01 + 2(0.1) + 3(0.89) = 2.88$.

Definition 1.20. If Y has pdf $f(y)$, then the *mean* or **expected value** of Y is

$$EY \equiv E(Y) = \int_{-\infty}^{\infty} y f(y) dy \tag{1.4}$$

and

$$E[g(Y)] = \int_{-\infty}^{\infty} g(y) f(y) dy \tag{1.5}$$

provided the integrals exist when y and $g(y)$ are replaced by $|y|$ and $|g(y)|$. If the modified integrals do not exist, then $E(Y)$ and $E[g(Y)]$ do not exist.

Definition 1.21. If $E(Y^2)$ exists, then the *variance* of a random variable Y is

$$\text{VAR}(Y) \equiv \text{Var}(Y) \equiv V\,Y \equiv V(Y) = E[(Y - E(Y))^2]$$

and the *standard deviation* of Y is $\text{SD}(Y) = \sqrt{V(Y)}$. If $E(Y^2)$ does not exist, then $V(Y)$ does not exist.

The notation $E(Y) = \infty$ or $V(Y) = \infty$ when the corresponding integral or sum diverges to ∞ can be useful. The following theorem is used over and over again, especially to find $E(Y^2) = V(Y) + (E(Y))^2$. The theorem is valid for all random variables that have a variance, including continuous and discrete random variables. If Y is a Cauchy (μ, σ) random variable (see Chap. 10), then neither $E(Y)$ nor $V(Y)$ exist.

Theorem 1.9: Short cut formula for variance.

$$V(Y) = E(Y^2) - (E(Y))^2. \tag{1.6}$$

If Y is a discrete random variable with sample space $S_Y = \{y_1, y_2, \ldots, y_k\}$, then

$$E(Y) = \sum_{i=1}^{k} y_i f(y_i) = y_1 f(y_1) + y_2 f(y_2) + \cdots + y_k f(y_k)$$

and $E[g(Y)] = \sum_{i=1}^{k} g(y_i)f(y_i) = g(y_1)f(y_1) + g(y_2)f(y_2) + \cdots + g(y_k)f(y_k)$. In particular,

$$E(Y^2) = y_1^2 f(y_1) + y_2^2 f(y_2) + \cdots + y_k^2 f(y_k).$$

Also

$$V(Y) = \sum_{i=1}^{k} (y_i - E(Y))^2 f(y_i) =$$

$$(y_1 - E(Y))^2 f(y_1) + (y_2 - E(Y))^2 f(y_2) + \cdots + (y_k - E(Y))^2 f(y_k).$$

For a continuous random variable Y with pdf $f(y)$, $V(Y) = \int_{-\infty}^{\infty} (y - E[Y])^2 f(y) dy$. Often using $V(Y) = E(Y^2) - (E(Y))^2$ is simpler.

Example 1.7: Common low level problem. i) Given a table of y and $f(y)$, find $E[g(Y)]$ and the standard deviation $\sigma = SD(Y)$. ii) Find $f(y)$ from $F(y)$. iii) Find $F(y)$ from $f(y)$. iv) Given that $f(y) = c\ g(y)$, find c. v) Given the pdf $f(y)$, find $P(a < Y < b)$, etc. vi) Given the pmf or pdf $f(y)$ find $E[Y]$, $V(Y)$, $SD(Y)$, and $E[g(Y)]$. The functions $g(y) = y$, $g(y) = y^2$, and $g(y) = e^{ty}$ are especially common.

Theorem 1.10. Let a and b be any constants and assume all relevant expectations exist.

i) $E(a) = a$.
ii) $E(aY + b) = aE(Y) + b$.
iii) $E(aX + bY) = aE(X) + bE(Y)$.
iv) $V(aY + b) = a^2 V(Y)$.

Definition 1.22. The **moment generating function** (mgf) of a random variable Y is

$$m(t) = E[e^{tY}] \tag{1.7}$$

if the expectation exists for t in some neighborhood of 0. Otherwise, the mgf does not exist. If Y is discrete, then $m(t) = \sum_y e^{ty} f(y)$, and if Y is continuous, then $m(t) = \int_{-\infty}^{\infty} e^{ty} f(y) dy$.

Definition 1.23. The **characteristic function** of a random variable Y is $c(t) = E[e^{itY}]$, where the complex number $i = \sqrt{-1}$.

Moment generating functions do not necessarily exist in a neighborhood of zero, but a characteristic function always exists. This text does not require much knowledge of theory of complex variables, but know that $i^2 = -1$, $i^3 = -i$ and $i^4 = 1$. Hence $i^{4k-3} = i$, $i^{4k-2} = -1$, $i^{4k-1} = -i$ and $i^{4k} = 1$ for $k = 1, 2, 3, \ldots$. To compute the characteristic function, the following result will be used.

Proposition 1.11. Suppose that Y is a random variable with an mgf $m(t)$ that exists for $|t| < b$ for some constant $b > 0$. Then the characteristic function of Y is $c(t) = m(it)$.

Definition 1.24. Random variables X and Y are *identically distributed*, written $X \sim Y$ or $Y \sim F_X$, if $F_X(y) = F_Y(y)$ for all real y.

Proposition 1.12. Let X and Y be random variables. Then X and Y are identically distributed, $X \sim Y$, if any of the following conditions hold.

a) $F_X(y) = F_Y(y)$ for all y,
b) $f_X(y) = f_Y(y)$ for all y,
c) $c_X(t) = c_Y(t)$ for all t or
d) $m_X(t) = m_Y(t)$ for all t in a neighborhood of zero.

Definition 1.25. For positive integers k, the *kth moment* of Y is $E[Y^k]$ while the *kth central moment* is $E[(Y - E[Y])^k]$.

Theorem 1.13. Suppose that the mgf $m(t)$ exists for $|t| < b$ for some constant $b > 0$, and suppose that the kth derivative $m^{(k)}(t)$ exists for $|t| < b$. Then $E[Y^k] = m^{(k)}(0)$ for positive integers k. In particular, $E[Y] = m'(0)$ and $E[Y^2] = m''(0)$.

Notation. The natural logarithm of y is $\log(y) = \ln(y)$. If another base is wanted, it will be given, e.g., $\log_{10}(y)$.

Example 1.8. Common problem. Let $h(y)$, $g(y)$, $n(y)$, and $d(y)$ be functions. Review how to find the derivative $g'(y)$ of $g(y)$ and how to find the kth derivative

$$g^{(k)}(y) = \frac{d^k}{dy^k} g(y)$$

for integers $k \geq 2$. Recall that the *product rule* is

$$(h(y)g(y))' = h'(y)g(y) + h(y)g'(y).$$

The **quotient rule** is

$$\left(\frac{n(y)}{d(y)}\right)' = \frac{d(y)n'(y) - n(y)d'(y)}{[d(y)]^2}.$$

The **chain rule** is

$$[h(g(y))]' = [h'(g(y))][g'(y)].$$

Know the derivative of $\log(y)$ and e^y and know the chain rule with these functions. Know the derivative of y^k.

Then given the mgf $m(t)$, find $E[Y] = m'(0)$, $E[Y^2] = m''(0)$ and $V(Y) = E[Y^2] - (E[Y])^2$.

Definition 1.26. Let $f(y) \equiv f_Y(y|\boldsymbol{\theta})$ be the pdf or pmf of a random variable Y. Then the set $\mathscr{Y}_{\boldsymbol{\theta}} = \{y | f_Y(y|\boldsymbol{\theta}) > 0\}$ is called the *sample space* or **support** of Y. Let the set Θ be the set of parameter values $\boldsymbol{\theta}$ of interest. Then Θ is the **parameter space** of Y. Use the notation $\mathscr{Y} = \{y | f(y|\boldsymbol{\theta}) > 0\}$ if the support does not depend on $\boldsymbol{\theta}$. So $\mathscr{Y}$ is the support of Y if $\mathscr{Y}_{\boldsymbol{\theta}} \equiv \mathscr{Y}\ \forall \boldsymbol{\theta} \in \Theta$.

Definition 1.27. The **indicator function** $I_A(x) \equiv I(x \in A) = 1$ if $x \in A$ and $I_A(x) = 0$, otherwise. Sometimes an indicator function such as $I_{(0,\infty)}(y)$ will be denoted by $I(y > 0)$.

Example 1.9. Often equations for functions such as the pmf, pdf, or cdf are given only on the support (or on the support plus points on the boundary of the support). For example, suppose

$$f(y) = P(Y = y) = \binom{k}{y} \rho^y (1-\rho)^{k-y}$$

for $y = 0, 1, \ldots, k$ where $0 < \rho < 1$. Then the support of Y is $\mathscr{Y} = \{0, 1, \ldots, k\}$, the parameter space is $\Theta = (0,1)$ and $f(y) = 0$ for y not $\in \mathscr{Y}$. Similarly, if $f(y) = 1$ and $F(y) = y$ for $0 \le y \le 1$, then the support $\mathscr{Y} = [0,1]$, $f(y) = 0$ for $y < 0$ and $y > 1$, $F(y) = 0$ for $y < 0$ and $F(y) = 1$ for $y > 1$.

Since the pmf and cdf are defined for all $y \in \mathbb{R} = (-\infty, \infty)$ and the pdf is defined for all but countably many y, it may be better to use indicator functions when giving the formula for $f(y)$. For example,

$$f(y) = 1I(0 \le y \le 1)$$

is defined for all $y \in \mathbb{R}$.

1.5 The Kernel Method

Notation. Notation such as $E(Y|\boldsymbol{\theta}) \equiv E_{\boldsymbol{\theta}}(Y)$ or $f_Y(y|\boldsymbol{\theta})$ is used to indicate that the formula for the expected value or pdf are for a family of distributions indexed by $\boldsymbol{\theta} \in \Theta$. A major goal of parametric inference is to collect data and estimate $\boldsymbol{\theta}$ from the data.

Example 1.10. If $Y \sim N(\mu, \sigma^2)$, then Y is a member of the normal family of distributions with $\boldsymbol{\theta} = (\mu, \sigma) \in \Theta = \{(\mu, \sigma) | -\infty < \mu < \infty \text{ and } \sigma > 0\}$. Then $E[Y|(\mu,\sigma)] = \mu$ and $V(Y|(\mu,\sigma)) = \sigma^2$. This family has uncountably many members.

The *kernel method* is a widely used technique for finding $E[g(Y)]$.

Definition 1.28. Let $f_Y(y)$ be the pdf or pmf of a random variable Y and suppose that $f_Y(y|\boldsymbol{\theta}) = c(\boldsymbol{\theta}) k(y|\boldsymbol{\theta})$. Then $k(y|\boldsymbol{\theta}) \ge 0$ is the **kernel** of f_Y and $c(\boldsymbol{\theta}) > 0$ is the constant term that makes f_Y sum or integrate to one. Thus $\int_{-\infty}^{\infty} k(y|\boldsymbol{\theta}) dy = 1/c(\boldsymbol{\theta})$ or $\sum_{y \in \mathscr{Y}} k(y|\boldsymbol{\theta}) = 1/c(\boldsymbol{\theta})$.

Often $E[g(Y)]$ is found using "tricks" tailored for a specific distribution. The word "kernel" means "essential part." Notice that if $f_Y(y)$ is a pdf, then $E[g(Y)] = \int_{-\infty}^{\infty} g(y) f(y|\boldsymbol{\theta}) dy = \int_{\mathscr{Y}} g(y) f(y|\boldsymbol{\theta}) dy$. Suppose that after algebra, it is found that

$$E[g(Y)] = a\, c(\boldsymbol{\theta}) \int_{-\infty}^{\infty} k(y|\boldsymbol{\tau}) dy$$

for some constant a where $\boldsymbol{\tau} \in \Theta$ and Θ is the parameter space. Then the kernel method says that

$$E[g(Y)] = a\, c(\boldsymbol{\theta}) \int_{-\infty}^{\infty} \frac{c(\boldsymbol{\tau})}{c(\boldsymbol{\tau})} k(y|\boldsymbol{\tau}) dy = \frac{a\, c(\boldsymbol{\theta})}{c(\boldsymbol{\tau})} \underbrace{\int_{-\infty}^{\infty} c(\boldsymbol{\tau}) k(y|\boldsymbol{\tau}) dy}_{1} = \frac{a\, c(\boldsymbol{\theta})}{c(\boldsymbol{\tau})}.$$

Similarly, if $f_Y(y)$ is a pmf, then

$$E[g(Y)] = \sum_{y: f(y) > 0} g(y) f(y|\boldsymbol{\theta}) = \sum_{y \in \mathscr{Y}} g(y) f(y|\boldsymbol{\theta}),$$

where $\mathscr{Y} = \{y : f_Y(y) > 0\}$ is the support of Y. Suppose that after algebra, it is found that

$$E[g(Y)] = a\, c(\boldsymbol{\theta}) \sum_{y \in \mathscr{Y}} k(y|\boldsymbol{\tau})$$

for some constant a, where $\boldsymbol{\tau} \in \Theta$. Then the kernel method says that

$$E[g(Y)] = a\, c(\boldsymbol{\theta}) \sum_{y \in \mathscr{Y}} \frac{c(\boldsymbol{\tau})}{c(\boldsymbol{\tau})} k(y|\boldsymbol{\tau}) = \frac{a\, c(\boldsymbol{\theta})}{c(\boldsymbol{\tau})} \underbrace{\sum_{y \in \mathscr{Y}} c(\boldsymbol{\tau}) k(y|\boldsymbol{\tau})}_{1} = \frac{a\, c(\boldsymbol{\theta})}{c(\boldsymbol{\tau})}.$$

The kernel method is often useful for finding $E[g(Y)]$, especially if $g(y) = y$, $g(y) = y^2$ or $g(y) = e^{ty}$. The kernel method is often easier than memorizing a trick specific to a distribution because the kernel method uses the same trick for every distribution: $\sum_{y \in \mathscr{Y}} f(y) = 1$ and $\int_{y \in \mathscr{Y}} f(y) dy = 1$. Of course sometimes tricks are needed to get the kernel $f(y|\boldsymbol{\tau})$ from $g(y) f(y|\boldsymbol{\theta})$. For example, complete the square for the normal (Gaussian) kernel.

Example 1.11. To use the kernel method to find the mgf of a gamma (ν, λ) distribution, refer to Chap. 10 and note that

$$m(t) = E(e^{tY}) = \int_0^{\infty} e^{ty} \frac{y^{\nu-1} e^{-y/\lambda}}{\lambda^{\nu} \Gamma(\nu)} dy = \frac{1}{\lambda^{\nu} \Gamma(\nu)} \int_0^{\infty} y^{\nu-1} \exp\left[-y\left(\frac{1}{\lambda} - t\right)\right] dy.$$

The integrand is the kernel of a gamma (ν, η) distribution with

$$\frac{1}{\eta} = \frac{1}{\lambda} - t = \frac{1 - \lambda t}{\lambda} \text{ so } \eta = \frac{\lambda}{1 - \lambda t}.$$

Now

$$\int_0^{\infty} y^{\nu-1} e^{-y/\lambda} dy = \frac{1}{c(\nu, \lambda)} = \lambda^{\nu} \Gamma(\nu).$$

Hence

$$m(t) = \frac{1}{\lambda^\nu \Gamma(\nu)} \int_0^\infty y^{\nu-1} \exp[-y/\eta] dy = c(\nu,\lambda) \frac{1}{c(\nu,\eta)} =$$

$$\frac{1}{\lambda^\nu \Gamma(\nu)} \eta^\nu \Gamma(\nu) = \left(\frac{\eta}{\lambda}\right)^\nu = \left(\frac{1}{1-\lambda t}\right)^\nu$$

for $t < 1/\lambda$.

Example 1.12. The zeta(ν) distribution has probability mass function

$$f(y) = P(Y = y) = \frac{1}{\zeta(\nu) y^\nu},$$

where $\nu > 1$ and $y = 1, 2, 3, \ldots$. Here the zeta function

$$\zeta(\nu) = \sum_{y=1}^{\infty} \frac{1}{y^\nu}$$

for $\nu > 1$. Hence

$$E(Y) = \sum_{y=1}^{\infty} y \frac{1}{\zeta(\nu)} \frac{1}{y^\nu}$$

$$= \frac{1}{\zeta(\nu)} \zeta(\nu-1) \underbrace{\sum_{y=1}^{\infty} \frac{1}{\zeta(\nu-1)} \frac{1}{y^{\nu-1}}}_{1=\text{sum of } zeta(\nu-1) \text{ pmf}} = \frac{\zeta(\nu-1)}{\zeta(\nu)}$$

if $\nu > 2$. Similarly

$$E(Y^k) = \sum_{y=1}^{\infty} y^k \frac{1}{\zeta(\nu)} \frac{1}{y^\nu}$$

$$= \frac{1}{\zeta(\nu)} \zeta(\nu-k) \underbrace{\sum_{y=1}^{\infty} \frac{1}{\zeta(\nu-k)} \frac{1}{y^{\nu-k}}}_{1=\text{sum of } zeta(\nu-k) \text{ pmf}} = \frac{\zeta(\nu-k)}{\zeta(\nu)}$$

if $\nu - k > 1$ or $\nu > k+1$. Thus if $\nu > 3$, then

$$V(Y) = E(Y^2) - [E(Y)]^2 = \frac{\zeta(\nu-2)}{\zeta(\nu)} - \left[\frac{\zeta(\nu-1)}{\zeta(\nu)}\right]^2.$$

Example 1.13. The generalized gamma distribution has pdf

$$f(y) = \frac{\phi y^{\phi\nu-1}}{\lambda^{\phi\nu}\Gamma(\nu)} \exp(-y^\phi/\lambda^\phi),$$

where ν, λ, ϕ and y are positive, and

$$E(Y^k) = \frac{\lambda^k \Gamma(\nu + \frac{k}{\phi})}{\Gamma(\nu)} \quad \text{if } k > -\phi\nu.$$

To prove this result using the kernel method, note that

$$E(Y^k) = \int_0^\infty y^k \frac{\phi y^{\phi\nu-1}}{\lambda^{\phi\nu}\Gamma(\nu)} \exp(-y^\phi/\lambda^\phi)dy = \int_0^\infty \frac{\phi y^{\phi\nu+k-1}}{\lambda^{\phi\nu}\Gamma(\nu)} \exp(-y^\phi/\lambda^\phi)dy.$$

This integrand looks much like a generalized gamma pdf with parameters ν_k, λ, and ϕ, where $\nu_k = \nu + (k/\phi)$ since

$$E(Y^k) = \int_0^\infty \frac{\phi y^{\phi(\nu+k/\phi)-1}}{\lambda^{\phi\nu}\Gamma(\nu)} \exp(-y^\phi/\lambda^\phi)dy.$$

Multiply the integrand by

$$1 = \frac{\lambda^k \Gamma(\nu + \frac{k}{\phi})}{\lambda^k \Gamma(\nu + \frac{k}{\phi})}$$

to get

$$E(Y^k) = \frac{\lambda^k \Gamma(\nu + \frac{k}{\phi})}{\Gamma(\nu)} \int_0^\infty \frac{\phi y^{\phi(\nu+k/\phi)-1}}{\lambda^{\phi(\nu+k/\phi)}\Gamma(\nu + \frac{k}{\phi})} \exp(-y^\phi/\lambda^\phi)dy.$$

Then the result follows since the integral of a generalized gamma pdf with parameters ν_k, λ, and ϕ over its support is 1. Notice that $\nu_k > 0$ implies $k > -\phi\nu$.

1.6 Mixture Distributions

Mixture distributions are often used as outlier models. The following two definitions and proposition are useful for finding the mean and variance of a mixture distribution. Parts a) and b) of Proposition 1.14 below show that the definition of expectation given in Definition 1.30 is the same as the usual definition for expectation if Y is a discrete or continuous random variable.

Definition 1.29. The distribution of a random variable Y is a *mixture distribution* if the cdf of Y has the form

$$F_Y(y) = \sum_{i=1}^{k} \alpha_i F_{W_i}(y) \tag{1.8}$$

where $0 < \alpha_i < 1$, $\sum_{i=1}^k \alpha_i = 1$, $k \geq 2$, and $F_{W_i}(y)$ is the cdf of a continuous or discrete random variable W_i, $i = 1, \ldots, k$.

Definition 1.30. Let Y be a random variable with cdf $F(y)$. Let h be a function such that the *expected value* $E[h(Y)]$ exists. Then

$$E[h(Y)] = \int_{-\infty}^{\infty} h(y)dF(y). \tag{1.9}$$

Proposition 1.14. Assume all expectations exist. a) If Y is a discrete random variable that has a pmf $f(y)$ with support $\mathscr{Y}$, then

$$E[h(Y)] = \int_{-\infty}^{\infty} h(y)dF(y) = \sum_{y \in \mathscr{Y}} h(y)f(y).$$

b) If Y is a continuous random variable that has a pdf $f(y)$, then

$$E[h(Y)] = \int_{-\infty}^{\infty} h(y)dF(y) = \int_{-\infty}^{\infty} h(y)f(y)dy.$$

c) If Y is a random variable that has a mixture distribution with cdf $F_Y(y) = \sum_{i=1}^{k} \alpha_i F_{W_i}(y)$, then

$$E[h(Y)] = \int_{-\infty}^{\infty} h(y)dF(y) = \sum_{i=1}^{k} \alpha_i E_{W_i}[h(W_i)],$$

where $E_{W_i}[h(W_i)] = \int_{-\infty}^{\infty} h(y)dF_{W_i}(y)$.

Example 1.14. Proposition 1.14c implies that the pmf or pdf of W_i is used to compute $E_{W_i}[h(W_i)]$. As an example, suppose the cdf of Y is $F(y) = (1-\varepsilon)\Phi(y) + \varepsilon\Phi(y/k)$, where $0 < \varepsilon < 1$ and $\Phi(y)$ is the cdf of $W_1 \sim N(0,1)$. Then $\Phi(x/k)$ is the cdf of $W_2 \sim N(0,k^2)$. To find $E[Y]$, use $h(y) = y$. Then

$$E[Y] = (1-\varepsilon)E[W_1] + \varepsilon E[W_2] = (1-\varepsilon)0 + \varepsilon 0 = 0.$$

To find $E[Y^2]$, use $h(y) = y^2$. Then

$$E[Y^2] = (1-\varepsilon)E[W_1^2] + \varepsilon E[W_2^2] = (1-\varepsilon)1 + \varepsilon k^2 = 1 - \varepsilon + \varepsilon k^2.$$

Thus $\text{VAR}(Y) = E[Y^2] - (E[Y])^2 = 1 - \varepsilon + \varepsilon k^2$. If $\varepsilon = 0.1$ and $k = 10$, then $EY = 0$, and $\text{VAR}(Y) = 10.9$.

Remark 1.2. Warning: Mixture distributions and linear combinations of random variables are very different quantities. As an example, let

$$W = (1-\varepsilon)W_1 + \varepsilon W_2,$$

where ε, W_1, and W_2 are as in the previous example and suppose that W_1 and W_2 are independent. Then W, a linear combination of W_1 and W_2, has a normal distribution with mean

$$E[W] = (1-\varepsilon)E[W_1] + \varepsilon E[W_2] = 0$$

and variance

$$\text{VAR}(W) = (1-\varepsilon)^2\text{VAR}(W_1) + \varepsilon^2\text{VAR}(W_2) = (1-\varepsilon)^2 + \varepsilon^2 k^2 < \text{VAR}(Y)$$

where Y is given in the example above. Moreover, W has a unimodal normal distribution while Y does not follow a normal distribution. In fact, if $W_1 \sim N(0,1)$, $W_2 \sim N(10,1)$, and W_1 and W_2 are independent, then $(W_1+W_2)/2 \sim N(5,0.5)$; however, if Y has a mixture distribution with cdf

$$F_Y(y) = 0.5F_{W_1}(y) + 0.5F_{W_2}(y) = 0.5\Phi(y) + 0.5\Phi(y-10),$$

then the pdf of Y is bimodal. See Fig. 1.1.

Remark 1.3. a) If all of the W_i are continuous random variables, then the pdf of Y is $f_Y(y) = \sum_{i=1}^k \alpha_i f_{W_i}(y)$, where $f_{W_i}(y)$ is the pdf corresponding to the random variable W_i.

This result can be proved by taking the derivative of both sides of Eq. (1.8).
b) If all of the W_i are discrete random variables, then the pmf of Y is $f_Y(y) = \sum_{i=1}^k \alpha_i f_{W_i}(y)$, where $f_{W_i}(y)$ is the pmf corresponding to the random variable W_i.

This result can be proved using Proposition 1.14c and the indicator function $h(x) = I(y=x) = 1$ if $y = x$ and $h(x) = I(y=x) = 0$ if $y \neq x$. Then $f(x) = P(Y=x) = E[h(Y)] = \sum_{i=1}^k \alpha_i E_{W_i}[h(W_i)] = \sum_{i=1}^k \alpha_i f_{W_i}(x)$. Replace the dummy variable x by y to get the result.

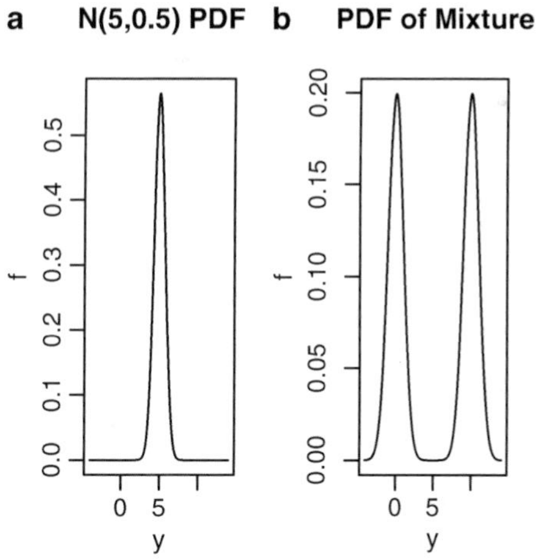

Fig. 1.1 PDF f of $(W_1+W_2)/2$ and $f = 0.5f_1(y) + 0.5f_2(y)$

Assume that all expectations exist. If each W_i is continuous, then by Remark 1.3 a), $E[h(Y)] = \int_{-\infty}^{\infty} h(y) f_Y(y) dy = \int_{-\infty}^{\infty} h(y) \left[\sum_{i=1}^k \alpha_i f_{W_i}(y)\right] dy =$
$\sum_{i=1}^k \alpha_i [\int_{-\infty}^{\infty} h(y) f_{W_i}(y) dy] = \sum_{i=1}^k \alpha_i E_{W_i}[h(W_i)]$. If each W_i is discrete, then by Remark 1.3 b), $E[h(Y)] = \sum_y h(y) f_Y(y) = \sum_{i=1}^k \alpha_i \left[\sum_y h(y) f_{W_i}(y)\right] = \sum_{i=1}^k \alpha_i E_{W_i}[h(W_i)]$, where the sum $\sum_y$ can be taken over all y such that at least one of $f_{W_i}(y) > 0$ for $i = 1, \ldots, k$.

1.7 Summary

Referring to Chap. 10, **memorize the pmf or pdf f, $E(Y)$ and $V(Y)$ for the following 10 random variables. You should recognize the mgf of the binomial, χ^2_p, exponential, gamma, normal and Poisson distributions. You should recognize the cdf of the exponential, normal and uniform distributions.** The Gamma function $\Gamma(x)$ is defined in Definition 10.3.

1) beta(δ, ν)

$$f(y) = \frac{\Gamma(\delta+\nu)}{\Gamma(\delta)\Gamma(\nu)} y^{\delta-1}(1-y)^{\nu-1}$$

where $\delta > 0$, $\nu > 0$ and $0 \le y \le 1$.

$$E(Y) = \frac{\delta}{\delta+\nu}.$$

$$\text{VAR}(Y) = \frac{\delta\nu}{(\delta+\nu)^2(\delta+\nu+1)}.$$

2) Bernoulli(ρ) = binomial$(k=1,\rho)$ $f(y) = \rho^y(1-\rho)^{1-y}$ for $y = 0, 1$.
$E(Y) = \rho$.
$\text{VAR}(Y) = \rho(1-\rho)$.

$$m(t) = [(1-\rho)+\rho e^t].$$

3) binomial(k,ρ)

$$f(y) = \binom{k}{y}\rho^y(1-\rho)^{k-y}$$

for $y = 0, 1, \ldots, k$ where $0 < \rho < 1$.
$E(Y) = k\rho$.
$\text{VAR}(Y) = k\rho(1-\rho)$.

$$m(t) = [(1-\rho)+\rho e^t]^k.$$

4) Cauchy(μ, σ)

$$f(y) = \frac{1}{\pi\sigma\left[1+\left(\frac{y-\mu}{\sigma}\right)^2\right]}$$

where y and μ are real numbers and $\sigma > 0$.
$E(Y) = \infty = \text{VAR}(Y)$.

5) chi-square(p) = gamma$(\nu = p/2, \lambda = 2)$

$$f(y) = \frac{y^{\frac{p}{2}-1}e^{-\frac{y}{2}}}{2^{\frac{p}{2}}\Gamma(\frac{p}{2})}$$

where $y > 0$ and p is a positive integer.
$E(Y) = p$.

$\text{VAR}(Y) = 2p.$

$$m(t) = \left(\frac{1}{1-2t}\right)^{p/2} = (1-2t)^{-p/2}$$

for $t < 1/2$.

6) exponential(λ)= gamma($\nu = 1, \lambda$)

$$f(y) = \frac{1}{\lambda} \exp\left(-\frac{y}{\lambda}\right) I(y \geq 0)$$

where $\lambda > 0$.
$E(Y) = \lambda$,
$\text{VAR}(Y) = \lambda^2$.

$$m(t) = 1/(1 - \lambda t)$$

for $t < 1/\lambda$.

$$F(y) = 1 - \exp(-y/\lambda),\ y \geq 0.$$

7) gamma(ν, λ)

$$f(y) = \frac{y^{\nu-1} e^{-y/\lambda}}{\lambda^\nu \Gamma(\nu)}$$

where ν, λ, and y are positive.
$E(Y) = \nu\lambda$.
$\text{VAR}(Y) = \nu\lambda^2$.

$$m(t) = \left(\frac{1}{1-\lambda t}\right)^{\nu}$$

for $t < 1/\lambda$.

8) $N(\mu, \sigma^2)$

$$f(y) = \frac{1}{\sqrt{2\pi\sigma^2}} \exp\left(\frac{-(y-\mu)^2}{2\sigma^2}\right)$$

where $\sigma > 0$ and μ and y are real.
$E(Y) = \mu$. $\text{VAR}(Y) = \sigma^2$.

$$m(t) = \exp(t\mu + t^2\sigma^2/2).$$

$$F(y) = \Phi\left(\frac{y-\mu}{\sigma}\right).$$

9) Poisson(θ)

$$f(y) = \frac{e^{-\theta}\theta^y}{y!}$$

for $y = 0, 1, \ldots$, where $\theta > 0$.
$E(Y) = \theta = \text{VAR}(Y)$.

$$m(t) = \exp(\theta(e^t - 1)).$$

10) uniform(θ_1, θ_2)

$$f(y) = \frac{1}{\theta_2 - \theta_1} I(\theta_1 \le y \le \theta_2).$$

$F(y) = (y - \theta_1)/(\theta_2 - \theta_1)$ for $\theta_1 \le y \le \theta_2$.
$E(Y) = (\theta_1 + \theta_2)/2$.
$\text{VAR}(Y) = (\theta_2 - \theta_1)^2/12$.

From an introductory course in statistics, the terms sample space S, events, disjoint, partition, probability function, sampling with and without replacement, conditional probability, Bayes' theorem, mutually independent events, random variable, cdf, continuous random variable, discrete random variable, identically distributed, pmf, and pdf are important.

I) Be able to find $E[g(Y)]$, especially $E(Y) = m'(0)$, $E(Y^2) = m''(0)$, $V(Y) = E(Y^2) - [E(Y)]^2$ and the mgf $m(t) = m_Y(t) = E[e^{tY}]$.
II) Let $f_Y(y|\boldsymbol{\theta}) = c(\boldsymbol{\theta})k(y|\boldsymbol{\theta})$ where $k(y|\boldsymbol{\theta})$ is the **kernel** of f_Y. Thus $\int_{-\infty}^{\infty} k(y|\boldsymbol{\theta})dy = 1/c(\boldsymbol{\theta})$. The kernel method is useful for finding $E[g(Y)]$ if $E[g(Y)] =$

$$a\, c(\boldsymbol{\theta}) \int_{-\infty}^{\infty} k(y|\boldsymbol{\tau})dy = a\, c(\boldsymbol{\theta}) \frac{1}{c(\boldsymbol{\tau})} \int_{-\infty}^{\infty} c(\boldsymbol{\tau})k(y|\boldsymbol{\tau})dy = \frac{a\, c(\boldsymbol{\theta})}{c(\boldsymbol{\tau})}$$

for some constant a. Replace the integral by a sum for a discrete distribution.
III) If the cdf of X is $F_X(x) = (1-\varepsilon)F_Z(x) + \varepsilon F_W(x)$ where $0 \le \varepsilon \le 1$ and F_Z and F_W are cdfs, then $E[g(X)] = (1-\varepsilon)E[g(Z)] + \varepsilon E[g(W)]$. In particular, $E(X^2) = (1-\varepsilon)E[Z^2] + \varepsilon E[W^2] = (1-\varepsilon)[V(Z) + (E[Z])^2] + \varepsilon[V(W) + (E[W])^2]$.

1.8 Complements

Kolmogorov's definition of a probability function makes a probability function a normed measure. Hence many of the tools of measure theory can be used for probability theory. See, for example, Ash and Doleans-Dade (1999), Billingsley (1995), Dudley (2002), Durrett (1995), Feller (1971), and Resnick (1999). Feller (1957) and Tucker (1984) are good references for combinatorics.

1.9 Problems

PROBLEMS WITH AN ASTERISK * ARE ESPECIALLY USEFUL. Refer to Chap. 10 for the pdf or pmf of the distributions in the problems below.

1.1*. Consider the Binomial(k, ρ) distribution.

a) Find E Y.
b) Find Var Y.
c) Find the mgf $m(t)$.

1.2*. Consider the Poisson(θ) distribution.

a) Find E Y.
b) Find Var Y. (Hint: Use the kernel method to find E $Y(Y-1)$.)
c) Find the mgf $m(t)$.

1.3*. Consider the Gamma(ν, λ) distribution.

a) Find E Y.
b) Find Var Y.
c) Find the mgf $m(t)$.

1.4*. Consider the Normal(μ, σ^2) (or Gaussian) distribution.

a) Find the mgf $m(t)$. (Hint: complete the square to get a Gaussian kernel.)
b) Use the mgf to find E Y.
c) Use the mgf to find Var Y.

1.5*. Consider the Uniform(θ_1, θ_2) distribution.

a) Find E Y.
b) Find Var Y.
c) Find the mgf $m(t)$.

1.6*. Consider the Beta(δ, ν) distribution.

a) Find E Y.
b) Find Var Y.

1.7*. See Mukhopadhyay (2000, p. 39). Recall integrals by u-substitution:

$$I = \int_a^b f(g(x))g'(x)dx = \int_{g(a)}^{g(b)} f(u)du = \int_c^d f(u)du =$$

$$F(u)|_c^d = F(d) - F(c) = F(u)|_{g(a)}^{g(b)} = F(g(x))|_a^b = F(g(b)) - F(g(a))$$

where $F'(x) = f(x)$, $u = g(x)$, $du = g'(x)dx$, $d = g(b)$, and $c = g(a)$.

This problem uses the Gamma function and u-substitution to show that the normal density integrates to 1 (usually shown with polar coordinates). When you perform the u-substitution, make sure you say what $u = g(x)$, $du = g'(x)dx$, $d = g(b)$, and $c = g(a)$ are.

a) Let $f(x)$ be the pdf of a $N(\mu, \sigma^2)$ random variable. Perform u-substitution on

$$I = \int_{-\infty}^{\infty} f(x)dx$$

with $u = (x - \mu)/\sigma$.

b) Break the result into two parts,

$$I = \frac{1}{\sqrt{2\pi}} \int_{-\infty}^{0} e^{-u^2/2} du + \frac{1}{\sqrt{2\pi}} \int_{0}^{\infty} e^{-u^2/2} du.$$

Then perform u-substitution on the first integral with $v = -u$.

c) Since the two integrals are now equal,

$$I = \frac{2}{\sqrt{2\pi}} \int_{0}^{\infty} e^{-v^2/2} dv = \frac{2}{\sqrt{2\pi}} \int_{0}^{\infty} e^{-v^2/2} \frac{1}{v} v dv.$$

Perform u-substitution with $w = v^2/2$.

d) Using the Gamma function, show that $I = \Gamma(1/2)/\sqrt{\pi} = 1$.

1.8. Let X be a $N(0,1)$ (standard normal) random variable. Use integration by parts to show that $EX^2 = 1$. Recall that integration by parts is used to evaluate $\int f(x)g'(x)dx = \int u dv = uv - \int v du$ where $u = f(x)$, $dv = g'(x)dx$, $du = f'(x)dx$, and $v = g(x)$. When you do the integration, clearly state what these four terms are (e.g., $u = x$).

1.9. Verify the formula for the cdf F for the following distributions. That is, either show that $F'(y) = f(y)$ or show that $\int_{-\infty}^{y} f(t)dt = F(y)\ \forall y \in \mathbb{R}$.

a) Cauchy (μ, σ).
b) Double exponential (θ, λ).
c) Exponential (λ).
d) Logistic (μ, σ).
e) Pareto (σ, λ).
f) Power (λ).
g) Uniform (θ_1, θ_2).
h) Weibull $W(\phi, \lambda)$.

1.10. Verify the formula for the expected value $E(Y)$ for the following distributions.

a) Double exponential (θ, λ).
b) Exponential (λ).
c) Logistic (μ, σ). (Hint from and deCani and Stine (1986): Let $Y = [\mu + \sigma W]$ so $E(Y) = \mu + \sigma E(W)$ where $W \sim L(0,1)$. Hence

$$E(W) = \int_{-\infty}^{\infty} y \frac{e^y}{[1+e^y]^2} dy.$$

Use substitution with

$$u = \frac{e^y}{1+e^y}.$$

Then

$$E(W^k) = \int_{0}^{1} [\log(u) - \log(1-u)]^k du.$$

Also use the fact that

$$\lim_{v\to 0} v\log(v) = 0$$

to show $E(W) = 0$.)
d) Lognormal (μ, σ^2).
e) Pareto (σ, λ).
f) Weibull (ϕ, λ).

1.11. Verify the formula for the variance VAR(Y) for the following distributions.

a) Double exponential (θ, λ).
b) Exponential (λ).
c) Logistic (μ, σ). (Hint from deCani and Stine (1986): Let $Y = [\mu + \sigma X]$ so $V(Y) = \sigma^2 V(X) = \sigma^2 E(X^2)$ where $X \sim L(0,1)$. Hence

$$E(X^2) = \int_{-\infty}^{\infty} y^2 \frac{e^y}{[1+e^y]^2} dy.$$

Use substitution with

$$v = \frac{e^y}{1+e^y}.$$

Then

$$E(X^2) = \int_0^1 [\log(v) - \log(1-v)]^2 dv.$$

Let $w = \log(v) - \log(1-v)$ and $du = [\log(v) - \log(1-v)]dv$. Then

$$E(X^2) = \int_0^1 w du = uw|_0^1 - \int_0^1 u dw.$$

Now

$$uw|_0^1 = [v\log(v) + (1-v)\log(1-v)] \; w|_0^1 = 0$$

since

$$\lim_{v\to 0} v\log(v) = 0.$$

Now

$$-\int_0^1 u dw = -\int_0^1 \frac{\log(v)}{1-v} dv - \int_0^1 \frac{\log(1-v)}{v} dv = 2\pi^2/6 = \pi^2/3$$

using

$$\int_0^1 \frac{\log(v)}{1-v} dv = \int_0^1 \frac{\log(1-v)}{v} dv = -\pi^2/6.)$$

d) Lognormal (μ, σ^2).
e) Pareto (σ, λ).
f) Weibull (ϕ, λ).

Problems from old quizzes and exams.

1.12. Suppose the random variable X has cdf $F_X(x) = 0.9\,\Phi(x-10) + 0.1\,F_W(x)$, where $\Phi(x-10)$ is the cdf of a normal $N(10,1)$ random variable with mean 10 and variance 1 and $F_W(x)$ is the cdf of the random variable W that satisfies $P(W=200)=1$.

a) Find $E\,W$.
b) Find $E\,X$.

1.13. Suppose the random variable X has cdf $F_X(x) = 0.9\,F_Z(x) + 0.1\,F_W(x)$, where F_Z is the cdf of a gamma($\alpha = 10, \beta = 1$) random variable with mean 10 and variance 10 and $F_W(x)$ is the cdf of the random variable W that satisfies $P(W=400)=1$.

a) Find $E\,W$.
b) Find $E\,X$.

1.14. Suppose the cdf $F_X(x) = (1-\varepsilon)F_Z(x) + \varepsilon F_W(x)$, where $0 \le \varepsilon \le 1$, F_Z is the cdf of a random variable Z, and F_W is the cdf of a random variable W. Then $E\,g(X) = (1-\varepsilon)E_Z\,g(Z) + \varepsilon E_W\,g(W)$, where $E_Z\,g(Z)$ means that the expectation should be computed using the pmf or pdf of Z. Suppose the random variable X has cdf $F_X(x) = 0.9\,F_Z(x) + 0.1\,F_W(x)$, where F_Z is the cdf of a gamma($\alpha = 20, \beta = 1$) random variable with mean 20 and variance 20 and $F_W(x)$ is the cdf of the random variable W that satisfies $P(W=400)=1$.

a) Find $E\,W$.
b) Find $E\,X$.

1.15. Let A and B be positive integers. A hypergeometric random variable $X = W_1 + W_2 + \cdots + W_n$, where the random variables W_i are identically distributed random variables with $P(W_i = 1) = A/(A+B)$ and $P(W_i = 0) = B/(A+B)$.

a) Find $E(W_1)$.
b) Find $E(X)$.

1.16. Suppose $P(X = x_o) = 1$ for some constant x_o.

a) Find $E\,g(X)$ in terms of x_o.
b) Find the moment generating function $m(t)$ of X.
c) Find $m^{(n)}(t) = \frac{d^n}{dt^n} m(t)$. (Hint: find $m^{(n)}(t)$ for $n = 1, 2,$ and 3. Then the pattern should be apparent.)

1.17. Suppose $P(X = 1) = 0.5$ and $P(X = -1) = 0.5$. Find the moment generating function of X.

1.18. Suppose that X is a discrete random variable with pmf $f(x) = P(X = x)$ for $x = 0, 1, \ldots, n$ so that the moment generating function of X is $m(t) = \sum_{x=0}^{n} e^{tx} f(x)$.

a) Find $\frac{d}{dt}m(t) = m'(t)$.
b) Find $m'(0)$.
c) Find $m''(t) = \frac{d^2}{dt^2}m(t)$.
d) Find $m''(0)$.
e) Find $m^{(k)}(t) = \frac{d^k}{dt^k}m(t)$. (Hint: you found $m^{(k)}(t)$ for $k = 1, 2$, and the pattern should be apparent.)

1.19. Suppose that the random variable $W = e^X$, where $X \sim N(\mu, \sigma^2)$. Find $E(W^r) = E[(e^X)^r]$ by recognizing the relationship of $E[(e^X)^r]$ with the moment generating function of a normal(μ, σ^2) random variable.

1.20. Let $X \sim N(\mu, \sigma^2)$ so that $EX = \mu$ and Var $X = \sigma^2$.

a) Find $E(X^2)$.
b) If $k \geq 2$ is an integer, then $E(X^k) = (k-1)\sigma^2 E(X^{k-2}) + \mu E(X^{k-1})$. Use this recursion relationship to find $E(X^3)$.

1.21*. Let $X \sim$ gamma(ν, λ). Using the kernel method, find EX^r where $r > -\nu$.

1.22. Find $\int_{-\infty}^{\infty} \exp(-\frac{1}{2}y^2)dy$.
(Hint: the integrand is a Gaussian kernel.)

1.23. Let X have a Pareto $(\sigma, \lambda = 1/\theta)$ pdf

$$f(x) = \frac{\theta\sigma^\theta}{x^{\theta+1}}$$

where $x > \sigma$, $\sigma > 0$ and $\theta > 0$. Using the kernel method, find EX^r where $\theta > r$.

1.24. Let $Y \sim$ beta (δ, ν). Using the kernel method, find EY^r where $r > -\delta$.

1.25. Use the kernel method to find the mgf of the logarithmic (θ) distribution.

1.26. Suppose that X has pdf

$$f(x) = \frac{h(x)e^{\theta x}}{\lambda(\theta)}$$

for $x \in \mathscr{X}$ and for $-\infty < \theta < \infty$ where $\lambda(\theta)$ is some positive function of θ and $h(x)$ is some nonnegative function of x. Find the moment generating function of X using the kernel method. Your final answer should be written in terms of λ, θ, and t.

1.27. Use the kernel method to find $E(Y^r)$ for the chi (p, σ) distribution.

1.28. Suppose the cdf $F_X(x) = (1-\varepsilon)F_Z(x) + \varepsilon F_W(x)$, where $0 \leq \varepsilon \leq 1$, F_Z is the cdf of a random variable Z, and F_W is the cdf of a random variable W. Then $E\,g(X) = (1-\varepsilon)E_Z\,g(Z) + \varepsilon E_W\,g(W)$, where $E_Z\,g(Z)$ means that the expectation should be computed using the pmf or pdf of Z.

Suppose the random variable X has cdf $F_X(x) = 0.9\ F_Z(x) + 0.1\ F_W(x)$, where F_Z is the cdf of a gamma $(\nu = 3, \lambda = 4)$ random variable and $F_W(x)$ is the cdf of a Poisson (10) random variable.

a) Find $E\ X$.
b) Find $E\ X^2$.

1.29. If Y has an exponential distribution truncated at 1, $Y \sim TEXP(\theta, 1)$, then the pdf of Y is

$$f(y) = \frac{\theta}{1 - e^{-\theta}} e^{-\theta y}$$

for $0 < y < 1$, where $\theta > 0$. Find the mgf of Y using the kernel method.

1.30. Following Morris (1982), let

$$f(y) = \frac{\cos(\theta)}{2\cosh(\pi y/2)} \exp(\theta y)$$

where y is real and $|\theta| < \pi/2$. Find the mgf of Y using the kernel method.

1.31. If Y has a log-gamma distribution, the pdf of Y is

$$f(y) = \frac{1}{\lambda^\nu \Gamma(\nu)} \exp\left(\nu y + \left(\frac{-1}{\lambda}\right) e^y\right)$$

where y is real, $\nu > 0$, and $\lambda > 0$. Find the mgf of Y using the kernel method.

1.32. If Y has an inverted gamma distribution, $Y \sim \text{INVG}(\nu, \lambda)$, then the pdf of Y is

$$f(y) = \frac{1}{y^{\nu+1}\Gamma(\nu)} I(y > 0) \frac{1}{\lambda^\nu} \exp\left(\frac{-1}{\lambda}\frac{1}{y}\right)$$

where λ, ν, and y are all positive. Using the kernel method, show

$$E(Y^r) = \frac{\Gamma(\nu - r)}{\lambda^r \Gamma(\nu)}$$

for $\nu > r$.

1.33. If Y has a zero truncated Poisson distribution, $Y \sim ZTP(\theta)$, then the pmf of Y is

$$f(y) = \frac{e^{-\theta}\ \theta^y}{(1 - e^{-\theta})\ y!}$$

for $y = 1, 2, 3, \ldots$, where $\theta > 0$. Find the mgf of Y using the kernel method.

1.34. If Y has a Zipf distribution, $Y \sim \text{Zipf}(\nu)$, then the pmf of Y is

$$f(y) = \frac{1}{y^\nu z(\nu)}$$

where $y \in \{1, \ldots, m\}$ and m is known, ν is real and

$$z(\nu) = \sum_{y=1}^{m} \frac{1}{y^{\nu}}.$$

Using the kernel method, show

$$E(Y^r) = \frac{z(\nu - r)}{z(\nu)}$$

for real r.

1.35. If Y has a Lindley distribution, then the pdf of Y is

$$f(y) = \frac{\theta^2}{1+\theta}(1+y)e^{-\theta y}$$

where $y > 0$ and $\theta > 0$. Using the kernel method, find the mgf of Y.

1.36. The Lindley distribution has cdf $F_Y(y) = (1-\varepsilon)F_Z(y) + \varepsilon F_W(y)$, where $\varepsilon = \theta/(1+\theta)$, $\theta > 0$, F_Z is the cdf of a gamma ($\nu = 2, \lambda = 1/\theta$) random variable Z, and F_W is the cdf of an EXP($1/\theta$) random variable W. Then $E\ g(Y) = (1-\varepsilon)E_Z\ g(Z) + \varepsilon E_W\ g(W)$, where $E_Z\ g(Z)$ means that the expectation should be computed using the pmf or pdf of Z.

a) Find $E\ Y$.
b) Find $E\ Y^2$.

1.37. According to and Consonni and Veronese (1992), if Y is a random variable with pdf

$$f(y) = \frac{1-\theta^2}{2}\ \exp(\theta y)$$

where $-\infty < y < \infty$ and $-1 < \theta < 1$, then Y is a one-parameter regular exponential family with an mgf that can be found using the kernel method.

a) Assuming $f(y)$ is a pdf, find the mgf of Y using the kernel method.
b) Show that $f(y)$ is not a pdf by showing $\int_{-\infty}^{\infty} f(y)dy \neq 1$.
(Problem 1.30 may have the correct pdf.)

Chapter 2
Multivariate Distributions and Transformations

This chapter continues the review of some tools from probability that are useful for statistics, and most of the material in Sects. 2.1–2.3, 2.5, and 2.6 should be familiar to the reader. The material on elliptically contoured distributions in Sect. 2.10 may be omitted when first reading this chapter.

2.1 Joint, Marginal, and Conditional Distributions

Often there are n random variables $Y_1, \ldots, Y_n$ that are of interest. For example, *age, blood pressure, weight, gender*, and *cholesterol level* might be some of the random variables of interest for patients suffering from heart disease.

Notation. Let $\mathbb{R}^n$ be the n-dimensional Euclidean space. Then the vector $\boldsymbol{y} = (y_1, \ldots, y_n) \in \mathbb{R}^n$ if y_i is an arbitrary real number for $i = 1, \ldots, n$.

Definition 2.1. If $Y_1, \ldots, Y_n$ are discrete random variables, then the **joint pmf** (probability mass function) of $Y_1, \ldots, Y_n$ is

$$f(y_1, \ldots, y_n) = P(Y_1 = y_1, \ldots, Y_n = y_n) \tag{2.1}$$

for any $(y_1, \ldots, y_n) \in \mathbb{R}^n$. A joint pmf f satisfies $f(\boldsymbol{y}) \equiv f(y_1, \ldots, y_n) \geq 0 \ \forall \boldsymbol{y} \in \mathbb{R}^n$ and

$$\sum_{\boldsymbol{y}\,:\, f(\boldsymbol{y}) > 0} \cdots \sum f(y_1, \ldots, y_n) = 1.$$

For any event $A \in \mathbb{R}^n$,

$$P[(Y_1, \ldots, Y_n) \in A] = \sum_{\boldsymbol{y}\,:\, \boldsymbol{y} \in A \text{ and } f(\boldsymbol{y}) > 0} \cdots \sum f(y_1, \ldots, y_n).$$

Definition 2.2. The **joint cdf** (cumulative distribution function) of $Y_1, \ldots, Y_n$ is $F(y_1, \ldots, y_n) = P(Y_1 \leq y_1, \ldots, Y_n \leq y_n)$ for any $(y_1, \ldots, y_n) \in \mathbb{R}^n$.

D.J. Olive, *Statistical Theory and Inference*, DOI 10.1007/978-3-319-04972-4_2,

Definition 2.3. If $Y_1,\ldots,Y_n$ are continuous random variables, then the **joint pdf** (probability density function) of $Y_1,\ldots,Y_n$ is a function $f(y_1,\ldots,y_n)$ that satisfies $F(y_1,\ldots,y_n) = \int_{-\infty}^{y_n} \cdots \int_{-\infty}^{y_1} f(t_1,\ldots,t_n)dt_1 \cdots dt_n$, where the y_i are any real numbers. A joint pdf f satisfies $f(\boldsymbol{y}) \equiv f(y_1,\ldots,y_n) \geq 0 \;\forall \boldsymbol{y} \in \mathbb{R}^n$ and $\int_{-\infty}^{\infty} \cdots \int_{-\infty}^{\infty} f(t_1,\ldots,t_n)dt_1 \cdots dt_n = 1$. For any event $A \in \mathbb{R}^n$,

$$P[(Y_1,\ldots,Y_n) \in A] = \underset{A}{\int \cdots \int} f(t_1,\ldots,t_n)dt_1 \cdots dt_n.$$

Definition 2.4. If $Y_1,\ldots,Y_n$ has a joint pdf or pmf f, then the *sample space* or **support** of $Y_1,\ldots,Y_n$ is

$$\mathscr{Y} = \{(y_1,\ldots,y_n) \in \mathbb{R}^n : f(y_1,\ldots,y_n) > 0\}.$$

If $\boldsymbol{Y}$ comes from a family of distributions $f(\boldsymbol{y}|\boldsymbol{\theta})$ for $\boldsymbol{\theta} \in \Theta$, then the support $\mathscr{Y}_{\boldsymbol{\theta}} = \{\boldsymbol{y} : f(\boldsymbol{y}|\boldsymbol{\theta}) > 0\}$ may depend on $\boldsymbol{\theta}$.

Theorem 2.1. Let $Y_1,\ldots,Y_n$ have joint cdf $F(y_1,\ldots,y_n)$ and joint pdf $f(y_1,\ldots,y_n)$. Then

$$f(y_1,\ldots,y_n) = \frac{\partial^n}{\partial y_1 \cdots \partial y_n} F(y_1,\ldots,y_n)$$

wherever the partial derivative exists.

Definition 2.5. The **marginal pmf** of any subset $Y_{i1},\ldots,Y_{ik}$ of the coordinates $(Y_1,\ldots,Y_n)$ is found by summing the joint pmf over all possible values of the other coordinates where the values $y_{i1},\ldots,y_{ik}$ are held fixed. For example,

$$f_{Y_1,\ldots,Y_k}(y_1,\ldots,y_k) = \sum_{y_{k+1}} \cdots \sum_{y_n} f(y_1,\ldots,y_n)$$

where $y_1,\ldots,y_k$ are held fixed. In particular, if Y_1 and Y_2 are discrete random variables with joint pmf $f(y_1,y_2)$, then the marginal pmf for Y_1 is

$$f_{Y_1}(y_1) = \sum_{y_2} f(y_1,y_2) \tag{2.2}$$

where y_1 is held fixed. The marginal pmf for Y_2 is

$$f_{Y_2}(y_2) = \sum_{y_1} f(y_1,y_2) \tag{2.3}$$

where y_2 is held fixed.

Remark 2.1. For $n = 2$, double integrals are used to find marginal pdfs (defined below) and to show that the joint pdf integrates to 1. If the region of integration Ω is bounded on top by the function $y_2 = \phi_T(y_1)$, on the bottom by the function $y_2 = \phi_B(y_1)$ and to the left and right by the lines $y_1 = a$ and $y_2 = b$, then $\int\int_\Omega f(y_1,y_2)dy_1dy_2 = \int\int_\Omega f(y_1,y_2)dy_2dy_2 =$

$$\int_a^b \left[\int_{\phi_B(y_1)}^{\phi_T(y_1)} f(y_1, y_2) dy_2 \right] dy_1.$$

Within the inner integral, treat y_2 as the variable, anything else, including y_1, is treated as a constant.

If the region of integration Ω is bounded on the left by the function $y_1 = \psi_L(y_2)$, on the right by the function $y_1 = \psi_R(y_2)$ and to the top and bottom by the lines $y_2 = c$ and $y_2 = d$, then $\int\int_\Omega f(y_1, y_2) dy_1 dy_2 = \int\int_\Omega f(y_1, y_2) dy_2 dy_2 =$

$$\int_c^d \left[\int_{\psi_L(y_2)}^{\psi_R(y_2)} f(y_1, y_2) dy_1 \right] dy_2.$$

Within the inner integral, treat y_1 as the variable, anything else, including y_2, is treated as a constant. See Example 2.3.

Definition 2.6. The **marginal pdf** of any subset $Y_{i1}, \ldots, Y_{ik}$ of the coordinates $(Y_1, \ldots, Y_n)$ is found by integrating the joint pdf over all possible values of the other coordinates where the values $y_{i1}, \ldots, y_{ik}$ are held fixed. For example, $f(y_1, \ldots, y_k) = \int_{-\infty}^{\infty} \cdots \int_{-\infty}^{\infty} f(t_1, \ldots, t_n) dt_{k+1} \cdots dt_n$, where $y_1, \ldots, y_k$ are held fixed. In particular, if Y_1 and Y_2 are continuous random variables with joint pdf $f(y_1, y_2)$, then the marginal pdf for Y_1 is

$$f_{Y_1}(y_1) = \int_{-\infty}^{\infty} f(y_1, y_2) dy_2 = \int_{\phi_B(y_1)}^{\phi_T(y_1)} f(y_1, y_2) dy_2 \tag{2.4}$$

where y_1 is held fixed (to get the region of integration, draw a line parallel to the y_2 axis, and use the functions $y_2 = \phi_B(y_1)$ and $y_2 = \phi_T(y_1)$ as the lower and upper limits of integration). The marginal pdf for Y_2 is

$$f_{Y_2}(y_2) = \int_{-\infty}^{\infty} f(y_1, y_2) dy_1 = \int_{\psi_L(y_2)}^{\psi_R(y_2)} f(y_1, y_2) dy_1 \tag{2.5}$$

where y_2 is held fixed (to get the region of integration, draw a line parallel to the y_1 axis, and use the functions $y_1 = \psi_L(y_2)$ and $y_1 = \psi_R(y_2)$ as the lower and upper limits of integration).

Definition 2.7. The **conditional pmf** of any subset $Y_{i1}, \ldots, Y_{ik}$ of the coordinates $(Y_1, \ldots, Y_n)$ is found by dividing the joint pmf by the marginal pmf of the remaining coordinates assuming that the values of the remaining coordinates are fixed and that the denominator > 0. For example,

$$f(y_1, \ldots, y_k | y_{k+1}, \ldots, y_n) = \frac{f(y_1, \ldots, y_n)}{f(y_{k+1}, \ldots, y_n)}$$

if $f(y_{k+1}, \ldots, y_n) > 0$. In particular, the conditional pmf of Y_1 given $Y_2 = y_2$ is a function of y_1 and

$$f_{Y_1|Y_2=y_2}(y_1|y_2) = \frac{f(y_1, y_2)}{f_{Y_2}(y_2)} \tag{2.6}$$

if $f_{Y_2}(y_2) > 0$, and the conditional pmf of Y_2 given $Y_1 = y_1$ is a function of y_2 and

$$f_{Y_2|Y_1=y_1}(y_2|y_1) = \frac{f(y_1,y_2)}{f_{Y_1}(y_1)} \tag{2.7}$$

if $f_{Y_1}(y_1) > 0$.

Definition 2.8. The **conditional pdf** of any subset $Y_{i1},\ldots,Y_{ik}$ of the coordinates $(Y_1,\ldots,Y_n)$ is found by dividing the joint pdf by the marginal pdf of the remaining coordinates assuming that the values of the remaining coordinates are fixed and that the denominator > 0. For example,

$$f(y_1,\ldots,y_k|y_{k+1},\ldots,y_n) = \frac{f(y_1,\ldots,y_n)}{f(y_{k+1},\ldots,y_n)}$$

if $f(y_{k+1},\ldots,y_n) > 0$. In particular, the conditional pdf of Y_1 given $Y_2 = y_2$ is a function of y_1 and

$$f_{Y_1|Y_2=y_2}(y_1|y_2) = \frac{f(y_1,y_2)}{f_{Y_2}(y_2)} \tag{2.8}$$

if $f_{Y_2}(y_2) > 0$, and the conditional pdf of Y_2 given $Y_1 = y_1$ is a function of y_2 and

$$f_{Y_2|Y_1=y_1}(y_2|y_1) = \frac{f(y_1,y_2)}{f_{Y_1}(y_1)} \tag{2.9}$$

if $f_{Y_1}(y_1) > 0$.

Example 2.1. Common Problem. If the joint pmf $f(y_1,y_2) = P(Y_1 = y_1, Y_2 = y_2)$ is given by a table, then the function $f(y_1,y_2)$ is a joint pmf if $f(y_1,y_2) \geq 0$, $\forall y_1, y_2$ and if

$$\sum_{(y_1,y_2):f(y_1,y_2)>0} f(y_1,y_2) = 1.$$

The marginal pmfs are found from the row sums and column sums using Definition 2.5, and the conditional pmfs are found with the formulas given in Definition 2.7. See Example 2.6b and 2.6f.

Example 2.2. Common Problem. Given the joint pdf $f(y_1,y_2) = kg(y_1,y_2)$ on its support, find k, find the marginal pdfs $f_{Y_1}(y_1)$ and $f_{Y_2}(y_2)$, and find the conditional pdfs $f_{Y_1|Y_2=y_2}(y_1|y_2)$ and $f_{Y_2|Y_1=y_1}(y_2|y_1)$. Also,
$P(a_1 < Y_1 < b_1, a_2 < Y_2 < b_2) = \int_{a_2}^{b_2} \int_{a_1}^{b_1} f(y_1,y_2) dy_1 dy_2$.

Tips: Often using **symmetry** helps.

The support of the marginal pdf does not depend on the second variable.

The *support* of the conditional pdf can depend on the second variable. For example, the support of $f_{Y_1|Y_2=y_2}(y_1|y_2)$ could have the form $0 \leq y_1 \leq y_2$.

The *support* of continuous random variables Y_1 and Y_2 is the region where $f(y_1,y_2) > 0$. The support is generally given by one to three inequalities such as $0 \le y_1 \le 1$, $0 \le y_2 \le 1$, and $0 \le y_1 \le y_2 \le 1$. For each variable, set the inequalities to equalities to get boundary lines. For example $0 \le y_1 \le y_2 \le 1$ yields 5 lines: $y_1 = 0, y_1 = 1$, $y_2 = 0, y_2 = 1$, and $y_2 = y_1$. Generally y_2 is on the vertical axis and y_1 is on the horizontal axis for pdfs.

To determine the **limits of integration**, examine the **dummy variable used in the inner integral**, say dy_1. Then within the region of integration, draw a line parallel to the same (y_1) axis as the dummy variable. The limits of integration will be functions of the other variable (y_2), never of the dummy variable (dy_1). See the following example.

Example 2.3. Suppose that the joint pdf of the random variables Y_1 and Y_2 is given by

$$f(y_1,y_2) = 2, \ \text{if } 0 < y_1 < y_2 < 1$$

and $f(y_1,y_2) = 0$, otherwise.

a) Show that $f(y_1,y_2)$ is a pdf.
b) Find the marginal pdf of Y_1. Include the support.
c) Find the marginal pdf of Y_2. Include the support.
d) Find the conditional pdf $f_{Y_1|Y_2=y_2}(y_1|y_2)$. Include the support.
e) Find the conditional pdf $f_{Y_2|Y_1=y_1}(y_2|y_1)$. Include the support.

Solution. Refer to Remark 2.1. The support is the region of integration Ω which is the triangle with vertices (0,0), (0,1), and (1,1). This triangle is bounded by the lines $y_1 = 0, y_2 = 1$, and $y_2 = y_1$. The latter line can also be written as $y_1 = y_2$.

a) Hence $\int_{-\infty}^{\infty}\int_{-\infty}^{\infty} f(y_1,y_2)dy_1dy_2 = \int_0^1[\int_0^{y_2} 2dy_1]dy_2 = \int_0^1 [2y_1|_0^{y_2}]dy_2 = \int_0^1 2y_2dy_2 = 2y_2^2/2|_0^1 = 1$. Here $\psi_L(y_2) \equiv 0$ and $\psi_R(y_2) = y_2$. Alternatively, $\int_{-\infty}^{\infty}\int_{-\infty}^{\infty} f(y_1,y_2)dy_2dy_1 = \int_0^1[\int_{y_1}^1 2dy_2]dy_1 = \int_0^1 [2y_2|_{y_1}^1]dy_1 = \int_0^1 2(1-y_1)dy_1 = 2(y_1 - y_1^2/2)|_0^1 = 2(1/2) = 1$. Here $\phi_B(y_1) = y_1$ and $\phi_T(y_1) \equiv 1$.
b) Now $f_{Y_1}(y_1) = \int_{-\infty}^{\infty} f(y_1,y_2)dy_2 = \int_{y_1}^1 2dy_2 = 2y_2|_{y_1}^1 = 2(1-y_1), 0 < y_1 < 1$.
c) Now $f_{Y_2}(y_2) = \int_{-\infty}^{\infty} f(y_1,y_2)dy_1 = \int_0^{y_2} 2dy_1 = 2y_1|_0^{y_2} = 2y_2, 0 < y_2 < 1$.
d) By Definition 2.8,

$$f_{Y_1|Y_2=y_2}(y_1|y_2) = \frac{f(y_1,y_2)}{f_{Y_2}(y_2)} = \frac{2}{2y_2} = \frac{1}{y_2}, \ 0 < y_1 < y_2.$$

Note that for fixed y_2, the variable y_1 can run from 0 to y_2.
e) By Definition 2.8,

$$f_{Y_2|Y_1=y_1}(y_2|y_1) = \frac{f(y_1,y_2)}{f_{Y_1}(y_1)} = \frac{2}{2(1-y_1)} = \frac{1}{1-y_1}, \ y_1 < y_2 < 1.$$

Note that for fixed y_1, the variable y_2 can run from y_1 to 1.

2.2 Expectation, Covariance, and Independence

For joint pmfs with $n = 2$ random variables Y_1 and Y_2, the marginal pmfs and conditional pmfs can provide important information about the data. For joint pdfs the integrals are usually too difficult for the joint, conditional and marginal pdfs to be of practical use unless the random variables are independent. (Exceptions are the multivariate normal distribution and the elliptically contoured distributions. See Sects. 2.9 and 2.10.)

For independent random variables, the joint cdf is the product of the marginal cdfs, the joint pmf is the product of the marginal pmfs, and the joint pdf is the product of the marginal pdfs. Recall that $\forall$ is read "for all."

Definition 2.9. i) The random variables $Y_1, Y_2, \ldots, Y_n$ are **independent** if $F(y_1, y_2, \ldots, y_n) = F_{Y_1}(y_1)F_{Y_2}(y_2)\cdots F_{Y_n}(y_n)\ \forall y_1, y_2, \ldots, y_n$.

ii) If the random variables have a joint pdf or pmf f, then the random variables $Y_1, Y_2, \ldots, Y_n$ are independent if $f(y_1, y_2, \ldots, y_n) = f_{Y_1}(y_1)f_{Y_2}(y_2)\cdots f_{Y_n}(y_n)$ $\forall y_1, y_2, \ldots, y_n$.

If the random variables are not independent, then they are **dependent**.

In particular random variables Y_1 and Y_2 are **independent**, written $Y_1 \perp\!\!\!\perp Y_2$, if either of the following conditions holds.

i) $F(y_1, y_2) = F_{Y_1}(y_1)F_{Y_2}(y_2)\ \ \forall y_1, y_2$.

ii) $f(y_1, y_2) = f_{Y_1}(y_1)f_{Y_2}(y_2)\ \ \forall y_1, y_2$. Otherwise, Y_1 and Y_2 are *dependent.*

Definition 2.10. Recall that the support $\mathscr{Y}$ of $(Y_1, Y_2, \ldots, Y_n)$ is $\mathscr{Y} = \{\boldsymbol{y} : f(\boldsymbol{y}) > 0\}$. The support is a **cross product** or **Cartesian product** if

$$\mathscr{Y} = \mathscr{Y}_1 \times \mathscr{Y}_2 \times \cdots \times \mathscr{Y}_n = \{\boldsymbol{y} : y_i \in \mathscr{Y}_i \text{ for } i = 1, \ldots, n\}$$

where $\mathscr{Y}_i$ is the support of Y_i. If f is a joint pdf then the support is **rectangular** if $\mathscr{Y}_i$ is an interval for each i. If f is a joint pmf then the support is rectangular if the points in $\mathscr{Y}_i$ are equally spaced for each i.

Example 2.4. In applications the support is often rectangular. For $n = 2$ the support is a cross product if

$$\mathscr{Y} = \mathscr{Y}_1 \times \mathscr{Y}_2 = \{(y_1, y_2) : y_1 \in \mathscr{Y}_1 \text{ and } y_2 \in \mathscr{Y}_2\}$$

where $\mathscr{Y}_i$ is the support of Y_i. The support is rectangular if $\mathscr{Y}_1$ and $\mathscr{Y}_2$ are intervals. For example, if

$$\mathscr{Y} = \{(y_1, y_2) : a < y_1 < \infty \text{ and } c \le y_2 \le d\},$$

then $\mathscr{Y}_1 = (a, \infty)$ and $\mathscr{Y}_2 = [c, d]$. For a joint pmf, the support is rectangular if the grid of points where $f(y_1, y_2) > 0$ is rectangular.

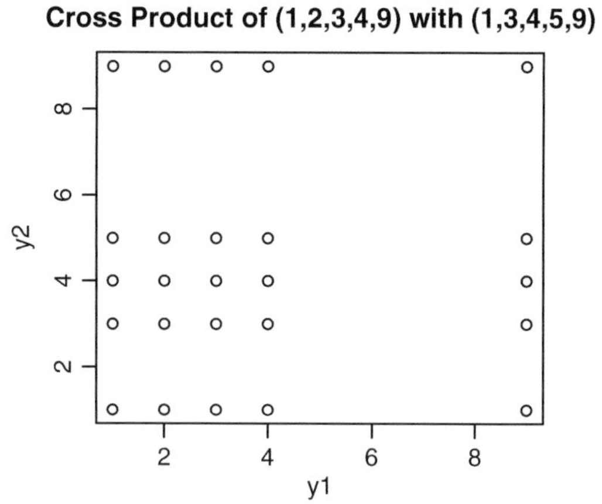

Fig. 2.1 Cross product for a joint PMF

Figure 2.1 shows the cross product of $\mathscr{Y}_1 \times \mathscr{Y}_2$ where $\mathscr{Y}_1 = \{1,2,3,4,9\}$ and $\mathscr{Y}_2 = \{1,3,4,5,9\}$. Each dot occurs where $f(y_1,y_2) > 0$. Notice that each point in $\mathscr{Y}_1$ occurs with each point in $\mathscr{Y}_2$. This support would not be a cross product if any point was deleted, but would be a cross product if any row of dots or column of dots was deleted. Note that the cross product support is not rectangular. The cross product of $\mathscr{Y}_1 = \{1,2,3,4\}$ with $\mathscr{Y}_2 = \{3,4,5\}$ is rectangular.

Theorem 2.2a below is useful because it is often immediate from the formula for the joint pdf or the table for the joint pmf that the support is not a cross product. Hence Y_1 and Y_2 are dependent. For example, if the support of Y_1 and Y_2 is a triangle, as in Example 2.3, then Y_1 and Y_2 are dependent. **A necessary condition for independence is that the support is a cross product.** Theorem 2.2b is useful because factorizing the joint pdf on cross product support is easier than using integration to find the marginal pdfs. Many texts give Theorem 2.2c, but 2.2b is easier to use. Recall that $\prod_{i=1}^n a_i = a_1 a_2 \cdots a_n$. For example, let $n = 3$ and $a_i = i$ for $i = 1,2,3$. Then $\prod_{i=1}^n a_i = a_1 a_2 a_3 = (1)(2)(3) = 6$.

Theorem 2.2. a) Random variables $Y_1,\ldots,Y_n$ with joint pdf or pmf f are dependent if their support $\mathscr{Y}$ is not a cross product. In particular, Y_1 and Y_2 are dependent if $\mathscr{Y}$ does not have the form $\mathscr{Y} = \mathscr{Y}_1 \times \mathscr{Y}_2$.

b) If random variables $Y_1,\ldots,Y_n$ with joint pdf or pmf f have support $\mathscr{Y}$ that is a cross product, then $Y_1,\ldots,Y_n$ are independent iff $f(y_1,y_2,\ldots,y_n) = h_1(y_1)h_2(y_2) \cdots h_n(y_n)$ for all $\boldsymbol{y} \in \mathscr{Y}$, where h_i is a positive function of y_i alone. In particular, if $\mathscr{Y} = \mathscr{Y}_1 \times \mathscr{Y}_2$, then $Y_1 \perp\!\!\!\perp Y_2$ iff $f(y_1,y_2) = h_1(y_1)h_2(y_2)$ for all $(y_1,y_2) \in \mathscr{Y}$ where $h_i(y_i) > 0$ for $y_i \in \mathscr{Y}_i$ and $i = 1,2$.

c) $Y_1,\ldots,Y_n$ are independent iff $f(y_1,y_2,\ldots,y_n) = g_1(y_1)g_2(y_2)\cdots g_n(y_n)$ for all $\boldsymbol{y}$ where g_i is a nonnegative function of y_i alone.

d) If discrete Y_1 and Y_2 have cross product support given by a table, find the row and column sums. If $f(y_1,y_2) \neq f_{Y_1}(y_1)f_{Y_2}(y_2)$ for **some entry** (y_1,y_2), then Y_1 and Y_2 are dependent. If $f(y_1,y_2) = f_{Y_1}(y_1)f_{Y_2}(y_2)$ for *all table entries, then Y_1 and Y_2 are independent.*

Proof. a) If the support is not a cross product, then there is a point $\boldsymbol{y}$ such that $f(\boldsymbol{y}) = 0$ but $f_{Y_i}(y_i) > 0$ for $i = 1, \ldots, n$. Hence $f(\boldsymbol{y}) \neq \prod_{i=1}^n f_{Y_i}(y_i)$ at the point $\boldsymbol{y}$ and $Y_1, \ldots, Y_n$ are dependent.

b) The proof for a joint pdf is given below. For a joint pmf, replace the integrals by appropriate sums. If $Y_1, \ldots, Y_n$ are independent, take $h_i(y_i) = f_{Y_i}(y_i) > 0$ for $y_i \in \mathscr{Y}_i$ and $i = 1, \ldots, n$.

If $f(\boldsymbol{y}) = h_1(y_1) \cdots h_n(y_n)$ for $\boldsymbol{y} \in \mathscr{Y} = \mathscr{Y}_1 \times \cdots \times \mathscr{Y}_n$ then $f(\boldsymbol{y}) = 0 = f_{Y_1}(y_1) \cdots f_{Y_n}(y_n)$ if $\boldsymbol{y}$ is not in $\mathscr{Y}$. Hence we need to show that $f(\boldsymbol{y}) = f_{Y_1}(y_1) \cdots f_{Y_n}(y_n) = h_1(y_1) \cdots h_n(y_n)$ if $\boldsymbol{y} \in \mathscr{Y}$. Since f is a joint pdf,

$$1 = \int \cdots \int_{\mathscr{Y}} f(\boldsymbol{y})\, d\boldsymbol{y} = \prod_{i=1}^n \int_{\mathscr{Y}_i} h_i(y_i)\, dy_i = \prod_{i=1}^n a_i$$

where $a_i = \int_{\mathscr{Y}_i} h_i(y_i)\, dy_i > 0$. For $y_i \in \mathscr{Y}_i$, the marginal pdfs $f_{Y_i}(y_i) =$

$$\int_{\mathscr{Y}_n} \cdots \int_{\mathscr{Y}_{i+1}} \int_{\mathscr{Y}_{i-1}} \cdots \int_{\mathscr{Y}_1} h_1(y_1) \cdots h_i(y_i) \cdots h(y_n)\, dy_1 \cdots dy_{i-1} dy_{i+1} \cdots dy_n$$

$$= h_i(y_i) \prod_{j=1, j\neq i}^n \int_{\mathscr{Y}_j} h_j(y_j)\, dy_j = h_i(y_i) \prod_{j=1, j\neq i}^n a_j = h_i(y_i) \frac{1}{a_i}.$$

Thus $a_i f_{Y_i}(y_i) = h_i(y_i)$ for $y_i \in \mathscr{Y}_i$. Since $\prod_{i=1}^n a_i = 1$,

$$f(\boldsymbol{y}) = \prod_{i=1}^n h_i(y_i) = \prod_{i=1}^n a_i f_{Y_i}(y_i) = \left(\prod_{i=1}^n a_i\right)\left(\prod_{i=1}^n f_{Y_i}(y_i)\right) = \prod_{i=1}^n f_{Y_i}(y_i)$$

if $\boldsymbol{y} \in \mathscr{Y}$.

c) Take

$$g_i(y_i) = \begin{cases} h_i(y_i), & \text{if } y_i \in \mathscr{Y}_i \\ 0, & \text{otherwise.} \end{cases}$$

Then the result follows from b).

d) Since $f(y_1, y_2) = 0 = f_{Y_1}(y_1) f_{Y_2}(y_2)$ if (y_1, y_2) is not in the support of Y_1 and Y_2, the result follows by the definition of independent random variables. □

The following theorem shows that finding the marginal and conditional pdfs or pmfs is simple if $Y_1, \ldots, Y_n$ are independent. Also **subsets of independent random variables are independent**: if $Y_1, \ldots, Y_n$ are independent and if $\{i_1, \ldots, i_k\} \subseteq \{1, \ldots, n\}$ for $k \geq 2$, then $Y_{i_1}, \ldots, Y_{i_k}$ are independent.

Theorem 2.3. Suppose that $Y_1, \ldots, Y_n$ are independent random variables with joint pdf or pmf $f(y_1, \ldots, y_n)$. Then

a) the marginal pdf or pmf of any subset $Y_{i_1}, \ldots, Y_{i_k}$ is $f(y_{i_1}, \ldots, y_{i_k}) = \prod_{j=1}^k f_{Y_{i_j}}(y_{i_j})$. Hence $Y_{i_1}, \ldots, Y_{i_k}$ are independent random variables for $k \geq 2$.

b) The conditional pdf or pmf of $Y_{i_1}, \ldots, Y_{i_k}$ given any subset of the remaining random variables $Y_{j_1} = y_{j_1}, \ldots, Y_{j_m} = y_{j_m}$ is equal to the marginal: $f(y_{i_1}, \ldots, y_{i_k} | y_{j_1}, \ldots, y1_{j_m}) = f(y_{i_1}, \ldots, y_{i_k}) = \prod_{j=1}^{k} f_{Y_{i_j}}(y_{i_j})$ if $f(y_{j_1}, \ldots, y_{j_m}) > 0$.

Proof. The proof for a joint pdf is given below. For a joint pmf, replace the integrals by appropriate sums. a) The marginal

$$f(y_{i_1}, \ldots, y_{i_k}) = \int_{-\infty}^{\infty} \cdots \int_{-\infty}^{\infty} \left[\prod_{j=1}^{n} f_{Y_{i_j}}(y_{i_j}) \right] dy_{i_{k+1}} \cdots dy_{i_n}$$

$$= \left[\prod_{j=1}^{k} f_{Y_{i_j}}(y_{i_j}) \right] \left[\prod_{j=k+1}^{n} \int_{-\infty}^{\infty} f_{Y_{i_j}}(y_{i_j})\, dy_{i_j} \right]$$

$$= \left[\prod_{j=1}^{k} f_{Y_{i_j}}(y_{i_j}) \right] (1)^{n-k} = \prod_{j=1}^{k} f_{Y_{i_j}}(y_{i_j}).$$

b) follows from a) and the definition of a conditional pdf assuming that $f(y_{j_1}, \ldots, y_{j_m}) > 0$. $\square$

Definition 2.11. Suppose that random variables $\boldsymbol{Y} = (Y_1, \ldots, Y_n)$ have support $\mathscr{Y}$ and joint pdf or pmf f. Then the **expected value** of the real valued function $h(\boldsymbol{Y}) = h(Y_1, \ldots, Y_n)$ is

$$E[h(\boldsymbol{Y})] = \int_{-\infty}^{\infty} \cdots \int_{-\infty}^{\infty} h(\boldsymbol{y}) f(\boldsymbol{y})\, d\boldsymbol{y} = \int \cdots \int_{\mathscr{Y}} h(\boldsymbol{y}) f(\boldsymbol{y})\, d\boldsymbol{y} \tag{2.10}$$

if f is a joint pdf and if

$$\int_{-\infty}^{\infty} \cdots \int_{-\infty}^{\infty} |h(\boldsymbol{y})| f(\boldsymbol{y})\, d\boldsymbol{y}$$

exists. Otherwise the expectation does not exist. The expected value is

$$E[h(\boldsymbol{Y})] = \sum_{y_1} \cdots \sum_{y_n} h(\boldsymbol{y}) f(\boldsymbol{y}) = \sum_{\boldsymbol{y} \in \mathbb{R}^n} h(\boldsymbol{y}) f(\boldsymbol{y}) = \sum_{\boldsymbol{y} \in \mathscr{Y}} h(\boldsymbol{y}) f(\boldsymbol{y}) \tag{2.11}$$

if f is a joint pmf and if $\sum_{\boldsymbol{y} \in \mathbb{R}^n} |h(\boldsymbol{y})| f(\boldsymbol{y})$ exists. Otherwise the expectation does not exist.

The notation $E[h(\boldsymbol{Y})] = \infty$ can be useful when the corresponding integral or sum diverges to ∞. The following theorem is useful since multiple integrals with smaller dimension are easier to compute than those with higher dimension.

Theorem 2.4. Suppose that $Y_1, \ldots, Y_n$ are random variables with joint pdf or pmf $f(y_1, \ldots, y_n)$. Let $\{i_1, \ldots, i_k\} \subset \{1, \ldots, n\}$, and let $f(y_{i_1}, \ldots, y_{i_k})$ be the marginal pdf or pmf of $Y_{i_1}, \ldots, Y_{i_k}$ with support $\mathscr{Y}_{Y_{i_1}, \ldots, Y_{i_k}}$. Assume that $E[h(Y_{i_1}, \ldots, Y_{i_k})]$ exists. Then

$$E[h(Y_{i_1},\ldots,Y_{i_k})] = \int_{-\infty}^{\infty}\cdots\int_{-\infty}^{\infty} h(y_{i_1},\ldots,y_{i_k})\, f(y_{i_1},\ldots,y_{i_k})\, dy_{i_1}\cdots dy_{i_k} =$$

$$\int\cdots\int_{\mathscr{Y}_{Y_{i_1},\ldots,Y_{i_k}}} h(y_{i_1},\ldots,y_{i_k})\, f(y_{i_1},\ldots,y_{i_k})\, dy_{i_1}\cdots dy_{i_k}$$

if f is a pdf, and

$$E[h(Y_{i_1},\ldots,Y_{i_k})] = \sum_{y_{i_1}}\cdots\sum_{y_{i_k}} h(y_{i_1},\ldots,y_{i_k})\, f(y_{i_1},\ldots,y_{i_k})$$

$$= \sum_{(y_{i_1},\ldots,y_{i_k})\in\mathscr{Y}_{Y_{i_1},\ldots,Y_{i_k}}} h(y_{i_1},\ldots,y_{i_k})\, f(y_{i_1},\ldots,y_{i_k})$$

if f is a pmf.

Proof. The proof for a joint pdf is given below. For a joint pmf, replace the integrals by appropriate sums. Let $g(Y_1,\ldots,Y_n) = h(Y_{i_1},\ldots,Y_{i_k})$. Then $E[g(\boldsymbol{Y})] =$

$$\int_{-\infty}^{\infty}\cdots\int_{-\infty}^{\infty} h(y_{i_1},\ldots,y_{i_k}) f(y_1,\ldots,y_n)\, dy_1\cdots dy_n =$$

$$\int_{-\infty}^{\infty}\cdots\int_{-\infty}^{\infty} h(y_{i_1},\ldots,y_{i_k})\left[\int_{-\infty}^{\infty}\cdots\int_{-\infty}^{\infty} f(y_1,\ldots,y_n)\, dy_{i_{k+1}}\cdots dy_{i_n}\right] dy_{i_1}\cdots dy_{i_k}$$

$$= \int_{-\infty}^{\infty}\cdots\int_{-\infty}^{\infty} h(y_{i_1},\ldots,y_{i_k}) f(y_{i_1},\ldots,y_{i_k})\, dy_{i_1}\cdots dy_{i_k}$$

since the term in the brackets gives the marginal. □

Example 2.5. Typically $E(Y_i), E(Y_i^2)$ and $E(Y_iY_j)$ for $i \neq j$ are of primary interest. Suppose that (Y_1,Y_2) has joint pdf $f(y_1,y_2)$. Then $E[h(Y_1,Y_2)]$

$$= \int_{-\infty}^{\infty}\int_{-\infty}^{\infty} h(y_1,y_2) f(y_1,y_2) dy_2 dy_1 = \int_{-\infty}^{\infty}\int_{-\infty}^{\infty} h(y_1,y_2) f(y_1,y_2) dy_1 dy_2$$

where $-\infty$ to ∞ could be replaced by the limits of integration for dy_i. **In particular,**

$$E(Y_1Y_2) = \int_{-\infty}^{\infty}\int_{-\infty}^{\infty} y_1y_2 f(y_1,y_2) dy_2 dy_1 = \int_{-\infty}^{\infty}\int_{-\infty}^{\infty} y_1y_2 f(y_1,y_2) dy_1 dy_2.$$

Since finding the marginal pdf is usually easier than doing the double integral, if h is a function of Y_i but not of Y_j, find the marginal for Y_i : $E[h(Y_1)] = \int_{-\infty}^{\infty}\int_{-\infty}^{\infty} h(y_1) f(y_1,y_2) dy_2 dy_1 = \int_{-\infty}^{\infty} h(y_1) f_{Y_1}(y_1) dy_1$. Similarly, $E[h(Y_2)] = \int_{-\infty}^{\infty} h(y_2) f_{Y_2}(y_2) dy_2$.

In particular, $E(Y_1) = \int_{-\infty}^{\infty} y_1 f_{Y_1}(y_1) dy_1$, and $E(Y_2) = \int_{-\infty}^{\infty} y_2 f_{Y_2}(y_2) dy_2$. See Example 2.8.

Suppose that (Y_1,Y_2) have a joint pmf $f(y_1,y_2)$. Then the **expectation** $E[h(Y_1,Y_2)] = \sum_{y_2}\sum_{y_1} h(y_1,y_2) f(y_1,y_2) = \sum_{y_1}\sum_{y_2} h(y_1,y_2) f(y_1,y_2)$. **In particular,**

$$E[Y_1Y_2] = \sum_{y_1}\sum_{y_2} y_1 y_2 f(y_1,y_2).$$

Since finding the marginal pmf is usually easier than doing the double summation, if h is a function of Y_i but not of Y_j, find the marginal for pmf for Y_i: $E[h(Y_1)] = \sum_{y_2}\sum_{y_1} h(y_1)f(y_1,y_2) = \sum_{y_1} h(y_1)f_{Y_1}(y_1)$. Similarly, $E[h(Y_2)] = \sum_{y_2} h(y_2)f_{Y_2}(y_2)$. **In particular**, $E(Y_1) = \sum_{y_1} y_1 f_{Y_1}(y_1)$ and $E(Y_2) = \sum_{y_2} y_2 f_{Y_2}(y_2)$. See Example 2.6.

For pdfs it is sometimes possible to find $E[h(Y_i)]$, but for $k \geq 2$ these expected values tend to be very difficult to compute unless $f(y_1,\ldots,y_k) = c\, y_1^{i_1}\cdots y_k^{i_k}$ for small integers i_j on rectangular or triangular support. Independence makes finding some expected values simple.

Theorem 2.5. Let $Y_1,\ldots,Y_n$ be independent random variables. If $h_i(Y_i)$ is a function of Y_i alone and if the relevant expected values exist, then

$$E[h_1(Y_1)h_2(Y_2)\cdots h_n(Y_n)] = E[h_1(Y_1)]\cdots E[h_n(Y_n)].$$

In particular, $E[Y_iY_j] = E[Y_i]E[Y_j]$ for $i \neq j$.

Proof. The result will be shown for the case where $\boldsymbol{Y} = (Y_1,\ldots,Y_n)$ has a joint pdf f. For a joint pmf, replace the integrals by appropriate sums. By independence, the support of $\boldsymbol{Y}$ is a cross product: $\mathscr{Y} = \mathscr{Y}_1 \times \cdots \times \mathscr{Y}_n$. Since $f(\boldsymbol{y}) = \prod_{i=1}^n f_{Y_i}(y_i)$, the expectation $E[h_1(Y_1)h_2(Y_2)\cdots h_n(Y_n)] =$

$$\int\cdots\int_{\mathscr{Y}} h_1(y_1)h_2(y_2)\cdots h_n(y_n)\, f(y_1,\ldots,y_n)\, dy_1\cdots dy_n$$

$$= \int_{\mathscr{Y}_n}\cdots\int_{\mathscr{Y}_1}\left[\prod_{i=1}^n h_i(y_i)f_{Y_i}(y_i)\right] dy_1\cdots dy_n$$

$$= \prod_{i=1}^n\left[\int_{\mathscr{Y}_i} h_i(y_i)f_{Y_i}(y_i)\, dy_i\right] = \prod_{i=1}^n E[h_i(Y_i)]. \ \square$$

Corollary 2.6. Let $Y_1,\ldots,Y_n$ be independent random variables. If $h_j(Y_{i_j})$ is a function of Y_{i_j} alone and if the relevant expected values exist, then

$$E[h_1(Y_{i_1})\cdots h_k(Y_{i_k})] = E[h_1(Y_{i_1})]\cdots E[h_k(Y_{i_k})].$$

Proof. Method 1: Take $X_j = Y_{i_j}$ for $j = 1,\ldots,k$. Then $X_1,\ldots,X_k$ are independent and Theorem 2.5 applies.
Method 2: Take $h_j(Y_{i_j}) \equiv 1$ for $j = k+1,\ldots,n$ and apply Theorem 2.5. $\square$

Theorem 2.7. Let $Y_1,\ldots,Y_n$ be independent random variables. If $h_i(Y_i)$ is a function of Y_i alone and $X_i = h_i(Y_i)$, then the random variables $X_1,\ldots,X_n$ are independent.

Definition 2.12. The **covariance** of Y_1 and Y_2 is

$$\text{Cov}(Y_1,Y_2) = E[(Y_1 - E(Y_1))(Y_2 - E(Y_2))]$$

provided the expectation exists. Otherwise the covariance does not exist.

Theorem 2.8: Short cut formula. If $\text{Cov}(Y_1,Y_2)$ exists then $\text{Cov}(Y_1,Y_2) = E(Y_1Y_2) - E(Y_1)E(Y_2)$.

Theorem 2.9. a) Let Y_1 and Y_2 be independent random variables. If $\text{Cov}(Y_1,Y_2)$ exists, then $\text{Cov}(Y_1,Y_2) = 0$.
b) **The converse is false**: $\text{Cov}(Y_1,Y_2) = 0$ does not imply $Y_1 \perp\!\!\!\perp Y_2$.

Example 2.6. When $f(y_1,y_2)$ is given by a table, a common problem is to determine whether Y_1 and Y_2 are independent or dependent, find the marginal pmfs $f_{Y_1}(y_1)$ and $f_{Y_2}(y_2)$ and find the conditional pmfs $f_{Y_1|Y_2=y_2}(y_1|y_2)$ and $f_{Y_2|Y_1=y_1}(y_2|y_1)$. Also find $E(Y_1), E(Y_2), V(Y_1), V(Y_2), E(Y_1Y_2)$, and $\text{Cov}(Y_1,Y_2)$.

Suppose that the joint probability mass function of Y_1 and Y_2 is $f(y_1,y_2)$ is tabled as shown.

$f(y_1,y_2)$		y_2		
		0	1	2
y_1	0	1/9	2/9	1/9
	1	2/9	2/9	0/9
	2	1/9	0/9	0/9

a) Are Y_1 and Y_2 independent? Explain.
b) Find the marginal pmfs.
c) Find $E(Y_1)$.
d) Find $E(Y_2)$.
e) Find $\text{Cov}(Y_1,Y_2)$.
f) Find $f_{Y_1|Y_2=y_2}(y_1|y_2)$.

Solution: a) No, the support is not a cross product. Alternatively, $f(2,2) = 0 < f_{Y_1}(2)f_{Y_2}(2)$.
b) Find $f_{Y_1}(y_1)$ by finding the row sums. Find $f_{Y_2}(y_2)$ by finding the column sums. In both cases, $f_{Y_i}(0) = f_{Y_i}(1) = 4/9$ and $f_{Y_i}(2) = 1/9$.
c) $E(Y_1) = \sum y_1 f_{Y_1}(y_1) = 0\frac{4}{9} + 1\frac{4}{9} + 2\frac{1}{9} = \frac{6}{9} \approx 0.6667$.
d) $E(Y_2) \approx 0.6667$ is found as in c) with y_2 replacing y_1.
e) $E(Y_1Y_2) = \sum\sum y_1y_2 f(y_1,y_2) =$
$0+0+0$
$+0+(1)(1)\frac{2}{9}+0$
$+0+0+0 = \frac{2}{9}$. Hence $\text{Cov}(Y_1,Y_2) = E(Y_1Y_2) - E(Y_1)E(Y_2) = \frac{2}{9} - (\frac{6}{9})(\frac{6}{9}) = -\frac{2}{9} \approx -0.2222$.
f) Now $f_{Y_1|Y_2=y_2}(y_1|y_2) = f(y_1,y_2)/f_{Y_2}(y_2)$. If $y_2 = 2$, then $f_{Y_1|Y_2=2}(y_1|2) = f(0,2)/f_{Y_2}(2) = 1$ for $y_1 = 0$. If $y_2 = 1$, then $f_{Y_1|Y_2=}(y_1|1) = f(y_1,2)/f_{Y_2}(1) = 1/2$ for $y_1 = 0,1$. If $y_2 = 0$, then $f_{Y_1|Y_2=0}(0|0) = 1/4$, $f_{Y_1|Y_2=0}(1|0) = 1/2$ and $f_{Y_1|Y_2=0}(2|0) = 1/4$.

Example 2.7. Given the joint pdf $f(y_1, y_2) = kg(y_1, y_2)$ on its support, a common problem is to find k, find the marginal pdfs $f_{Y_1}(y_1)$ and $f_{Y_2}(y_2)$ and find the conditional pdfs $f_{Y_1|Y_2=y_2}(y_1|y_2)$ and $f_{Y_2|Y_1=y_1}(y_2|y_1)$. Also determine whether Y_1 and Y_2 are independent or dependent, and find $E(Y_1), E(Y_2), V(Y_1), V(Y_2), E(Y_1Y_2)$, and $\text{Cov}(Y_1, Y_2)$.

Suppose that the joint pdf of the random variables Y_1 and Y_2 is given by

$$f(y_1, y_2) = 10y_1y_2^2, \text{ if } 0 < y_1 < y_2 < 1$$

and $f(y_1, y_2) = 0$, otherwise. a) Find the marginal pdf of Y_1. Include the support. b) Is $Y_1 \perp\!\!\!\perp Y_2$?

Solution: a) Notice that for a given value of y_1, the joint pdf is positive for $y_1 < y_2 < 1$. Thus

$$f_{Y_1}(y_1) = \int_{y_1}^{1} 10y_1y_2^2 dy_2 = 10y_1 \frac{y_2^3}{3}\Bigg|_{y_1}^{1} = \frac{10y_1}{3}(1 - y_1^3), 0 < y_1 < 1.$$

b) No, the support is not a cross product.

Example 2.8. Suppose that the joint pdf of the random variables Y_1 and Y_2 is given by

$$f(y_1, y_2) = 4y_1(1 - y_2), \text{ if } 0 \le y_1 \le 1, 0 \le y_2 \le 1$$

and $f(y_1, y_2) = 0$, otherwise.

a) Find the marginal pdf of Y_1. Include the support.
b) Find $E(Y_1)$.
c) Find $V(Y_1)$.
d) Are Y_1 and Y_2 independent? Explain.

Solution: a) $f_{Y_1}(y_1) = \int_0^1 4y_1(1 - y_2)dy_2 = 4y_1\left(y_2 - \frac{y_2^2}{2}\right)\Big|_0^1 = 4y_1(1 - \frac{1}{2}) = 2y_1, 0 < y_1 < 1$.

b) $E(Y_1) = \int_0^1 y_1 f_{Y_1}(y_1)dy_1 = \int_0^1 y_1 2y_1 dy_1 = 2\int_0^1 y_1^2 dy_1 = 2\frac{y_1^3}{3}\Big|_0^1 = 2/3$.

c) $E(Y_1^2) = \int_0^1 y_1^2 f_{Y_1}(y_1)dy_1 = \int_0^1 y_1^2 2y_1 dy_1 = 2\int_0^1 y_1^3 dy_1 = 2\frac{y_1^4}{4}\Big|_0^1 = 1/2$. So $V(Y_1) = E(Y_1^2) - [E(Y_1)]^2 = \frac{1}{2} - \frac{4}{9} = \frac{1}{18} \approx 0.0556$.

d) Yes, use Theorem 2.2b with $f(y_1, y_2) = (4y_1)(1 - y_2) = h_1(y_1)h_2(y_2)$ on cross product support.

2.3 Conditional Expectation and Variance

Notation. $Y|X = x$ is a single conditional distribution while $Y|X$ is a family of distributions. For example, if $Y|X = x \sim N(c + dx, \sigma^2)$, then $Y|X \sim N(c + dX, \sigma^2)$ is the family of normal distributions with variance σ^2 and mean $\mu_{Y|X=x} = c + dx$.

Think of Y = weight and X = height. There is a distribution of weights for each value x of height where $X = x$, and weights of people who are $x = 60$ in. tall will on average be less than weights of people who are $x = 70$ in. tall. This notation will be useful for defining $E[Y|X]$ and VAR$[Y|X]$ in Definition 2.15.

Definition 2.13. Suppose that $f(y|x)$ is the conditional pmf or pdf of $Y|X = x$ and that $h(Y)$ is a function of Y. Then the *conditional expected value* $E[h(Y)|X = x]$ of $h(Y)$ given $X = x$ is

$$E[h(Y)|X = x] = \sum_y h(y)f(y|x) \tag{2.12}$$

if $f(y|x)$ is a pmf and if the sum exists when $h(y)$ is replaced by $|h(y)|$. In particular,

$$E[Y|X = x] = \sum_y y f(y|x). \tag{2.13}$$

Similarly,

$$E[h(Y)|X = x] = \int_{-\infty}^{\infty} h(y)f(y|x)dy \tag{2.14}$$

if $f(y|x)$ is a pdf and if the integral exists when $h(y)$ is replaced by $|h(y)|$. In particular,

$$E[Y|X = x] = \int_{-\infty}^{\infty} y f(y|x)dy. \tag{2.15}$$

Definition 2.14. Suppose that $f(y|x)$ is the conditional pmf or pdf of $Y|X = x$. Then the *conditional variance*

$$\text{VAR}(Y|X = x) = E(Y^2|X = x) - [E(Y|X = x)]^2$$

whenever $E(Y^2|X = x)$ exists.

Recall that the pmf or pdf $f(y|x)$ is a function of y with x fixed, but $E(Y|X = x) \equiv m(x)$ is a function of x. In the definition below, both $E(Y|X)$ and VAR$(Y|X)$ are random variables since $m(X)$ and $v(X)$ are random variables. Now think of Y = weight and X = height. Young children 36 in. tall have weights that are less variable than the weights of adults who are 72 in. tall.

Definition 2.15. If $E(Y|X = x) = m(x)$, then the random variable $E(Y|X) = m(X)$. Similarly if VAR$(Y|X = x) = v(x)$, then the random variable VAR$(Y|X) = v(X) = E(Y^2|X) - [E(Y|X)]^2$.

Example 2.9. Suppose that Y = *weight* and X = *height* of college students. Then $E(Y|X = x)$ is a function of x. For example, the weight of 5 ft tall students is less than the weight of 6 ft tall students, on average.

Notation. When computing $E(h(Y))$, the marginal pdf or pmf $f(y)$ is used. When computing $E[h(Y)|X = x]$, the conditional pdf or pmf $f(y|x)$ is used. In a formula such as $E[E(Y|X)]$ the inner expectation uses $f(y|x)$ but the outer expectation uses $f(x)$ since $E(Y|X)$ is a function of X. In the formula below, we could write $E_Y(Y) = E_X[E_{Y|X}(Y|X)]$, but such notation is usually omitted.

Theorem 2.10. Iterated Expectations. Assume the relevant expected values exist. Then

$$E(Y) = E[E(Y|X)].$$

Proof: The result will be shown for the case where (Y,X) has a joint pmf f. For a joint pdf, replace the sums by appropriate integrals. Now

$$E(Y) = \sum_x \sum_y y f(x,y) = \sum_x \sum_y y f_{Y|X}(y|x) f_X(x)$$

$$= \sum_x \left[\sum_y y f_{Y|X}(y|x) \right] f_X(x) = \sum_x E(Y|X=x) f_X(x) = E[E(Y|X)]$$

since the term in brackets is $E(Y|X=x)$. □

Theorem 2.11: Steiner's Formula or the Conditional Variance Identity. Assume the relevant expectations exist. Then

$$\text{VAR}(Y) = E[\text{VAR}(Y|X)] + \text{VAR}[E(Y|X)].$$

Proof: Following Rice (1988, p. 132), since $\text{VAR}(Y|X) = E(Y^2|X) - [E(Y|X)]^2$ is a random variable,

$$E[\text{VAR}(Y|X)] = E[E(Y^2|X)] - E([E(Y|X)]^2).$$

If W is a random variable, then $E(W) = E[E(W|X)]$ by Theorem 2.10 and $\text{VAR}(W) = E(W^2) - [E(W)]^2$ by the shortcut formula. Letting $W = E(Y|X)$ gives

$$\text{VAR}(E(Y|X)) = E([E(Y|X)]^2) - (E[E(Y|X)])^2.$$

Since $E(Y^2) = E[E(Y^2|X)]$ and since $E(Y) = E[E(Y|X)]$,

$$\text{VAR}(Y) = E(Y^2) - [E(Y)]^2 = E[E(Y^2|X)] - (E[E(Y|X)])^2.$$

Adding 0 to $\text{VAR}(Y)$ gives

$$\text{VAR}(Y) = E[E(Y^2|X)] - E([E(Y|X)]^2) + E([E(Y|X)]^2) - (E[E(Y|X)])^2$$

$$= E[\text{VAR}(Y|X)] + \text{VAR}[E(Y|X)].$$ □

A *hierarchical model* models a complicated process by a sequence of models placed in a hierarchy. Interest might be in the marginal expectation $E(Y)$ and marginal variance $\text{VAR}(Y)$. One could find the joint pmf from $f(x,y) = f(y|x)f(x)$, then find the marginal distribution $f_Y(y)$ and then find $E(Y)$ and $\text{VAR}(Y)$. Alternatively, use Theorems 2.10 and 2.11. Hierarchical models are also used in Bayesian applications. See Chap. 11.

Example 2.10. Suppose $Y|X \sim \text{BIN}(X,\rho)$ and $X \sim$ Poisson (λ). Then $E(Y|X) = X\rho$, $\text{VAR}(Y|X) = X\rho(1-\rho)$, and $E(X) = \text{VAR}(X) = \lambda$. Hence $E(Y) = E[E(Y|X)] = E(X\rho) = \rho E(X) = \rho\lambda$ and $\text{VAR}(Y) = E[\text{VAR}(Y|X)] + \text{VAR}[E(Y|X)] = E[X\rho(1-\rho)] + \text{VAR}(X\rho) = \lambda\rho(1-\rho) + \rho^2\text{VAR}(X) = \lambda\rho(1-\rho) + \rho^2\lambda = \lambda\rho$.

2.4 Location–Scale Families

Many univariate distributions are location, scale, or location–scale families. Assume that the random variable Y has a pdf $f_Y(y)$.

Definition 2.16. Let $f_Y(y)$ be the pdf of Y. Then the family of pdfs $f_W(w) = f_Y(w-\mu)$ indexed by the *location parameter* μ, $-\infty < \mu < \infty$, is the *location family* for the random variable $W = \mu + Y$ with *standard pdf* $f_Y(y)$.

Definition 2.17. Let $f_Y(y)$ be the pdf of Y. Then the family of pdfs $f_W(w) = (1/\sigma)f_Y(w/\sigma)$ indexed by the *scale parameter* $\sigma > 0$ is the *scale family* for the random variable $W = \sigma Y$ with *standard pdf* $f_Y(y)$.

Definition 2.18. Let $f_Y(y)$ be the pdf of Y. Then the family of pdfs $f_W(w) = (1/\sigma)f_Y((w-\mu)/\sigma)$ indexed by the *location and scale parameters* μ, $-\infty < \mu < \infty$, and $\sigma > 0$ is the *location–scale family* for the random variable $W = \mu + \sigma Y$ with *standard pdf* $f_Y(y)$.

The most important scale family is the exponential EXP(λ) distribution. Other scale families from Chap. 10 include the chi (p,σ) distribution if p is known, the Gamma G(ν,λ) distribution if ν is known, the lognormal (μ,σ^2) distribution with scale parameter $\tau = e^{\mu}$ if σ^2 is known, the one-sided stable OSS(σ) distribution, the Pareto PAR(σ,λ) distribution if λ is known, and the Weibull $W(\phi,\lambda)$ distribution with scale parameter $\sigma = \lambda^{1/\phi}$ if ϕ is known.

A location family can be obtained from a location–scale family by fixing the scale parameter while a scale family can be obtained by fixing the location parameter. The most important location–scale families are the Cauchy C(μ,σ), double exponential DE(θ,λ), logistic L(μ,σ), normal N(μ,σ^2), and uniform U(θ_1,θ_2) distributions. Other location–scale families from Chap. 10 include the two-parameter exponential EXP(θ,λ), half Cauchy HC(μ,σ), half logistic HL(μ,σ), half normal HN(μ,σ), largest extreme value LEV(θ,σ), Maxwell Boltzmann MB(μ,σ), Rayleigh R(μ,σ), and smallest extreme value SEV(θ,σ) distributions.

2.5 Transformations

Transformations for univariate distributions are important because many "brand name" random variables are transformations of other brand name distributions.

These transformations will also be useful for finding the distribution of the complete sufficient statistic for a one-parameter exponential family. See Chap. 10.

Example 2.11: Common problem. Suppose that X is a discrete random variable with pmf $f_X(x)$ given by a table. Let the **transformation** $Y = t(X)$ for some function t and find the probability function $f_Y(y)$.

Solution: Step 1) Find $t(x)$ for each value of x.
Step 2) Collect $x : t(x) = y$, and sum the corresponding probabilities:
$f_Y(y) = \sum_{x:t(x)=y} f_X(x)$, and table the resulting pmf $f_Y(y)$ of Y.
For example, if $Y = X^2$ and $f_X(-1) = 1/3, f_X(0) = 1/3$, and $f_X(1) = 1/3$, then $f_Y(0) = 1/3$ and $f_Y(1) = 2/3$.

Definition 2.19. Let $h : D \to \mathbb{R}$ be a real valued function with domain D. Then h is **increasing** if $h(y_1) < h(y_2)$, *nondecreasing* if $h(y_1) \leq h(y_2)$, **decreasing** if $h(y_1) > h(y_2)$ and *nonincreasing* if $h(y_1) \geq h(y_2)$ provided that y_1 and y_2 are any two numbers in D with $y_1 < y_2$. The function h is a monotone function if h is either increasing or decreasing.

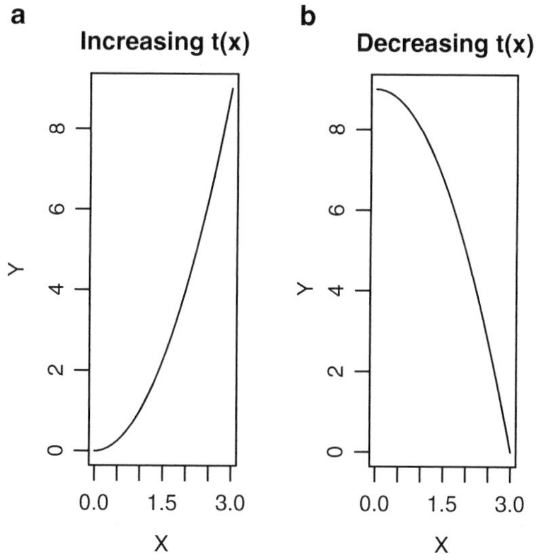

Fig. 2.2 Increasing and decreasing $t(x)$. (**a**) Increasing $t(x)$; (**b**) decreasing $t(x)$

Recall that if h is differentiable on an open interval D or continuous on a closed interval D and differentiable on the interior of D, then h is increasing if $h'(y) > 0$ for all y in the interior of D and h is decreasing if $h'(y) < 0$ for all y in the interior of D. Also if h is increasing then $-h$ is decreasing. Similarly, if h is decreasing then $-h$ is increasing.

Suppose that X is a continuous random variable with pdf $f_X(x)$ on support $\mathscr{X}$. Let the transformation $Y = t(X)$ for some monotone function t. Then there are two ways to find the support $\mathscr{Y}$ of $Y = t(X)$ if the support $\mathscr{X}$ of X is an interval with endpoints $a < b$ where $a = -\infty$ and $b = \infty$ are possible. Let $t(a) \equiv \lim_{y \downarrow a} t(y)$ and let $t(b) \equiv \lim_{y \uparrow b} t(y)$. A graph can help. If t is an increasing function, then $\mathscr{Y}$ is an interval with endpoints $t(a) < t(b)$. If t is a decreasing function, then $\mathscr{Y}$ is an interval with endpoints $t(b) < t(a)$. The second method is to find $x = t^{-1}(y)$. Then if $\mathscr{X} = [a,b]$, say, solve $a \le t^{-1}(y) \le b$ in terms of y.

If $t(x)$ is increasing then $P(\{Y \le y\}) = P(\{X \le t^{-1}(y)\})$ while if $t(x)$ is decreasing $P(\{Y \le y\}) = P(\{X \ge t^{-1}(y)\})$. To see this, look at Fig. 2.2. Suppose the support of Y is $[0,9]$ and the support of X is $[0,3]$. Now the height of the curve is $y = t(x)$. Mentally draw a horizontal line from y to $t(x)$ and then drop a vertical line to the x-axis. The value on the x-axis is $t^{-1}(y)$ since $t(t^{-1}(y)) = y$. Hence in Fig. 2.2a $t^{-1}(4) = 2$ and in Fig. 2.2b $t^{-1}(8) = 1$. If $w < y$ then $t^{-1}(w) < t^{-1}(y)$ if $t(x)$ is increasing as in Fig. 2.2a, but $t^{-1}(w) > t^{-1}(y)$ if $t(x)$ is decreasing as in Fig. 2.2b. Hence $P(Y \le y) = P(t^{-1}(Y) \ge t^{-1}(y)) = P(X \ge t^{-1}(y))$.

Theorem 2.12: The CDF Method or Method of Distributions: Suppose that the continuous cdf $F_X(x)$ is known and that $Y = t(X)$. Let $\mathscr{Y}$ be the support of Y.

i) If t is an increasing function, then $F_Y(y) = P(Y \le y) = P(t(X) \le y) =$
$P(X \le t^{-1}(y)) = F_X(t^{-1}(y))$.
ii) If t is a decreasing function, then $F_Y(y) = P(Y \le y) = P(t(X) \le y) =$
$P(X \ge t^{-1}(y)) = 1 - P(X < t^{-1}(y)) = 1 - P(X \le t^{-1}(y)) = 1 - F_X(t^{-1}(y))$.
iii) The special case $Y = X^2$ is important. If the support of X is positive, use i). If the support of X is negative, use ii). If the support of X is $(-a,a)$ (where $a = \infty$ is allowed), then $F_Y(y) = P(Y \le y) =$
$P(X^2 \le y) = P(-\sqrt{y} \le X \le \sqrt{y}) =$

$$\int_{-\sqrt{y}}^{\sqrt{y}} f_X(x)dx = F_X(\sqrt{y}) - F_X(-\sqrt{y}),\ 0 \le y < a^2.$$

After finding the cdf $F_Y(y)$, the pdf of Y is $f_Y(y) = \dfrac{d}{dy} F_Y(y)$ for $y \in \mathscr{Y}$.

Example 2.12. Suppose X has a pdf with support on the real line and that the pdf is symmetric about μ so $f_X(\mu - w) = f_X(\mu + w)$ for all real w. It can be shown that X has a symmetric distribution about μ if $Z = X - \mu$ and $-Z = \mu - X$ have the same distribution. Several named right skewed distributions with support $y \ge \mu$ are obtained by the transformation $Y = \mu + |X - \mu|$. Similarly, let U be a U(0,1) random variable that is independent of Y, then a symmetric random variable X can be obtained from Y by letting $X = Y$ if $U \le 0.5$ and $X = 2\mu - Y$ if $U > 0.5$. Pairs of such distributions include the exponential and double exponential, normal and half normal, Cauchy and half Cauchy, and logistic and half logistic distributions. Figure 2.3 shows the $N(0,1)$ and HN$(0,1)$ pdfs.

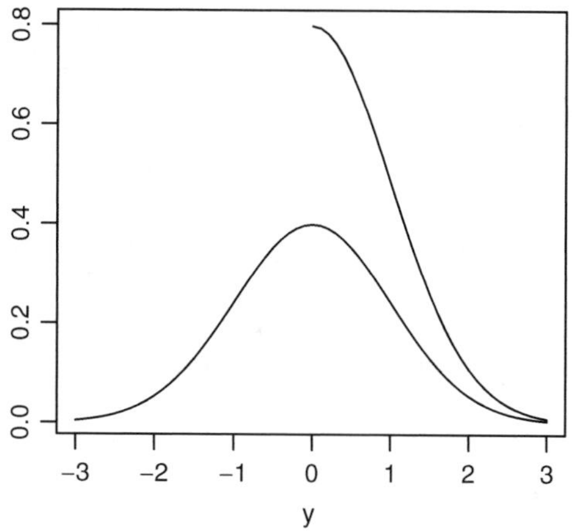

Fig. 2.3 Pdfs for $N(0,1)$ and HN(0, 1) distributions

Notice that for $y \geq \mu$,

$$F_Y(y) = P(Y \leq y) = P(\mu + |X - \mu| \leq y) = P(|X - \mu| \leq y - \mu) =$$

$$P(\mu - y \leq X - \mu \leq y - \mu) = P(2\mu - y \leq X \leq y) = F_X(y) - F_X(2\mu - y).$$

Taking derivatives and using the symmetry of f_X gives $f_Y(y) =$

$$f_X(y) + f_X(2\mu - y) = f_X(\mu + (y - \mu)) + f_X(\mu - (y - \mu)) = 2f_X(\mu + (y - \mu))$$

$= 2f_X(y)$ for $y \geq \mu$. Hence $f_Y(y) = 2f_X(y)I(y \geq \mu)$.

Then X has pdf

$$f_X(x) = \frac{1}{2} f_Y(\mu + |x - \mu|)$$

for all real x, since this pdf is symmetric about μ and $f_X(x) = 0.5 f_Y(x)$ if $x \geq \mu$.

Example 2.13. Often the rules of differentiation such as the multiplication, quotient, and chain rules are needed. For example if the support of X is $[-a, a]$ and if $Y = X^2$, then

$$f_Y(y) = \frac{1}{2\sqrt{y}} [f_X(\sqrt{y}) + f_X(-\sqrt{y})]$$

for $0 \leq y \leq a^2$.

Theorem 2.13: The Transformation Method. Assume that X has pdf $f_X(x)$ and support $\mathscr{X}$. Let $\mathscr{Y}$ be the support of $Y = t(X)$. If $t(x)$ is either increasing or decreasing on $\mathscr{X}$ and if $t^{-1}(y)$ has a continuous derivative on $\mathscr{Y}$, then $Y = t(X)$ has pdf

$$f_Y(y) = f_X(t^{-1}(y)) \left| \frac{dt^{-1}(y)}{dy} \right| \tag{2.16}$$

for $y \in \mathscr{Y}$. As always, $f_Y(y) = 0$ for y not in $\mathscr{Y}$.

Proof: Examining Theorem 2.12, if t is increasing then $F_Y(y) = F_X(t^{-1}(y))$ and

$$f_Y(y) = \frac{d}{dy} F_Y(y)$$

$$= \frac{d}{dy} F_X(t^{-1}(y)) = f_X(t^{-1}(y)) \frac{d}{dy} t^{-1}(y) = f_X(t^{-1}(y)) \left| \frac{dt^{-1}(y)}{dy} \right|$$

for $y \in \mathscr{Y}$ since the derivative of a differentiable increasing function is positive.

If t is a decreasing function, then from Theorem 2.12, $F_Y(y) = 1 - F_X(t^{-1}(x))$. Hence

$$f_Y(y) = \frac{d}{dy}[1 - F_X(t^{-1}(y))] = -f_X(t^{-1}(y)) \frac{d}{dy} t^{-1}(y) = f_X(t^{-1}(y)) \left| \frac{dt^{-1}(y)}{dy} \right|$$

for $y \in \mathscr{Y}$ since the derivative of a differentiable decreasing function is negative. □

Tips: To be useful, formula (2.16) should be simplified as much as possible.

(a) The pdf of Y will often be that of a gamma random variable. In particular, the pdf of Y is often the pdf of an exponential(λ) random variable.
(b) To find the inverse function $x = t^{-1}(y)$, solve the equation $y = t(x)$ for x.
(c) The log transformation is often used. Know how to sketch $\log(x)$ and e^x for $x > 0$. Recall that in this text, $\log(x)$ is the natural logarithm of x.
(d) If $\mathscr{X}$ is an interval with endpoints a and b, find

$$\mathscr{Y} = (\min\{t(a), t(b)\}, \max\{t(a), t(b)\})$$

as described two paragraphs above Theorem 2.12.

Example 2.14. Let X be a random variable with pdf

$$f_X(x) = \frac{1}{x\sqrt{2\pi\sigma^2}} \exp\left(\frac{-(\log(x) - \mu)^2}{2\sigma^2}\right)$$

where $x > 0$, μ is real and $\sigma > 0$. Let $Y = \log(X)$ and find the distribution of Y.

Solution: $X = e^Y = t^{-1}(Y)$. So

$$\left| \frac{dt^{-1}(y)}{dy} \right| = |e^y| = e^y,$$

and

$$f_Y(y) = f_X(t^{-1}(y)) \left| \frac{dt^{-1}(y)}{dy} \right| = f_X(e^y) e^y$$

$$= \frac{1}{e^y\sqrt{2\pi\sigma^2}} \exp\left(\frac{-(\log(e^y) - \mu)^2}{2\sigma^2}\right) e^y$$

$$= \frac{1}{\sqrt{2\pi\sigma^2}} \exp\left(\frac{-(y-\mu)^2}{2\sigma^2}\right)$$

for $y \in (-\infty, \infty)$ since $x > 0$ implies that $y = \log(x) \in (-\infty, \infty)$. Notice that X is lognormal (μ, σ^2) and $Y \sim N(\mu, \sigma^2)$.

Example 2.15. If Y has a Topp–Leone distribution, then pdf of Y is

$$f(y) = \nu(2-2y)(2y-y^2)^{\nu-1}$$

for $\nu > 0$ and $0 < y < 1$. Notice that $F(y) = (2y-y^2)^\nu$ for $0 < y < 1$ since $F'(y) = f(y)$. Then the distribution of $W = -\log(2Y-Y^2)$ will be of interest for later chapters.

Let $X = Y - 1$. Then the support of X is $(-1, 0)$ and $F_X(x) =$
$P(X \le x) = P(Y-1 \le x) = P(Y \le x+1) = F_Y(x+1)$

$$= (2(x+1)-(x+1)^2)^\nu = ((x+1)(2-(x+1)))^\nu = [(x+1)(1-x)]^\nu = (1-x^2)^\nu.$$

So $F_X(x) = (1-x^2)^\nu$ for $-1 < x < 0$. Now the support of W is $w > 0$ and $F_W(w) =$
$P(W \le w) = P(-\log(2Y-Y^2) \le w) = P(\log(2Y-Y^2) \ge -w) =$
$P(2Y-Y^2 \ge e^{-w}) = P(2Y-Y^2-1 \ge e^{-w}-1) = P(-(Y-1)^2 \ge e^{-w}-1) =$
$P((Y-1)^2 \le 1-e^{-w})$. So $F_W(w) = P(X^2 \le 1-e^{-w}) =$
$P(-\sqrt{a} \le X \le \sqrt{a})$ where $a = 1-e^{-w} \in (0,1)$. So $F_W(w) =$
$F_X(\sqrt{a}) - F_X(-\sqrt{a}) = 1 - F_X(-\sqrt{a}) = 1 - F_X(-\sqrt{1-e^{-w}})$

$$= 1 - [1-(-\sqrt{1-e^{-w}})^2]^\nu = 1 - [1-(1-e^{-w})]^\nu = 1 - e^{-w\nu}$$

for $w > 0$. Thus $W = -\log(2Y-Y^2) \sim \text{EXP}(1/\nu)$.

Transformations for vectors are often less useful in applications because the transformation formulas tend to be impractical to compute. For the theorem below, typically $n = 2$. If $Y_1 = t_1(X_1, X_2)$ is of interest, choose $Y_2 = t_2(X_1, X_2)$ such that the determinant J is easy to compute. For example, $Y_2 = X_2$ may work. Finding the support $\mathscr{Y}$ can be difficult, but if the joint pdf of X_1, X_2 is $g(x_1, x_2) = h(x_1, x_2)\, I[(x_1, x_2) \in \mathscr{X}]$, then the joint pdf of Y_1, Y_2 is

$$f(y_1, y_2) = h(t_1^{-1}(\boldsymbol{y}), t_2^{-1}(\boldsymbol{y}))\, I[(t_1^{-1}(\boldsymbol{y}), t_2^{-1}(\boldsymbol{y})) \in \mathscr{X}]\, |J|,$$

and using $I[(t_1^{-1}(\boldsymbol{y}), t_2^{-1}(\boldsymbol{y})) \in \mathscr{X}]$ can be useful for finding $\mathscr{Y}$. The fact that $\prod_{j=1}^k I_{A_j}(\boldsymbol{y}) = I_{\cap_{j=1}^k A_j}(\boldsymbol{y})$ can also be useful. Also sketch $\mathscr{X}$ with x_1 on the horizontal axis and x_2 on the vertical axis, and sketch $\mathscr{Y}$ with y_1 on the horizontal axis and y_2 on the vertical axis. See the following two examples and Problem 2.67.

Theorem 2.14. The Multivariate Transformation Method. Let $X_1, \ldots, X_n$ be random variables with joint pdf $g(x_1, \ldots, x_n)$ and support $\mathscr{X}$. Let $Y_i = t_i(X_1, \ldots, X_n)$ for $i = 1, \ldots, n$. Suppose that $f(y_1, \ldots, y_n)$ is the joint pdf of $Y_1, \ldots, Y_n$ and that the multivariate transformation is one to one. Hence the transformation is invertible and can be solved for the equations $x_i = t_i^{-1}(y_1, \ldots, y_n)$ for $i = 1, \ldots, n$. Then the Jacobian of this multivariate transformation is

$$J = \det \begin{bmatrix} \frac{\partial t_1^{-1}}{\partial y_1} & \cdots & \frac{\partial t_1^{-1}}{\partial y_n} \\ \vdots & & \vdots \\ \frac{\partial t_n^{-1}}{\partial y_1} & \cdots & \frac{\partial t_n^{-1}}{\partial y_n} \end{bmatrix}.$$

Let $|J|$ denote the absolute value of the determinant J. Then the pdf of $Y_1, \ldots, Y_n$ is

$$f(y_1, \ldots, y_n) = g(t_1^{-1}(\mathbf{y}), \ldots, t_n^{-1}(\mathbf{y}))\ |J|. \tag{2.17}$$

Example 2.16. Let X_1 and X_2 have joint pdf

$$g(x_1, x_2) = 2e^{-(x_1+x_2)}$$

for $0 < x_1 < x_2 < \infty$. Let $Y_1 = X_1$ and $Y_2 = X_1 + X_2$. An important step is finding the support $\mathscr{Y}$ of (Y_1, Y_2) from the support of (X_1, X_2)

$$= \mathscr{X} = \{(x_1, x_2) | 0 < x_1 < x_2 < \infty\}.$$

Now $x_1 = y_1 = t_1^{-1}(y_1, y_2)$ and $x_2 = y_2 - y_1 = t_2^{-1}(y_1, y_2)$. Hence $x_1 < x_2$ implies $y_1 < y_2 - y_1$ or $2y_1 < y_2$, and

$$\mathscr{Y} = \{(y_1, y_2) | 0 < 2y_1 < y_2\}.$$

Now

$$\frac{\partial t_1^{-1}}{\partial y_1} = 1, \quad \frac{\partial t_1^{-1}}{\partial y_2} = 0,$$

$$\frac{\partial t_2^{-1}}{\partial y_1} = -1, \quad \frac{\partial t_2^{-1}}{\partial y_2} = 1,$$

and the Jacobian

$$J = \begin{vmatrix} 1 & 0 \\ -1 & 1 \end{vmatrix} = 1.$$

Hence $|J| = 1$. Using indicators,

$$g_{X_1,X_2}(x_1, x_2) = 2e^{-(x_1+x_2)} I(0 < x_1 < x_2 < \infty),$$

and

$$f_{Y_1,Y_2}(y_1, y_2) = g_{X_1,X_2}(y_1, y_2 - y_1)|J| = 2e^{-(y_1+y_2-y_1)} I(0 < y_1 < y_2 - y_1)1 =$$

$$2e^{-y_2} I(0 < 2y_1 < y_2).$$

Notice that Y_1 and Y_2 are not independent since the support $\mathscr{Y}$ is not a cross product. The marginals

$$f_{Y_1}(y_1) = \int_{-\infty}^{\infty} 2e^{-y_2} I(0 < 2y_1 < y_2) dy_2 = \int_{2y_1}^{\infty} 2e^{-y_2} dy_2$$

$$= -2e^{-y_2}\Big|_{y_2=2y_1}^{\infty} = 0 - -2e^{-2y_1} = 2e^{-2y_1}$$

for $0 < y_1 < \infty$, and

$$f_{Y_2}(y_2) = \int_{-\infty}^{\infty} 2e^{-y_2} I(0 < 2y_1 < y_2) dy_1 = \int_0^{y_2/2} 2e^{-y_2} dy_1$$

$$= 2e^{-y_2} y_1 \Big|_{y_1=0}^{y_1=y_2/2} = y_2 e^{-y_2}$$

for $0 < y_2 < \infty$.

Example 2.17. Following Bickel and Doksum (2007, pp. 489–490) , let X_1 and X_2 be independent gamma (ν_i, λ) random variables for $i = 1, 2$. Then X_1 and X_2 have joint pdf $g(x_1, x_2) = g_1(x_1) g_2(x_2) =$

$$\frac{x_1^{\nu_1 - 1} e^{-x_1/\lambda}}{\lambda^{\nu_1} \Gamma(\nu_1)} \frac{x_2^{\nu_2 - 1} e^{-x_2/\lambda}}{\lambda^{\nu_2} \Gamma(\nu_2)} = \frac{1}{\lambda^{\nu_1 + \nu_2} \Gamma(\nu_1) \Gamma(\nu_2)} x_1^{\nu_1 - 1} x_2^{\nu_2 - 1} \exp[-(x_1 + x_2)/\lambda]$$

for $0 < x_1$ and $0 < x_2$. Let $Y_1 = X_1 + X_2$ and $Y_2 = X_1/(X_1 + X_2)$. An important step is finding the support $\mathscr{Y}$ of (Y_1, Y_2) from the support of (X_1, X_2)

$$= \mathscr{X} = \{(x_1, x_2) | 0 < x_1 \text{ and } 0 < x_2\}.$$

Now $y_2 = x_1/y_1$, so $x_1 = y_1 y_2 = t_1^{-1}(y_1, y_2)$ and $x_2 = y_1 - x_1 = y_1 - y_1 y_2 = t_2^{-1}(y_1, y_2)$. Notice that $0 < y_1$ and $0 < x_1 < x_1 + x_2$. Thus $0 < y_2 < 1$, and

$$\mathscr{Y} = \{(y_1, y_2) | 0 < y_1 \text{ and } 0 < y_2 < 1\}.$$

Now

$$\frac{\partial t_1^{-1}}{\partial y_1} = y_2, \quad \frac{\partial t_1^{-1}}{\partial y_2} = y_1,$$

$$\frac{\partial t_2^{-1}}{\partial y_1} = 1 - y_2, \quad \frac{\partial t_2^{-1}}{\partial y_2} = -y_1,$$

and the Jacobian

$$J = \begin{vmatrix} y_2 & y_1 \\ 1 - y_2 & -y_1 \end{vmatrix} = -y_1 y_2 - (y_1 - y_1 y_2) = -y_1,$$

and $|J| = y_1$. So the joint pdf

$$f(y_1,y_2) = g(t_1^{-1}(\boldsymbol{y}),\, t_2^{-1}(\boldsymbol{y}))\ |J| = g(y_1y_2, y_1 - y_1y_2)y_1 =$$

$$\frac{1}{\lambda^{\nu_1+\nu_2}\Gamma(\nu_1)\Gamma(\nu_2)} y_1^{\nu_1-1} y_2^{\nu_1-1} y_1^{\nu_2-1}(1-y_2)^{\nu_2-1}\exp[-(y_1y_2+y_1-y_1y_2)/\lambda]y_1 =$$

$$\frac{1}{\lambda^{\nu_1+\nu_2}\Gamma(\nu_1)\Gamma(\nu_2)} y_1^{\nu_1+\nu_2-1} y_2^{\nu_1-1}(1-y_2)^{\nu_2-1}e^{-y_1/\lambda} =$$

$$\frac{1}{\lambda^{\nu_1+\nu_2}\Gamma(\nu_1+\nu_2)}\, y_1^{\nu_1+\nu_2-1}\, e^{-y_1/\lambda}\ \frac{\Gamma(\nu_1+\nu_2)}{\Gamma(\nu_1)\Gamma(\nu_2)}\, y_2^{\nu_1-1}\, (1-y_2)^{\nu_2-1}.$$

Thus $f(y_1,y_2) = f_1(y_1)f_2(y_2)$ on $\mathscr{Y}$, and $Y_1 \sim \text{gamma}(\nu_1+\nu_2,\lambda) \perp\!\!\!\perp Y_2 \sim \text{beta}(\nu_1,\nu_2)$ by Theorem 2.2b.

2.6 Sums of Random Variables

An important multivariate transformation of the random variables $\boldsymbol{Y} = (Y_1,\ldots,Y_n)$ is $T(Y_1,\ldots,Y_n) = \sum_{i=1}^n Y_i$. Some properties of sums are given below.

Theorem 2.15. Assume that all relevant expectations exist. Let a, $a_1,\ldots,a_n$ and $b_1,\ldots,b_m$ be constants. Let $Y_1,\ldots,Y_n$, and $X_1,\ldots,X_m$ be random variables. Let $g_1,\ldots,g_k$ be functions of $Y_1,\ldots,Y_n$.

i) $E(a) = a$.
ii) $E[aY] = aE[Y]$
iii) $V(aY) = a^2V(Y)$.
iv) $E[g_1(Y_1,\ldots,Y_n)+\cdots+g_k(Y_1,\ldots,Y_n)] = \sum_{i=1}^k E[g_i(Y_1,\ldots,Y_n)]$.
Let $W_1 = \sum_{i=1}^n a_iY_i$ and $W_2 = \sum_{i=1}^m b_iX_i$.

v) $E(W_1) = \sum_{i=1}^n a_iE(Y_i)$.

vi) $V(W_1) = \text{Cov}(W_1,W_1) = \sum_{i=1}^n a_i^2V(Y_i) + 2\sum_{i=1}^{n-1}\sum_{j=i+1}^n a_ia_j\text{Cov}(Y_i,Y_j)$.

vii) $\text{Cov}(W_1,W_2) = \sum_{i=1}^n\sum_{j=1}^m a_ib_j\text{Cov}(Y_i,X_j)$.

viii) $E(\sum_{i=1}^n Y_i) = \sum_{i=1}^n E(Y_i)$.

ix) If $Y_1,\ldots,Y_n$ are independent, $V(\sum_{i=1}^n Y_i) = \sum_{i=1}^n V(Y_i)$.

Let $Y_1,\ldots,Y_n$ be iid random variables with $E(Y_i) = \mu$ and $V(Y_i) = \sigma^2$, then the **sample mean** $\overline{Y} = \frac{1}{n}\sum_{i=1}^n Y_i$. Then

x) $E(\overline{Y}) = \mu$ and
xi) $V(\overline{Y}) = \sigma^2/n$.

Hence the expected value of the sum is the sum of the expected values, the variance of the sum is the sum of the variances for independent random variables, and the covariance of two sums is the double sum of the covariances. Note that ix) follows from vi) with $a_i \equiv 1$, viii) follows from iv) with $g_i(\boldsymbol{Y}) = Y_i$ or from v) with $a_i \equiv 1$, x) follows from v) with $a_i \equiv 1/n$, and xi) can be shown using iii) and ix) using $\overline{Y} = \sum_{i=1}^n (Y_i/n)$.

The assumption that the data are iid or a random sample is often used in a first course in statistics. The assumption will also be often used in this text. The iid assumption is useful for finding the joint pdf or pmf, and the exact or large sample distribution of many important statistics.

Definition 2.20. $Y_1, \ldots, Y_n$ are a **random sample** or **iid** if $Y_1, \ldots, Y_n$ are independent and identically distributed (all of the Y_i have the same distribution).

Example 2.18: Common problem. Let $Y_1, \ldots, Y_n$ be independent random variables with $E(Y_i) = \mu_i$ and $V(Y_i) = \sigma_i^2$. Let $W = \sum_{i=1}^n Y_i$. Then
a) $E(W) = E(\sum_{i=1}^n Y_i) = \sum_{i=1}^n E(Y_i) = \sum_{i=1}^n \mu_i$, and
b) $V(W) = V(\sum_{i=1}^n Y_i) = \sum_{i=1}^n V(Y_i) = \sum_{i=1}^n \sigma_i^2$.

A **statistic** is a function of the data (often a random sample) and known constants. A statistic is a random variable and the **sampling distribution** of a statistic is the distribution of the statistic. Important statistics are $\sum_{i=1}^n Y_i$, $\overline{Y} = \frac{1}{n}\sum_{i=1}^n Y_i$ and $\sum_{i=1}^n a_i Y_i$, where $a_1, \ldots, a_n$ are constants. The following theorem shows how to find the mgf and characteristic function of such statistics.

Theorem 2.16. a) The characteristic function uniquely determines the distribution.

b) If the moment generating function exists, then it uniquely determines the distribution.

c) Assume that $Y_1, \ldots, Y_n$ are independent with characteristic functions $\phi_{Y_i}(t)$. Then the characteristic function of $W = \sum_{i=1}^n Y_i$ is

$$\phi_W(t) = \prod_{i=1}^n \phi_{Y_i}(t). \tag{2.18}$$

d) Assume that $Y_1, \ldots, Y_n$ are iid with characteristic functions $\phi_Y(t)$. Then the characteristic function of $W = \sum_{i=1}^n Y_i$ is

$$\phi_W(t) = [\phi_Y(t)]^n. \tag{2.19}$$

e) Assume that $Y_1, \ldots, Y_n$ are independent with mgfs $m_{Y_i}(t)$. Then the mgf of $W = \sum_{i=1}^n Y_i$ is

$$m_W(t) = \prod_{i=1}^n m_{Y_i}(t). \tag{2.20}$$

f) Assume that $Y_1, \ldots, Y_n$ are iid with mgf $m_Y(t)$. Then the mgf of $W = \sum_{i=1}^n Y_i$ is

$$m_W(t) = [m_Y(t)]^n. \tag{2.21}$$

g) Assume that $Y_1,\ldots,Y_n$ are independent with characteristic functions $\phi_{Y_i}(t)$. Then the characteristic function of $W = \sum_{j=1}^n (a_j + b_j Y_j)$ is

$$\phi_W(t) = \exp\left(it \sum_{j=1}^n a_j\right) \prod_{j=1}^n \phi_{Y_j}(b_j t). \tag{2.22}$$

h) Assume that $Y_1,\ldots,Y_n$ are independent with mgfs $m_{Y_i}(t)$. Then the mgf of $W = \sum_{i=1}^n (a_i + b_i Y_i)$ is

$$m_W(t) = \exp\left(t \sum_{i=1}^n a_i\right) \prod_{i=1}^n m_{Y_i}(b_i t). \tag{2.23}$$

Proof of g): Recall that $\exp(w) = e^w$ and $\exp(\sum_{j=1}^n d_j) = \prod_{j=1}^n \exp(d_j)$. It can be shown that for the purposes of this proof, that the complex constant i in the characteristic function (cf) can be treated in the same way as if it were a real constant. Now

$$\phi_W(t) = E(e^{itW}) = E\left(\exp\left[it \sum_{j=1}^n (a_j + b_j Y_j)\right]\right)$$

$$= \exp\left(it \sum_{j=1}^n a_j\right) E\left(\exp\left[\sum_{j=1}^n itb_j Y_j\right]\right)$$

$$= \exp\left(it \sum_{j=1}^n a_j\right) E\left(\prod_{i=1}^n \exp[itb_j Y_j]\right)$$

$$= \exp\left(it \sum_{j=1}^n a_j\right) \prod_{i=1}^n E[\exp(itb_j Y_j)]$$

since by Theorem 2.5 the expected value of a product of independent random variables is the product of the expected values of the independent random variables. Now in the definition of a cf, the t is a dummy variable as long as t is real. Hence $\phi_Y(t) = E[\exp(itY)]$ and $\phi_Y(s) = E[\exp(isY)]$. Taking $s = tb_j$ gives $E[\exp(itb_j Y_j)] = \phi_{Y_j}(tb_j)$. Thus

$$\phi_W(t) = \exp\left(it \sum_{j=1}^n a_j\right) \prod_{i=1}^n \phi_{Y_j}(tb_j). \quad \square$$

The distribution of $W = \sum_{i=1}^n Y_i$ is known as the convolution of $Y_1,\ldots,Y_n$. Even for $n = 2$, convolution formulas tend to be hard; however, the following two theorems suggest that to find the distribution of $W = \sum_{i=1}^n Y_i$, first find the mgf or characteristic function of W using Theorem 2.16. If the mgf or cf is that of a brand name distribution, then W has that distribution. For example, if the mgf of W is a normal (ν, τ^2) mgf, then W has a normal (ν, τ^2) distribution, written $W \sim N(\nu, \tau^2)$. This technique is useful for several brand name distributions. Chapter 10 will show that many of these distributions are exponential families.

Theorem 2.17. a) If $Y_1, \ldots, Y_n$ are independent binomial BIN(k_i, ρ) random variables, then

$$\sum_{i=1}^{n} Y_i \sim \text{BIN}\left(\sum_{i=1}^{n} k_i, \rho\right).$$

Thus if $Y_1, \ldots, Y_n$ are iid BIN(k, ρ) random variables, then $\sum_{i=1}^{n} Y_i \sim \text{BIN}(nk, \rho)$.

b) Denote a chi-square χ^2_p random variable by $\chi^2(p)$. If $Y_1, \ldots, Y_n$ are independent chi-square $\chi^2_{p_i}$, then

$$\sum_{i=1}^{n} Y_i \sim \chi^2\left(\sum_{i=1}^{n} p_i\right).$$

Thus if $Y_1, \ldots, Y_n$ are iid χ^2_p, then

$$\sum_{i=1}^{n} Y_i \sim \chi^2_{np}.$$

c) If $Y_1, \ldots, Y_n$ are iid exponential EXP(λ), then

$$\sum_{i=1}^{n} Y_i \sim G(n, \lambda).$$

d) If $Y_1, \ldots, Y_n$ are independent Gamma $G(\nu_i, \lambda)$ then

$$\sum_{i=1}^{n} Y_i \sim G\left(\sum_{i=1}^{n} \nu_i, \lambda\right).$$

Thus if $Y_1, \ldots, Y_n$ are iid $G(\nu, \lambda)$, then

$$\sum_{i=1}^{n} Y_i \sim G(n\nu, \lambda).$$

e) If $Y_1, \ldots, Y_n$ are independent normal $N(\mu_i, \sigma_i^2)$, then

$$\sum_{i=1}^{n} (a_i + b_i Y_i) \sim N\left(\sum_{i=1}^{n} (a_i + b_i \mu_i), \sum_{i=1}^{n} b_i^2 \sigma_i^2\right).$$

Here a_i and b_i are fixed constants. Thus if $Y_1, \ldots, Y_n$ are iid $N(\mu, \sigma^2)$, then $\overline{Y} \sim N(\mu, \sigma^2/n)$.

f) If $Y_1, \ldots, Y_n$ are independent Poisson POIS(θ_i), then

$$\sum_{i=1}^{n} Y_i \sim \text{POIS}\left(\sum_{i=1}^{n} \theta_i\right).$$

Thus if $Y_1, \ldots, Y_n$ are iid $POIS(\theta)$, then

$$\sum_{i=1}^{n} Y_i \sim \text{POIS}(n\theta).$$

Theorem 2.18. a) If $Y_1, \ldots, Y_n$ are independent Cauchy $C(\mu_i, \sigma_i)$, then

$$\sum_{i=1}^{n} (a_i + b_i Y_i) \sim C\left(\sum_{i=1}^{n} (a_i + b_i \mu_i), \sum_{i=1}^{n} |b_i| \sigma_i\right).$$

Thus if $Y_1, \ldots, Y_n$ are iid $C(\mu, \sigma)$, then $\overline{Y} \sim C(\mu, \sigma)$.
b) If $Y_1, \ldots, Y_n$ are iid geometric geom(p), then

$$\sum_{i=1}^{n} Y_i \sim \text{NB}(n, p).$$

c) If $Y_1, \ldots, Y_n$ are iid inverse Gaussian $IG(\theta, \lambda)$, then

$$\sum_{i=1}^{n} Y_i \sim \text{IG}(n\theta, n^2\lambda).$$

Also

$$\overline{Y} \sim \text{IG}(\theta, n\lambda).$$

d) If $Y_1, \ldots, Y_n$ are independent negative binomial NB(r_i, ρ), then

$$\sum_{i=1}^{n} Y_i \sim \text{NB}\left(\sum_{i=1}^{n} r_i, \rho\right).$$

Thus if $Y_1, \ldots, Y_n$ are iid $NB(r, \rho)$, then

$$\sum_{i=1}^{n} Y_i \sim \text{NB}(nr, \rho).$$

Example 2.19: Common problem. Given that $Y_1, \ldots, Y_n$ are independent random variables from one of the distributions in Theorem 2.17, find the distribution of $W = \sum_{i=1}^{n} Y_i$ or $W = \sum_{i=1}^{n} b_i Y_i$ by finding the mgf or characteristic function of W and recognizing that it comes from a brand name distribution.

Tips: a) in the product, anything that does not depend on the product index i is treated as a constant.

b) $\exp(a) = e^a$ and $\log(y) = \ln(y) = \log_e(y)$ is the **natural logarithm**.
c)

$$\prod_{i=1}^{n} a^{b\theta_i} = a^{\sum_{i=1}^{n} b\theta_i} = a^{b\sum_{i=1}^{n} \theta_i}.$$

In particular, $\prod_{i=1}^{n} \exp(b\theta_i) = \exp\left(\sum_{i=1}^{n} b\theta_i\right) = \exp\left(b\sum_{i=1}^{n}\theta_i\right)$.

Example 2.20. Suppose $Y_1,\ldots,Y_n$ are iid $IG(\theta,\lambda)$ where the mgf

$$m_{Y_i}(t) = m(t) = \exp\left[\frac{\lambda}{\theta}\left(1-\sqrt{1-\frac{2\theta^2 t}{\lambda}}\right)\right]$$

for $t < \lambda/(2\theta^2)$. Then

$$m_{\sum_{i=1}^n Y_i}(t) = \prod_{i=1}^{n} m_{Y_i}(t) = [m(t)]^n = \exp\left[\frac{n\lambda}{\theta}\left(1-\sqrt{1-\frac{2\theta^2 t}{\lambda}}\right)\right]$$

$$= \exp\left[\frac{n^2\lambda}{n\,\theta}\left(1-\sqrt{1-\frac{2(n\theta)^2\,t}{n^2\lambda}}\right)\right]$$

which is the mgf of an $IG(n\theta, n^2\lambda)$ random variable. The last equality was obtained by multiplying $\frac{n\lambda}{\theta}$ by $1 = n/n$ and by multiplying $\frac{2\theta^2 t}{\lambda}$ by $1 = n^2/n^2$. Hence $\sum_{i=1}^n Y_i \sim \text{IG}(n\theta, n^2\lambda)$.

2.7 Random Vectors

Definition 2.21. $\boldsymbol{Y} = (Y_1,\ldots,Y_p)$ is a $1\times p$ **random vector** if Y_i is a random variable for $i = 1,\ldots,p$. $\boldsymbol{Y}$ is a discrete random vector if each Y_i is discrete, and $\boldsymbol{Y}$ is a continuous random vector if each Y_i is continuous. A random variable Y_1 is the special case of a random vector with $p = 1$.

In the previous sections each $\boldsymbol{Y} = (Y_1,\ldots,Y_n)$ was a random vector. In this section we will consider n random vectors $\boldsymbol{Y}_1,\ldots,\boldsymbol{Y}_n$. Often double subscripts will be used: $\boldsymbol{Y}_i = (Y_{i,1},\ldots,Y_{i,p_i})$ for $i = 1,\ldots,n$.

Notation. The notation for random vectors is rather awkward. In most of the statistical inference literature, $\boldsymbol{Y}$ is a row vector, but in most of the multivariate analysis literature $\boldsymbol{Y}$ is a column vector. In this text, if $\boldsymbol{X}$ and $\boldsymbol{Y}$ are both vectors, a phrase with $\boldsymbol{Y}$ and $\boldsymbol{X}^T$ means that $\boldsymbol{Y}$ is a column vector and $\boldsymbol{X}^T$ is a row vector where T stands for transpose. Hence in the definition below, first $E(\boldsymbol{Y})$ is a $p\times 1$ row vector, but in the definition of $\text{Cov}(\boldsymbol{Y})$ below, $E(\boldsymbol{Y})$ and $\boldsymbol{Y} - E(\boldsymbol{Y})$ are $p\times 1$ column vectors and $(\boldsymbol{Y} - E(\boldsymbol{Y}))^T$ is a $1\times p$ row vector.

Definition 2.22. The *population mean* or **expected value** of a random $1\times p$ random vector $(Y_1,\ldots,Y_p)$ is

$$E(\boldsymbol{Y}) = (E(Y_1),\ldots,E(Y_p))$$

provided that $E(Y_i)$ exists for $i = 1, \ldots, p$. Otherwise the expected value does not exist. Now let $\boldsymbol{Y}$ be a $p \times 1$ column vector. The $p \times p$ *population covariance matrix*

$$\text{Cov}(\boldsymbol{Y}) = E(\boldsymbol{Y} - E(\boldsymbol{Y}))(\boldsymbol{Y} - E(\boldsymbol{Y}))^T = (\sigma_{i,j})$$

where the ij entry of Cov($\boldsymbol{Y}$) is Cov(Y_i, Y_j) $= \sigma_{i,j}$ provided that each $\sigma_{i,j}$ exists. Otherwise Cov($\boldsymbol{Y}$) does not exist.

The covariance matrix is also called the variance–covariance matrix and variance matrix. Sometimes the notation Var($\boldsymbol{Y}$) is used. Note that Cov($\boldsymbol{Y}$) is a symmetric positive semi-definite matrix. If $\boldsymbol{X}$ and $\boldsymbol{Y}$ are $p \times 1$ random vectors, $\boldsymbol{a}$ a conformable constant vector and $\boldsymbol{A}$ and $\boldsymbol{B}$ are conformable constant matrices, then

$$E(\boldsymbol{a} + \boldsymbol{X}) = \boldsymbol{a} + E(\boldsymbol{X}) \text{ and } E(\boldsymbol{X} + \boldsymbol{Y}) = E(\boldsymbol{X}) + E(\boldsymbol{Y}) \tag{2.24}$$

and

$$E(\boldsymbol{AX}) = \boldsymbol{A}E(\boldsymbol{X}) \text{ and } E(\boldsymbol{AXB}) = \boldsymbol{A}E(\boldsymbol{X})\boldsymbol{B}. \tag{2.25}$$

Thus

$$\text{Cov}(\boldsymbol{a} + \boldsymbol{AX}) = \text{Cov}(\boldsymbol{AX}) = \boldsymbol{A}\text{Cov}(\boldsymbol{X})\boldsymbol{A}^T. \tag{2.26}$$

Definition 2.23. Let $\boldsymbol{Y}_1, \ldots, \boldsymbol{Y}_n$ be random vectors with joint pdf or pmf $f(\boldsymbol{y}_1, \ldots, \boldsymbol{y}_n)$. Let $f_{\boldsymbol{Y}_i}(\boldsymbol{y}_i)$ be the marginal pdf or pmf of $\boldsymbol{Y}_i$. Then $\boldsymbol{Y}_1, \ldots, \boldsymbol{Y}_n$ are **independent random vectors** if

$$f(\boldsymbol{y}_1, \ldots, \boldsymbol{y}_n) = f_{\boldsymbol{Y}_1}(\boldsymbol{y}_1) \cdots f_{\boldsymbol{Y}_n}(\boldsymbol{y}_n) = \prod_{i=1}^n f_{\boldsymbol{Y}_i}(\boldsymbol{y}_i).$$

The following theorem is a useful generalization of Theorem 2.7.

Theorem 2.19. Let $\boldsymbol{Y}_1, \ldots, \boldsymbol{Y}_n$ be independent random vectors where $\boldsymbol{Y}_i$ is a $1 \times p_i$ vector for $i = 1, \ldots, n$. and let $\boldsymbol{h}_i : \mathbb{R}^{p_i} \to \mathbb{R}^{p_{j_i}}$ be vector valued functions and suppose that $\boldsymbol{h}_i(\boldsymbol{y}_i)$ is a function of $\boldsymbol{y}_i$ alone for $i = 1, \ldots, n$. Then the random vectors $\boldsymbol{X}_i = \boldsymbol{h}_i(\boldsymbol{Y}_i)$ are independent. There are three important special cases.

i) If $p_{j_i} = 1$ so that each h_i is a real valued function, then the random variables $X_i = h_i(\boldsymbol{Y}_i)$ are independent.
ii) If $p_i = p_{j_i} = 1$ so that each Y_i and each $X_i = h(Y_i)$ are random variables, then $X_1, \ldots, X_n$ are independent.
iii) Let $\boldsymbol{Y} = (Y_1, \ldots, Y_n)$ and $\boldsymbol{X} = (X_1, .., X_m)$ and assume that $\boldsymbol{Y} \perp\!\!\!\perp \boldsymbol{X}$. If $\boldsymbol{h}(\boldsymbol{Y})$ is a vector valued function of $\boldsymbol{Y}$ alone and if $\boldsymbol{g}(\boldsymbol{X})$ is a vector valued function of $\boldsymbol{X}$ alone, then $\boldsymbol{h}(\boldsymbol{Y})$ and $\boldsymbol{g}(\boldsymbol{X})$ are independent random vectors.

Definition 2.24. The **characteristic function** (cf) of a random vector $\boldsymbol{Y}$ is

$$\phi_{\boldsymbol{Y}}(\boldsymbol{t}) = E(e^{i\boldsymbol{t}^T\boldsymbol{Y}})$$

$\forall \boldsymbol{t} \in \mathbb{R}^n$ where the complex number $i = \sqrt{-1}$.

Definition 2.25. The **moment generating function** (mgf) of a random vector $\boldsymbol{Y}$ is

$$m_{\boldsymbol{Y}}(\boldsymbol{t}) = E(e^{\boldsymbol{t}^T \boldsymbol{Y}})$$

provided that the expectation exists for all $\boldsymbol{t}$ in some neighborhood of the origin $\boldsymbol{0}$.

Theorem 2.20. If $Y_1, \ldots, Y_n$ have mgf $m(\boldsymbol{t})$, then moments of all orders exist and for any nonnegative integers $k_1, \ldots, k_j$,

$$E(Y_{i_1}^{k_1} \cdots Y_{i_j}^{k_j}) = \left. \frac{\partial^{k_1+\cdots+k_j}}{\partial t_{i_1}^{k_1} \cdots \partial t_{i_j}^{k_j}} m(\boldsymbol{t}) \right|_{\boldsymbol{t}=\boldsymbol{0}}.$$

In particular,

$$E(Y_i) = \left. \frac{\partial m(\boldsymbol{t})}{\partial t_i} \right|_{\boldsymbol{t}=\boldsymbol{0}}$$

and

$$E(Y_i Y_j) = \left. \frac{\partial^2 m(\boldsymbol{t})}{\partial t_i \partial t_j} \right|_{\boldsymbol{t}=\boldsymbol{0}}.$$

Theorem 2.21. If $Y_1, \ldots, Y_n$ have a cf $\phi_{\boldsymbol{Y}}(\boldsymbol{t})$ and mgf $m_{\boldsymbol{Y}}(\boldsymbol{t})$ then the marginal cf and mgf for $Y_{i_1}, \ldots, Y_{i_k}$ are found from the joint cf and mgf by replacing t_{i_j} by 0 for $j = k+1, \ldots, n$. In particular, if $\boldsymbol{Y} = (\boldsymbol{Y}_1, \boldsymbol{Y}_2)$ and $\boldsymbol{t} = (\boldsymbol{t}_1, \boldsymbol{t}_2)$, then

$$\phi_{\boldsymbol{Y}_1}(\boldsymbol{t}_1) = \phi_{\boldsymbol{Y}}(\boldsymbol{t}_1, \boldsymbol{0}) \text{ and } m_{\boldsymbol{Y}_1}(\boldsymbol{t}_1) = m_{\boldsymbol{Y}}(\boldsymbol{t}_1, \boldsymbol{0}).$$

Proof. Use the definition of the cf and mgf. For example, if $\boldsymbol{Y}_1 = (Y_1, \ldots, Y_k)$ and $\boldsymbol{s} = \boldsymbol{t}_1$, then $m(\boldsymbol{t}_1, \boldsymbol{0}) =$

$$E[\exp(t_1 Y_1 + \cdots + t_k Y_k + 0 Y_{k+1} + \cdots + 0 Y_n)] = E[\exp(t_1 Y_1 + \cdots + t_k Y_k)] =$$

$E[\exp(\boldsymbol{s}^T \boldsymbol{Y}_1)] = m_{\boldsymbol{Y}_1}(\boldsymbol{s})$, which is the mgf of $\boldsymbol{Y}_1$. □

Theorem 2.22. Partition the $1 \times n$ vectors $\boldsymbol{Y}$ and $\boldsymbol{t}$ as $\boldsymbol{Y} = (\boldsymbol{Y}_1, \boldsymbol{Y}_2)$ and $\boldsymbol{t} = (\boldsymbol{t}_1, \boldsymbol{t}_2)$. Then the random vectors $\boldsymbol{Y}_1$ and $\boldsymbol{Y}_2$ are independent iff their joint cf factors into the product of their marginal cfs:

$$\phi_{\boldsymbol{Y}}(\boldsymbol{t}) = \phi_{\boldsymbol{Y}_1}(\boldsymbol{t}_1) \phi_{\boldsymbol{Y}_2}(\boldsymbol{t}_2) \ \forall \boldsymbol{t} \in \mathbb{R}^n.$$

If the joint mgf exists, then the random vectors $\boldsymbol{Y}_1$ and $\boldsymbol{Y}_2$ are independent iff their joint mgf factors into the product of their marginal mgfs:

$$m_{\boldsymbol{Y}}(\boldsymbol{t}) = m_{\boldsymbol{Y}_1}(\boldsymbol{t}_1) m_{\boldsymbol{Y}_2}(\boldsymbol{t}_2)$$

$\forall \boldsymbol{t}$ in some neighborhood of $\boldsymbol{0}$.

2.8 The Multinomial Distribution

Definition 2.26. Assume that there are m iid trials with n outcomes. Let Y_i be the number of the m trials that resulted in the ith outcome and let ρ_i be the probability of the ith outcome for $i = 1,\ldots,n$ where $0 \le \rho_i \le 1$. Thus $\sum_{i=1}^n Y_i = m$ and $\sum_{i=1}^n \rho_i = 1$. Then $\boldsymbol{Y} = (Y_1,\ldots,Y_n)$ has a *multinomial distribution*, written $\boldsymbol{Y} \sim M_n(m,\rho_1,\ldots,\rho_n)$, if the joint pmf of $\boldsymbol{Y}$ is
$f(y_1,\ldots,y_n) = P(Y_1 = y_1,\ldots,Y_n = y_n)$

$$= \frac{m!}{y_1!\cdots y_n!}\rho_1^{y_1}\rho_2^{y_2}\cdots\rho_n^{y_n} = m!\prod_{i=1}^n \frac{\rho_i^{y_i}}{y_i!}. \tag{2.27}$$

The support of $\boldsymbol{Y}$ is $\mathscr{Y} = \{\boldsymbol{y} : \sum_{i=1}^n y_i = m \text{ and } 0 \le y_i \le m \text{ for } i = 1,\ldots,n\}$.

The **multinomial theorem** states that for real x_i and positive integers m and n,

$$(x_1 + \cdots + x_n)^m = \sum_{\boldsymbol{y}\in\mathscr{Y}} \frac{m!}{y_1!\cdots y_n!}x_1^{y_1}x_2^{y_2}\cdots x_n^{y_n}. \tag{2.28}$$

Taking $x_i = \rho_i$ shows that (2.27) is a pmf.

Since Y_n and ρ_n are known if $Y_1,\ldots,Y_{n-1}$ and $\rho_1,\ldots,\rho_{n-1}$ are known, it is convenient to act as if $n-1$ of the outcomes $Y_1,\ldots,Y_{n-1}$ are important and the nth outcome means that none of the $n-1$ important outcomes occurred. With this reasoning, suppose that $\{i_1,\ldots,i_{k-1}\} \subset \{1,\ldots,n\}$. Let $W_j = Y_{i_j}$, and let W_k count the number of times that none of $Y_{i_1},\ldots,Y_{i_{k-1}}$ occurred. Then $W_k = m - \sum_{j=1}^{k-1} Y_{i_j}$ and $P(W_k) = 1 - \sum_{j=1}^{k-1}\rho_{i_j}$. Here W_k represents the unimportant outcomes and the joint distribution of $W_1,\ldots,W_{k-1},W_k$ is multinomial $M_k(m,\rho_{i_1},\ldots,\rho_{i_{k-1}},1-\sum_{j=1}^{k-1}\rho_{i_j})$.

Notice that $\sum_{j=1}^k Y_{i_j}$ counts the number of times that the outcome "one of the outcomes $i_1,\ldots,i_k$ occurred," an outcome with probability $\sum_{j=1}^k \rho_{i_j}$. Hence $\sum_{j=1}^k Y_{i_j} \sim \text{BIN}(m,\sum_{j=1}^k \rho_{i_j})$.

Now consider conditional distributions. If it is known that $Y_{i_j} = y_{i_j}$ for $j = k+1,\ldots,n$, then there are $m - \sum_{j=k+1}^n y_{i_j}$ outcomes left to distribute among $Y_{i_1},\ldots,Y_{i_k}$. The conditional probabilities of Y_i remains proportional to ρ_i, but the conditional probabilities must sum to one. Hence the conditional distribution is again multinomial. These results prove the following theorem.

Theorem 2.23. Assume that $(Y_1,\ldots,Y_n)$ has an $M_n(m,\rho_1,\ldots,\rho_n)$ distribution and that $\{i_1,\ldots,i_k\} \subset \{1,\ldots,n\}$ with $k < n$ and $1 \le i_1 < i_2 < \cdots < i_k \le n$.

a) $(Y_{i_1},\ldots,Y_{i_{k-1}},m-\sum_{j=1}^{k-1}Y_{i_j})$ has an $M_k(m,\rho_{i_1},\ldots,\rho_{i_{k-1}},1-\sum_{j=1}^{k-1}\rho_{i_j})$ distribution.
b) $\sum_{j=1}^k Y_{i_j} \sim \text{BIN}(m,\sum_{j=1}^k \rho_{i_j})$. In particular, $Y_i \sim \text{BIN}(m,\rho_i)$.
c) Suppose that $0 \le y_{i_j} < m$ for $j = k+1,\ldots,n$ and that $0 \le \sum_{j=k+1}^n y_{i_j} < m$.

Let $t = m - \sum_{j=k+1}^n y_{i_j}$ and let $\pi_{i_j} = \rho_{i_j}/\sum_{j=1}^k \rho_{i_j}$ for $j = 1,\ldots,k$. Then the conditional distribution of $Y_{i_1},\ldots,Y_{i_k}|Y_{i_{k+1}} = y_{i_{k+1}},\ldots,Y_{i_n} = y_{i_n}$ is the $M_k(t,\pi_{i_1},\ldots,\pi_{i_k})$

distribution. The support of this conditional distribution is
$\{(y_{i_1},\ldots,y_{i_k}) : \sum_{j=1}^k y_{i_j} = t, \text{ and } 0 \le y_{i_j} \le t \text{ for } j=1,\ldots,k\}$.

Theorem 2.24. Assume that $(Y_1,\ldots,Y_n)$ has an $M_n(m,\rho_1,\ldots,\rho_n)$ distribution. Then the mgf is

$$m(\boldsymbol{t}) = (\rho_1 e^{t_1} + \cdots + \rho_n e^{t_n})^m, \tag{2.29}$$

$E(Y_i) = m\rho_i$, $\text{VAR}(Y_i) = m\rho_i(1-\rho_i)$ and $\text{Cov}(Y_i,Y_j) = -m\rho_i\rho_j$ for $i \ne j$.

Proof. $E(Y_i)$ and $V(Y_i)$ follow from Theorem 2.23b, and $m(\boldsymbol{t}) =$

$$E[\exp(t_1Y_1+\cdots+t_nY_n)] = \sum_{\mathscr{Y}} \exp(t_1y_1+\cdots+t_ny_n)\frac{m!}{y_1!\cdots y_n!}\rho_1^{y_1}\rho_2^{y_2}\cdots\rho_n^{y_n}$$

$$= \sum_{\mathscr{Y}} \frac{m!}{y_1!\cdots y_n!}(\rho_1e^{t_1})^{y_1}\cdots(\rho_ne^{t_n})^{y_n} = (\rho_1e^{t_1}+\cdots+\rho_ne^{t_n})^m$$

by the multinomial theorem (2.28). By Theorem 2.20,

$$E(Y_iY_j) = \frac{\partial^2}{\partial t_i \partial t_j}(\rho_1e^{t_1}+\cdots+\rho_ne^{t_n})^m\bigg|_{\boldsymbol{t}=\boldsymbol{0}} =$$

$$\frac{\partial}{\partial t_j} m(\rho_1e^{t_1}+\cdots+\rho_ne^{t_n})^{m-1}\rho_ie^{t_i}\bigg|_{\boldsymbol{t}=\boldsymbol{0}} =$$

$$m(m-1)(\rho_1e^{t_1}+\cdots+\rho_ne^{t_n})^{m-2}\rho_ie^{t_i}\rho_je^{t_j}\bigg|_{\boldsymbol{t}=\boldsymbol{0}} = m(m-1)\rho_i\rho_j.$$

Hence $\text{Cov}(Y_i,Y_j) = E(Y_iY_j) - E(Y_i)E(Y_j) = m(m-1)\rho_i\rho_j - m\rho_i m\rho_j$
$= -m\rho_i\rho_j$. □

2.9 The Multivariate Normal Distribution

Definition 2.27 (Rao 1973, p. 437). A $p \times 1$ random vector $\boldsymbol{X}$ has a $p-$dimensional *multivariate normal distribution* $N_p(\boldsymbol{\mu},\boldsymbol{\Sigma})$ iff $\boldsymbol{t}^T\boldsymbol{X}$ has a univariate normal distribution for any $p \times 1$ vector $\boldsymbol{t}$.

If $\boldsymbol{\Sigma}$ is positive definite, then $\boldsymbol{X}$ has a joint pdf

$$f(\boldsymbol{z}) = \frac{1}{(2\pi)^{p/2}|\boldsymbol{\Sigma}|^{1/2}} e^{-(1/2)(\boldsymbol{z}-\boldsymbol{\mu})^T\boldsymbol{\Sigma}^{-1}(\boldsymbol{z}-\boldsymbol{\mu})} \tag{2.30}$$

where $|\boldsymbol{\Sigma}|^{1/2}$ is the square root of the determinant of $\boldsymbol{\Sigma}$. Note that if $p=1$, then the quadratic form in the exponent is $(z-\mu)(\sigma^2)^{-1}(z-\mu)$ and X has the univariate $N(\mu,\sigma^2)$ pdf. If $\boldsymbol{\Sigma}$ is positive semi-definite but not positive definite, then $\boldsymbol{X}$ has a degenerate distribution. For example, the univariate $N(0, 0^2)$ distribution is degenerate (the point mass at 0).

Some important properties of MVN distributions are given in the following three propositions. These propositions can be proved using results from Johnson and Wichern (1988, pp. 127–132).

Proposition 2.25. a) If $\boldsymbol{X} \sim N_p(\boldsymbol{\mu}, \boldsymbol{\Sigma})$, then $E(\boldsymbol{X}) = \boldsymbol{\mu}$ and

$$\text{Cov}(\boldsymbol{X}) = \boldsymbol{\Sigma}.$$

b) If $\boldsymbol{X} \sim N_p(\boldsymbol{\mu}, \boldsymbol{\Sigma})$, then any linear combination $\boldsymbol{t}^T\boldsymbol{X} = t_1X_1 + \cdots + t_pX_p \sim N_1(\boldsymbol{t}^T\boldsymbol{\mu}, \boldsymbol{t}^T\boldsymbol{\Sigma}\boldsymbol{t})$. Conversely, if $\boldsymbol{t}^T\boldsymbol{X} \sim N_1(\boldsymbol{t}^T\boldsymbol{\mu}, \boldsymbol{t}^T\boldsymbol{\Sigma}\boldsymbol{t})$ for every $p \times 1$ vector $\boldsymbol{t}$, then $\boldsymbol{X} \sim N_p(\boldsymbol{\mu}, \boldsymbol{\Sigma})$.

c) **The joint distribution of independent normal random variables is MVN.** If $X_1, \ldots, X_p$ are independent univariate normal $N(\mu_i, \sigma_i^2)$ random vectors, then $\boldsymbol{X} = (X_1, \ldots, X_p)^T$ is $N_p(\boldsymbol{\mu}, \boldsymbol{\Sigma})$ where $\boldsymbol{\mu} = (\mu_1, \ldots, \mu_p)^T$ and $\boldsymbol{\Sigma} = diag(\sigma_1^2, \ldots, \sigma_p^2)$ (so the off diagonal entries $\sigma_{i,j} = 0$ while the diagonal entries of $\boldsymbol{\Sigma}$ are $\sigma_{i,i} = \sigma_i^2$.)

d) If $\boldsymbol{X} \sim N_p(\boldsymbol{\mu}, \boldsymbol{\Sigma})$ and if $\boldsymbol{A}$ is a $q \times p$ matrix, then $\boldsymbol{A}\boldsymbol{X} \sim N_q(\boldsymbol{A}\boldsymbol{\mu}, \boldsymbol{A}\boldsymbol{\Sigma}\boldsymbol{A}^T)$. If $\boldsymbol{a}$ is a $p \times 1$ vector of constants, then $\boldsymbol{a} + \boldsymbol{X} \sim N_p(\boldsymbol{a} + \boldsymbol{\mu}, \boldsymbol{\Sigma})$.

It will be useful to partition $\boldsymbol{X}$, $\boldsymbol{\mu}$, and $\boldsymbol{\Sigma}$. Let $\boldsymbol{X}_1$ and $\boldsymbol{\mu}_1$ be $q \times 1$ vectors, let $\boldsymbol{X}_2$ and $\boldsymbol{\mu}_2$ be $(p-q) \times 1$ vectors, let $\boldsymbol{\Sigma}_{11}$ be a $q \times q$ matrix, let $\boldsymbol{\Sigma}_{12}$ be a $q \times (p-q)$ matrix, let $\boldsymbol{\Sigma}_{21}$ be a $(p-q) \times q$ matrix, and let $\boldsymbol{\Sigma}_{22}$ be a $(p-q) \times (p-q)$ matrix. Then

$$\boldsymbol{X} = \begin{pmatrix} \boldsymbol{X}_1 \\ \boldsymbol{X}_2 \end{pmatrix}, \ \boldsymbol{\mu} = \begin{pmatrix} \boldsymbol{\mu}_1 \\ \boldsymbol{\mu}_2 \end{pmatrix}, \ \text{and } \boldsymbol{\Sigma} = \begin{pmatrix} \boldsymbol{\Sigma}_{11} & \boldsymbol{\Sigma}_{12} \\ \boldsymbol{\Sigma}_{21} & \boldsymbol{\Sigma}_{22} \end{pmatrix}.$$

Proposition 2.26. a) **All subsets of a MVN are MVN:** $(X_{k_1}, \ldots, X_{k_q})^T \sim N_q(\tilde{\boldsymbol{\mu}}, \tilde{\boldsymbol{\Sigma}})$ where $\tilde{\boldsymbol{\mu}}_i = E(X_{k_i})$ and $\tilde{\boldsymbol{\Sigma}}_{ij} = \text{Cov}(X_{k_i}, X_{k_j})$. In particular, $\boldsymbol{X}_1 \sim N_q(\boldsymbol{\mu}_1, \boldsymbol{\Sigma}_{11})$ and $\boldsymbol{X}_2 \sim N_{p-q}(\boldsymbol{\mu}_2, \boldsymbol{\Sigma}_{22})$.

b) If $\boldsymbol{X}_1$ and $\boldsymbol{X}_2$ are independent, then $\text{Cov}(\boldsymbol{X}_1, \boldsymbol{X}_2) = \boldsymbol{\Sigma}_{12} = E[(\boldsymbol{X}_1 - E(\boldsymbol{X}_1))(\boldsymbol{X}_2 - E(\boldsymbol{X}_2))^T] = \boldsymbol{0}$, a $q \times (p-q)$ matrix of zeroes.

c) If $\boldsymbol{X} \sim N_p(\boldsymbol{\mu}, \boldsymbol{\Sigma})$, then $\boldsymbol{X}_1$ and $\boldsymbol{X}_2$ are independent iff $\boldsymbol{\Sigma}_{12} = \boldsymbol{0}$.

d) If $\boldsymbol{X}_1 \sim N_q(\boldsymbol{\mu}_1, \boldsymbol{\Sigma}_{11})$ and $\boldsymbol{X}_2 \sim N_{p-q}(\boldsymbol{\mu}_2, \boldsymbol{\Sigma}_{22})$ are independent, then

$$\begin{pmatrix} \boldsymbol{X}_1 \\ \boldsymbol{X}_2 \end{pmatrix} \sim N_p\left(\begin{pmatrix} \boldsymbol{\mu}_1 \\ \boldsymbol{\mu}_2 \end{pmatrix}, \begin{pmatrix} \boldsymbol{\Sigma}_{11} & \boldsymbol{0} \\ \boldsymbol{0} & \boldsymbol{\Sigma}_{22} \end{pmatrix} \right).$$

Proposition 2.27. The conditional distribution of a MVN is MVN. If $\boldsymbol{X} \sim N_p(\boldsymbol{\mu}, \boldsymbol{\Sigma})$, then the conditional distribution of $\boldsymbol{X}_1$ given that $\boldsymbol{X}_2 = \boldsymbol{x}_2$ is multivariate normal with mean $\boldsymbol{\mu}_1 + \boldsymbol{\Sigma}_{12}\boldsymbol{\Sigma}_{22}^{-1}(\boldsymbol{x}_2 - \boldsymbol{\mu}_2)$ and covariance matrix $\boldsymbol{\Sigma}_{11} - \boldsymbol{\Sigma}_{12}\boldsymbol{\Sigma}_{22}^{-1}\boldsymbol{\Sigma}_{21}$. That is,

$$\boldsymbol{X}_1 | \boldsymbol{X}_2 = \boldsymbol{x}_2 \sim N_q(\boldsymbol{\mu}_1 + \boldsymbol{\Sigma}_{12}\boldsymbol{\Sigma}_{22}^{-1}(\boldsymbol{x}_2 - \boldsymbol{\mu}_2), \boldsymbol{\Sigma}_{11} - \boldsymbol{\Sigma}_{12}\boldsymbol{\Sigma}_{22}^{-1}\boldsymbol{\Sigma}_{21}).$$

Example 2.21. Let $p = 2$ and let $(Y,X)^T$ have a bivariate normal distribution. That is,

$$\begin{pmatrix} Y \\ X \end{pmatrix} \sim N_2\left(\begin{pmatrix} \mu_Y \\ \mu_X \end{pmatrix}, \begin{pmatrix} \sigma_Y^2 & \text{Cov}(Y,X) \\ \text{Cov}(X,Y) & \sigma_X^2 \end{pmatrix} \right).$$

Also recall that the population correlation between X and Y is given by

$$\rho(X,Y) = \frac{\text{Cov}(X,Y)}{\sqrt{\text{VAR}(X)}\sqrt{\text{VAR}(Y)}} = \frac{\sigma_{X,Y}}{\sigma_X \sigma_Y}$$

if $\sigma_X > 0$ and $\sigma_Y > 0$. Then $Y|X = x \sim N(E(Y|X=x), \text{VAR}(Y|X=x))$ where the conditional mean

$$E(Y|X=x) = \mu_Y + \text{Cov}(Y,X)\frac{1}{\sigma_X^2}(x - \mu_X) = \mu_Y + \rho(X,Y)\sqrt{\frac{\sigma_Y^2}{\sigma_X^2}}(x - \mu_X)$$

and the conditional variance

$$\text{VAR}(Y|X=x) = \sigma_Y^2 - \text{Cov}(X,Y)\frac{1}{\sigma_X^2}\text{Cov}(X,Y)$$

$$= \sigma_Y^2 - \rho(X,Y)\sqrt{\frac{\sigma_Y^2}{\sigma_X^2}}\rho(X,Y)\sqrt{\sigma_X^2}\sqrt{\sigma_Y^2}$$

$$= \sigma_Y^2 - \rho^2(X,Y)\sigma_Y^2 = \sigma_Y^2[1 - \rho^2(X,Y)].$$

Also $aX + bY$ is univariate normal with mean $a\mu_X + b\mu_Y$ and variance

$$a^2\sigma_X^2 + b^2\sigma_Y^2 + 2ab\,\text{Cov}(X,Y).$$

Remark 2.2. There are several common misconceptions. First, **it is not true that every linear combination $t^T X$ of normal random variables is a normal random variable,** and **it is not true that all uncorrelated normal random variables are independent**. The key condition in Proposition 2.25b and Proposition 2.26c is that the joint distribution of $\boldsymbol{X}$ is MVN. It is possible that $X_1, X_2, \ldots, X_p$ each has a marginal distribution that is univariate normal, but the joint distribution of $\boldsymbol{X}$ is not MVN. Examine the following example from Rohatgi (1976, p. 229). Suppose that the joint pdf of X and Y is a mixture of two bivariate normal distributions both with $EX = EY = 0$ and $\text{VAR}(X) = \text{VAR}(Y) = 1$, but $\text{Cov}(X,Y) = \pm\rho$. Hence

$$f(x,y) = \frac{1}{2}\frac{1}{2\pi\sqrt{1-\rho^2}}\exp\left(\frac{-1}{2(1-\rho^2)}(x^2 - 2\rho xy + y^2)\right) +$$

$$\frac{1}{2}\frac{1}{2\pi\sqrt{1-\rho^2}}\exp\left(\frac{-1}{2(1-\rho^2)}(x^2 + 2\rho xy + y^2)\right) \equiv \frac{1}{2}f_1(x,y) + \frac{1}{2}f_2(x,y)$$

where x and y are real and $0 < \rho < 1$. Since both marginal distributions of $f_i(x,y)$ are $N(0,1)$ for $i = 1$ and 2 by Proposition 2.26a, the marginal distributions of X and Y are $N(0,1)$. Since $\int\int xyf_i(x,y)dxdy = \rho$ for $i = 1$ and $-\rho$ for $i = 2$, X and Y are uncorrelated, but X and Y are not independent since $f(x,y) \neq f_X(x)f_Y(y)$.

Remark 2.3. In Proposition 2.27, suppose that $\boldsymbol{X} = (Y, X_2, \ldots, X_p)^T$. Let $X_1 = Y$ and $\boldsymbol{X}_2 = (X_2, \ldots, X_p)^T$. Then $E[Y|\boldsymbol{X}_2] = \beta_1 + \beta_2 X_2 + \cdots + \beta_p X_p$ and $\text{VAR}[Y|\boldsymbol{X}_2]$ is a constant that does not depend on $\boldsymbol{X}_2$. Hence $Y|\boldsymbol{X}_2 = \beta_1 + \beta_2 X_2 + \cdots + \beta_p X_p + e$ follows the multiple linear regression model.

2.10 Elliptically Contoured Distributions

Definition 2.28 (Johnson 1987, pp. 107–108). A $p \times 1$ random vector has an *elliptically contoured distribution,* also called an *elliptically symmetric distribution,* if $\boldsymbol{X}$ has joint pdf

$$f(\boldsymbol{z}) = k_p |\boldsymbol{\Sigma}|^{-1/2} g[(\boldsymbol{z} - \boldsymbol{\mu})^T \boldsymbol{\Sigma}^{-1} (\boldsymbol{z} - \boldsymbol{\mu})], \tag{2.31}$$

and we say $\boldsymbol{X}$ has an elliptically contoured $EC_p(\boldsymbol{\mu}, \boldsymbol{\Sigma}, g)$ distribution.

If $\boldsymbol{X}$ has an elliptically contoured (EC) distribution, then the characteristic function of $\boldsymbol{X}$ is

$$\phi_{\boldsymbol{X}}(\boldsymbol{t}) = \exp(i\boldsymbol{t}^T \boldsymbol{\mu}) \psi(\boldsymbol{t}^T \boldsymbol{\Sigma} \boldsymbol{t}) \tag{2.32}$$

for some function ψ. If the second moments exist, then

$$E(\boldsymbol{X}) = \boldsymbol{\mu} \tag{2.33}$$

and

$$\text{Cov}(\boldsymbol{X}) = c_X \boldsymbol{\Sigma} \tag{2.34}$$

where

$$c_X = -2\psi'(0).$$

Definition 2.29. The *population squared Mahalanobis distance*

$$U \equiv D^2 = D^2(\boldsymbol{\mu}, \boldsymbol{\Sigma}) = (\boldsymbol{X} - \boldsymbol{\mu})^T \boldsymbol{\Sigma}^{-1} (\boldsymbol{X} - \boldsymbol{\mu}) \tag{2.35}$$

has pdf

$$h(u) = \frac{\pi^{p/2}}{\Gamma(p/2)} k_p u^{p/2-1} g(u). \tag{2.36}$$

For $c > 0$, an $EC_p(\boldsymbol{\mu}, c\boldsymbol{I}, g)$ distribution is *spherical about* $\boldsymbol{\mu}$ where $\boldsymbol{I}$ is the $p \times p$ identity matrix. The *multivariate normal distribution* $N_p(\boldsymbol{\mu}, \boldsymbol{\Sigma})$ has $k_p = (2\pi)^{-p/2}$, $\psi(u) = g(u) = \exp(-u/2)$, and $h(u)$ is the χ^2_p density.

The following lemma is useful for proving properties of EC distributions without using the characteristic function (2.32). See Eaton (1986) and Cook (1998, pp. 57, 130).

Lemma 2.28. Let $\boldsymbol{X}$ be a $p \times 1$ random vector with 1st moments; i.e., $E(\boldsymbol{X})$ exists. Let $\boldsymbol{B}$ be any constant full rank $p \times r$ matrix where $1 \leq r \leq p$. Then $\boldsymbol{X}$ is elliptically contoured iff for all such conforming matrices $\boldsymbol{B}$,

$$E(\boldsymbol{X}|\boldsymbol{B}^T \boldsymbol{X}) = \boldsymbol{\mu} + \boldsymbol{M}_B \boldsymbol{B}^T(\boldsymbol{X} - \boldsymbol{\mu}) = \boldsymbol{a}_B + \boldsymbol{M}_B \boldsymbol{B}^T \boldsymbol{X} \tag{2.37}$$

where the $p \times 1$ constant vector $\boldsymbol{a}_B$ and the $p \times r$ constant matrix $\boldsymbol{M}_B$ both depend on $\boldsymbol{B}$.

A useful fact is that $\boldsymbol{a}_B$ and $\boldsymbol{M}_B$ do not depend on g:

$$\boldsymbol{a}_B = \boldsymbol{\mu} - \boldsymbol{M}_B \boldsymbol{B}^T \boldsymbol{\mu} = (\boldsymbol{I}_p - \boldsymbol{M}_B \boldsymbol{B}^T)\boldsymbol{\mu},$$

and

$$\boldsymbol{M}_B = \boldsymbol{\Sigma} \boldsymbol{B}(\boldsymbol{B}^T \boldsymbol{\Sigma} \boldsymbol{B})^{-1}.$$

Notice that in the formula for $\boldsymbol{M}_B$, $\boldsymbol{\Sigma}$ can be replaced by $c\boldsymbol{\Sigma}$ where $c > 0$ is a constant. In particular, if the EC distribution has second moments, $\text{Cov}(\boldsymbol{X})$ can be used instead of $\boldsymbol{\Sigma}$.

To use Lemma 2.28 to prove interesting properties, partition $\boldsymbol{X}$, $\boldsymbol{\mu}$, and $\boldsymbol{\Sigma}$. Let $\boldsymbol{X}_1$ and $\boldsymbol{\mu}_1$ be $q \times 1$ vectors, let $\boldsymbol{X}_2$ and $\boldsymbol{\mu}_2$ be $(p-q) \times 1$ vectors. Let $\boldsymbol{\Sigma}_{11}$ be a $q \times q$ matrix, let $\boldsymbol{\Sigma}_{12}$ be a $q \times (p-q)$ matrix, let $\boldsymbol{\Sigma}_{21}$ be a $(p-q) \times q$ matrix, and let $\boldsymbol{\Sigma}_{22}$ be a $(p-q) \times (p-q)$ matrix. Then

$$\boldsymbol{X} = \begin{pmatrix} \boldsymbol{X}_1 \\ \boldsymbol{X}_2 \end{pmatrix}, \ \boldsymbol{\mu} = \begin{pmatrix} \boldsymbol{\mu}_1 \\ \boldsymbol{\mu}_2 \end{pmatrix}, \text{ and } \boldsymbol{\Sigma} = \begin{pmatrix} \boldsymbol{\Sigma}_{11} & \boldsymbol{\Sigma}_{12} \\ \boldsymbol{\Sigma}_{21} & \boldsymbol{\Sigma}_{22} \end{pmatrix}.$$

Also assume that the $(p+1) \times 1$ vector $(Y, \boldsymbol{X}^T)^T$ is $EC_{p+1}(\boldsymbol{\mu}, \boldsymbol{\Sigma}, g)$ where Y is a random variable, $\boldsymbol{X}$ is a $p \times 1$ vector, and use

$$\begin{pmatrix} Y \\ \boldsymbol{X} \end{pmatrix}, \ \boldsymbol{\mu} = \begin{pmatrix} \mu_Y \\ \boldsymbol{\mu}_X \end{pmatrix}, \text{ and } \boldsymbol{\Sigma} = \begin{pmatrix} \Sigma_{YY} & \boldsymbol{\Sigma}_{YX} \\ \boldsymbol{\Sigma}_{XY} & \boldsymbol{\Sigma}_{XX} \end{pmatrix}.$$

Proposition 2.29. Let $\boldsymbol{X} \sim EC_p(\boldsymbol{\mu}, \boldsymbol{\Sigma}, g)$ and assume that $E(\boldsymbol{X})$ exists.

a) Any subset of $\boldsymbol{X}$ is EC, in particular $\boldsymbol{X}_1$ is EC.
b) (Cook 1998 p. 131, Kelker 1970). If $\text{Cov}(\boldsymbol{X})$ is nonsingular,

$$\text{Cov}(\boldsymbol{X}|\boldsymbol{B}^T \boldsymbol{X}) = d_g(\boldsymbol{B}^T \boldsymbol{X})[\boldsymbol{\Sigma} - \boldsymbol{\Sigma} \boldsymbol{B}(\boldsymbol{B}^T \boldsymbol{\Sigma} \boldsymbol{B})^{-1} \boldsymbol{B}^T \boldsymbol{\Sigma}]$$

where the real valued function $d_g(\boldsymbol{B}^T \boldsymbol{X})$ is constant iff $\boldsymbol{X}$ is MVN.

Proof of a). Let $\boldsymbol{A}$ be an arbitrary full rank $q \times r$ matrix where $1 \leq r \leq q$. Let

$$\boldsymbol{B} = \begin{pmatrix} \boldsymbol{A} \\ \boldsymbol{0} \end{pmatrix}.$$

Then $\boldsymbol{B}^T\boldsymbol{X} = \boldsymbol{A}^T\boldsymbol{X}_1$, and

$$E[\boldsymbol{X}|\boldsymbol{B}^T\boldsymbol{X}] = E\left[\begin{pmatrix}\boldsymbol{X}_1\\ \boldsymbol{X}_2\end{pmatrix}|\boldsymbol{A}^T\boldsymbol{X}_1\right] =$$

$$\begin{pmatrix}\boldsymbol{\mu}_1\\ \boldsymbol{\mu}_2\end{pmatrix} + \begin{pmatrix}\boldsymbol{M}_{1B}\\ \boldsymbol{M}_{2B}\end{pmatrix}\left(\boldsymbol{A}^T\ \boldsymbol{0}^T\right)\begin{pmatrix}\boldsymbol{X}_1-\boldsymbol{\mu}_1\\ \boldsymbol{X}_2-\boldsymbol{\mu}_2\end{pmatrix}$$

by Lemma 2.28. Hence $E[\boldsymbol{X}_1|\boldsymbol{A}^T\boldsymbol{X}_1] = \boldsymbol{\mu}_1 + \boldsymbol{M}_{1B}\boldsymbol{A}^T(\boldsymbol{X}_1-\boldsymbol{\mu}_1)$. Since $\boldsymbol{A}$ was arbitrary, $\boldsymbol{X}_1$ is EC by Lemma 2.28. Notice that $\boldsymbol{M}_B = \boldsymbol{\Sigma}\boldsymbol{B}(\boldsymbol{B}^T\boldsymbol{\Sigma}\boldsymbol{B})^{-1} =$

$$\begin{pmatrix}\boldsymbol{\Sigma}_{11} & \boldsymbol{\Sigma}_{12}\\ \boldsymbol{\Sigma}_{21} & \boldsymbol{\Sigma}_{22}\end{pmatrix}\begin{pmatrix}\boldsymbol{A}\\ \boldsymbol{0}\end{pmatrix}\left[\left(\boldsymbol{A}^T\ \boldsymbol{0}^T\right)\begin{pmatrix}\boldsymbol{\Sigma}_{11} & \boldsymbol{\Sigma}_{12}\\ \boldsymbol{\Sigma}_{21} & \boldsymbol{\Sigma}_{22}\end{pmatrix}\begin{pmatrix}\boldsymbol{A}\\ \boldsymbol{0}\end{pmatrix}\right]^{-1}$$

$$= \begin{pmatrix}\boldsymbol{M}_{1B}\\ \boldsymbol{M}_{2B}\end{pmatrix}.$$

Hence

$$\boldsymbol{M}_{1B} = \boldsymbol{\Sigma}_{11}\boldsymbol{A}(\boldsymbol{A}^T\boldsymbol{\Sigma}_{11}\boldsymbol{A})^{-1}$$

and $\boldsymbol{X}_1$ is EC with location and dispersion parameters $\boldsymbol{\mu}_1$ and $\boldsymbol{\Sigma}_{11}$. □

Proposition 2.30. Let $(Y, \boldsymbol{X}^T)^T$ be $EC_{p+1}(\boldsymbol{\mu}, \boldsymbol{\Sigma}, g)$ where Y is a random variable.

a) Assume that $E[(Y, \boldsymbol{X}^T)^T]$ exists. Then $E(Y|\boldsymbol{X}) = \alpha + \boldsymbol{\beta}^T\boldsymbol{X}$ where $\alpha = \mu_Y - \boldsymbol{\beta}^T\boldsymbol{\mu}_X$ and

$$\boldsymbol{\beta} = \boldsymbol{\Sigma}_{XX}^{-1}\boldsymbol{\Sigma}_{XY}.$$

b) Even if the first moment does not exist, the conditional median

$$\text{MED}(Y|\boldsymbol{X}) = \alpha + \boldsymbol{\beta}^T\boldsymbol{X}$$

where α and $\boldsymbol{\beta}$ are given in a).

Proof. a) The trick is to choose $\boldsymbol{B}$ so that Lemma 2.28 applies. Let

$$\boldsymbol{B} = \begin{pmatrix}\boldsymbol{0}^T\\ \boldsymbol{I}_p\end{pmatrix}.$$

Then $\boldsymbol{B}^T\boldsymbol{\Sigma}\boldsymbol{B} = \boldsymbol{\Sigma}_{XX}$ and

$$\boldsymbol{\Sigma}\boldsymbol{B} = \begin{pmatrix}\boldsymbol{\Sigma}_{YX}\\ \boldsymbol{\Sigma}_{XX}\end{pmatrix}.$$

Now

$$E\left[\begin{pmatrix}Y\\ \boldsymbol{X}\end{pmatrix}|\boldsymbol{X}\right] = E\left[\begin{pmatrix}Y\\ \boldsymbol{X}\end{pmatrix}|\boldsymbol{B}^T\begin{pmatrix}Y\\ \boldsymbol{X}\end{pmatrix}\right]$$

$$= \boldsymbol{\mu} + \boldsymbol{\Sigma}\boldsymbol{B}(\boldsymbol{B}^T\boldsymbol{\Sigma}\boldsymbol{B})^{-1}\boldsymbol{B}^T\begin{pmatrix}Y-\mu_Y\\ \boldsymbol{X}-\boldsymbol{\mu}_X\end{pmatrix}$$

by Lemma 2.28. The right-hand side of the last equation is equal to

$$\boldsymbol{\mu} + \begin{pmatrix} \boldsymbol{\Sigma}_{YX} \\ \boldsymbol{\Sigma}_{XX} \end{pmatrix} \boldsymbol{\Sigma}_{XX}^{-1}(\boldsymbol{X} - \boldsymbol{\mu}_X) = \begin{pmatrix} \mu_Y - \boldsymbol{\Sigma}_{YX}\boldsymbol{\Sigma}_{XX}^{-1}\boldsymbol{\mu}_X + \boldsymbol{\Sigma}_{YX}\boldsymbol{\Sigma}_{XX}^{-1}\boldsymbol{X} \\ \boldsymbol{X} \end{pmatrix}$$

and the result follows since

$$\boldsymbol{\beta}^T = \boldsymbol{\Sigma}_{YX}\boldsymbol{\Sigma}_{XX}^{-1}.$$

b) See and Croux et al. (2001) for references.

Example 2.22. This example illustrates another application of Lemma 2.28. Suppose that $\boldsymbol{X}$ comes from a mixture of two multivariate normals with the same mean and proportional covariance matrices. That is, let

$$\boldsymbol{X} \sim (1-\gamma)N_p(\boldsymbol{\mu}, \boldsymbol{\Sigma}) + \gamma N_p(\boldsymbol{\mu}, c\boldsymbol{\Sigma})$$

where $c > 0$ and $0 < \gamma < 1$. Since the multivariate normal distribution is elliptically contoured (and see Proposition 1.14c),

$$E(\boldsymbol{X}|\boldsymbol{B}^T\boldsymbol{X}) = (1-\gamma)[\boldsymbol{\mu} + \boldsymbol{M}_1\boldsymbol{B}^T(\boldsymbol{X}-\boldsymbol{\mu})] + \gamma[\boldsymbol{\mu} + \boldsymbol{M}_2\boldsymbol{B}^T(\boldsymbol{X}-\boldsymbol{\mu})]$$

$$= \boldsymbol{\mu} + [(1-\gamma)\boldsymbol{M}_1 + \gamma\boldsymbol{M}_2]\boldsymbol{B}^T(\boldsymbol{X}-\boldsymbol{\mu}) \equiv \boldsymbol{\mu} + \boldsymbol{M}\boldsymbol{B}^T(\boldsymbol{X}-\boldsymbol{\mu}).$$

Since $\boldsymbol{M}_B$ only depends on $\boldsymbol{B}$ and $\boldsymbol{\Sigma}$, it follows that $\boldsymbol{M}_1 = \boldsymbol{M}_2 = \boldsymbol{M} = \boldsymbol{M}_B$. Hence $\boldsymbol{X}$ has an elliptically contoured distribution by Lemma 2.28.

2.11 Summary

1. Y_1 and Y_2 are dependent if the support $\mathscr{Y} = \{(y_1,y_2)|f(y_1,y_2) > 0\}$ is not a cross product.
2. If the support is a cross product, then Y_1 and Y_2 are independent iff $f(y_1,y_2) = h_1(y_1)h_2(y_2)$ for all $(y_1,y_2) \in \mathscr{Y}$ where $h_i(y_i)$ is a positive function of y_i alone. If no such factorization exists, then Y_1 and Y_2 are dependent.
3. If $Y_1,\ldots,Y_n$ are independent, then the functions $h_1(Y_1),\ldots,h_n(Y_n)$ are independent.
4. Given $f(y_1,y_2)$, find $E[h(Y_i)]$ by finding the marginal pdf or pmf $f_{Y_i}(y_i)$ and using the marginal distribution in the expectation.
5. $E[Y] = E[E(Y|X)]$ and $V(Y) = E[V(Y|X)] + V[E(Y|X)]$.
6. Find the pmf of $Y = t(X)$ and the (sample space =) support $\mathscr{Y}$ given the pmf of X by collecting terms $x : y = t(x)$.
7. For increasing or decreasing t, the pdf of $Y = t(X)$ is

$$f_Y(y) = f_X(t^{-1}(y)) \left| \frac{dt^{-1}(y)}{dy} \right|$$

for $y \in \mathscr{Y}$. Also be able to find the support $\mathscr{Y}$.

8. Find the joint pdf of $Y_1 = t_1(X_1, X_2)$ and $Y_2 = t_2(X_1, X_2)$: $f_{Y_1,Y_2}(y_1, y_2)$ $= f_{X_1,X_2}(t_1^{-1}(y_1, y_2), t_2^{-1}(y_1, y_2))|J|$. Finding the support $\mathscr{Y}$ is crucial. Using indicator functions can help. Know that $\prod_{j=1}^k I_{A_j}(\boldsymbol{y}) = I_{\cap_{j=1}^k A_j}(\boldsymbol{y})$. The Jacobian of the bivariate transformation is

$$J = \det \begin{bmatrix} \frac{\partial t_1^{-1}}{\partial y_1} & \frac{\partial t_1^{-1}}{\partial y_2} \\ \frac{\partial t_2^{-1}}{\partial y_1} & \frac{\partial t_2^{-1}}{\partial y_2} \end{bmatrix},$$

and $|J|$ is the absolute value of the determinant J. Recall that

$$\det \begin{bmatrix} a & b \\ c & d \end{bmatrix} = \begin{vmatrix} a & b \\ c & d \end{vmatrix} = ad - bc.$$

To find $t_i^{-1}(y_1, y_2)$, use $y_i = t_i(x_1, x_2)$ and solve for x_1 and x_2 where $i = 1, 2$.
9. If $Y_1, \ldots, Y_n$ are independent with mgfs $m_{Y_i}(t)$, then the mgf of $W = \sum_{i=1}^n Y_i$ is

$$m_W(t) = \prod_{i=1}^n m_{Y_i}(t).$$

10. If $Y_1, \ldots, Y_n$ are iid with mgf $m_Y(t)$, then the mgf of $W = \sum_{i=1}^n Y_i$ is

$$m_W(t) = [m_Y(t)]^n,$$

and the mgf of $\overline{Y}$ is

$$m_{\overline{Y}}(t) = [m_Y(t/n)]^n.$$

11. Know that if $Y_1, \ldots, Y_n$ are iid with $E(Y) = \mu$ and $\mathrm{V}(Y) = \sigma^2$, then $E(\overline{Y}) = \mu$ and $\mathrm{V}(\overline{Y}) = \sigma^2/n$.
12. Suppose $W = \sum_{i=1}^n Y_i$ or $W = \overline{Y}$ where $Y_1, \ldots, Y_n$ are independent. For several distributions (especially Y_i iid gamma(ν, λ) and Y_i independent $N(\mu_i, \sigma_i^2)$), be able to find the distribution of W, the mgf of W, $E(W)$, Var(W), and $E(W^2) = V(W) + [E(W)]^2$.
13. If $\boldsymbol{X} \sim N_p(\boldsymbol{\mu}, \boldsymbol{\Sigma})$, then $\boldsymbol{t}^T \boldsymbol{X} = t_1 X_1 + \ldots + t_p X_p \sim N(\boldsymbol{t}^T \boldsymbol{\mu}, \boldsymbol{t}^T \boldsymbol{\Sigma} \boldsymbol{t})$.
14. If $\boldsymbol{X} \sim N_p(\boldsymbol{\mu}, \boldsymbol{\Sigma})$ and if $\boldsymbol{A}$ is a $q \times p$ matrix, then $\boldsymbol{AX} \sim N_q(\boldsymbol{A\mu}, \boldsymbol{A\Sigma A}^T)$. If $\boldsymbol{a}$ is a $p \times 1$ vector of constants, then $\boldsymbol{a} + \boldsymbol{X} \sim N_p(\boldsymbol{a} + \boldsymbol{\mu}, \boldsymbol{\Sigma})$.
Suppose $\boldsymbol{X}_1$ is $q \times 1$ and

$$\boldsymbol{X} = \begin{pmatrix} \boldsymbol{X}_1 \\ \boldsymbol{X}_2 \end{pmatrix}, \ \boldsymbol{\mu} = \begin{pmatrix} \boldsymbol{\mu}_1 \\ \boldsymbol{\mu}_2 \end{pmatrix}, \text{ and } \boldsymbol{\Sigma} = \begin{pmatrix} \boldsymbol{\Sigma}_{11} & \boldsymbol{\Sigma}_{12} \\ \boldsymbol{\Sigma}_{21} & \boldsymbol{\Sigma}_{22} \end{pmatrix}.$$

15. $\boldsymbol{X}_1 \sim N_q(\boldsymbol{\mu}_1, \boldsymbol{\Sigma}_{11})$.
16. If $\boldsymbol{X} \sim N_p(\boldsymbol{\mu}, \boldsymbol{\Sigma})$, then the conditional distribution of $\boldsymbol{X}_1$ given that $\boldsymbol{X}_2 = \boldsymbol{x}_2$ is multivariate normal with mean $\boldsymbol{\mu}_1 + \boldsymbol{\Sigma}_{12}\boldsymbol{\Sigma}_{22}^{-1}(\boldsymbol{x}_2 - \boldsymbol{\mu}_2)$ and covariance matrix $\boldsymbol{\Sigma}_{11} - \boldsymbol{\Sigma}_{12}\boldsymbol{\Sigma}_{22}^{-1}\boldsymbol{\Sigma}_{21}$. That is,

$$X_1|X_2 = x_2 \sim N_q(\mu_1 + \Sigma_{12}\Sigma_{22}^{-1}(x_2 - \mu_2), \Sigma_{11} - \Sigma_{12}\Sigma_{22}^{-1}\Sigma_{21}).$$

17.
$$\rho(X_i, X_j) = \frac{\sigma_{i,j}}{\sqrt{\sigma_{ii}\sigma_{jj}}} = \text{Cov}(X_i, X_j)/\sqrt{V(X_i)V(X_j)}.$$

18. Know that (X,Y) can have a joint distribution that is not multivariate normal, yet the marginal distributions of X and Y are both univariate normal. Hence X and Y can be normal, but $aX + bY$ is not normal. (Need the joint distribution of (X,Y) to be MVN for all linear combinations to be univariate normal.)

2.12 Complements

Panjer (1969) provides generalizations of Steiner's formula.

Johnson and Wichern (1988), Mardia et al. (1979) and Press (2005) are good references for multivariate statistical analysis based on the multivariate normal distribution. The elliptically contoured distributions generalize the multivariate normal distribution and are discussed (in increasing order of difficulty) in Johnson (1987), Fang et al. (1990), Fang and Anderson (1990), and Gupta and Varga (1993). Fang et al. (1990) sketch the history of elliptically contoured distributions while Gupta and Varga (1993) discuss matrix valued elliptically contoured distributions. Cambanis et al. (1981), Chmielewski (1981) and Eaton (1986) are also important references. Also see Muirhead (1982, pp. 30–42).

Broffitt (1986), Kowalski (1973), Melnick and Tenebien (1982) and Seber and Lee (2003, p. 23) give examples of dependent marginally normal random variables that have 0 correlation. The example in Remark 2.1 appears in Rohatgi (1976, p. 229) and Lancaster (1959).

See Abuhassan (2007) for more information about the distributions in Problems 2.52– 2.59.

2.13 Problems

PROBLEMS WITH AN ASTERISK * ARE ESPECIALLY USEFUL.

Refer to Chap. 10 for the pdf or pmf of the distributions in the problems below.

Theorem 2.16 is useful for Problems 2.1*– 2.7*.

2.1*. Let $X_1, \ldots, X_n$ be independent Poisson(λ_i). Let $W = \sum_{i=1}^n X_i$. Find the mgf of W and find the distribution of W.

2.2*. Let $X_1, \ldots, X_n$ be iid Bernoulli(ρ). Let $W = \sum_{i=1}^n X_i$. Find the mgf of W and find the distribution of W.

2.3*. Let $X_1,\ldots,X_n$ be iid exponential (λ). Let $W = \sum_{i=1}^n X_i$. Find the mgf of W and find the distribution of W.

2.4*. Let $X_1,\ldots,X_n$ be independent $N(\mu_i, \sigma_i^2)$. Let $W = \sum_{i=1}^n (a_i + b_i X_i)$ where a_i and b_i are fixed constants. Find the mgf of W and find the distribution of W.

2.5*. Let $X_1,\ldots,X_n$ be iid negative binomial (1, ρ). Let $W = \sum_{i=1}^n X_i$. Find the mgf of W and find the distribution of W.

2.6*. Let $X_1,\ldots,X_n$ be independent gamma (ν_i, λ). Let $W = \sum_{i=1}^n X_i$. Find the mgf of W and find the distribution of W.

2.7*. Let $X_1,\ldots,X_n$ be independent $\chi^2_{p_i}$. Let $W = \sum_{i=1}^n X_i$. Find the mgf of W and find the distribution of W.

2.8. a) Let $f_Y(y)$ be the pdf of Y. If $W = \mu + Y$ where $-\infty < \mu < \infty$, show that the pdf of W is $f_W(w) = f_Y(w - \mu)$.
b) Let $f_Y(y)$ be the pdf of Y. If $W = \sigma Y$ where $\sigma > 0$, show that the pdf of W is $f_W(w) = (1/\sigma) f_Y(w/\sigma)$.
c) Let $f_Y(y)$ be the pdf of Y. If $W = \mu + \sigma Y$ where $-\infty < \mu < \infty$ and $\sigma > 0$, show that the pdf of W is $f_W(w) = (1/\sigma) f_Y((w - \mu)/\sigma)$.

2.9. a) If Y is lognormal $LN(\mu, \sigma^2)$, show that $W = \log(Y)$ is a normal $N(\mu, \sigma^2)$ random variable.
b) If Y is a normal $N(\mu, \sigma^2)$ random variable, show that $W = e^Y$ is a lognormal $LN(\mu, \sigma^2)$ random variable.

2.10. a) If Y is uniform (0,1), Show that $W = -\log(Y)$ is exponential (1).
b) If Y is exponential (1), show that $W = \exp(-Y)$ is uniform (0,1).

2.11. If $Y \sim N(\mu, \sigma^2)$, find the pdf of

$$W = \left(\frac{Y - \mu}{\sigma}\right)^2.$$

2.12. If Y has a half normal distribution, $Y \sim \text{HN}(\mu, \sigma^2)$, show that $W = (Y - \mu)^2 \sim G(1/2, 2\sigma^2)$.

2.13. a) Suppose that Y has a Weibull (ϕ, λ) distribution with pdf

$$f(y) = \frac{\phi}{\lambda} y^{\phi - 1} e^{-\frac{y^\phi}{\lambda}}$$

where $\lambda, y,$ and ϕ are all positive. Show that $W = \log(Y)$ has a smallest extreme value SEV($\theta = \log(\lambda^{1/\phi}), \sigma = 1/\phi$) distribution.
b) If Y has a SEV($\theta = \log(\lambda^{1/\phi}), \sigma = 1/\phi$) distribution, show that $W = e^Y$ has a Weibull (ϕ, λ) distribution.

2.14. a) Suppose that Y has a Pareto(σ, λ) distribution with pdf

$$f(y) = \frac{\frac{1}{\lambda}\sigma^{1/\lambda}}{y^{1+1/\lambda}}$$

where $y \geq \sigma$, $\sigma > 0$, and $\lambda > 0$. Show that $W = \log(Y) \sim \text{EXP}(\theta = \log(\sigma), \lambda)$.
b) If Y as an EXP$(\theta = \log(\sigma), \lambda)$ distribution, show that $W = e^Y$ has a Pareto(σ, λ) distribution.

2.15. a) If Y is chi χ_p, then the pdf of Y is

$$f(y) = \frac{y^{p-1}e^{-y^2/2}}{2^{\frac{p}{2}-1}\Gamma(p/2)}$$

where $y \geq 0$ and p is a positive integer. Show that the pdf of $W = Y^2$ is the χ^2_p pdf.
b) If Y is a chi-square χ^2_p random variable, show that $W = \sqrt{Y}$ is a chi χ_p random variable.

2.16. a) If Y is power POW(λ), then the pdf of Y is

$$f(y) = \frac{1}{\lambda}y^{\frac{1}{\lambda}-1},$$

where $\lambda > 0$ and $0 < y < 1$. Show that $W = -\log(Y)$ is an exponential (λ) random variable.
b) If Y is an exponential(λ) random variable, show that $W = e^{-Y}$ is a power $POW(\lambda)$ random variable.

2.17. a) If Y is truncated extreme value TEV(λ) then the pdf of Y is

$$f(y) = \frac{1}{\lambda}\exp\left(y - \frac{e^y - 1}{\lambda}\right)$$

where $y > 0$, and $\lambda > 0$. Show that $W = e^Y - 1$ is an exponential (λ) random variable.
b) If Y is an exponential(λ) random variable, show that $W = \log(Y + 1)$ is a truncated extreme value TEV(λ) random variable.
c) If Y has an inverse exponential distribution, $Y \sim \text{IEXP}(\theta)$, show that $W = 1/Y \sim \text{EXP}(1/\theta)$.
d) If Y has an inverse Weibull distribution, $Y \sim \text{IW}(\phi, \lambda)$, show that $1/Y \sim W(\phi, \lambda)$, the Weibull distribution with parameters ϕ and λ.
e) If Y has a log-gamma distribution, $Y \sim \text{LG}(\nu, \lambda)$, show that $W = e^Y \sim$ gamma (ν, λ).
f) If Y has a two-parameter power distribution, $Y \sim \text{power}(\tau, \lambda)$, show that $W = -\log(Y) \sim \text{EXP}(-\log(\tau), \lambda)$.

2.18. a) If Y is BurrXII(ϕ,λ), show that $W = \log(1+Y^{\phi})$ is an exponential(λ) random variable.

b) If Y is an exponential(λ) random variable, show that $W = (e^Y - 1)^{1/\phi}$ is a BurrXII(ϕ,λ) random variable.

2.19. a) If Y is Pareto PAR(σ,λ), show that $W = \log(Y/\sigma)$ is an exponential(λ) random variable.

b) If Y is an exponential(λ) random variable, show that $W = \sigma e^Y$ is a Pareto $PAR(\sigma,\lambda)$ random variable.

2.20. a) If Y is Weibull $W(\phi,\lambda)$, show that $W = Y^{\phi}$ is an exponential (λ) random variable.

b) If Y is an exponential(λ) random variable, show that $W = Y^{1/\phi}$ is a Weibull $W(\phi,\lambda)$ random variable.

2.21. If Y is double exponential (θ,λ), show that $W = |Y-\theta| \sim \text{EXP}(\lambda)$.

2.22. If Y has a generalized gamma distribution, $Y \sim \text{GG}(\nu,\lambda,\phi)$, show that $W = Y^{\phi} \sim G(\nu,\lambda^{\phi})$.

2.23. If Y has an inverted gamma distribution, $Y \sim \text{INV}G(\nu,\lambda)$, show that $W = 1/Y \sim G(\nu,\lambda)$.

2.24. a) If Y has a largest extreme value distribution $Y \sim \text{LEV}(\theta,\sigma)$, show that $W = \exp(-(Y-\theta)/\sigma) \sim \text{EXP}(1)$.

b) If $Y \sim \text{EXP}(1)$, show that $W = \theta - \sigma\log(Y) \sim \text{LEV}(\theta,\sigma)$.

2.25. a) If Y has a log–Cauchy distribution, $Y \sim \text{LC}(\mu,\sigma)$, show that $W = \log(Y)$ has a Cauchy(μ,σ) distribution.

b) If $Y \sim C(\mu,\sigma)$ show that $W = e^Y \sim \text{LC}(\mu,\sigma)$.

2.26. a) If Y has a log–logistic distribution, $Y \sim \text{LL}(\phi,\tau)$, show that $W = \log(Y)$ has a logistic $L(\mu = -\log(\phi), \sigma = 1/\tau)$ distribution.

b) If $Y \sim L(\mu = -\log(\phi), \sigma = 1/\tau)$, show that $W = e^Y \sim \text{LL}(\phi,\tau)$.

2.27. If Y has a Maxwell–Boltzmann distribution, $Y \sim \text{MB}(\mu,\sigma)$, show that $W = (Y-\mu)^2 \sim G(3/2, 2\sigma^2)$.

2.28. If Y has a one-sided stable distribution, $Y \sim \text{OSS}(\sigma)$, show that $W = 1/Y \sim G(1/2, 2/\sigma)$.

2.29. a) If Y has a Rayleigh distribution, $Y \sim R(\mu,\sigma)$, show that $W = (Y-\mu)^2 \sim \text{EXP}(2\sigma^2)$.

b) If $Y \sim \text{EXP}(2\sigma^2)$, show that $W = \sqrt{Y} + \mu \sim R(\mu,\sigma)$.

2.30. If Y has a smallest extreme value distribution, $Y \sim \text{SEV}(\theta,\sigma)$, show that $W = -Y$ has an LEV$(-\theta,\sigma)$ distribution.

2.31. Let $Y \sim C(0,1)$. Show that the Cauchy distribution is a location–scale family by showing that $W = \mu + \sigma Y \sim C(\mu,\sigma)$ where μ is real and $\sigma > 0$.

2.32. Let Y have a chi distribution, $Y \sim \text{chi}(p,1)$ where p is known. Show that the $\text{chi}(p,\sigma)$ distribution is a scale family for p known by showing that $W = \sigma Y \sim \text{chi}(p,\sigma)$ for $\sigma > 0$.

2.33. Let $Y \sim \text{DE}(0,1)$. Show that the double exponential distribution is a location–scale family by showing that $W = \theta + \lambda Y \sim \text{DE}(\theta,\lambda)$ where θ is real and $\lambda > 0$.

2.34. Let $Y \sim \text{EXP}(1)$. Show that the exponential distribution is a scale family by showing that $W = \lambda Y \sim \text{EXP}(\lambda)$ for $\lambda > 0$.

2.35. Let $Y \sim \text{EXP}(0,1)$. Show that the two-parameter exponential distribution is a location–scale family by showing that $W = \theta + \lambda Y \sim \text{EXP}(\theta,\lambda)$ where θ is real and $\lambda > 0$.

2.36. Let $Y \sim \text{LEV}(0,1)$. Show that the largest extreme value distribution is a location–scale family by showing that $W = \theta + \sigma Y \sim \text{LEV}(\theta,\sigma)$ where θ is real and $\sigma > 0$.

2.37. Let $Y \sim G(\nu,1)$ where ν is known. Show that the gamma (ν,λ) distribution is a scale family for ν known by showing that $W = \lambda Y \sim G(\nu,\lambda)$ for $\lambda > 0$.

2.38. Let $Y \sim \text{HC}(0,1)$. Show that the half Cauchy distribution is a location–scale family by showing that $W = \mu + \sigma Y \sim \text{HC}(\mu,\sigma)$ where μ is real and $\sigma > 0$.

2.39. Let $Y \sim \text{HL}(0,1)$. Show that the half logistic distribution is a location–scale family by showing that $W = \mu + \sigma Y \sim \text{HL}(\mu,\sigma)$ where μ is real and $\sigma > 0$.

2.40. Let $Y \sim \text{HN}(0,1)$. Show that the half normal distribution is a location–scale family by showing that $W = \mu + \sigma Y \sim \text{HN}(\mu,\sigma^2)$ where μ is real and $\sigma > 0$.

2.41. Let $Y \sim L(0,1)$. Show that the logistic distribution is a location–scale family by showing that $W = \mu + \sigma Y \sim L(\mu,\sigma)$ where μ is real and $\sigma > 0$.

2.42. Let $Y \sim \text{MB}(0,1)$. Show that the Maxwell–Boltzmann distribution is a location–scale family by showing that $W = \mu + \sigma Y \sim \text{MB}(\mu,\sigma)$ where μ is real and $\sigma > 0$.

2.43. Let $Y \sim N(0,1)$. Show that the normal distribution is a location–scale family by showing that $W = \mu + \sigma Y \sim N(\mu,\sigma^2)$ where μ is real and $\sigma > 0$.

2.44. Let $Y \sim \text{OSS}(1)$. Show that the one-sided stable distribution is a scale family by showing that $W = \sigma Y \sim \text{OSS}(\sigma)$ for $\sigma > 0$.

2.45. Let $Y \sim \text{PAR}(1,\lambda)$ where λ is known. Show that the Pareto (σ,λ) distribution is a scale family for λ known by showing that $W = \sigma Y \sim \text{PAR}(\sigma,\lambda)$ for $\sigma > 0$.

2.46. Let $Y \sim R(0,1)$. Show that the Rayleigh distribution is a location–scale family by showing that $W = \mu + \sigma Y \sim R(\mu,\sigma)$ where μ is real and $\sigma > 0$.

2.47. Let $Y \sim U(0,1)$. Show that the uniform distribution is a location–scale family by showing that $W = \mu + \sigma Y \sim U(\theta_1, \theta_2)$ where $\mu = \theta_1$ is real and $\sigma = \theta_2 - \theta_1 > 0$.

2.48. Examine the proof of Theorem 2.2b for a joint pdf and prove the result for a joint pmf by replacing the integrals by appropriate sums.

2.49. Examine the proof of Theorem 2.3 for a joint pdf and prove the result for a joint pmf by replacing the integrals by appropriate sums.

2.50. Examine the proof of Theorem 2.4 for a joint pdf and prove the result for a joint pmf by replacing the integrals by appropriate sums.

2.51. Examine the proof of Theorem 2.5 for a joint pdf and prove the result for a joint pmf by replacing the integrals by appropriate sums.

2.52. If $Y \sim$ hburr(ϕ, λ), then the pdf of Y is

$$f(y) = \frac{2}{\lambda\sqrt{2\pi}} \frac{\phi y^{\phi-1}}{(1+y^{\phi})} \exp\left(\frac{-[\log(1+y^{\phi})]^2}{2\lambda^2}\right) I(y > 0)$$

where ϕ and λ are positive.

a) Show that $W = \log(1 + Y^{\phi}) \sim \text{HN}(0, \lambda)$, the half normal distribution with parameters 0 and λ.
b) If $W \sim \text{HN}(0, \lambda)$, then show $Y = [e^W - 1]^{1/\phi} \sim$ hburr(ϕ, λ).

2.53. If $Y \sim$ hlev(θ, λ), then the pdf of Y is

$$f(y) = \frac{2}{\lambda\sqrt{2\pi}} \exp\left(\frac{-(y-\theta)}{\lambda}\right) \exp\left[-\frac{1}{2}\left[\exp\left(\frac{-(y-\theta)}{\lambda}\right)\right]^2\right]$$

where y and θ are real and $\lambda > 0$.

a) Show that $W = \exp(-(Y-\theta)/\lambda) \sim \text{HN}(0,1)$, the half normal distribution with parameters 0 and 1.
b) If $W \sim \text{HN}(0,1)$, then show $Y = -\lambda \log(W) + \theta \sim$ hlev(θ, λ).

2.54. If $Y \sim$ hpar(θ, λ), then the pdf of Y is

$$f(y) = \frac{2}{\lambda\sqrt{2\pi}} \frac{1}{y} I[y \geq \theta] \exp\left[\frac{-(\log(y) - \log(\theta))^2}{2\lambda^2}\right]$$

where $\theta > 0$ and $\lambda > 0$.

a) Show that $W = \log(Y) \sim \text{HN}(\mu = \log(\theta), \sigma = \lambda)$. (See the half normal distribution in Chap. 10.)
b) If $W \sim \text{HN}(\mu, \sigma)$, then show $Y = e^W \sim$ hpar$(\theta = e^{\mu}, \lambda = \sigma)$.

2.55. If $Y \sim \text{hpow}(\lambda)$, then the pdf of Y is

$$f(y) = \frac{2}{\lambda\sqrt{2\pi}} \frac{1}{y} I_{[0,1]}(y) \exp\left[\frac{-(\log(y))^2}{2\lambda^2}\right]$$

where $\lambda > 0$.

a) Show that $W = -\log(Y) \sim \text{HN}(0, \sigma = \lambda)$, the half normal distribution with parameters 0 and λ.
b) If $W \sim \text{HN}(0, \sigma)$, then show $Y = e^{-W} \sim \text{hpow}(\lambda = \sigma)$.

2.56. If $Y \sim \text{hray}(\theta, \lambda)$, then the pdf of Y is

$$f(y) = \frac{4}{\lambda\sqrt{2\pi}}(y-\theta)I[y \geq \theta]\exp\left[\frac{-(y-\theta)^4}{2\lambda^2}\right]$$

where $\lambda > 0$ and θ is real.

a) Show that $W = (Y-\theta)^2 \sim \text{HN}(0, \sigma = \lambda)$, the half normal distribution with parameters 0 and λ.
b) If $W \sim \text{HN}(0, \sigma)$, then show $Y = \sqrt{W} + \theta \sim \text{hray}(\theta, \lambda = \sigma)$.

2.57. If $Y \sim \text{hsev}(\theta, \lambda)$, then the pdf of Y is

$$f(y) = \frac{2}{\lambda\sqrt{2\pi}} \exp\left(\frac{y-\theta}{\lambda}\right) \exp\left(\frac{-1}{2}\left[\exp\left(\frac{y-\theta}{\lambda}\right)\right]^2\right)$$

where y and θ are real and $\lambda > 0$.

a) Show that $W = \exp[(y-\theta)/\lambda] \sim \text{HN}(0,1)$.
b) If $W \sim \text{HN}(0,1)$, then show $Y = \lambda\log(W) + \theta \sim \text{hsev}(\theta, \lambda)$.

2.58. If $Y \sim \text{htev}(\lambda)$, then the pdf of Y is

$$f(y) = \frac{2}{\lambda\sqrt{2\pi}} \exp\left(y - \frac{(e^y-1)^2}{2\lambda^2}\right) = \frac{2}{\lambda\sqrt{2\pi}} e^y \exp\left(\frac{-(e^y-1)^2}{2\lambda^2}\right)$$

where $y > 0$ and $\lambda > 0$.

a) Show that $W = e^Y - 1 \sim \text{HN}(0, \sigma = \lambda)$, the half normal distribution with parameters 0 and λ.
b) If $W \sim \text{HN}(0, \sigma)$, then show $Y = \log(W+1) \sim \text{htev}(\lambda = \sigma)$.

2.59. If $Y \sim \text{hweib}(\phi, \lambda)$, then the pdf of Y is

$$f(y) = \frac{2}{\lambda\sqrt{2\pi}} \phi\, y^{\phi-1}\, I[y > 0] \exp\left(\frac{-y^{2\phi}}{2\lambda^2}\right)$$

where λ and ϕ are positive.

a) Show that $W = Y^{\phi} \sim \text{HN}(0, \sigma = \lambda)$, the half normal distribution with parameters 0 and λ.
b) If $W \sim \text{HN}(0, \sigma)$, then show $Y = W^{1/\phi} \sim \text{hweib}(\phi, \lambda = \sigma)$.

Problems from old quizzes and exams. Problems from old qualifying exams are marked with a Q since these problems take longer than quiz and exam problems.

2.60. If Y is a random variable with pdf

$$f(y) = \lambda y^{\lambda - 1} \text{ for } 0 < y < 1$$

where $\lambda > 0$, show that $W = -\log(Y)$ is an exponential$(1/\lambda)$ random variable.

2.61. If Y is an exponential$(1/\lambda)$ random variable, show that $W = e^{-Y}$ has pdf

$$f_W(w) = \lambda w^{\lambda - 1} \text{ for } 0 < w < 1.$$

2.62. If $Y \sim \text{EXP}(\lambda)$, find the pdf of $W = 2\lambda Y$.

2.63*. (Mukhopadhyay 2000, p. 113): Suppose that $X|Y \sim N(\beta_0 + \beta_1 Y, Y^2)$, and that $Y \sim N(3, 10)$. That is, the conditional distribution of X given that $Y = y$ is normal with mean $\beta_0 + \beta_1 y$ and variance y^2 while the (marginal) distribution of Y is normal with mean 3 and variance 10.

a) Find EX.
b) Find Var X.

2.64*. Suppose that

$$\begin{pmatrix} X_1 \\ X_2 \\ X_3 \\ X_4 \end{pmatrix} \sim N_4 \left(\begin{pmatrix} 49 \\ 100 \\ 17 \\ 7 \end{pmatrix}, \begin{pmatrix} 3 & 1 & -1 & 0 \\ 1 & 6 & 1 & -1 \\ -1 & 1 & 4 & 0 \\ 0 & -1 & 0 & 2 \end{pmatrix} \right).$$

a) Find the distribution of X_2.
b) Find the distribution of $(X_1, X_3)^T$.
c) Which pairs of random variables X_i and X_j are independent?
d) Find the correlation $\rho(X_1, X_3)$.

2.65*. Recall that if $\boldsymbol{X} \sim N_p(\boldsymbol{\mu}, \boldsymbol{\Sigma})$, then the conditional distribution of $\boldsymbol{X}_1$ given that $\boldsymbol{X}_2 = \boldsymbol{x}_2$ is multivariate normal with mean $\boldsymbol{\mu}_1 + \boldsymbol{\Sigma}_{12}\boldsymbol{\Sigma}_{22}^{-1}(\boldsymbol{x}_2 - \boldsymbol{\mu}_2)$ and covariance matrix $\boldsymbol{\Sigma}_{11} - \boldsymbol{\Sigma}_{12}\boldsymbol{\Sigma}_{22}^{-1}\boldsymbol{\Sigma}_{21}$.

Let $\sigma_{12} = \text{Cov}(Y, X)$ and suppose Y and X follow a bivariate normal distribution

$$\begin{pmatrix} Y \\ X \end{pmatrix} \sim N_2 \left(\begin{pmatrix} 49 \\ 100 \end{pmatrix}, \begin{pmatrix} 16 & \sigma_{12} \\ \sigma_{12} & 25 \end{pmatrix} \right).$$

a) If $\sigma_{12} = 0$, find $Y|X$. Explain your reasoning.
b) If $\sigma_{12} = 10$ find $E(Y|X)$.

c) If $\sigma_{12} = 10$, find $\text{Var}(Y|X)$.

2.66. Let $\sigma_{12} = \text{Cov}(Y,X)$ and suppose Y and X follow a bivariate normal distribution

$$\begin{pmatrix} Y \\ X \end{pmatrix} \sim N_2\left(\begin{pmatrix} 15 \\ 20 \end{pmatrix}, \begin{pmatrix} 64 & \sigma_{12} \\ \sigma_{12} & 81 \end{pmatrix} \right).$$

a) If $\sigma_{12} = 10$ find $E(Y|X)$.
b) If $\sigma_{12} = 10$, find $\text{Var}(Y|X)$.
c) If $\sigma_{12} = 10$, find $\rho(Y,X)$, the correlation between Y and X.

2.67*. (Mukhopadhyay 2000, p. 197): Suppose that X_1 and X_2 have a joint pdf given by

$$f(x_1,x_2) = 3(x_1+x_2)I(0<x_1<1)I(0<x_2<1)I(0<x_1+x_2<1).$$

Consider the transformation $Y_1 = X_1 + X_2$ and $Y_2 = X_1 - X_2$.

a) Find the Jacobian J for the transformation.
b) Find the support $\mathscr{Y}$ of Y_1 and Y_2.
c) Find the joint density $f_{Y_1,Y_2}(y_1,y_2)$.
d) Find the marginal pdf $f_{Y_1}(y_1)$.
e) Find the marginal pdf $f_{Y_2}(y_2)$.

Hint for d) and e): $I_{A_1}(\boldsymbol{y})I_{A_2}(\boldsymbol{y})I_{A_3}(\boldsymbol{y}) = I_{\cap_{j=1}^3 A_j}(\boldsymbol{y}) = I_{\mathscr{Y}}(\boldsymbol{y})$ where $\mathscr{Y}$ is a triangle.

2.68*Q. The number of defects per yard Y of a certain fabric is known to have a Poisson distribution with parameter λ. However, λ is a random variable with pdf

$$f(\lambda) = e^{-\lambda}I(\lambda > 0).$$

a) Find E(Y).
b) Find Var(Y).

2.69. Let A and B be positive integers. A hypergeometric random variable $X = W_1 + W_2 + \cdots + W_n$ where the random variables W_i are identically distributed random variables with $P(W_i = 1) = A/(A+B)$ and $P(W_i = 0) = B/(A+B)$. You may use the fact that $E(W_1) = A/(A+B)$ and that $E(X) = nA/(A+B)$.

a) Find $\text{Var}(W_1)$.

b) If $i \neq j$, then $\text{Cov}(W_i, W_j) = \dfrac{-AB}{(A+B)^2(A+B-1)}$. Find Var(X) using the formula

$$\text{Var}\left(\sum_{i=1}^n W_i\right) = \sum_{i=1}^n \text{Var}(W_i) + 2\sum_{i=1}^{n-1}\sum_{j=i+1}^n \text{Cov}(W_i, W_j).$$

(Hint: the sum $\sum_{i=1}^{n-1}\sum_{j=i+1}^n$ has $(n-1)n/2$ terms.)

2.70. Let $X = W_1 + W_2 + \cdots + W_n$ where the joint distribution of the random variables W_i is an n-dimensional multivariate normal distribution with $E(W_i) = 1$ and $\text{Var}(W_i) = 100$ for $i = 1, \ldots, n$.

a) Find $E(X)$.
b) Suppose that if $i \neq j$, then $\text{Cov}(W_i, W_j) = 10$. Find Var(X) using the formula

$$\text{Var}\left(\sum_{i=1}^{n} W_i\right) = \sum_{i=1}^{n} \text{Var}(W_i) + 2\sum_{i=1}^{n-1} \sum_{j=i+1}^{n} \text{Cov}(W_i, W_j).$$

(Hint: the sum $\sum_{i=1}^{n-1} \sum_{j=i+1}^{n}$ has $(n-1)n/2$ terms.)

2.71. Find the moment generating function for Y_1 if the joint probability mass function $f(y_1, y_2)$ of Y_1 and Y_2 is tabled as shown.

$f(y_1,y_2)$	y_2: 0	1	2
y_1 = 0	0.38	0.14	0.24
y_1 = 1	0.17	0.02	0.05

2.72. Suppose that the joint pdf of X and Y is $f(x,y) =$

$$\frac{1}{2}\frac{1}{2\pi\sqrt{1-\rho^2}} \exp\left(\frac{-1}{2(1-\rho^2)}(x^2 - 2\rho xy + y^2)\right)$$

$$+\frac{1}{2}\frac{1}{2\pi\sqrt{1-\rho^2}} \exp\left(\frac{-1}{2(1-\rho^2)}(x^2 + 2\rho xy + y^2)\right)$$

where x and y are real and $0 < \rho < 1$. It can be shown that the marginal pdfs are

$$f_X(x) = \frac{1}{\sqrt{2\pi}} \exp\left(\frac{-1}{2}x^2\right)$$

for x real and

$$f_Y(y) = \frac{1}{\sqrt{2\pi}} \exp\left(\frac{-1}{2}y^2\right)$$

for y real. Are X and Y independent? Explain briefly.

2.73*. Suppose that the conditional distribution of $Y|P = \rho$ is the binomial(k, ρ) distribution and that the random variable P has a beta$(\delta = 4, \nu = 6)$ distribution.

a) Find E(Y).
b) Find Var(Y).

2.74*. Suppose that the joint probability mass function $f(y_1, y_2)$ of Y_1 and Y_2 is given in the following table.

$f(y_1,y_2)$		y_2 0	1	2
y_1	0	0.38	0.14	0.24
	1	0.17	0.02	0.05

a) Find the marginal probability function $f_{Y_2}(y_2)$ for Y_2.
b) Find the conditional probability function $f(y_1|y_2)$ of Y_1 given $Y_2 = 2$.

2.75*. Find the pmf of $Y = X^2 + 4$ where the pmf of X is given below.

X	-2	-1	0	1	2
Probability	0.1	0.2	0.4	0.2	0.1

2.76. Suppose that X_1 and X_2 are independent with $X_1 \sim N(0,1)$ and $X_2 \sim N(0,4)$ so $\text{Var}(X_2) = 4$. Consider the transformation $Y_1 = X_1 + X_2$ and $Y_2 = X_1 - X_2$.

a) Find the Jacobian J for the transformation.
b) Find the joint pdf $f(y_1, y_2)$ of Y_1 and Y_2.
c) Are Y_1 and Y_2 independent? Explain briefly. Hint: can you factor the joint pdf so that $f(y_1, y_2) = g(y_1)h(y_2)$ for every real y_1 and y_2?

2.77. This problem follows Severini (2005, p. 236). Let $W \sim N(\mu_W, \sigma_W^2)$ and let $\boldsymbol{X} \sim N_p(\boldsymbol{\mu}, \boldsymbol{\Sigma})$.

a) Write down the moment generating function (mgf) $m_W(t)$ of W.
b) Suppose $W = \boldsymbol{t}^T\boldsymbol{X}$. Then $W \sim N(\mu_W, \sigma_W^2)$. What are μ_W and σ_W?
c) The mgf of $\boldsymbol{X}$ is $m_{\boldsymbol{X}}(\boldsymbol{t}) = E(e^{\boldsymbol{t}^T\boldsymbol{X}}) = E(e^W) = m_W(1)$. Using a) and b), find $m_{\boldsymbol{X}}(\boldsymbol{t})$.

2.78. Consider k insect eggs. Eggs may not hatch. If the egg hatches into a juvenile, the juvenile may not survive long enough to turn into an adult. Let ρ be the probability that the egg hatches into a juvenile that eventually turns into an adult. Let X_1 be the number of eggs that turn into a juvenile, and let X_2 be the number of juveniles that turn into adults = the number of eggs that turn into juveniles that turn into adults. Assuming that such events are iid, then $X_2 \sim$ binomial (k, ρ). Let ρ_1 be the probability that an egg hatches into a juvenile, and let ρ_2 be the probability that a juvenile turns into an adult. Then $X_2|X_1 \sim \text{binomial}(X_1, \rho_2)$ and $X_1 \sim \text{binomial}(k, \rho_1)$. Part a) below will show that $\rho = \rho_1\rho_2$.

a) Find $E(X_2)$.
b) Find $V(X_2)$ using Steiner's formula.

Chapter 3
Exponential Families

Suppose the data is a random sample from some parametric brand name distribution with parameters $\boldsymbol{\theta}$. This brand name distribution comes from a family of distributions parameterized by $\boldsymbol{\theta} \in \Theta$. Each different value of $\boldsymbol{\theta}$ in the parameter space Θ gives a distribution that is a member of the family of distributions. Often the brand name family of distributions is from an exponential family.

The theory of exponential families will be used in the following chapters to study some of the most important topics in statistical inference such as minimal and complete sufficient statistics, maximum likelihood estimators (MLEs), uniform minimum variance estimators (UMVUEs), and the Fréchet–Cramér–Rao lower bound (FCRLB), uniformly most powerful (UMP) tests and large sample theory.

3.1 Regular Exponential Families

Often a "brand name distribution" such as the normal distribution will have three useful parameterizations: the *usual parameterization* with parameter space Θ_U is simply the formula for the probability distribution or mass function (pdf or pmf, respectively) given when the distribution is first defined. The *k-parameter exponential family parameterization* with parameter space Θ, given in Definition 3.1 below, provides a simple way to determine if the distribution is an exponential family, while the *natural parameterization* with parameter space Ω, given in Definition 3.2 below, is used for *theory* that requires a complete sufficient statistic. See Chaps. 4 and 6.

Definition 3.1. A *family* of joint pdfs or joint pmfs $\{f(\boldsymbol{y}|\boldsymbol{\theta}) : \boldsymbol{\theta} = (\theta_1, \ldots, \theta_j) \in \Theta\}$ for a random vector $\boldsymbol{Y}$ is an **exponential family** if

$$f(\boldsymbol{y}|\boldsymbol{\theta}) = h(\boldsymbol{y})c(\boldsymbol{\theta})\exp\left[\sum_{i=1}^{k} w_i(\boldsymbol{\theta})t_i(\boldsymbol{y})\right] \tag{3.1}$$

D.J. Olive, *Statistical Theory and Inference*, DOI 10.1007/978-3-319-04972-4_3,

for all $\boldsymbol{y}$ where $c(\boldsymbol{\theta}) \geq 0$ and $h(\boldsymbol{y}) \geq 0$. The functions c, h, t_i, and w_i are real valued functions. The parameter $\boldsymbol{\theta}$ can be a scalar and $\boldsymbol{y}$ can be a scalar. It is crucial that $c, w_1, \ldots, w_k$ do not depend on $\boldsymbol{y}$ and that $h, t_1, \ldots, t_k$ do not depend on $\boldsymbol{\theta}$. The support of the distribution is $\mathscr{Y}$ and the parameter space is Θ. The family is a **k-parameter exponential family** if k is the smallest integer where (3.1) holds.

Notice that the distribution of Y is an exponential family if

$$f(y|\boldsymbol{\theta}) = h(y)c(\boldsymbol{\theta}) \exp\left[\sum_{i=1}^{k} w_i(\boldsymbol{\theta}) t_i(y)\right] \tag{3.2}$$

and the distribution is a one-parameter exponential family if

$$f(y|\theta) = h(y)c(\theta) \exp[w(\theta)t(y)]. \tag{3.3}$$

The parameterization is not unique since, for example, w_i could be multiplied by a nonzero constant a if t_i is divided by a. Many other parameterizations are possible. If $h(y) = g(y) I_{\mathscr{Y}}(y)$, then usually $c(\boldsymbol{\theta})$ and $g(y)$ are positive, so another parameterization is

$$f(y|\boldsymbol{\theta}) = \exp\left[\sum_{i=1}^{k} w_i(\boldsymbol{\theta}) t_i(y) + d(\boldsymbol{\theta}) + S(y)\right] I_{\mathscr{Y}}(y) \tag{3.4}$$

where $S(y) = \log(g(y))$, $d(\boldsymbol{\theta}) = \log(c(\boldsymbol{\theta}))$, and $\mathscr{Y}$ does not depend on $\boldsymbol{\theta}$.

To demonstrate that $\{f(\boldsymbol{y}|\boldsymbol{\theta}) : \boldsymbol{\theta} \in \Theta\}$ is an exponential family, find $h(\mathbf{y}), c(\boldsymbol{\theta})$, $w_i(\boldsymbol{\theta})$ and $t_i(\boldsymbol{y})$ such that (3.1), (3.2), (3.3) or (3.4) holds.

Theorem 3.1. Suppose that $\boldsymbol{Y}_1, \ldots, \boldsymbol{Y}_n$ are iid random vectors from an exponential family. Then the joint distribution of $\boldsymbol{Y}_1, \ldots, \boldsymbol{Y}_n$ follows an exponential family.

Proof. Suppose that $f_{\boldsymbol{Y}_i}(\boldsymbol{y}_i)$ has the form of (3.1). Then by independence,

$$f(\boldsymbol{y}_1, \ldots, \boldsymbol{y}_n) = \prod_{i=1}^{n} f_{\boldsymbol{Y}_i}(\boldsymbol{y}_i) = \prod_{i=1}^{n} h(\boldsymbol{y}_i) c(\boldsymbol{\theta}) \exp\left[\sum_{j=1}^{k} w_j(\boldsymbol{\theta}) t_j(\boldsymbol{y}_i)\right]$$

$$= \left[\prod_{i=1}^{n} h(\boldsymbol{y}_i)\right] [c(\boldsymbol{\theta})]^n \prod_{i=1}^{n} \exp\left[\sum_{j=1}^{k} w_j(\boldsymbol{\theta}) t_j(\boldsymbol{y}_i)\right]$$

$$= \left[\prod_{i=1}^{n} h(\boldsymbol{y}_i)\right] [c(\boldsymbol{\theta})]^n \exp\left(\sum_{i=1}^{n}\left[\sum_{j=1}^{k} w_j(\boldsymbol{\theta}) t_j(\boldsymbol{y}_i)\right]\right)$$

$$= \left[\prod_{i=1}^{n} h(\boldsymbol{y}_i)\right] [c(\boldsymbol{\theta})]^n \exp\left[\sum_{j=1}^{k} w_j(\boldsymbol{\theta}) \left(\sum_{i=1}^{n} t_j(\boldsymbol{y}_i)\right)\right].$$

To see that this has the form (3.1), take $h^*(\boldsymbol{y}_1, \ldots, \boldsymbol{y}_n) = \prod_{i=1}^{n} h(\boldsymbol{y}_i)$, $c^*(\boldsymbol{\theta}) = [c(\boldsymbol{\theta})]^n$, $w_j^*(\boldsymbol{\theta}) = w_j(\boldsymbol{\theta})$ and $t_j^*(\boldsymbol{y}_1, \ldots, \boldsymbol{y}_n) = \sum_{i=1}^{n} t_j(\boldsymbol{y}_i)$. □

The parameterization that uses the **natural parameter** $\boldsymbol{\eta}$ is especially useful for theory. See Definition 3.3 for the natural parameter space Ω.

Definition 3.2. Let Ω be the natural parameter space for $\boldsymbol{\eta}$. The **natural parameterization for an exponential family** is

$$f(\boldsymbol{y}|\boldsymbol{\eta}) = h(\boldsymbol{y})b(\boldsymbol{\eta})\exp\left[\sum_{i=1}^{k}\eta_i t_i(\boldsymbol{y})\right] \tag{3.5}$$

where $h(\boldsymbol{y})$ and $t_i(\boldsymbol{y})$ are the same as in Eq. (3.1) and $\boldsymbol{\eta} \in \Omega$. The natural parameterization for a random variable Y is

$$f(y|\boldsymbol{\eta}) = h(y)b(\boldsymbol{\eta})\exp\left[\sum_{i=1}^{k}\eta_i t_i(y)\right] \tag{3.6}$$

where $h(y)$ and $t_i(y)$ are the same as in Eq. (3.2) and $\boldsymbol{\eta} \in \Omega$. Again, the parameterization is not unique. If $a \neq 0$, then $a\eta_i$ and $t_i(y)/a$ would also work.

Notice that the natural parameterization (3.6) has the same form as (3.2) with $\boldsymbol{\theta}^* = \boldsymbol{\eta}$, $c^*(\boldsymbol{\theta}^*) = b(\boldsymbol{\eta})$ and $w_i(\boldsymbol{\theta}^*) = w_i(\boldsymbol{\eta}) = \eta_i$. In applications often $\boldsymbol{\eta}$ and Ω are of interest while $b(\boldsymbol{\eta})$ is not computed.

The next important idea is that of a regular exponential family (and of a full exponential family). Let $d_i(x)$ denote $t_i(y)$, $w_i(\boldsymbol{\theta})$ or η_i. A *linearity constraint* is satisfied by $d_1(x),\ldots,d_k(x)$ if $\sum_{i=1}^k a_i d_i(x) = c$ for some constants a_i and c and for all x (or η_i) in the sample or parameter space where not all of the $a_i = 0$. If $\sum_{i=1}^k a_i d_i(x) = c$ for all x only if $a_1 = \cdots = a_k = 0$, then the $d_i(x)$ do not satisfy a linearity constraint. In linear algebra, we would say that the $d_i(x)$ are *linearly independent* if they do not satisfy a linearity constraint.

For $k = 2$, a linearity constraint is satisfied if a plot of $d_1(x)$ versus $d_2(x)$ falls on a line as x varies. If the parameter space for the η_1 and η_2 is a nonempty open set, then the plot of η_1 versus η_2 is that nonempty open set, and the η_i cannot satisfy a linearity constraint since the plot is not a line.

Let $\tilde{\Omega}$ be the set where the integral of the kernel function is finite:

$$\tilde{\Omega} = \left\{\boldsymbol{\eta} = (\eta_1,\ldots,\eta_k) : \frac{1}{b(\boldsymbol{\eta})} \equiv \int_{-\infty}^{\infty} h(y)\exp\left[\sum_{i=1}^{k}\eta_i t_i(y)\right] dy < \infty\right\}. \tag{3.7}$$

Replace the integral by a sum for a pmf. An interesting fact is that $\tilde{\Omega}$ is a convex set. If the parameter space Θ of the exponential family is not a convex set, then the exponential family cannot be regular. Example 3.2 shows that the χ^2_p distribution is not regular since the set of positive integers is not convex.

Definition 3.3. Condition E1: the natural parameter space $\Omega = \tilde{\Omega}$.
Condition E2: assume that in the natural parameterization, neither the η_i nor the t_i satisfy a linearity constraint.
Condition E3: Ω is a k-dimensional nonempty open set.

If conditions E1), E2) and E3) hold then the exponential family is a k-parameter **regular exponential family** (REF).
If conditions E1) and E2) hold then the exponential family is a k-parameter *full exponential family*.

Notation. A kP-REF is a k-parameter regular exponential family. So a 1P-REF is a one-parameter REF and a 2P-REF is a two-parameter REF.

Notice that every REF is full. Any k-dimensional nonempty open set will contain a k-dimensional nonempty rectangle. A k-fold cross product of nonempty open intervals is a k-dimensional nonempty open set. For a one-parameter exponential family, a one-dimensional rectangle is just an interval, and the only type of function of one variable that satisfies a linearity constraint is a constant function. In the definition of an exponential family and in the usual parameterization, $\boldsymbol{\theta}$ is a $1 \times j$ vector. Typically $j = k$ if the family is a kP-REF. If $j < k$ and k is as small as possible, the family will usually not be regular. For example, a $N(\theta, \theta^2)$ family has $\boldsymbol{\theta} = \theta$ with $j = 1 < 2 = k$, and is not regular. See Example 3.8 for more details.

Some care has to be taken with the definitions of Θ and Ω since formulas (3.1) and (3.6) need to hold for every $\boldsymbol{\theta} \in \Theta$ and for every $\boldsymbol{\eta} \in \Omega$. Let Θ_U be the usual parameter space given for the distribution. For a continuous random variable or vector, the pdf needs to exist. Hence all degenerate distributions need to be deleted from Θ_U to form Θ and Ω. For continuous and discrete distributions, the natural parameter needs to exist (and often does not exist for discrete degenerate distributions). As a rule of thumb, remove values from Θ_U that cause the pmf to have the form 0^0. For example, for the binomial(k, ρ) distribution with k known, the natural parameter $\eta = \log(\rho/(1-\rho))$. Hence instead of using $\Theta_U = [0, 1]$, use $\rho \in \Theta = (0, 1)$, so that $\eta \in \Omega = (-\infty, \infty)$.

These conditions have some redundancy. If Ω contains a k-dimensional rectangle (e.g., if the family is a kP-REF, then Ω is a k-dimensional open set and contains a k-dimensional open ball which contains a k-dimensional rectangle), no η_i is completely determined by the remaining $\eta_j's$. In particular, the η_i cannot satisfy a linearity constraint. If the η_i do satisfy a linearity constraint, then the η_i lie on a hyperplane of dimension at most k, and such a surface cannot contain a k-dimensional rectangle. For example, if $k = 2$, a line cannot contain an open box. If $k = 2$ and $\eta_2 = \eta_1^2$, then the parameter space is not a two-dimensional open set and does not contain a two-dimensional rectangle. Thus the family is not a 2P-REF although η_1 and η_2 do not satisfy a linearity constraint. Again, see Example 3.8.

The most important 1P-REFs are the binomial (k, ρ) distribution with k known, the exponential (λ) distribution, and the Poisson (θ) distribution.

Other 1P-REFs are discussed in Chap. 10, including the Burr type III (λ, ϕ) distribution with ϕ known, the Burr Type X (τ) distribution, the Burr type XII (ϕ, λ) distribution with ϕ known, the double exponential (θ, λ) distribution with θ known, the two-parameter exponential (θ, λ) distribution with θ known, the generalized negative binomial (μ, κ) distribution if κ is known, the geometric (ρ) distribution, the Gompertz (θ, ν) distribution with $\theta > 0$ known, the half normal (μ, σ^2) distribution with μ known, the inverse exponential (θ) distribution, the inverse Weibull (ϕ, λ) distribution with ϕ known, the largest extreme value (θ, σ) distribution if σ is known, the smallest extreme value (θ, σ) distribution if σ is known, the inverted gamma (ν, λ) distribution if ν is known, the logarithmic (θ) distribution, the Maxwell–Boltzmann (μ, σ) distribution if μ is known, the modified DeMoivre's law (θ, ϕ) distribution if θ is known, the negative binomial (r, ρ) distribution if r is known, the one-sided stable (σ) distribution, the Pareto (σ, λ) distribution if σ is known, the power (λ) distribution, the Rayleigh (μ, σ) distribution if μ is known, the Topp–Leone (ν) distribution, the two-parameter power (τ, λ) distribution with τ known, the truncated extreme value (λ) distribution, the Weibull (ϕ, λ) distribution if ϕ is known, the zero truncated Poisson (θ) distribution, the Zeta (ν) distribution, and the Zipf (ν) distribution.

A one-parameter exponential family can often be obtained from a k-parameter exponential family by holding $k-1$ of the parameters fixed. Hence a normal (μ, σ^2) distribution is a 1P-REF if σ^2 is known. When data is modeled with an exponential family, often the scale, location, and shape parameters are unknown. For example, the mean and standard deviation are usually both unknown.

The most important 2P-REFs are the beta (δ, ν) distribution, the gamma (ν, λ) distribution and the normal (μ, σ^2) distribution. The chi (p, σ) distribution, the inverted gamma (ν, λ) distribution, the log-gamma (ν, λ) distribution, and the lognormal (μ, σ^2) distribution are also 2P-REFs. Example 3.9 will show that the inverse Gaussian distribution is full but not regular. The two-parameter Cauchy distribution is not an exponential family because its pdf cannot be put into the form of Eq. (3.1).

The natural parameterization can result in a family that is much larger than the family defined by the usual parameterization. See the definition of $\Omega = \tilde{\Omega}$ given by Eq. (3.7). Casella and Berger (2002, p. 114) remarks that

$$\{\boldsymbol{\eta} : \boldsymbol{\eta} = (w_1(\boldsymbol{\theta}), \ldots, w_k(\boldsymbol{\theta})) | \boldsymbol{\theta} \in \Theta\} \subseteq \Omega, \tag{3.8}$$

but often Ω is a strictly larger set.

Remark 3.1. For the families in Chap. 10 other than the χ^2_p and inverse Gaussian distributions, make the following assumptions if $\dim(\Theta) = k = \dim(\Omega)$. Assume that $\eta_i = w_i(\boldsymbol{\theta})$. Assume that the usual parameter space Θ_U is as big as possible (replace the integral by a sum for a pmf):

$$\Theta_U = \left\{ \boldsymbol{\theta} \in \mathbb{R}^k : \int f(y|\boldsymbol{\theta}) dy = 1 \right\},$$

and let

$$\Theta = \{\boldsymbol{\theta} \in \Theta_U : w_1(\boldsymbol{\theta}), \ldots, w_k(\boldsymbol{\theta}) \text{ are defined } \}.$$

Then assume that the natural parameter space satisfies condition E1) with

$$\Omega = \{(\eta_1, \ldots, \eta_k) : \eta_i = w_i(\boldsymbol{\theta}) \text{ for } \boldsymbol{\theta} \in \Theta\}.$$

In other words, simply define $\eta_i = w_i(\boldsymbol{\theta})$. For many common distributions, $\boldsymbol{\eta}$ is a one-to-one function of $\boldsymbol{\theta}$, and the above map is correct, especially if Θ_U is an open interval or cross product of open intervals.

Remark 3.2. Chapter 10 has many examples showing that a distribution is a 1P-REF or 2P-REF.

Example 3.1. Let $f(x|\mu,\sigma)$ be the $N(\mu,\sigma^2)$ family of pdfs. Then $\boldsymbol{\theta} = (\mu,\sigma)$ where $-\infty < \mu < \infty$ and $\sigma > 0$. Recall that μ is the mean and σ is the standard deviation (SD) of the distribution. The usual parameterization is

$$f(x|\boldsymbol{\theta}) = \frac{1}{\sqrt{2\pi}\sigma} \exp\left(\frac{-(x-\mu)^2}{2\sigma^2}\right) I_{\mathbb{R}}(x)$$

where $\mathbb{R} = (-\infty,\infty)$ and the indicator $I_A(x) = 1$ if $x \in A$ and $I_A(x) = 0$ otherwise. Notice that $I_{\mathbb{R}}(x) = 1 \;\; \forall x$. Since

$$f(x|\mu,\sigma) = \underbrace{\frac{1}{\sqrt{2\pi}\sigma} \exp\left(\frac{-\mu^2}{2\sigma^2}\right)}_{c(\mu,\sigma)\geq 0} \exp\left(\underbrace{\frac{-1}{2\sigma^2}}_{w_1(\boldsymbol{\theta})} \underbrace{x^2}_{t_1(x)} + \underbrace{\frac{\mu}{\sigma^2}}_{w_2(\boldsymbol{\theta})} \underbrace{x}_{t_2(x)}\right) \underbrace{I_{\mathbb{R}}(x)}_{h(x)\geq 0},$$

this family is a two-parameter exponential family. Hence $\eta_1 = -0.5/\sigma^2$ and $\eta_2 = \mu/\sigma^2$ if $\sigma > 0$, and $\Omega = (-\infty,0) \times (-\infty,\infty)$. Plotting η_1 on the horizontal axis and η_2 on the vertical axis yields the left half plane which certainly contains a two-dimensional rectangle. Since t_1 and t_2 lie on a quadratic rather than a line, the family is a 2P-REF. Notice that if $X_1, \ldots, X_n$ are iid $N(\mu,\sigma^2)$ random variables, then the joint pdf $f(\boldsymbol{x}|\boldsymbol{\theta}) = f(x_1, \ldots, x_n|\mu,\sigma) =$

$$\underbrace{\left[\frac{1}{\sqrt{2\pi}\sigma} \exp\left(\frac{-\mu^2}{2\sigma^2}\right)\right]^n}_{C(\mu,\sigma)\geq 0} \exp\left(\underbrace{\frac{-1}{2\sigma^2}}_{w_1(\boldsymbol{\theta})} \underbrace{\sum_{i=1}^n x_i^2}_{T_1(\boldsymbol{x})} + \underbrace{\frac{\mu}{\sigma^2}}_{w_2(\boldsymbol{\theta})} \underbrace{\sum_{i=1}^n x_i}_{T_2(\boldsymbol{x})}\right) \underbrace{1}_{h(\boldsymbol{x})\geq 0},$$

and is thus a 2P-REF.

Example 3.2. The χ^2_p distribution is not a 1P-REF since the usual parameter space Θ_U for the χ^2_p distribution is the set of positive integers, which is neither an open set nor a convex set. Nevertheless, the natural parameterization is the gamma$(\nu, \lambda = 2)$ family which is a 1P-REF. Note that this family has uncountably many members while the χ^2_p family does not.

Example 3.3. The binomial(k,ρ) pmf is

$$f(x|\rho) = \binom{k}{x}\rho^x(1-\rho)^{k-x}\, I_{\{0,\dots,k\}}(x)$$

$$= \underbrace{\binom{k}{x} I_{\{0,\dots,k\}}(x)}_{h(x)\geq 0}\ \underbrace{(1-\rho)^k}_{c(\rho)\geq 0} \exp\left[\underbrace{\log\left(\frac{\rho}{1-\rho}\right)}_{w(\rho)}\underbrace{x}_{t(x)}\right]$$

where $\Theta_U = [0,1]$. Since the pmf and $\eta = \log(\rho/(1-\rho))$ are undefined for $\rho = 0$ and $\rho = 1$, we have $\Theta = (0,1)$. Notice that $\Omega = (-\infty, \infty)$.

Example 3.4. The uniform$(0,\theta)$ family is not an exponential family since the support $\mathscr{Y}_\theta = (0,\theta)$ depends on the unknown parameter θ.

Example 3.5. If Y has a half normal distribution, $Y \sim \text{HN}(\mu,\sigma)$, then the pdf of Y is

$$f(y) = \frac{2}{\sqrt{2\pi}\,\sigma}\exp\left(\frac{-(y-\mu)^2}{2\sigma^2}\right)$$

where $\sigma > 0$ and $y \geq \mu$ and μ is real. Notice that

$$f(y) = \frac{2}{\sqrt{2\pi}\,\sigma} I(y \geq \mu)\exp\left[\left(\frac{-1}{2\sigma^2}\right)(y-\mu)^2\right]$$

is a 1P-REF if μ is known. Hence $\Theta = (0,\infty)$, $\eta = -1/(2\sigma^2)$ and $\Omega = (-\infty, 0)$. Notice that a different 1P-REF is obtained for each value of μ when μ is known with support $\mathscr{Y}_\mu = [\mu,\infty)$. If μ is not known, then this family is not an exponential family since the support depends on μ.

The following two examples are important examples of REFs where $\dim(\Theta) > \dim(\Omega)$.

Example 3.6. If the t_i or η_i satisfy a linearity constraint, then the number of terms in the exponent of Eq. (3.1) can be reduced. Suppose that $Y_1,\dots,Y_n$ follow the multinomial $M_n(m,\rho_1,\dots,\rho_n)$ distribution which has $\dim(\Theta) = n$ if m is known. Then $\sum_{i=1}^n Y_i = m$, $\sum_{i=1}^n \rho_i = 1$ and the joint pmf of $\boldsymbol{Y}$ is

$$f(\boldsymbol{y}) = m! \prod_{i=1}^n \frac{\rho_i^{y_i}}{y_i!}.$$

The support of $\boldsymbol{Y}$ is $\mathscr{Y} = \{\boldsymbol{y} : \sum_{i=1}^n y_i = m \text{ and } 0 \leq y_i \leq m \text{ for } i = 1,\dots,n\}$.

Since Y_n and ρ_n are known if $Y_1,\dots,Y_{n-1}$ and $\rho_1,\dots,\rho_{n-1}$ are known, we can use an equivalent joint pmf f_{EF} in terms of $Y_1,\dots,Y_{n-1}$. Let

$$h(y_1,\ldots,y_{n-1}) = \left[\frac{m!}{\prod_{i=1}^n y_i!}\right] \; I[(y_1,\ldots,y_{n-1},y_n) \in \mathscr{Y}].$$

(This is a function of $y_1,\ldots,y_{n-1}$ since $y_n = m - \sum_{i=1}^{n-1} y_i$.) Then $Y_1,\ldots,Y_{n-1}$ have a $M_n(m,\rho_1,\ldots,\rho_n)$ distribution if the joint pmf of $Y_1,\ldots,Y_{n-1}$ is

$$f_{EF}(y_1,\ldots,y_{n-1}) = \exp\left[\sum_{i=1}^{n-1} y_i \log(\rho_i) + \left(m - \sum_{i=1}^{n-1} y_i\right)\log(\rho_n)\right] \; h(y_1,\ldots,y_{n-1})$$

$$= \exp[m\log(\rho_n)] \; \exp\left[\sum_{i=1}^{n-1} y_i \log(\rho_i/\rho_n)\right] \; h(y_1,\ldots,y_{n-1}). \tag{3.9}$$

Since $\rho_n = 1 - \sum_{j=1}^{n-1} \rho_j$, this is an $n-1$ dimensional REF with

$$\eta_i = \log(\rho_i/\rho_n) = \log\left(\frac{\rho_i}{1 - \sum_{j=1}^{n-1} \rho_j}\right)$$

and $\Omega = \mathbb{R}^{n-1}$.

Example 3.7. Similarly, let $\boldsymbol{\mu}$ be a $1 \times j$ row vector and let $\boldsymbol{\Sigma}$ be a $j \times j$ positive definite matrix. Then the usual parameterization of the multivariate normal $\text{MVN}_j(\boldsymbol{\mu}, \boldsymbol{\Sigma})$ distribution has $\dim(\Theta) = j + j^2$ but is a $j + j(j+1)/2$ parameter REF.

A **curved exponential family** is a k-parameter exponential family where the elements of $\boldsymbol{\theta} = (\theta_1,\ldots,\theta_k)$ are completely determined by $d < k$ of the elements. For example if $\boldsymbol{\theta} = (\theta, \theta^2)$, then the elements of $\boldsymbol{\theta}$ are completely determined by $\theta_1 = \theta$. A curved exponential family is not regular since it places a restriction on

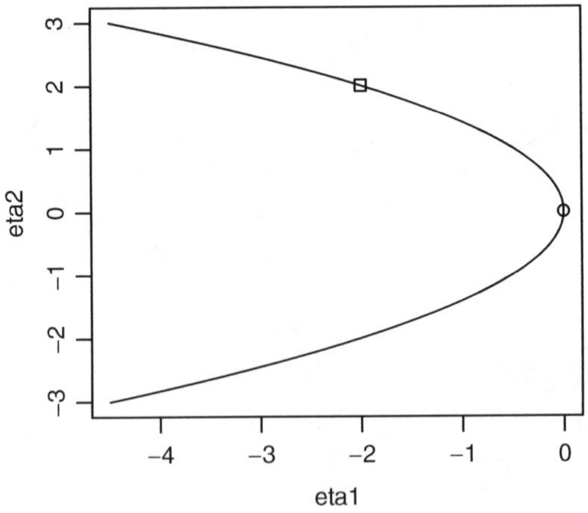

Fig. 3.1 The parameter space is a quadratic, not a two-dimensional open set

the parameter space Ω resulting in a new "natural parameter space" Ω_C where Ω_C does not contain a k-dimensional rectangle and is not a k-dimensional open set.

Example 3.8. The $N(\theta,\theta^2)$ distribution is a two-parameter exponential family with $\eta_1 = -1/(2\theta^2)$ and $\eta_2 = 1/\theta$. To see this, note that

$$f(y|\theta) = \frac{1}{\theta\sqrt{2\pi}}\exp\left(\frac{-(y-\theta)^2}{2\theta^2}\right) = \frac{1}{\theta\sqrt{2\pi}}\exp\left[\frac{-1}{2\theta^2}(y^2-2\theta y+\theta^2)\right] =$$

$$\frac{1}{\theta\sqrt{2\pi}}\exp(-1/2)\exp\left[\frac{-1}{2\theta^2}y^2+\frac{1}{\theta}y\right].$$

Thus the "natural parameter space" is

$$\Omega_C = \{(\eta_1,\eta_2)|\eta_1 = -0.5\eta_2^2, -\infty < \eta_1 < 0, -\infty < \eta_2 < \infty, \eta_2 \neq 0\}.$$

To be a 2P-REF, the parameter space needs to be a two-dimensional open set. A k-dimensional open set contains a k-dimensional open ball which contains a k-dimensional open set. The graph of the "natural parameter space" is a quadratic and cannot contain a two-dimensional rectangle. (Any rectangle will contain points that are not on the quadratic, so Ω_C is not a two-dimensional open set.) See Fig. 3.1 where the small rectangle centered at (−2,2) contains points that are not in the parameter space. Hence this two-parameter curved exponential family is not a 2P-REF.

3.2 Properties of $(t_1(Y),\ldots,t_k(Y))$

This section follows Lehmann (1983, pp. 29–35) closely, and several of the results will be used in later chapters. Write the *natural parameterization for the exponential family* as

$$f(y|\boldsymbol{\eta}) = h(y)b(\boldsymbol{\eta})\exp\left[\sum_{i=1}^k \eta_i t_i(y)\right]$$

$$= h(y)\exp\left[\sum_{i=1}^k \eta_i t_i(y) \quad - \quad a(\boldsymbol{\eta})\right] \tag{3.10}$$

where $a(\boldsymbol{\eta}) = -\log(b(\boldsymbol{\eta}))$. The kernel function of this pdf or pmf is

$$h(y)\exp\left[\sum_{i=1}^k \eta_i t_i(y)\right].$$

Lemma 3.2. Suppose that Y comes from an exponential family (3.10) and that $g(y)$ is any function with $E_{\boldsymbol{\eta}}[|g(Y)|] < \infty$. Then for any $\boldsymbol{\eta}$ in the interior of Ω, the integral $\int g(y)f(y|\boldsymbol{\eta})dy$ is continuous and has derivatives of all orders. These derivatives can be obtained by interchanging the derivative and integral operators. If f is a pmf, replace the integral by a sum.

Proof. See Lehmann (1986, p. 59).

Hence

$$\frac{\partial}{\partial \eta_i} \int g(y) f(y|\boldsymbol{\eta}) dy = \int g(y) \frac{\partial}{\partial \eta_i} f(y|\boldsymbol{\eta}) dy \tag{3.11}$$

if f is a pdf and

$$\frac{\partial}{\partial \eta_i} \sum g(y) f(y|\boldsymbol{\eta}) = \sum g(y) \frac{\partial}{\partial \eta_i} f(y|\boldsymbol{\eta}) \tag{3.12}$$

if f is a pmf.

Remark 3.3. If $\boldsymbol{Y}$ comes from an exponential family (3.1), then the derivative and integral (or sum) operators can be interchanged. Hence

$$\frac{\partial}{\partial \theta_i} \int \cdots \int g(\boldsymbol{y}) f(\boldsymbol{y}|\boldsymbol{\theta}) d\boldsymbol{y} = \int \cdots \int g(\boldsymbol{y}) \frac{\partial}{\partial \theta_i} f(\boldsymbol{y}|\boldsymbol{\theta}) d\boldsymbol{y}$$

for any function $g(\boldsymbol{y})$ with $E_{\theta}|g(\boldsymbol{Y})| < \infty$.

The behavior of $(t_1(Y), \ldots, t_k(Y))$ will be of considerable interest in later chapters. The following result is in Lehmann (1983, pp. 29–30). Also see Johnson et al. (1979).

Theorem 3.3. Suppose that Y comes from an exponential family (3.10). Then a)

$$E(t_i(Y)) = \frac{\partial}{\partial \eta_i} a(\boldsymbol{\eta}) = -\frac{\partial}{\partial \eta_i} \log(b(\boldsymbol{\eta})) \tag{3.13}$$

and b)

$$\text{Cov}(t_i(Y), t_j(Y)) = \frac{\partial^2}{\partial \eta_i \partial \eta_j} a(\boldsymbol{\eta}) = -\frac{\partial^2}{\partial \eta_i \partial \eta_j} \log(b(\boldsymbol{\eta})). \tag{3.14}$$

Notice that $i = j$ gives the formula for $\text{VAR}(t_i(Y))$.

Proof. The proof will be for pdfs. For pmfs replace the integrals by sums. Use Lemma 3.2 with $g(y) = 1\ \forall y$. a) Since $1 = \int f(y|\boldsymbol{\eta}) dy$,

$$0 = \frac{\partial}{\partial \eta_i} 1 = \frac{\partial}{\partial \eta_i} \int h(y) \exp\left[\sum_{m=1}^{k} \eta_m t_m(y) \quad - \quad a(\boldsymbol{\eta})\right] dy$$

$$= \int h(y) \frac{\partial}{\partial \eta_i} \exp\left[\sum_{m=1}^{k} \eta_m t_m(y) \quad - \quad a(\boldsymbol{\eta})\right] dy$$

$$= \int h(y)\exp\left[\sum_{m=1}^{k}\eta_m t_m(y) \quad - \quad a(\boldsymbol{\eta})\right](t_i(y) - \frac{\partial}{\partial \eta_i}a(\boldsymbol{\eta}))dy$$

$$= \int (t_i(y) - \frac{\partial}{\partial \eta_i}a(\boldsymbol{\eta}))f(y|\boldsymbol{\eta})dy$$

$$= E(t_i(Y)) - \frac{\partial}{\partial \eta_i}a(\boldsymbol{\eta}).$$

b) Similarly,

$$0 = \int h(y)\frac{\partial^2}{\partial \eta_i \partial \eta_j}\exp\left[\sum_{m=1}^{k}\eta_m t_m(y) \quad - \quad a(\boldsymbol{\eta})\right]dy.$$

From the proof of a),

$$0 = \int h(y)\frac{\partial}{\partial \eta_j}\left[\exp\left[\sum_{m=1}^{k}\eta_m t_m(y) \quad - \quad a(\boldsymbol{\eta})\right](t_i(y) - \frac{\partial}{\partial \eta_i}a(\boldsymbol{\eta}))\right]dy$$

$$= \int h(y)\exp\left[\sum_{m=1}^{k}\eta_m t_m(y) \quad - \quad a(\boldsymbol{\eta})\right](t_i(y) - \frac{\partial}{\partial \eta_i}a(\boldsymbol{\eta}))(t_j(y) - \frac{\partial}{\partial \eta_j}a(\boldsymbol{\eta}))dy$$

$$- \int h(y)\exp\left[\sum_{m=1}^{k}\eta_m t_m(y) \quad - \quad a(\boldsymbol{\eta})\right]\left(\frac{\partial^2}{\partial \eta_i \partial \eta_j}a(\boldsymbol{\eta})\right)dy$$

$$= \text{Cov}(t_i(Y),t_j(Y)) - \frac{\partial^2}{\partial \eta_i \partial \eta_j}a(\boldsymbol{\eta})$$

since $\frac{\partial}{\partial \eta_j}a(\boldsymbol{\eta}) = E(t_j(Y))$ by a). □

Theorem 3.4. Suppose that Y comes from an exponential family (3.10), and let $\boldsymbol{T} = (t_1(Y),\ldots,t_k(Y))$. Then for any $\boldsymbol{\eta}$ in the interior of Ω, the moment generating function of $\boldsymbol{T}$ is

$$m_{\boldsymbol{T}}(\boldsymbol{s}) = \exp[a(\boldsymbol{\eta}+\boldsymbol{s}) - a(\boldsymbol{\eta})] = \exp[a(\boldsymbol{\eta}+\boldsymbol{s})]/\exp[a(\boldsymbol{\eta})].$$

Proof. The proof will be for pdfs. For pmfs replace the integrals by sums. Since $\boldsymbol{\eta}$ is in the interior of Ω there is a neighborhood of $\boldsymbol{\eta}$ such that if $\boldsymbol{s}$ is in that neighborhood, then $\boldsymbol{\eta}+\boldsymbol{s} \in \Omega$. (Hence there exists a $\delta > 0$ such that if $\|\boldsymbol{s}\| < \delta$, then $\boldsymbol{\eta}+\boldsymbol{s} \in \Omega$.) For such $\boldsymbol{s}$ (see Definition 2.25),

$$m_{\boldsymbol{T}}(\boldsymbol{s}) = E\left[\exp\left(\sum_{i=1}^{k} s_i t_i(Y)\right)\right] \equiv E(g(Y)).$$

It is important to notice that we are finding the mgf of $\boldsymbol{T}$, not the mgf of Y. Hence we can use the kernel method of Sect. 1.5 to find $E(g(Y)) = \int g(y)f(y)dy$ without finding the joint distribution of $\boldsymbol{T}$. So

$$m_{\boldsymbol{T}}(\boldsymbol{s}) = \int \exp\left(\sum_{i=1}^{k} s_i t_i(y)\right) h(y) \exp\left[\sum_{i=1}^{k} \eta_i t_i(y) \; - \; a(\boldsymbol{\eta})\right] dy$$

$$= \int h(y) \exp\left[\sum_{i=1}^{k} (\eta_i + s_i) t_i(y) \; - \; a(\boldsymbol{\eta}+\boldsymbol{s}) \; + \; a(\boldsymbol{\eta}+\boldsymbol{s}) \; - \; a(\boldsymbol{\eta})\right] dy$$

$$= \exp[a(\boldsymbol{\eta}+\boldsymbol{s}) \; - \; a(\boldsymbol{\eta})] \int h(y) \exp\left[\sum_{i=1}^{k} (\eta_i + s_i) t_i(y) \; - \; a(\boldsymbol{\eta}+\boldsymbol{s})\right] dy$$

$$= \exp[a(\boldsymbol{\eta}+\boldsymbol{s}) \; - \; a(\boldsymbol{\eta})] \int f(y|[\boldsymbol{\eta}+\boldsymbol{s}]) dy = \exp[a(\boldsymbol{\eta}+\boldsymbol{s}) \; - \; a(\boldsymbol{\eta})]$$

since the pdf $f(y|[\boldsymbol{\eta}+\boldsymbol{s}])$ integrates to one. □

Theorem 3.5. Suppose that Y comes from an exponential family (3.10), and let $\boldsymbol{T} = (t_1(Y), \ldots, t_k(Y)) = (T_1, \ldots, T_k)$. Then the distribution of $\boldsymbol{T}$ is an exponential family with

$$f(\boldsymbol{t}|\boldsymbol{\eta}) = h^*(\boldsymbol{t}) \exp\left[\sum_{i=1}^{k} \eta_i t_i \; - \; a(\boldsymbol{\eta})\right].$$

Proof. See Lehmann (1986, p. 58).

The main point of this section is that $\boldsymbol{T}$ is well behaved even if Y is not. For example, if Y follows a one-sided stable distribution, then Y is from an exponential family, but $E(Y)$ does not exist. However the mgf of T exists, so all moments of T exist. If $Y_1, \ldots, Y_n$ are iid from a one-parameter exponential family, then $T \equiv T_n = \sum_{i=1}^n t(Y_i)$ is from a one-parameter exponential family. One way to find the distribution function of T is to find the distribution of $t(Y)$ using the transformation method, then find the distribution of $\sum_{i=1}^n t(Y_i)$ using moment generating functions or Theorems 2.17 and 2.18. This technique results in the following two theorems. Notice that T often has a gamma distribution.

Theorem 3.6. Let $Y_1, \ldots, Y_n$ be iid from the given one-parameter exponential family and let $T \equiv T_n = \sum_{i=1}^n t(Y_i)$.

a) If Y_i is from a binomial (k, ρ) distribution with k known, then $t(Y) = Y \sim \text{BIN}(k, \rho)$ and $T_n = \sum_{i=1}^n Y_i \sim \text{BIN}(nk, \rho)$.
b) If Y is from an exponential (λ) distribution then, $t(Y) = Y \sim \text{EXP}(\lambda)$ and $T_n = \sum_{i=1}^n Y_i \sim G(n, \lambda)$.
c) If Y is from a gamma (ν, λ) distribution with ν known, then $t(Y) = Y \sim G(\nu, \lambda)$ and $T_n = \sum_{i=1}^n Y_i \sim G(n\nu, \lambda)$.
d) If Y is from a geometric (ρ) distribution, then $t(Y) = Y \sim \text{geom}(\rho)$ and $T_n = \sum_{i=1}^n Y_i \sim \text{NB}(n, \rho)$ where NB stands for negative binomial.
e) If Y is from a negative binomial (r, ρ) distribution with r known, then $t(Y) = Y \sim \text{NB}(r, \rho)$ and $T_n = \sum_{i=1}^n Y_i \sim \text{NB}(nr, \rho)$.

f) If Y is from a normal (μ,σ^2) distribution with σ^2 known, then $t(Y) = Y \sim N(\mu,\sigma^2)$ and $T_n = \sum_{i=1}^n Y_i \sim N(n\mu, n\sigma^2)$.
g) If Y is from a normal (μ,σ^2) distribution with μ known, then $t(Y) = (Y-\mu)^2 \sim G(1/2, 2\sigma^2)$ and $T_n = \sum_{i=1}^n (Y_i - \mu)^2 \sim G(n/2, 2\sigma^2)$.
h) If Y is from a Poisson (θ) distribution, then $t(Y) = Y \sim \text{POIS}(\theta)$ and $T_n = \sum_{i=1}^n Y_i \sim \text{POIS}(n\theta)$.

Theorem 3.7. Let $Y_1,\ldots,Y_n$ be iid from the given one-parameter exponential family and let $T \equiv T_n = \sum_{i=1}^n t(Y_i)$.

a) If Y_i is from a Burr type XII (ϕ,λ) distribution with ϕ known, then $t(Y) = \log(1+Y^\phi) \sim \text{EXP}(\lambda)$ and $T_n = \sum \log(1+Y_i^\phi) \sim G(n,\lambda)$.
b) If Y is from a chi(p,σ) distribution with p known, then $t(Y) = Y^2 \sim G(p/2, 2\sigma^2)$ and $T_n = \sum Y_i^2 \sim G(np/2, 2\sigma^2)$.
c) If Y is from a double exponential (θ,λ) distribution with θ known, then $t(Y) = |Y-\theta| \sim \text{EXP}(\lambda)$ and $T_n = \sum_{i=1}^n |Y_i - \theta| \sim G(n,\lambda)$.
d) If Y is from a two-parameter exponential (θ,λ) distribution with θ known, then $t(Y) = Y_i - \theta \sim \text{EXP}(\lambda)$ and $T_n = \sum_{i=1}^n (Y_i - \theta) \sim G(n,\lambda)$.
e) If Y is from a generalized negative binomial GNB(μ,κ) distribution with κ known, then $T_n = \sum_{i=1}^n Y_i \sim \text{GNB}(n\mu, n\kappa)$
f) If Y is from a half normal (μ,σ^2) distribution with μ known, then $t(Y) = (Y-\mu)^2 \sim G(1/2, 2\sigma^2)$ and $T_n = \sum_{i=1}^n (Y_i-\mu)^2 \sim G(n/2, 2\sigma^2)$.
g) If Y is from an inverse Gaussian IG(θ,λ) distribution with λ known, then $T_n = \sum_{i=1}^n Y_i \sim \text{IG}(n\theta, n^2\lambda)$.
h) If Y is from an inverted gamma (ν,λ) distribution with ν known, then $t(Y) = 1/Y \sim G(\nu,\lambda)$ and $T_n = \sum_{i=1}^n 1/Y_i \sim G(n\nu,\lambda)$.
i) If Y is from a lognormal (μ,σ^2) distribution with μ known, then $t(Y) = (\log(Y)-\mu)^2 \sim G(1/2, 2\sigma^2)$ and $T_n = \sum_{i=1}^n (\log(Y_i)-\mu)^2 \sim G(n/2, 2\sigma^2)$.
j) If Y is from a lognormal (μ,σ^2) distribution with σ^2 known, then $t(Y) = \log(Y) \sim N(\mu,\sigma^2)$ and $T_n = \sum_{i=1}^n \log(Y_i) \sim N(n\mu, n\sigma^2)$.
k) If Y is from a Maxwell–Boltzmann (μ,σ) distribution with μ known, then $t(Y) = (Y-\mu)^2 \sim G(3/2, 2\sigma^2)$ and $T_n = \sum_{i=1}^n (Y_i-\mu)^2 \sim G(3n/2, 2\sigma^2)$.
l) If Y is from a one-sided stable (σ) distribution, then $t(Y) = 1/Y \sim G(1/2, 2/\sigma)$ and $T_n = \sum_{i=1}^n 1/Y_i \sim G(n/2, 2/\sigma)$.
m) If Y is from a Pareto (σ,λ) distribution with σ known, then $t(Y) = \log(Y/\sigma) \sim \text{EXP}(\lambda)$ and $T_n = \sum_{i=1}^n \log(Y_i/\sigma) \sim G(n,\lambda)$.
n) If Y is from a power (λ) distribution, then $t(Y) = -\log(Y) \sim \text{EXP}(\lambda)$ and $T_n = \sum_{i=1}^n [-\log(Y_i)] \sim G(n,\lambda)$.
o) If Y is from a Rayleigh (μ,σ) distribution with μ known, then $t(Y) = (Y-\mu)^2 \sim \text{EXP}(2\sigma^2)$ and $T_n = \sum_{i=1}^n (Y_i-\mu)^2 \sim G(n, 2\sigma^2)$.
p) If Y is from a Topp–Leone (ν) distribution, then $t(Y) =$
$-\log(2Y - Y^2) \sim \text{EXP}(1/\nu)$ and $T_n = \sum_{i=1}^n [-\log(2Y_i - Y_i^2)] \sim G(n, 1/\nu)$.
q) If Y is from a truncated extreme value (λ) distribution, then $t(Y) = e^Y - 1 \sim \text{EXP}(\lambda)$ and $T_n = \sum_{i=1}^n (e^{Y_i} - 1) \sim G(n,\lambda)$.
r) If Y is from a Weibull (ϕ,λ) distribution with ϕ known, then $t(Y) = Y^\phi \sim \text{EXP}(\lambda)$ and $T_n = \sum_{i=1}^n Y_i^\phi \sim G(n,\lambda)$.

3.3 Summary

1. Given the pmf or pdf of Y or that Y is a brand name distribution from Sect. 1.7, know how to show whether Y belongs to an exponential family or not using

$$f(y|\boldsymbol{\theta}) = h(y)c(\boldsymbol{\theta})\exp\left[\sum_{i=1}^{k} w_i(\boldsymbol{\theta})t_i(y)\right]$$

or

$$f(y|\theta) = h(y)c(\theta)\exp[w(\theta)t(y)].$$

Tips: a) The F, Cauchy, logistic, t and uniform distributions cannot be put in form 1) and so are not exponential families.
b) If the support depends on an unknown parameter, then the family is not an exponential family.

2. If Y belongs to an exponential family, set $\eta_i = w_i(\boldsymbol{\theta})$ and typically set the natural parameter space Ω equal to the cross product of the ranges of η_i. For a kP-REF, Ω is an open set that is typically a cross product of k open intervals. For a 1P-REF, Ω is an open interval.

3. If Y belongs to an exponential family, know how to show whether Y belongs to a k-dimensional regular exponential family. Suppose Θ is found after deleting values of $\boldsymbol{\theta}$ from Θ_U such that the pdf or pmf is undefined and such that $w_i(\boldsymbol{\theta})$ is undefined. Assume $\dim(\Theta) = k = \dim(\Omega)$. Typically we assume that condition E1) is true: that is, Ω is given by (3.7). Then typically Ω is the cross product of the ranges of $\eta_i = w_i(\boldsymbol{\theta})$. If Ω contains a k-dimensional rectangle, e.g., if Ω is a k-dimensional open set, then $\eta_1, \ldots, \eta_k$ do not satisfy a linearity constraint. In particular, if $p = 2$ you should be able to show whether η_1 and η_2 satisfy a linearity constraint and whether t_1 and t_2 satisfy a linearity constraint, to plot Ω and to determine whether the natural parameter space Ω contains a two-dimensional rectangle.
Tips: a) If $\dim(\Theta) = k = \dim(\Omega)$, then the usual parameterization has k parameters, and so does the kP-REF parameterization.
b) If one of the two parameters is known for the usual parameterization, then the family will often be a 1P-REF with $\eta = w(\theta)$ where θ is the unknown parameter.
c) If the family is a two-dimensional exponential family with natural parameters η_1 and η_2, but the usual parameterization is determined by one-parameter θ, then the family is probably not regular. The $N(a\mu, b\mu^2)$ and $N(a\sigma, b\sigma^2)$ families are typical examples. If $\dim(\Theta) = j < k = \dim(\Omega)$, the family is usually not regular. If η_1 is a simple function of η_2, then the "natural parameter space" is not a cross product of the ranges of η_i. For example, if $\eta_1 = \eta_2^2$, then the "natural parameter space" is a parabola and is not a two-dimensional open set, and does not contain a two-dimensional rectangle.

3.4 Complements

Theorem 3.8. Suppose Y has a pdf that is a kP-REF, and that $X = t_0(Y)$ where $t_0(y)$ is a parameter free monotone differentiable transformation. Then X has a pdf that is a kP-REF.

Proof. Let $f_Y(y|\boldsymbol{\eta}) = h(y)b(\boldsymbol{\eta})\exp\left[\sum_{i=1}^k \eta_i t_i(y)\right]$ where the η_i and the $t_i(y)$ do not satisfy a linearity constraint. Then

$$f_X(x|\boldsymbol{\eta}) = h(t_0^{-1}(x))\left|\frac{dt_0^{-1}(x)}{dx}\right| b(\boldsymbol{\eta})\exp\left[\sum_{i=1}^k \eta_i t_i(t_0^{-1}(x))\right]$$

$= h^*(x)b(\boldsymbol{\eta})\exp\left[\sum_{i=1}^k \eta_i t_i^*(x)\right]$ which is an exponential family. The family is a kP-REF since the η_i and natural parameter space Ω are the same as for Y, and neither the η_i nor the $t_i^*(x)$ satisfy a linearity constraint. To see this, note that if the $t_i^*(x)$ satisfy a linearity constraint, then $\sum_{i=1}^k a_i t_i(t_0^{-1}(x)) = \sum_{i=1}^k a_i t_i(y) = c$ where not all $a_i = 0$. But this contradicts the fact that the $t_i(y)$ do not satisfy a linearity constraint. $\square$

If Y is a k-parameter exponential family, then $X = t_0(Y)$ is an exponential family of dimension no larger than k since

$$f_X(x|\boldsymbol{\theta}) = h(t_0^{-1}(x))\left|\frac{dt_0^{-1}(x)}{dx}\right| c(\boldsymbol{\theta})\exp\left[\sum_{i=1}^k w_i(\boldsymbol{\theta})t_i(t_0^{-1}(x))\right].$$

Notice that for a one-parameter exponential family with $t_0(y) \equiv t(y)$, the above result implies that $t(Y)$ is a one-parameter exponential family. This result is a special case of Theorem 3.5. The chi distribution and log gamma distribution are 2P-REFs since they are transformations of the gamma distribution. The lognormal distribution is a 2P-REF since it is a transformation of the normal distribution. The power and truncated extreme value distributions are 1P-REFs since they are transformations of the exponential distributions.

Example 3.9. Following Barndorff–Nielsen (1978, p. 117), if Y has an inverse Gaussian distribution, $Y \sim \text{IG}(\theta, \lambda)$, then the pdf of Y is

$$f(y) = \sqrt{\frac{\lambda}{2\pi y^3}}\exp\left[\frac{-\lambda(y-\theta)^2}{2\theta^2 y}\right]$$

where $y, \theta, \lambda > 0$.

Notice that

$$f(y) = \sqrt{\frac{\lambda}{2\pi}}e^{\lambda/\theta}\sqrt{\frac{1}{y^3}}I(y>0)\exp\left[\frac{-\lambda}{2\theta^2}y - \frac{\lambda}{2}\frac{1}{y}\right]$$

is a two-parameter exponential family.

Another parameterization of the inverse Gaussian distribution takes $\theta = \sqrt{\lambda/\psi}$ so that

$$f(y) = \sqrt{\frac{\lambda}{2\pi}} e^{\sqrt{\lambda\psi}} \sqrt{\frac{1}{y^3}} I[y>0] \exp\left[\frac{-\psi}{2} y - \frac{\lambda}{2}\frac{1}{y}\right],$$

where $\lambda > 0$ and $\psi \geq 0$. Here $\Theta = (0,\infty) \times [0,\infty)$, $\eta_1 = -\psi/2$, $\eta_2 = -\lambda/2$ and $\Omega = (-\infty, 0] \times (-\infty, 0)$. Since Ω is not an open set, this is a **two-parameter full exponential family that is not regular**. If ψ is known then Y is a 1P-REF, but if λ is known, then Y is a one-parameter full exponential family. When $\psi = 0$, Y has a one- sided stable distribution.

The following chapters show that exponential families can be used to simplify the theory of sufficiency, MLEs, UMVUEs, UMP tests, and large sample theory. Barndorff–Nielsen (1982) and Olive (2008) are useful introductions to exponential families. Also see Bühler and Sehr (1987). Interesting subclasses of exponential families are given by Rahman and Gupta (1993), and Sankaran and Gupta (2005). Most statistical inference texts at the same level as this text also cover exponential families. History and references for additional topics (such as finding conjugate priors in Bayesian statistics) can be found in Lehmann (1983, p. 70), Brown (1986) and Barndorff–Nielsen (1978, 1982).

Barndorff–Nielsen (1982), Brown (1986), and Johanson (1979) are post-PhD treatments and hence very difficult. Mukhopadhyay (2000) and Brown (1986) place restrictions on the exponential families that make their theory less useful. For example, Brown (1986) covers linear exponential distributions. See Johnson and Kotz (1972).

3.5 Problems

PROBLEMS WITH AN ASTERISK * ARE ESPECIALLY USEFUL.

Refer to Chap. 10 for the pdf or pmf of the distributions in the problems below.

3.1*. Show that each of the following families is a 1P-REF by writing the pdf or pmf as a one-parameter exponential family, finding $\eta = w(\theta)$ and by showing that the natural parameter space Ω is an open interval.

a) The binomial (k, ρ) distribution with k known and $\rho \in \Theta = (0,1)$.
b) The exponential (λ) distribution with $\lambda \in \Theta = (0,\infty)$.
c) The Poisson (θ) distribution with $\theta \in \Theta = (0,\infty)$.
d) The half normal (μ, σ^2) distribution with μ known and $\sigma^2 \in \Theta = (0,\infty)$.

3.2*. Show that each of the following families is a 2P-REF by writing the pdf or pmf as a two-parameter exponential family, finding $\eta_i = w_i(\theta)$ for $i = 1,2$ and by showing that the natural parameter space Ω is a cross product of two open intervals.

a) The beta (δ, ν) distribution with $\Theta = (0, \infty) \times (0, \infty)$.
b) The chi (p, σ) distribution with $\Theta = (0, \infty) \times (0, \infty)$.
c) The gamma (ν, λ) distribution with $\Theta = (0, \infty) \times (0, \infty)$.
d) The lognormal (μ, σ^2) distribution with $\Theta = (-\infty, \infty) \times (0, \infty)$.
e) The normal (μ, σ^2) distribution with $\Theta = (-\infty, \infty) \times (0, \infty)$.

3.3. Show that each of the following families is a 1P-REF by writing the pdf or pmf as a one-parameter exponential family, finding $\eta = w(\theta)$ and by showing that the natural parameter space Ω is an open interval.

a) The generalized negative binomial (μ, κ) distribution if κ is known.
b) The geometric (ρ) distribution.
c) The logarithmic (θ) distribution.
d) The negative binomial (r, ρ) distribution if r is known.
e) The one-sided stable (σ) distribution.
f) The power (λ) distribution.
g) The truncated extreme value (λ) distribution.
h) The Zeta (ν) distribution.

3.4*. Show that each of the following families is a 1P-REF by writing the pdf or pmf as a one-parameter exponential family, finding $\eta = w(\theta)$ and by showing that the natural parameter space Ω is an open interval.

a) The $N(\mu, \sigma^2)$ family with $\sigma > 0$ known.
b) The $N(\mu, \sigma^2)$ family with μ known and $\sigma > 0$.
(See Problem 3.12 for a common error.)
c) The gamma (ν, λ) family with ν known.
d) The gamma (ν, λ) family with λ known.
e) The beta (δ, ν) distribution with δ known.
f) The beta (δ, ν) distribution with ν known.

3.5. Show that each of the following families is a 1P-REF by writing the pdf or pmf as a one-parameter exponential family, finding $\eta = w(\theta)$ and by showing that the natural parameter space Ω is an open interval.

a) The Burr Type XII (ϕ, λ) distribution with ϕ known.
b) The double exponential (θ, λ) distribution with θ known.
c) The two-parameter exponential (θ, λ) distribution with θ known.
d) The largest extreme value (θ, σ) distribution if σ is known.
e) The smallest extreme value (θ, σ) distribution if σ is known.
f) The inverted gamma (ν, λ) distribution if ν is known.
g) The Maxwell–Boltzmann (μ, σ) distribution if μ is known.
h) The Pareto (σ, λ) distribution if σ is known.
i) The Rayleigh (μ, σ) distribution if μ is known.
j) The Weibull (ϕ, λ) distribution if ϕ is known.

3.6*. Determine whether the Pareto (σ, λ) distribution is an exponential family or not.

3.7. Following Kotz and van Dorp (2004, pp. 35–36), if Y has a Topp–Leone distribution, $Y \sim \text{TL}(\nu)$, then the cdf of Y is $F(y) = (2y - y^2)^\nu$ for $\nu > 0$ and $0 < y < 1$. The pdf of Y is

$$f(y) = \nu(2-2y)(2y-y^2)^{\nu-1}$$

for $0 < y < 1$. Determine whether this distribution is an exponential family or not.

3.8. In Spiegel (1975, p. 210), Y has pdf

$$f_Y(y) = \frac{2\gamma^{3/2}}{\sqrt{\pi}} y^2 \exp(-\gamma y^2)$$

where $\gamma > 0$ and y is real. Is Y a 1P-REF?

3.9. Let Y be a (one-sided) truncated exponential $TEXP(\lambda, b)$ random variable. Then the pdf of Y is

$$f_Y(y|\lambda, b) = \frac{\frac{1}{\lambda} e^{-y/\lambda}}{1 - \exp(-\frac{b}{\lambda})}$$

for $0 < y \le b$ where $\lambda > 0$. If b is known, is Y a 1P-REF? (Also see O'Reilly and Rueda (2007).)

3.10. Suppose Y has a Birnbaum Saunders distribution. If $t(Y) = \frac{Y}{\theta} + \frac{\theta}{Y} - 2$, then $t(Y) \sim \nu^2 \chi^2_1$. If θ is known, is this distribution a regular exponential family?

3.11. If Y has a Burr Type X distribution, then the pdf of Y is

$$f(y) = I(y>0)\; 2\;\tau\; y\; e^{-y^2}\; (1-e^{-y^2})^{\tau-1} =$$

$$I(y>0)\; 2y\; e^{-y^2}\; \tau\; \exp[(1-\tau)(-\log(1-e^{-y^2}))]$$

where $\tau > 0$ and $-\log(1-e^{-Y^2}) \sim \text{EXP}(1/\tau)$. Is this distribution a regular exponential family?

Problems from old quizzes and exams.

3.12*. Suppose that X has a $\text{N}(\mu, \sigma^2)$ distribution where $\sigma > 0$ and μ **is known**. Then

$$f(x) = \frac{1}{\sqrt{2\pi}\sigma} e^{-\mu^2/(2\sigma^2)} \exp\left[-\frac{1}{2\sigma^2}x^2 + \frac{1}{\sigma^2}\mu x\right].$$

Let $\eta_1 = -1/(2\sigma^2)$ and $\eta_2 = 1/\sigma^2$. Why is this parameterization not the regular exponential family parameterization? (Hint: show that η_1 and η_2 satisfy a linearity constraint.)

3.13. Let $X_1, \ldots, X_n$ be iid $N(\mu, \gamma_o^2 \mu^2)$ random variables where $\gamma_o^2 > 0$ is **known** and $\mu > 0$.

a) Find the distribution of $\sum_{i=1}^n X_i$.
b) Find $E[(\sum_{i=1}^n X_i)^2]$.
c) The pdf of X is

$$f_X(x|\mu) = \frac{1}{\gamma_o \mu \sqrt{2\pi}} \exp\left[\frac{-(x-\mu)^2}{2\gamma_o^2 \mu^2}\right].$$

Show that the family $\{f(x|\mu) : \mu > 0\}$ is a two-parameter exponential family.
d) Show that the "natural parameter space" is a parabola. You may assume that $\eta_i = w_i(\mu)$. Is this family a regular exponential family?

3.14. Let $X_1, \ldots, X_n$ be iid $N(\alpha\sigma, \sigma^2)$ random variables where α is a **known** real number and $\sigma > 0$.

a) Find $E[\sum_{i=1}^n X_i^2]$.
b) Find $E[(\sum_{i=1}^n X_i)^2]$.
c) Show that the family $\{f(x|\sigma) : \sigma > 0\}$ is a two-parameter exponential family.
d) Show that the "natural parameter space" is a parabola. You may assume that $\eta_i = w_i(\sigma)$. Is this family a regular exponential family?

3.15. If Y has a Lindley distribution, then the pdf of Y is

$$f(y) = \frac{\theta^2}{1+\theta}(1+y)e^{-\theta y}$$

where $y > 0$ and $\theta > 0$. Is Y a 1P-REF?

3.16. Suppose the pdf of Y is

$$f(y) = \frac{\theta}{(1+y)^{\theta+1}}$$

where $y > 0$ and $\theta > 0$. Is Y a 1P-REF?

3.17. Suppose the pdf of Y is

$$f(y) = \frac{\theta}{2(1+|y|)^{\theta+1}}$$

where $-\infty < y < \infty$ and $\theta > 0$. Is Y a 1P-REF?

Chapter 4
Sufficient Statistics

A statistic is a function of the data that does not depend on any unknown parameters, and a statistic is a random variable that has a distribution called the sampling distribution. Suppose the data is a random sample from a distribution with unknown parameters $\boldsymbol{\theta}$. Heuristically, a sufficient statistic for $\boldsymbol{\theta}$ contains all of the information from the data about $\boldsymbol{\theta}$. Since the data contains all of the information from the data, the data $(Y_1, \ldots, Y_n)$ is a sufficient statistic of dimension n. Heuristically, a minimal sufficient statistic is a sufficient statistic with the smallest dimension k, where $1 \le k \le n$. If k is small and does not depend on n, then there is considerable dimension reduction.

The Factorization Theorem is used to find a sufficient statistic. The Lehmann–Scheffé Theorem and a theorem for exponential families are useful for finding a minimal sufficient statistic. Complete sufficient statistics are useful for UMVUE theory in Chap. 6.

4.1 Statistics and Sampling Distributions

Suppose that the data $Y_1, \ldots, Y_n$ is drawn from some population. The observed data is $Y_1 = y_1, \ldots, Y_n = y_n$ where $y_1, \ldots, y_n$ are numbers. Let $\boldsymbol{y} = (y_1, \ldots, y_n)$. Real valued functions $T(y_1, \ldots, y_n) = T(\boldsymbol{y})$ are of interest as are vector valued functions $\boldsymbol{T}(\boldsymbol{y}) = (T_1(\boldsymbol{y}), \ldots, T_k(\boldsymbol{y}))$. Sometimes the data $Y_1, \ldots, Y_n$ are random vectors. Again interest is in functions of the data. Typically the data has a joint pdf or pmf $f(y_1, \ldots, y_n | \boldsymbol{\theta})$ where the vector of unknown parameters is $\boldsymbol{\theta} = (\theta_1, \ldots, \theta_k)$. (In the joint pdf or pmf, the $y_1, \ldots, y_n$ are dummy variables, not the observed data.)

Definition 4.1. A **statistic** is a function of the data that does not depend on any unknown parameters. The probability distribution of the statistic is called the **sampling distribution** of the statistic.

D.J. Olive, *Statistical Theory and Inference*, DOI 10.1007/978-3-319-04972-4_4,

Let the data $\boldsymbol{Y} = (Y_1, \ldots, Y_n)$ where the Y_i are random variables. If $T(y_1, \ldots, y_n)$ is a real valued function whose domain includes the sample space $\mathscr{Y}$ of $\boldsymbol{Y}$, then $W = T(Y_1, \ldots, Y_n)$ is a statistic provided that T does not depend on any unknown parameters. The data comes from some probability distribution and the statistic is a random variable and hence also comes from some probability distribution. To avoid confusing the distribution of the statistic with the distribution of the data, the distribution of the statistic is called the sampling distribution of the statistic. If the observed data is $Y_1 = y_1, \ldots, Y_n = y_n$, then the observed value of the statistic is $W = w = T(y_1, \ldots, y_n)$. Similar remarks apply when the statistic $\boldsymbol{T}$ is vector valued and when the data $\boldsymbol{Y}_1, \ldots, \boldsymbol{Y}_n$ are random vectors.

Often $Y_1, \ldots, Y_n$ will be iid and statistics of the form

$$\sum_{i=1}^{n} a_i Y_i \text{ and } \sum_{i=1}^{n} t(Y_i)$$

are especially important. Chapter 10 and Theorems 2.17, 2.18, 3.6, and 3.7 are useful for finding the sampling distributions of some of these statistics when the Y_i are iid from a given brand name distribution that is usually an exponential family. The following example lists some important statistics.

Example 4.1. Let $Y_1, \ldots, Y_n$ be the data.

a) The *sample mean*

$$\overline{Y} = \frac{\sum_{i=1}^{n} Y_i}{n}. \tag{4.1}$$

b) The *sample variance*

$$S^2 \equiv S_n^2 = \frac{\sum_{i=1}^{n}(Y_i - \overline{Y})^2}{n-1} = \frac{\sum_{i=1}^{n} Y_i^2 - n(\overline{Y})^2}{n-1}. \tag{4.2}$$

c) The *sample standard deviation* $S \equiv S_n = \sqrt{S_n^2}$.

d) If the data $Y_1, \ldots, Y_n$ is arranged in ascending order from smallest to largest and written as $Y_{(1)} \le \cdots \le Y_{(n)}$, then $Y_{(i)}$ is the ith order statistic and the $Y_{(i)}$'s are called the *order statistics*.

e) The *sample median*

$$\text{MED}(n) = Y_{((n+1)/2)} \text{ if n is odd,} \tag{4.3}$$

$$\text{MED}(n) = \frac{Y_{(n/2)} + Y_{((n/2)+1)}}{2} \text{ if n is even.}$$

f) The *sample median absolute deviation* is

$$\text{MAD}(n) = \text{MED}(|Y_i - \text{MED}(n)|,\ i = 1, \ldots, n). \tag{4.4}$$

g) The *sample maximum*

$$\max(n) = Y_{(n)} \tag{4.5}$$

and the observed max $y_{(n)}$ is the largest value of the observed data.

h) The *sample minimum*

$$\min(n) = Y_{(1)} \tag{4.6}$$

and the observed min $y_{(1)}$ is the smallest value of the observed data.

Example 4.2. Usually the term "observed" is dropped. Hence below "data" is "observed data," "observed order statistics" is "order statistics," and "observed value of MED(n)" is "MED(n)."
Let the data be $9,2,7,4,1,6,3,8,5$ (so $Y_1 = y_1 = 9, \ldots, Y_9 = y_9 = 5$). Then the order statistics are $1,2,3,4,5,6,7,8,9$. Then MED$(n) = 5$ and MAD$(n) = 2 =$ MED$\{0,1,1,2,2,3,3,4,4\}$.

Example 4.3. Let $Y_1, \ldots, Y_n$ be iid $N(\mu, \sigma^2)$. Then

$$T_n = \frac{\sum_{i=1}^n (Y_i - \mu)^2}{n}$$

is a statistic iff μ is known.

The following theorem is extremely important and the proof follows Rice (1988, pp. 171–173) closely.

Theorem 4.1. Let the $Y_1, \ldots, Y_n$ be iid $N(\mu, \sigma^2)$.

a) The sample mean $\overline{Y} \sim N(\mu, \sigma^2/n)$.
b) $\overline{Y}$ and S^2 are independent.
c) $(n-1)S^2/\sigma^2 \sim \chi^2_{n-1}$. Hence $\sum_{i=1}^n (Y_i - \overline{Y})^2 \sim \sigma^2 \chi^2_{n-1}$.
d) $\frac{\sqrt{n}(\overline{Y}-\mu)}{S} = \frac{(\overline{Y}-\mu)}{S/\sqrt{n}} \sim t_{n-1}$.

Proof. a) This result follows from Theorem 2.17e.
b) The moment generating function of $(\overline{Y}, Y_1 - \overline{Y}, \ldots, Y_n - \overline{Y})$ is

$$m(s, t_1, \ldots, t_n) = E(\exp[s\overline{Y} + t_1(Y_1 - \overline{Y}) + \cdots + t_n(Y_n - \overline{Y})]).$$

By Theorem 2.22, $\overline{Y}$ and $(Y_1 - \overline{Y}, \ldots, Y_n - \overline{Y})$ are independent if

$$m(s, t_1, \ldots, t_n) = m_{\overline{Y}}(s)\; m(t_1, \ldots, t_n)$$

where $m_{\overline{Y}}(s)$ is the mgf of $\overline{Y}$ and $m(t_1, \ldots, t_n)$ is the mgf of $(Y_1 - \overline{Y}, \ldots, Y_n - \overline{Y})$.
Now

$$\sum_{i=1}^n t_i(Y_i - \overline{Y}) = \sum_{i=1}^n t_i Y_i - \overline{Y} n \overline{t} = \sum_{i=1}^n t_i Y_i - \sum_{i=1}^n \overline{t} Y_i$$

and thus

$$s\overline{Y} + \sum_{i=1}^{n} t_i(Y_i - \overline{Y}) = \sum_{i=1}^{n} \left[\frac{s}{n} + (t_i - \bar{t}\,)\right] Y_i = \sum_{i=1}^{n} a_i Y_i.$$

Now $\sum_{i=1}^{n} a_i = \sum_{i=1}^{n} [\frac{s}{n} + (t_i - \bar{t}\,)] = s$ and

$$\sum_{i=1}^{n} a_i^2 = \sum_{i=1}^{n} \left[\frac{s^2}{n^2} + 2\frac{s}{n}(t_i - \bar{t}\,) + (t_i - \bar{t}\,)^2\right] = \frac{s^2}{n} + \sum_{i=1}^{n} (t_i - \bar{t}\,)^2.$$

Hence

$$m(s, t_1, \ldots, t_n) = E\left(\exp\left[s\overline{Y} + \sum_{i=1}^{n} t_i(Y_i - \overline{Y})\right]\right) = E\left[\exp\left(\sum_{i=1}^{n} a_i Y_i\right)\right]$$

$$= m_{Y_1,\ldots,Y_n}(a_1, \ldots, a_n) = \prod_{i=1}^{n} m_{Y_i}(a_i)$$

since the Y_i are independent. Now

$$\prod_{i=1}^{n} m_{Y_i}(a_i) = \prod_{i=1}^{n} \exp\left(\mu a_i + \frac{\sigma^2}{2} a_i^2\right) = \exp\left(\mu \sum_{i=1}^{n} a_i + \frac{\sigma^2}{2} \sum_{i=1}^{n} a_i^2\right)$$

$$= \exp\left[\mu s + \frac{\sigma^2}{2}\frac{s^2}{n} + \frac{\sigma^2}{2} \sum_{i=1}^{n} (t_i - \bar{t}\,)^2\right]$$

$$= \exp\left[\mu s + \frac{\sigma^2}{2n} s^2\right] \exp\left[\frac{\sigma^2}{2} \sum_{i=1}^{n} (t_i - \bar{t}\,)^2\right].$$

Now the first factor is the mgf of $\overline{Y}$ and the second factor is $m(t_1, \ldots, t_n) = m(0, t_1, \ldots, t_n)$ since the mgf of the marginal is found from the mgf of the joint distribution by setting all terms not in the marginal to 0 (i.e., set $s = 0$ in $m(s, t_1, \ldots, t_n)$ to find $m(t_1, \ldots, t_n)$). Hence the mgf factors and

$$\overline{Y} \perp\!\!\!\perp (Y_1 - \overline{Y}, \ldots, Y_n - \overline{Y}).$$

Since S^2 is a function of $(Y_1 - \overline{Y}, \ldots, Y_n - \overline{Y})$, it is also true that $\overline{Y} \perp\!\!\!\perp S^2$.

c) $(Y_i - \mu)/\sigma \sim N(0, 1)$ so $(Y_i - \mu)^2/\sigma^2 \sim \chi_1^2$ and

$$\frac{1}{\sigma^2} \sum_{i=1}^{n} (Y_i - \mu)^2 \sim \chi_n^2.$$

Now

$$\sum_{i=1}^{n} (Y_i - \mu)^2 = \sum_{i=1}^{n} (Y_i - \overline{Y} + \overline{Y} - \mu)^2 = \sum_{i=1}^{n} (Y_i - \overline{Y})^2 + n(\overline{Y} - \mu)^2.$$

Hence

$$W = \frac{1}{\sigma^2}\sum_{i=1}^n (Y_i - \mu)^2 = \frac{1}{\sigma^2}\sum_{i=1}^n (Y_i - \overline{Y})^2 + \left(\frac{\overline{Y}-\mu}{\sigma/\sqrt{n}}\right)^2 = U + V.$$

Since $U \perp\!\!\!\perp V$ by b), $m_W(t) = m_U(t)\, m_V(t)$. Since $W \sim \chi^2_n$ and $V \sim \chi^2_1$,

$$m_U(t) = \frac{m_W(t)}{m_V(t)} = \frac{(1-2t)^{-n/2}}{(1-2t)^{-1/2}} = (1-2t)^{-(n-1)/2}$$

which is the mgf of a χ^2_{n-1} distribution.

d)

$$Z = \frac{\overline{Y}-\mu}{\sigma/\sqrt{n}} \sim N(0,1),$$

and

$$S^2/\sigma^2 = \frac{\frac{(n-1)S^2}{\sigma^2}}{n-1} \sim \frac{1}{n-1}\chi^2_{n-1}.$$

Suppose $Z \sim N(0,1)$, $X \sim \chi^2_{n-1}$ and $Z \perp\!\!\!\perp X$. Then $Z/\sqrt{X/(n-1)} \sim t_{n-1}$. Hence

$$\frac{\sqrt{n}(\overline{Y}-\mu)}{S} = \frac{\sqrt{n}(\overline{Y}-\mu)/\sigma}{\sqrt{S^2/\sigma^2}} \sim t_{n-1}.$$

□

Theorem 4.2. Let $Y_1, \ldots, Y_n$ be iid with cdf F_Y and pdf f_Y.

a) The pdf of $T = Y_{(n)}$ is

$$f_{Y_{(n)}}(t) = n[F_Y(t)]^{n-1} f_Y(t).$$

b) The pdf of $T = Y_{(1)}$ is

$$f_{Y_{(1)}}(t) = n[1 - F_Y(t)]^{n-1} f_Y(t).$$

c) Let $2 \le r \le n$. Then the joint pdf of $Y_{(1)}, Y_{(2)}, \ldots, Y_{(r)}$ is

$$f_{Y_{(1)},\ldots,Y_{(r)}}(t_1,\ldots,t_r) = \frac{n!}{(n-r)!}[1 - F_Y(t_r)]^{n-r}\prod_{i=1}^r f_Y(t_i).$$

Proof of a) and b). a) The cdf of $Y_{(n)}$ is

$$F_{Y_{(n)}}(t) = P(Y_{(n)} \le t) = P(Y_1 \le t, \ldots, Y_n \le t) = \prod_{i=1}^n P(Y_i \le t) = [F_Y(t)]^n.$$

Hence the pdf of $Y_{(n)}$ is

$$\frac{d}{dt}F_{Y_{(n)}}(t) = \frac{d}{dt}[F_Y(t)]^n = n[F_Y(t)]^{n-1}f_Y(t).$$

b) The cdf of $Y_{(1)}$ is

$$F_{Y_{(1)}}(t) = P(Y_{(1)} \le t) = 1 - P(Y_{(1)} > t) = 1 - P(Y_1 > t, \ldots, Y_n > t)$$

$$= 1 - \prod_{i=1}^{n} P(Y_i > t) = 1 - [1 - F_Y(t)]^n.$$

Hence the pdf of $Y_{(n)}$ is

$$\frac{d}{dt}F_{Y_{(n)}}(t) = \frac{d}{dt}(1 - [1 - F_Y(t)]^n) = n[1 - F_Y(t)]^{n-1}f_Y(t). \quad \square$$

To see that c) may be true, consider the following argument adapted from Mann et al. (1974, p. 93). Let Δt_i be a small positive number and notice that $P(E) \equiv$

$$P(t_1 < Y_{(1)} < t_1 + \Delta t_1, t_2 < Y_{(2)} < t_2 + \Delta t_2, \ldots, t_r < Y_{(r)} < t_r + \Delta t_r)$$

$$= \int_{t_r}^{t_r+\Delta t_r} \cdots \int_{t_1}^{t_1+\Delta t_1} f_{Y_{(1)},\ldots,Y_{(r)}}(w_1, \ldots, w_r) dw_1 \cdots dw_r$$

$$\approx f_{Y_{(1)},\ldots,Y_{(r)}}(t_1, \ldots, t_r) \prod_{i=1}^{r} \Delta t_i.$$

Since the event E denotes the occurrence of no observations before t_1, exactly one occurrence between t_1 and $t_1 + \Delta t_1$, no observations between $t_1 + \Delta t_1$ and t_2 and so on, and finally the occurrence of $n - r$ observations after $t_r + \Delta t_r$, using the multinomial pmf shows that

$$P(E) = \frac{n!}{0!1!\cdots 0!1!(n-r)!} \rho_1^0 \rho_2^1 \rho_3^0 \rho_4^1 \cdots \rho_{2r-1}^0 \rho_{2r}^1 \rho_{2r+1}^{n-r}$$

where

$$\rho_{2i} = P(t_i < Y < t_i + \Delta t_i) \approx f(t_i)\Delta t_i$$

for $i = 1, \ldots, r$ and

$$\rho_{2r+1} = P(n - r\, Y's > t_r + \Delta t_r) \approx (1 - F(t_r))^{n-r}.$$

Hence

$$P(E) \approx \frac{n!}{(n-r)!}(1 - F(t_r))^{n-r} \prod_{i=1}^{r} f(t_i) \prod_{i=1}^{r} \Delta t_i$$

$$\approx f_{Y_{(1)},\ldots,Y_{(r)}}(t_1, \ldots, t_r) \prod_{i=1}^{r} \Delta t_i,$$

and result c) seems reasonable.

Example 4.4. Let $Y_1,\ldots,Y_n$ be iid from the following distributions.

a) Bernoulli(ρ): Then $Y_{(1)} \sim$ Bernoulli(ρ^n).
b) Geometric(ρ): Then $Y_{(1)} \sim$ Geometric($1-(1-\rho)^n$).
c) Burr Type XII (ϕ,λ): Then $Y_{(1)} \sim$ Burr Type XII ($\phi,\lambda/n$).
d) EXP(λ): Then $Y_{(1)} \sim$ EXP(λ/n).
e) EXP(θ,λ): Then $Y_{(1)} \sim$ EXP($\theta,\lambda/n$).
f) Pareto PAR(σ,λ): Then $Y_{(1)} \sim$ PAR($\sigma,\lambda/n$).
g) Gompertz Gomp(θ,ν): Then $Y_{(1)} \sim$ Gomp($\theta,n\nu$).
h) Rayleigh $R(\mu,\sigma)$: Then $Y_{(1)} \sim R(\mu,\sigma/\sqrt{n})$.
i) Truncated Extreme Value TEV(λ) : Then $Y_{(1)} \sim$ TEV(λ/n).
j) Weibull $W(\phi,\lambda)$: Then $Y_{(1)} \sim W(\phi,\lambda/n)$.

Proof: a) $Y_i \in \{0,1\}$ so $Y_{(1)} \in \{0,1\}$. Hence $Y_{(1)}$ is Bernoulli, and $P(Y_{(1)} = 1) = P(Y_1 = 1,\ldots,Y_n = 1) = [P(Y_1 = 1)]^n = \rho^n$.
b) $P(Y_{(1)} \le y) = 1 - P(Y_{(1)} > y) = 1 - [1-F(y)]^n =$
$1-[1-(1-(1-\rho)^{\lfloor y \rfloor+1}]^n = 1-[(1-\rho)^n]^{\lfloor y \rfloor+1} = 1-[1-(1-(1-\rho)^n)]^{\lfloor y \rfloor+1}$
for $y \ge 0$ which is the cdf of a Geometric($1-(1-\rho)^n$) random variable.
Parts c)-j) follow from Theorem 4.2b. For example, suppose $Y_1,\ldots,Y_n$ are iid EXP(λ) with cdf $F(y) = 1-\exp(-y/\lambda)$ for $y > 0$. Then $F_{Y_{(1)}}(t) = 1-[1-(1-\exp(-t/\lambda))]^n = 1-[\exp(-t/\lambda)]^n = 1-\exp[-t/(\lambda/n)]$ for $t > 0$. Hence $Y_{(1)} \sim$ EXP(λ/n).

4.2 Minimal Sufficient Statistics

For parametric inference, the pmf or pdf of a random variable Y is $f_{\boldsymbol{\theta}}(y)$ where $\boldsymbol{\theta} \in \Theta$ is unknown. Hence Y comes from a family of distributions indexed by $\boldsymbol{\theta}$, and quantities such as $E_{\boldsymbol{\theta}}(g(Y))$ depend on $\boldsymbol{\theta}$. Since the parametric distribution is completely specified by $\boldsymbol{\theta}$, an important goal of parametric inference is finding good estimators of $\boldsymbol{\theta}$. For example, if $Y_1,\ldots,Y_n$ are iid $N(\mu,\sigma^2)$, then $\boldsymbol{\theta} = (\mu,\sigma)$ is fixed but unknown, $\boldsymbol{\theta} \in \Theta = (-\infty,\infty) \times (0,\infty)$ and $E_{\boldsymbol{\theta}}(\overline{Y}) \equiv E_{(\mu,\sigma)}(\overline{Y}) = \mu$. Since $V_{\boldsymbol{\theta}}(\overline{Y}) \equiv V_{(\mu,\sigma)}(\overline{Y}) = \sigma^2/n$, $\overline{Y}$ is a good estimator for μ if n is large. The notation $f_{\boldsymbol{\theta}}(\boldsymbol{y}) \equiv f(\boldsymbol{y}|\boldsymbol{\theta})$ is also used.

The basic idea of a sufficient statistic $\boldsymbol{T}(\boldsymbol{Y})$ for $\boldsymbol{\theta}$ is that all of the information needed for inference from the data $Y_1,\ldots,Y_n$ about the parameter $\boldsymbol{\theta}$ is contained in the statistic $\boldsymbol{T}(\boldsymbol{Y})$. For example, suppose that $Y_1,\ldots,Y_n$ are iid binomial($1,\rho$) random variables. Hence each observed Y_i is a 0 or a 1 and the observed data is an n-tuple of 0's and 1's, e.g., 0,0,1,...,0,0,1. It will turn out that $\sum_{i=1}^n Y_i$, the number of 1's in the n-tuple, is a sufficient statistic for ρ. From Theorem 2.17a, $\sum_{i=1}^n Y_i \sim$ BIN(n,ρ). The importance of a sufficient statistic is *dimension reduction*: the statistic $\sum_{i=1}^n Y_i$ has all of the information from the data needed to perform inference about ρ, and the statistic is one-dimensional and thus much easier to understand than the n dimensional n-tuple of 0's and 1's. Also notice that all n-tuples with the same number of 1's have the same amount of information needed for inference about ρ: the n-tuples 1,1,1,0,0,0,0 and 0,1,0,0,1,0,1 both give $\sum_{i=1}^{n} Y_i = 3$. The ordering of the observed sequences of 0's and 1's does not contain any information about the parameter ρ because the observations are iid.

Definition 4.2. Suppose that $(\boldsymbol{Y}_1, \ldots, \boldsymbol{Y}_n)$ have a joint distribution that depends on a vector of parameters $\boldsymbol{\theta}$ for $\boldsymbol{\theta} \in \Theta$ where Θ is the parameter space. A statistic $\boldsymbol{T}(\boldsymbol{Y}_1, \ldots, \boldsymbol{Y}_n)$ is a **sufficient statistic** for $\boldsymbol{\theta}$ if the conditional distribution of $(\boldsymbol{Y}_1, \ldots, \boldsymbol{Y}_n)$ given $\boldsymbol{T} = \boldsymbol{t}$ does not depend on $\boldsymbol{\theta}$ for any value of $\boldsymbol{t}$ in the support of $\boldsymbol{T}$.

Example 4.5. Let the random vector $\boldsymbol{Y}$ denote the data.

a) Suppose $T(\boldsymbol{y}) \equiv 7 \; \forall \boldsymbol{y}$. Then T is a constant and any constant is independent of a random vector $\boldsymbol{Y}$. Hence the conditional distribution $f_{\boldsymbol{\theta}}(\boldsymbol{y}|T) = f_{\boldsymbol{\theta}}(\boldsymbol{y})$ is not independent of $\boldsymbol{\theta}$. Thus T is not a sufficient statistic.
b) Let $T(\boldsymbol{Y}) \equiv \boldsymbol{Y}$, and let $\boldsymbol{W}$ have the same distribution as $\boldsymbol{Y}|\boldsymbol{Y} = \boldsymbol{y}$. Since $P(\boldsymbol{Y} = \boldsymbol{y}|\boldsymbol{Y} = \boldsymbol{y}) = 1$, the pmf $f_{\boldsymbol{W}}(\boldsymbol{w})$ of $\boldsymbol{W}$ is equal to 1 if $\boldsymbol{w} = \boldsymbol{y}$ and is equal to 0, otherwise. Hence the distribution of $\boldsymbol{Y}||\boldsymbol{Y} = \boldsymbol{y}$ is independent of $\boldsymbol{\theta}$, and the data $\boldsymbol{Y}$ is a sufficient statistic. Of course there is no dimension reduction when the data is used as the sufficient statistic.

Often $\boldsymbol{T}$ and $\boldsymbol{Y}_i$ are real valued. Then $T(Y_1, \ldots, Y_n)$ is a sufficient statistic if the conditional distribution of $\boldsymbol{Y} = (Y_1, \ldots, Y_n)$ given $T = t$ does not depend on $\boldsymbol{\theta}$. The following theorem provides such an effective method for showing that a statistic is a sufficient statistic that the definition should rarely be used to prove that the statistic is a sufficient statistic.

Regularity Condition F.1: If $f(\boldsymbol{y}|\boldsymbol{\theta})$ is a family of pmfs for $\boldsymbol{\theta} \in \Theta$, assume that there exists a set $\{\boldsymbol{y}_i\}_{i=1}^{\infty}$ that does not depend on $\boldsymbol{\theta} \in \Theta$ such that $\sum_{i=1}^{\infty} f(\boldsymbol{y}_i|\boldsymbol{\theta}) = 1$ for all $\boldsymbol{\theta} \in \Theta$. (This condition is usually satisfied. For example, F.1 holds if the support $\mathscr{Y}$ is free of $\boldsymbol{\theta}$ or if $\boldsymbol{y} = (y_1, \ldots, y_n)$ and y_i takes on values on a lattice such as $y_i \in \{1, \ldots, \theta\}$ for $\theta \in \{1, 2, 3, \ldots\}$.)

Theorem 4.3: Factorization Theorem. Let $f(\boldsymbol{y}|\boldsymbol{\theta})$ for $\boldsymbol{\theta} \in \Theta$ denote a family of pdfs or pmfs for $\boldsymbol{Y}$. For a family of pmfs, assume condition F.1 holds. A statistic $\boldsymbol{T}(\boldsymbol{Y})$ is a sufficient statistic for $\boldsymbol{\theta}$ iff for all sample points $\boldsymbol{y}$ and for all $\boldsymbol{\theta}$ in the parameter space Θ,

$$f(\boldsymbol{y}|\boldsymbol{\theta}) = g(\boldsymbol{T}(\boldsymbol{y})|\boldsymbol{\theta})\, h(\boldsymbol{y})$$

where both g and h are nonnegative functions. The function h does not depend on $\boldsymbol{\theta}$ and the function g depends on $\boldsymbol{y}$ only through $\boldsymbol{T}(\boldsymbol{y})$.

Proof for pmfs. If $\boldsymbol{T}(\boldsymbol{Y})$ is a sufficient statistic, then the conditional distribution of $\boldsymbol{Y}$ given $\boldsymbol{T}(\boldsymbol{Y}) = \boldsymbol{t}$ does not depend on $\boldsymbol{\theta}$ for any $\boldsymbol{t}$ in the support of $\boldsymbol{T}$. Taking $\boldsymbol{t} = \boldsymbol{T}(\boldsymbol{y})$ gives

$$P_{\boldsymbol{\theta}}(\boldsymbol{Y} = \boldsymbol{y}|\boldsymbol{T}(\boldsymbol{Y}) = \boldsymbol{T}(\boldsymbol{y})) \equiv P(\boldsymbol{Y} = \boldsymbol{y}|\boldsymbol{T}(\boldsymbol{Y}) = \boldsymbol{T}(\boldsymbol{y}))$$

for all θ in the parameter space. Now

$$\{\boldsymbol{Y} = \boldsymbol{y}\} \subseteq \{\boldsymbol{T}(\boldsymbol{Y}) = \boldsymbol{T}(\boldsymbol{y})\} \tag{4.7}$$

and $P(A) = P(A \cap B)$ if $A \subseteq B$. Hence

$$f(\boldsymbol{y}|\boldsymbol{\theta}) = P_{\boldsymbol{\theta}}(\boldsymbol{Y} = \boldsymbol{y}) = P_{\boldsymbol{\theta}}(\boldsymbol{Y} = \boldsymbol{y} \text{ and } \boldsymbol{T}(\boldsymbol{Y}) = \boldsymbol{T}(\boldsymbol{y}))$$

$$= P_{\boldsymbol{\theta}}(\boldsymbol{T}(\boldsymbol{Y}) = \boldsymbol{T}(\boldsymbol{y}))P(\boldsymbol{Y} = \boldsymbol{y}|\boldsymbol{T}(\boldsymbol{Y}) = \boldsymbol{T}(\boldsymbol{y})) = g(\boldsymbol{T}(\boldsymbol{y})|\boldsymbol{\theta})h(\boldsymbol{y}).$$

Now suppose

$$f(\boldsymbol{y}|\boldsymbol{\theta}) = g(\boldsymbol{T}(\boldsymbol{y})|\boldsymbol{\theta})\, h(\boldsymbol{y})$$

for all $\boldsymbol{y}$ and for all $\boldsymbol{\theta} \in \Theta$. Now

$$P_{\boldsymbol{\theta}}(\boldsymbol{T}(\boldsymbol{Y}) = \boldsymbol{t}) = \sum_{\{\boldsymbol{y}:\boldsymbol{T}(\boldsymbol{y}) = \boldsymbol{t}\}} f(\boldsymbol{y}|\boldsymbol{\theta}) = g(\boldsymbol{t}|\boldsymbol{\theta}) \sum_{\{\boldsymbol{y}:\boldsymbol{T}(\boldsymbol{y}) = \boldsymbol{t}\}} h(\boldsymbol{y}).$$

If $\boldsymbol{Y} = \boldsymbol{y}$ and $\boldsymbol{T}(\boldsymbol{Y}) = \boldsymbol{t}$, then $\boldsymbol{T}(\boldsymbol{y}) = \boldsymbol{t}$ and $\{\boldsymbol{Y} = \boldsymbol{y}\} \subseteq \{\boldsymbol{T}(\boldsymbol{Y}) = \boldsymbol{t}\}$. Thus

$$P_{\boldsymbol{\theta}}(\boldsymbol{Y} = \boldsymbol{y}|\boldsymbol{T}(\boldsymbol{Y}) = \boldsymbol{t}) = \frac{P_{\boldsymbol{\theta}}(\boldsymbol{Y} = \boldsymbol{y}, \boldsymbol{T}(\boldsymbol{Y}) = \boldsymbol{t})}{P_{\boldsymbol{\theta}}(\boldsymbol{T}(\boldsymbol{Y}) = \boldsymbol{t})} = \frac{P_{\boldsymbol{\theta}}(\boldsymbol{Y} = \boldsymbol{y})}{P_{\boldsymbol{\theta}}(\boldsymbol{T}(\boldsymbol{Y}) = \boldsymbol{t})}$$

$$= \frac{g(\boldsymbol{t}|\boldsymbol{\theta})\, h(\boldsymbol{y})}{g(\boldsymbol{t}|\boldsymbol{\theta}) \sum_{\{\boldsymbol{y}:\boldsymbol{T}(\boldsymbol{y}) = \boldsymbol{t}\}} h(\boldsymbol{y})} = \frac{h(\boldsymbol{y})}{\sum_{\{\boldsymbol{y}:\boldsymbol{T}(\boldsymbol{y}) = \boldsymbol{t}\}} h(\boldsymbol{y})}$$

which does not depend on $\boldsymbol{\theta}$ since the terms in the sum do not depend on $\boldsymbol{\theta}$ by condition F.1. Hence $\boldsymbol{T}$ is a sufficient statistic. □

Remark 4.1. If no such factorization exists for $\boldsymbol{T}$, then $\boldsymbol{T}$ is not a sufficient statistic. The "iff" in the Factorization Theorem is important.

Example 4.6. a) To use factorization to show that the data $\boldsymbol{Y} = (Y_1, \ldots, Y_n)$ is a sufficient statistic, take $\boldsymbol{T}(\boldsymbol{Y}) = \boldsymbol{Y}$, $g(\boldsymbol{T}(\boldsymbol{y})|\boldsymbol{\theta}) = f(\boldsymbol{y}|\boldsymbol{\theta})$, and $h(\boldsymbol{y}) = 1 \forall\, \boldsymbol{y}$.
b) For iid data with $n > 1$, the statistic Y_1 is not a sufficient statistic by the Factorization Theorem since $f(\boldsymbol{y}|\boldsymbol{\theta}) = f(y_1|\boldsymbol{\theta}) \prod_{i=2}^n f(y_i|\boldsymbol{\theta})$ cannot be written as $f(\boldsymbol{y}|\boldsymbol{\theta}) = g(y_1|\boldsymbol{\theta})h(\boldsymbol{y})$ where g depends on $\boldsymbol{y}$ only through y_1 and $h(\boldsymbol{y})$ does not depend on $\boldsymbol{\theta}$.

Example 4.7. Let $X_1, \ldots, X_n$ be iid $N(\mu, \sigma^2)$. Then

$$f(x_1, \ldots, x_n) = \prod_{i=1}^n f(x_i) = \left[\frac{1}{\sqrt{2\pi}\sigma} \exp\left(\frac{-\mu^2}{2\sigma^2}\right)\right]^n \exp\left(\frac{-1}{2\sigma^2} \sum_{i=1}^n x_i^2 + \frac{\mu}{\sigma^2} \sum_{i=1}^n x_i\right)$$

$$= g(T(\boldsymbol{x})|\boldsymbol{\theta})h(\boldsymbol{x})$$

where $\boldsymbol{\theta} = (\mu, \sigma)$ and $h(\boldsymbol{x}) = 1$. Hence $T(\boldsymbol{X}) = (\sum_{i=1}^n X_i^2, \sum_{i=1}^n X_i)$ is a sufficient statistic for (μ, σ) or equivalently for (μ, σ^2) by the Factorization Theorem.

Example 4.8. Let $Y_1, \ldots, Y_n$ be iid binomial(k, ρ) with k known and pmf

$$f(y|\rho) = \binom{k}{y} \rho^y (1-\rho)^{k-y} I_{\{0,\ldots,k\}}(y).$$

Then

$$f(\boldsymbol{y}|\rho) = \prod_{i=1}^n f(y_i|\rho) = \prod_{i=1}^n \left[\binom{k}{y_i} I_{\{0,\ldots,k\}}(y_i) \right] (1-\rho)^{nk} \left(\frac{\rho}{1-\rho} \right)^{\sum_{i=1}^n y_i}.$$

Hence by the Factorization Theorem, $\sum_{i=1}^n Y_i$ is a sufficient statistic.

Example 4.9. Suppose $X_1, \ldots, X_n$ are iid uniform observations on the interval $(\theta, \theta+1)$, $-\infty < \theta < \infty$. Notice that

$$\prod_{i=1}^n I_A(x_i) = I(\text{all } x_i \in A) \text{ and } \prod_{i=1}^n I_{A_n}(x) = I_{\cap_{i=1}^n A_i}(x)$$

where the latter holds since both terms are 1 if x is in all sets A_i for $i = 1, \ldots, n$ and both terms are 0 otherwise. Hence $f(\boldsymbol{x}|\theta) =$

$$\prod_{i=1}^n f(x_i|\theta) = \prod_{i=1}^n 1 I(x_i \ge \theta) I(x_i \le \theta+1) = 1 I(\min(x_i) \ge \theta) I(\max(x_i) \le \theta+1).$$

Then $h(\boldsymbol{x}) \equiv 1$ and $g(\boldsymbol{T}(\boldsymbol{x})|\theta) = I(\min(x_i) \ge \theta) I(\max(x_i) \le \theta+1)$, and $\boldsymbol{T}(\boldsymbol{X}) = (X_{(1)}, X_{(n)})$ is a sufficient statistic by the Factorization Theorem.

Remark 4.2. i) Suppose $Y_1, \ldots, Y_n$ are iid from a distribution with support $\mathscr{Y}_i \equiv \mathscr{Y}^*$ and pdf or pmf $f(y|\boldsymbol{\theta}) = k(y|\boldsymbol{\theta}) I(y \in \mathscr{Y}^*)$. Then $f(\boldsymbol{y}|\boldsymbol{\theta}) = \prod_{i=1}^n k(y_i|\boldsymbol{\theta}) \prod_{i=1}^n I(y_i \in \mathscr{Y}^*)$. Now the support of $\boldsymbol{Y}$ is the n-fold cross product $\mathscr{Y} = \mathscr{Y}^* \times \cdots \times \mathscr{Y}^*$, and $I(\boldsymbol{y} \in \mathscr{Y}) = \prod_{i=1}^n I(y_i \in \mathscr{Y}^*) =$ $I(\text{ all } y_i \in \mathscr{Y}^*)$. Thus $f(\boldsymbol{y}|\boldsymbol{\theta}) = \prod_{i=1}^n k(y_i|\boldsymbol{\theta}) I(\text{ all } y_i \in \mathscr{Y}^*)$.

ii) If $\mathscr{Y}^*$ does not depend on $\boldsymbol{\theta}$, then $I(\text{ all } y_i \in \mathscr{Y}^*)$ is part of $h(\boldsymbol{y})$. If $\mathscr{Y}^*$ does depend on unknown $\boldsymbol{\theta}$, then $I(\text{ all } y_i \in \mathscr{Y}^*)$ could be placed in $g(\boldsymbol{T}(\boldsymbol{y})|\boldsymbol{\theta})$. Typically $\mathscr{Y}^*$ is an interval with endpoints a and b, not necessarily finite. For pdfs, $\prod_{i=1}^n I(y_i \in [a,b]) = I(a \le y_{(1)} < y_{(n)} \le b) = I[a \le y_{(1)}] I[y_{(n)} \le b]$. If both a and b are unknown parameters, put the middle term in $g(\boldsymbol{T}(\boldsymbol{y})|\boldsymbol{\theta})$. If both a and b are known, put the middle term in $h(\boldsymbol{y})$. If a is an unknown parameter and b is known, put $I[a \le y_{(1)}]$ in $g(\boldsymbol{T}(\boldsymbol{y})|\boldsymbol{\theta})$ and $I[y_{(n)} \le b]$ in $h(\boldsymbol{y})$.

iii) $\prod_{i=1}^n I(y_i \in (-\infty, b)) = I(y_{(n)} < b)$.
$\prod_{i=1}^n I(y_i \in [a, \infty)) = I(a \le y_{(1)})$, et cetera.

iv) Another useful fact is that $\prod_{j=1}^k I(\boldsymbol{y} \in A_j) = I(\boldsymbol{y} \in \cap_{j=1}^k A_j)$.

Example 4.10. Try to place any part of $f(\boldsymbol{y}|\boldsymbol{\theta})$ that depends on $\boldsymbol{y}$ but not on $\boldsymbol{\theta}$ into $h(\boldsymbol{y})$. For example, if $Y_1, \ldots, Y_n$ are iid $U(\theta_1, \theta_2)$, then $f(\boldsymbol{y}|\boldsymbol{\theta}) =$

$$\prod_{i=1}^n f(y_i|\boldsymbol{\theta}) = \prod_{i=1}^n \frac{1}{\theta_2 - \theta_1} I(\theta_1 \le y_i \le \theta_2) = \frac{1}{(\theta_2 - \theta_1)^n} I(\theta_1 \le y_{(1)} < y_{(n)} \le \theta_2).$$

Then $h(\boldsymbol{y}) \equiv 1$ and $\boldsymbol{T}(\boldsymbol{Y}) = (Y_{(1)}, Y_{(n)})$ is a sufficient statistic for (θ_1, θ_2) by factorization.

If θ_1 or θ_2 is known, then the above factorization works, but it is better to make the dimension of the sufficient statistic as small as possible. If θ_1 is known, then

$$f(\boldsymbol{y}|\boldsymbol{\theta}) = \frac{1}{(\theta_2 - \theta_1)^n} I(y_{(n)} \leq \theta_2) I(\theta_1 \leq y_{(1)})$$

where the first two terms are $g(T(\boldsymbol{y})|\theta_2)$ and the third term is $h(\boldsymbol{y})$. Hence $T(\boldsymbol{Y}) = Y_{(n)}$ is a sufficient statistic for θ_2 by factorization. If θ_2 is known, then

$$f(\boldsymbol{y}|\boldsymbol{\theta}) = \frac{1}{(\theta_2 - \theta_1)^n} I(\theta_1 \leq y_{(1)}) I(y_{(n)} \leq \theta_2)$$

where the first two terms are $g(T(\boldsymbol{y})|\theta_1)$ and the third term is $h(\boldsymbol{y})$. Hence $T(\boldsymbol{Y}) = Y_{(1)}$ is a sufficient statistic for θ_1 by factorization.

There are infinitely many sufficient statistics (see Theorem 4.8 below), but typically we want the dimension of the sufficient statistic to be as small as possible since lower dimensional statistics are easier to understand and to use for inference than higher dimensional statistics. Dimension reduction is extremely important and the following definition is useful.

Definition 4.3. Suppose that $Y_1, \ldots, Y_n$ have a joint distribution that depends on a vector of parameters $\boldsymbol{\theta}$ for $\boldsymbol{\theta} \in \Theta$ where Θ is the parameter space. A sufficient statistic $\boldsymbol{T}(\boldsymbol{Y})$ for $\boldsymbol{\theta}$ is a **minimal sufficient statistic** for $\boldsymbol{\theta}$ if $\boldsymbol{T}(\boldsymbol{Y})$ is a function of $\boldsymbol{S}(\boldsymbol{Y})$ for any other sufficient statistic $\boldsymbol{S}(\boldsymbol{Y})$ for $\boldsymbol{\theta}$.

Remark 4.3. A useful mnemonic is that $\boldsymbol{S} = \boldsymbol{Y}$ is a sufficient statistic, and $\boldsymbol{T} \equiv \boldsymbol{T}(\boldsymbol{Y})$ is a function of $\boldsymbol{S}$.

A minimal sufficient statistic $\boldsymbol{T}(\boldsymbol{Y}) = \boldsymbol{g_S}(\boldsymbol{S}(\boldsymbol{Y}))$ for some function $\boldsymbol{g_S}$ where $\boldsymbol{S}(\boldsymbol{Y})$ is a sufficient statistic. If $\boldsymbol{S}(\boldsymbol{Y})$ is not a minimal sufficient statistic, then $\boldsymbol{S}(\boldsymbol{Y})$ is not a function of the minimal sufficient statistic $\boldsymbol{T}(\boldsymbol{Y})$. To see this, note that there exists a sufficient statistic $\boldsymbol{W}(\boldsymbol{Y})$ such that $\boldsymbol{S}(\boldsymbol{Y})$ is not a function of $\boldsymbol{W}(\boldsymbol{Y})$. Suppose $\boldsymbol{S}(\boldsymbol{Y}) = \boldsymbol{h}[\boldsymbol{T}(\boldsymbol{Y})]$ for some function $\boldsymbol{h}$. Then $\boldsymbol{S}(\boldsymbol{Y}) = \boldsymbol{h}[\boldsymbol{g_W}(\boldsymbol{W}(\boldsymbol{Y}))]$, a function of $\boldsymbol{W}(\boldsymbol{Y})$, which is a contradiction. If $\boldsymbol{T}_1$ and $\boldsymbol{T}_2$ are both minimal sufficient statistics, then $\boldsymbol{T}_1 = \boldsymbol{g}(\boldsymbol{T}_2)$ and $\boldsymbol{T}_2 = \boldsymbol{h}(\boldsymbol{T}_1)$. Hence $\boldsymbol{g}(\boldsymbol{h}(\boldsymbol{T}_1)) = \boldsymbol{T}_1$ and $\boldsymbol{h}(\boldsymbol{g}(\boldsymbol{T}_2)) = \boldsymbol{T}_2$. Hence $\boldsymbol{h}$ and $\boldsymbol{g}$ are inverse functions which are one to one and onto, and $\boldsymbol{T}_1$ and $\boldsymbol{T}_2$ are equivalent statistics. Following Lehmann (1983, p. 41), if the minimal sufficient statistic $\boldsymbol{T}(\boldsymbol{Y}) = \boldsymbol{g}(\boldsymbol{S}(\boldsymbol{Y}))$ where $\boldsymbol{g}$ is not one to one, then $\boldsymbol{T}(\boldsymbol{Y})$ provides greater reduction of the data than the sufficient statistic $\boldsymbol{S}(\boldsymbol{Y})$. Hence minimal sufficient statistics provide the greatest possible reduction of the data.

Complete sufficient statistics, defined below, are primarily used for the theory of uniformly minimum variance estimators covered in Chap. 6.

Definition 4.4. Suppose that a *statistic* $\boldsymbol{T}(\boldsymbol{Y})$ has a pmf or pdf $f(\boldsymbol{t}|\boldsymbol{\theta})$. Then $\boldsymbol{T}(\boldsymbol{Y})$ is a *complete sufficient statistic* for $\boldsymbol{\theta}$ if $E_{\boldsymbol{\theta}}[g(\boldsymbol{T}(\boldsymbol{Y}))] = 0$ for all $\boldsymbol{\theta}$ implies that $P_{\boldsymbol{\theta}}[g(\boldsymbol{T}(\boldsymbol{Y})) = 0] = 1$ for all $\boldsymbol{\theta}$. The function g cannot depend on any unknown parameters.

The statistic $\boldsymbol{T}(\boldsymbol{Y})$ has a sampling distribution that depends on n and on $\boldsymbol{\theta} \in \Theta$. Hence the property of being a complete sufficient statistic depends on the family of distributions with pdf or pmf $f(\boldsymbol{t}|\boldsymbol{\theta})$. Regular exponential families will have a complete sufficient statistic $\boldsymbol{T}(\boldsymbol{Y})$. The criterion used to show that a statistic is complete places a strong restriction on g, and the larger the family of distributions, the greater is the restriction on g. Following Casella and Berger (2002, p. 285), suppose $n = 1$ and $T(Y) = Y \sim N(\theta, 1)$. The family of $N(\theta, 1)$ distributions for $\theta \in \Theta = (-\infty, \infty)$ is a 1P-REF, and it will turn out that Y is a complete sufficient statistic for this family when $n = 1$. Suppose instead that the only member of the family of distributions is the $N(0, 1)$ distribution. Then $\Theta = \{0\}$. Using $g(Y) = Y$ gives $E_0(Y) = 0$ but $P_0(Y = 0) = 0$, not 1. Hence Y is not complete when the family only contains the N(0,1) distribution.

The following two theorems are useful for finding minimal sufficient statistics.

Theorem 4.4: Lehmann–Scheffé Theorem for Minimal Sufficient Statistics (LSM). Let $f(\boldsymbol{y}|\boldsymbol{\theta})$ be the pmf or pdf of an iid sample $\boldsymbol{Y}$. Let $c_{\boldsymbol{x},\boldsymbol{y}}$ be a constant. Suppose there exists a function $\boldsymbol{T}(\boldsymbol{y})$ such that for any two sample points $\boldsymbol{x}$ and $\boldsymbol{y}$, the ratio $R_{\boldsymbol{x},\boldsymbol{y}}(\boldsymbol{\theta}) = f(\boldsymbol{x}|\boldsymbol{\theta})/f(\boldsymbol{y}|\boldsymbol{\theta}) = c_{\boldsymbol{x},\boldsymbol{y}}$ for all $\boldsymbol{\theta}$ in Θ iff $\boldsymbol{T}(\boldsymbol{x}) = \boldsymbol{T}(\boldsymbol{y})$. Then $\boldsymbol{T}(\boldsymbol{Y})$ is a minimal sufficient statistic for $\boldsymbol{\theta}$.

In the Lehmann–Scheffé Theorem, for R to be constant as a function of $\boldsymbol{\theta}$, define $0/0 = c_{\boldsymbol{x},\boldsymbol{y}}$. Alternatively, replace $R_{\boldsymbol{x},\boldsymbol{y}}(\boldsymbol{\theta}) = f(\boldsymbol{x}|\boldsymbol{\theta})/f(\boldsymbol{y}|\boldsymbol{\theta}) = c_{\boldsymbol{x},\boldsymbol{y}}$ by $f(\boldsymbol{x}|\boldsymbol{\theta}) = c_{\boldsymbol{x},\boldsymbol{y}} f(\boldsymbol{y}|\boldsymbol{\theta})$ in the above definition.

Finding sufficient, minimal sufficient, and complete sufficient statistics is often simple for a kP-REF (k-parameter regular exponential family). **If the family given by Eq. (4.8) is a kP-REF, then the conditions for Theorem 4.5a–d are satisfied as are the conditions for e)** if $\boldsymbol{\eta}$ is a one-to-one function of $\boldsymbol{\theta}$. In a), k does not need to be as small as possible. In Corollary 4.6 below, assume that Eqs. (4.8) and (4.9) hold.

Note that any one-to-one function is onto its range. Hence if $\boldsymbol{\eta} = \boldsymbol{\tau}(\boldsymbol{\theta})$ for any $\boldsymbol{\eta} \in \Omega$ where $\boldsymbol{\tau}$ is a one-to-one function, then $\boldsymbol{\tau} : \Theta \to \Omega$ is one to one and onto. Thus there is a one-to-one (and onto) inverse function $\boldsymbol{\tau}^{-1}$ such that $\boldsymbol{\theta} = \boldsymbol{\tau}^{-1}(\boldsymbol{\eta})$ for any $\boldsymbol{\theta} \in \Theta$.

Theorem 4.5: Sufficiency, Minimal Sufficiency, and Completeness of Exponential Families. Suppose that $Y_1, \ldots, Y_n$ are iid from an exponential family

$$f(y|\boldsymbol{\theta}) = h(y)c(\boldsymbol{\theta})\exp[w_1(\boldsymbol{\theta})t_1(y) + \cdots + w_k(\boldsymbol{\theta})t_k(y)] \tag{4.8}$$

with the natural parameterization

$$f(y|\boldsymbol{\eta}) = h(y)b(\boldsymbol{\eta})\exp[\eta_1 t_1(y) + \cdots + \eta_k t_k(y)] \tag{4.9}$$

so that the joint pdf or pmf is given by

$$f(y_1,\ldots,y_n|\boldsymbol{\eta}) = \left(\prod_{j=1}^n h(y_j)\right)[b(\boldsymbol{\eta})]^n \exp\left[\eta_1 \sum_{j=1}^n t_1(y_j) + \cdots + \eta_k \sum_{j=1}^n t_k(y_j)\right]$$

which is a k-parameter exponential family. Then

$$\boldsymbol{T}(\boldsymbol{Y}) = \left(\sum_{j=1}^n t_1(Y_j), \ldots, \sum_{j=1}^n t_k(Y_j)\right) \quad \text{is}$$

a) a sufficient statistic for $\boldsymbol{\theta}$ and for $\boldsymbol{\eta}$,
b) a minimal sufficient statistic for $\boldsymbol{\eta}$ if $\eta_1,\ldots,\eta_k$ do not satisfy a linearity constraint,
c) a minimal sufficient statistic for $\boldsymbol{\theta}$ if $w_1(\boldsymbol{\theta}),\ldots,w_k(\boldsymbol{\theta})$ do not satisfy a linearity constraint,
d) a complete sufficient statistic for $\boldsymbol{\eta}$ if Ω contains a k-dimensional rectangle,
e) a complete sufficient statistic for $\boldsymbol{\theta}$ if $\boldsymbol{\eta}$ is a one-to-one function of $\boldsymbol{\theta}$ and if Ω contains a k-dimensional rectangle.

Proof. a) Use the Factorization Theorem.

b) The proof expands on remarks given in Johanson (1979, p. 3) and Lehmann (1983, p. 44). The ratio

$$\frac{f(\boldsymbol{x}|\boldsymbol{\eta})}{f(\boldsymbol{y}|\boldsymbol{\eta})} = \frac{\prod_{j=1}^n h(x_j)}{\prod_{j=1}^n h(y_j)} \exp\left[\sum_{i=1}^k \eta_i (T_i(\boldsymbol{x}) - T_i(\boldsymbol{y}))\right]$$

is equal to a constant with respect to $\boldsymbol{\eta}$ iff

$$\sum_{i=1}^k \eta_i [T_i(\boldsymbol{x}) - T_i(\boldsymbol{y})] = \sum_{i=1}^k \eta_i a_i = d$$

for all η_i where d is some constant and where $a_i = T_i(\boldsymbol{x}) - T_i(\boldsymbol{y})$ and $T_i(\boldsymbol{x}) = \sum_{j=1}^n t_i(x_j)$. Since the η_i do not satisfy a linearity constraint, $\sum_{i=1}^k \eta_i a_i = d$ for all $\boldsymbol{\eta}$ iff all of the $a_i = 0$. Hence

$$\boldsymbol{T}(\boldsymbol{Y}) = (T_1(\boldsymbol{Y}),\ldots,T_k(\boldsymbol{Y}))$$

is a minimal sufficient statistic by the Lehmann–Scheffé LSM Theorem.

c) Use almost the same proof as b) with $w_i(\boldsymbol{\theta})$ in the place of η_i and $\boldsymbol{\theta}$ in the place of $\boldsymbol{\eta}$. (In particular, the result holds if $\eta_i = w_i(\boldsymbol{\theta})$ for $i = 1,\ldots,k$ provided that the η_i do not satisfy a linearity constraint.)

d) See Lehmann (1986, p. 142).

e) If $\boldsymbol{\eta} = \tau(\boldsymbol{\theta})$ then $\boldsymbol{\theta} = \tau^{-1}(\boldsymbol{\eta})$ and the parameters have just been renamed. Hence $E_{\boldsymbol{\theta}}[g(\boldsymbol{T})] = 0$ for all $\boldsymbol{\theta}$ implies that $E_{\boldsymbol{\eta}}[g(\boldsymbol{T})] = 0$ for all $\boldsymbol{\eta}$, and thus

$P_{\boldsymbol{\eta}}[g(\boldsymbol{T}(\boldsymbol{Y}))=0]=1$ for all $\boldsymbol{\eta}$ since $\boldsymbol{T}$ is a complete sufficient statistic for $\boldsymbol{\eta}$ by d). Thus $P_{\boldsymbol{\theta}}[g(\boldsymbol{T}(\boldsymbol{Y}))=0]=1$ for all $\boldsymbol{\theta}$, and $\boldsymbol{T}$ is a complete sufficient statistic for $\boldsymbol{\theta}$.

Corollary 4.6: Completeness of a kP-REF. Suppose that $Y_1,\ldots,Y_n$ are iid from a kP-REF (k-parameter regular exponential family)

$$f(y|\boldsymbol{\theta}) = h(y)c(\boldsymbol{\theta})\exp[w_1(\boldsymbol{\theta})t_1(y)+\cdots+w_k(\boldsymbol{\theta})t_k(y)]$$

with $\boldsymbol{\theta}\in\Theta$ and natural parameterization given by (4.9) with $\boldsymbol{\eta}\in\Omega$. Then

$$\boldsymbol{T}(\boldsymbol{Y}) = \left(\sum_{j=1}^n t_1(Y_j),\ldots,\sum_{j=1}^n t_k(Y_j)\right) \quad \text{is}$$

a) a minimal sufficient statistic for $\boldsymbol{\theta}$ and for $\boldsymbol{\eta}$,
b) a complete sufficient statistic for $\boldsymbol{\theta}$ and for $\boldsymbol{\eta}$ if $\boldsymbol{\eta}$ is a one-to-one function of $\boldsymbol{\theta}$.

Proof. The result follows by Theorem 4.5 since for a kP-REF, the $w_i(\boldsymbol{\theta})$ and η_i do not satisfy a linearity constraint and Ω contains a k-dimensional rectangle. □

Theorem 4.7: Bahadur's Theorem. A finite dimensional complete sufficient statistic is also minimal sufficient.

Theorem 4.8. A one-to-one function of a sufficient, minimal sufficient, or complete sufficient statistic is sufficient, minimal sufficient, or complete sufficient, respectively.

If $\boldsymbol{T}_1(\boldsymbol{Y}) = \boldsymbol{g}(\boldsymbol{T}_2(\boldsymbol{Y}))$ where $\boldsymbol{g}$ is a one-to-one function, then $\boldsymbol{T}_2(\boldsymbol{Y}) = \boldsymbol{g}^{-1}(\boldsymbol{T}_1(\boldsymbol{Y}))$ where $\boldsymbol{g}^{-1}$ is the inverse function of $\boldsymbol{g}$. Hence $\boldsymbol{T}_1$ and $\boldsymbol{T}_2$ provide an equivalent amount reduction of the data. Also see the discussion below Remark 4.3. Note that in a kP-REF, the statistic $\boldsymbol{T}$ is k-dimensional and thus $\boldsymbol{T}$ is minimal sufficient by Theorem 4.7 if $\boldsymbol{T}$ is complete sufficient. Corollary 4.6 is useful because often you know or can show that the given family is a REF. The theorem gives a particularly simple way to find complete sufficient statistics for one-parameter exponential families and for any family that is known to be a REF. If it is known that the distribution is regular, find the exponential family parameterization given by Eq. (4.8) or (4.9). These parameterizations give $t_1(y),\ldots,t_k(y)$. Then $\boldsymbol{T}(\boldsymbol{Y}) = (\sum_{j=1}^n t_1(Y_j),\ldots,\sum_{j=1}^n t_k(Y_j))$.

Example 4.11. Let $X_1,\ldots,X_n$ be iid $N(\mu,\sigma^2)$. Then the $N(\mu,\sigma^2)$ pdf is

$$f(x|\mu,\sigma) = \underbrace{\frac{1}{\sqrt{2\pi}\sigma}\exp\left(\frac{-\mu^2}{2\sigma^2}\right)}_{c(\mu,\sigma)\geq 0}\exp\left(\underbrace{\frac{-1}{2\sigma^2}}_{w_1(\boldsymbol{\theta})}\underbrace{x^2}_{t_1(x)}+\underbrace{\frac{\mu}{\sigma^2}}_{w_2(\boldsymbol{\theta})}\underbrace{x}_{t_2(x)}\right)\underbrace{I_{\mathbb{R}}(x)}_{h(x)\geq 0},$$

with $\eta_1=-0.5/\sigma^2$ and $\eta_2=\mu/\sigma^2$ if $\sigma>0$. As shown in Example 3.1, this is a 2P–REF. By Corollary 4.6, $\boldsymbol{T}=(\sum_{i=1}^n X_i,\sum_{i=1}^n X_i^2)$ is a complete sufficient statistic for (μ,σ^2). The one-to-one functions

$$\boldsymbol{T}_2 = (\overline{X}, S^2) \text{ and } \boldsymbol{T}_3 = (\overline{X}, S)$$

of $\boldsymbol{T}$ are also complete sufficient where $\overline{X}$ is the sample mean and S is the sample standard deviation. $\boldsymbol{T}, \boldsymbol{T}_2$ and $\boldsymbol{T}_3$ are minimal sufficient by Corollary 4.6 or by Theorem 4.7 since the statistics are two dimensional.

Example 4.12. Let $Y_1, \ldots, Y_n$ be iid binomial(k, ρ) with k known and pmf

$$f(y|\rho) = \binom{k}{y} \rho^y (1-\rho)^{k-y} I_{\{0,\ldots,k\}}(y)$$

$$= \underbrace{\binom{k}{y} I_{\{0,\ldots,k\}}(y)}_{h(y) \geq 0} \underbrace{(1-\rho)^k}_{c(\rho) \geq 0} \exp\left[\underbrace{\log\left(\frac{\rho}{1-\rho}\right)}_{w(\rho)} \underbrace{y}_{t(y)} \right]$$

where $\Theta = (0,1)$ and $\Omega = (-\infty, \infty)$. Notice that $\eta = \log(\frac{\rho}{1-\rho})$ is an increasing and hence one-to-one function of ρ. Since this family is a 1P-REF, $T_n = \sum_{i=1}^n t(Y_i) = \sum_{i=1}^n Y_i$ is complete sufficient statistic for ρ.

Compare Examples 4.7 and 4.8 with Examples 4.11 and 4.12. The exponential family theorem gives more powerful results than the Factorization Theorem, but often the Factorization Theorem is useful for suggesting a potential minimal sufficient statistic.

Example 4.13. In testing theory, a single sample is often created by combining two independent samples of iid data. Let $X_1, \ldots, X_n$ be iid exponential (θ) and $Y_1, \ldots, Y_m$ iid exponential $(\theta/2)$. If the two samples are independent, then the joint pdf $f(\boldsymbol{x}, \boldsymbol{y}|\theta)$ belongs to a regular one-parameter exponential family with complete sufficient statistic $T = \sum_{i=1}^n X_i + 2\sum_{i=1}^m Y_i$. (Let $W_i = 2Y_i$. Then the W_i and X_i are iid and Corollary 4.6 applies.)

Rule of thumb 4.1: A k-parameter minimal sufficient statistic for a d-dimensional parameter where $d < k$ will not be complete. In the following example $d = 1 < 2 = k$. (A rule of thumb is something that is frequently true but cannot be used to rigorously prove something. Hence this rule of thumb cannot be used to prove that the minimal sufficient statistic is not complete.)

Warning: Showing that a minimal sufficient statistic is not complete is a problem that often appears in qualifying exams on statistical inference.

Example 4.14 (Cox and Hinkley 1974, p. 31). Let $X_1, \ldots, X_n$ be iid $N(\mu, \gamma_o^2 \mu^2)$ random variables where $\gamma_o^2 > 0$ is *known* and $\mu > 0$. Then this family has a one-dimensional parameter μ, but

$$f(x|\mu) = \frac{1}{\sqrt{2\pi\gamma_o^2\mu^2}} \exp\left(\frac{-1}{2\gamma_o^2}\right) \exp\left(\frac{-1}{2\gamma_o^2\mu^2} x^2 + \frac{1}{\gamma_o^2\mu} x\right)$$

is a two-parameter exponential family with $\Theta = (0,\infty)$ (which contains a one-dimensional rectangle), and $(\sum_{i=1}^n X_i, \sum_{i=1}^n X_i^2)$ is a minimal sufficient statistic. (Theorem 4.5 applies since the functions $1/\mu$ and $1/\mu^2$ do not satisfy a linearity constraint.) However, since $E_\mu(X^2) = \gamma_o^2\mu^2 + \mu^2$ and $\sum_{i=1}^n X_i \sim N(n\mu, n\gamma_o^2\mu^2)$ implies that

$$E_\mu\left[\left(\sum_{i=1}^n X_i\right)^2\right] = n\gamma_o^2\mu^2 + n^2\mu^2,$$

we find that

$$E_\mu\left[\frac{n+\gamma_o^2}{1+\gamma_o^2}\sum_{i=1}^n X_i^2 - \left(\sum_{i=1}^n X_i\right)^2\right] = \frac{n+\gamma_o^2}{1+\gamma_o^2} n\mu^2(1+\gamma_o^2) - (n\gamma_o^2\mu^2 + n^2\mu^2) = 0$$

for all μ so the minimal sufficient statistic is not complete. Notice that

$$\Omega = \left\{(\eta_1, \eta_2) : \eta_1 = \frac{-1}{2}\gamma_o^2\eta_2^2\right\}$$

and a plot of η_1 versus η_2 is a quadratic function which cannot contain a two-dimensional rectangle. Notice that (η_1, η_2) is a one-to-one function of μ, and thus this example illustrates that the rectangle needs to be contained in Ω rather than Θ.

Example 4.15. The theory does not say that any sufficient statistic from a REF is complete. Let Y be a random variable from a normal $N(0,\sigma^2)$ distribution with $\sigma^2 > 0$. This family is a REF with complete minimal sufficient statistic Y^2. The data Y is also a sufficient statistic, but Y is not a function of Y^2. Hence Y is not minimal sufficient and (by Bahadur's Theorem) not complete. Alternatively $E_{\sigma^2}(Y) = 0$ but $P_{\sigma^2}(Y = 0) = 0 < 1$, so Y is not complete.

Theorem 4.9. Let $Y_1, \ldots, Y_n$ be iid.

a) If $Y_i \sim U(\theta_1, \theta_2)$, then $(Y_{(1)}, Y_{(n)})$ is a complete sufficient statistic for (θ_1, θ_2). See David (1981, p. 123.)
b) If $Y_i \sim U(\theta_1, \theta_2)$ with θ_1 known, then $Y_{(n)}$ is a complete sufficient statistic for θ_2.
c) If $Y_i \sim U(\theta_1, \theta_2)$ with θ_2 known, then $Y_{(1)}$ is a complete sufficient statistic for θ_1.
d) If $Y_i \sim U(-\theta, \theta)$, then $\max(|Y_i|)$ is a complete sufficient statistic for θ.
e) If $Y_i \sim \text{EXP}(\theta, \lambda)$, then $(Y_{(1)}, \overline{Y})$ is a complete sufficient statistic for (θ, λ). See David (1981, pp. 153–154).
f) If $Y_i \sim \text{EXP}(\theta, \lambda)$ with λ known, then $Y_{(1)}$ is a complete sufficient statistic for θ.
g) If $Y_i \sim \text{Cauchy}(\mu, \sigma)$ with σ known, then the order statistics are minimal sufficient.
h) If $Y_i \sim$ Double Exponential(θ, λ) with λ known, then the order statistics $(Y_{(1)}, \ldots, Y_{(n)})$ are minimal sufficient.
i) If $Y_i \sim \text{logistic}(\mu, \sigma)$, then the order statistics are minimal sufficient.
j) If $Y_i \sim \text{Weibull}(\phi, \lambda)$, then the order statistics $(Y_{(1)}, \ldots, Y_{(n)})$ are minimal sufficient.

A **common midterm, final, and qual question** takes $X_1,\ldots,X_n$ iid $U(h_l(\theta),h_u(\theta))$ where h_l and h_u are functions of θ such that $h_l(\theta) < h_u(\theta)$. The function h_l and h_u are chosen so that the min = $X_{(1)}$ and the max = $X_{(n)}$ form the two-dimensional minimal sufficient statistic by the LSM theorem. Since θ is one dimensional, the rule of thumb suggests that the minimal sufficient statistic is not complete. State this fact, but if you have time find $E_\theta[X_{(1)}]$ and $E_\theta[X_{(n)}]$. Then show that $E_\theta[aX_{(1)}+bX_{(n)}+c] \equiv 0$ so that $\boldsymbol{T} = (X_{(1)},X_{(n)})$ is not complete.

Example 4.16. The uniform distribution is tricky since usually $(X_{(1)},X_{(n)})$ is minimal sufficient by the LSM theorem, since

$$f(\boldsymbol{x}) = \frac{1}{(\theta_2-\theta_1)^n} I(\theta_1 < x_{(1)} < x_{(n)} < \theta_2)$$

if $n > 1$. But occasionally θ_1 and θ_2 are functions of the one-dimensional parameter θ such that the indicator can be written as $I(\theta > T)$ or $I(\theta < T)$ where the minimal sufficient statistic T is a one-dimensional function of $(X_{(1)},X_{(n)})$. If $X \sim U(c_1+d_1\theta, c_2+d_2\theta)$ where d_1,d_2,c_1,c_2 are known and $d_1 < 0$ and $d_2 > \theta$, then

$$T = \max\left(\frac{X_{(1)}-c_1}{d_1}, \frac{X_{(n)}-c_2}{d_2}\right)$$

is minimal sufficient.

Let $X_1,\ldots,X_n$ be iid $U(1-\theta,1+\theta)$ where $\theta > 0$ is unknown. Hence

$$f_X(x) = \frac{1}{2\theta} I(1-\theta < x < 1+\theta)$$

and

$$\frac{f(\boldsymbol{x})}{f(\boldsymbol{y})} = \frac{I(1-\theta < x_{(1)} \le x_{(n)} < 1+\theta)}{I(1-\theta < y_{(1)} \le y_{(n)} < 1+\theta)}.$$

This ratio may look to be constant for all $\theta > 0$ iff $(x_{(1)},x_{(n)}) = (y_{(1)},y_{(n)})$, but it is not. To show that $\boldsymbol{T}_1 = (X_{(1)},X_{(n)})$ is not a minimal sufficient statistic, note that

$$f_X(x) = \frac{1}{2\theta} I(\theta > 1-x) I(\theta > x-1).$$

Hence

$$\frac{f(\boldsymbol{x})}{f(\boldsymbol{y})} = \frac{I(\theta > \max(1-x_{(1)},x_{(n)}-1))}{I(\theta > \max(1-y_{(1)},y_{(n)}-1))}$$

which is constant for all $\theta > 0$ iff $T_2(\boldsymbol{x}) = T_2(\boldsymbol{y})$ where $T_2(\boldsymbol{x}) = \max(1-x_{(1)},x_{(n)}-1)$. Hence $T_2 = T_2(\boldsymbol{X}) = \max(1-X_{(1)},X_{(n)}-1)$ is minimal sufficient by the LSM theorem. Thus $\boldsymbol{T}_1$ is not a minimal sufficient statistic (and so not complete) since $\boldsymbol{T}_1$ is not a function of T_2.

To show that $\boldsymbol{T}_1$ is not complete using the definition of complete statistic, first find $E(\boldsymbol{T}_1)$. Now

$$F_X(t) = \int_{1-\theta}^{t} \frac{1}{2\theta} dx = \frac{t+\theta-1}{2\theta}$$

for $1-\theta < t < 1+\theta$. Hence by Theorem 4.2a),

$$f_{X_{(n)}}(t) = \frac{n}{2\theta} \left(\frac{t+\theta-1}{2\theta} \right)^{n-1}$$

for $1-\theta < t < 1+\theta$ and

$$E_\theta(X_{(n)}) = \int x f_{X_{(n)}}(x) dx = \int_{1-\theta}^{1+\theta} x \frac{n}{2\theta} \left(\frac{x+\theta-1}{2\theta} \right)^{n-1} dx.$$

Use u–substitution with $u = (x+\theta-1)/2\theta$ and $x = 2\theta u + 1 - \theta$. Hence $x = 1+\theta$ implies $u = 1$, and $x = 1-\theta$ implies $u = 0$ and $dx = 2\theta du$. Thus

$$E_\theta(X_{(n)}) = n \int_0^1 \frac{2\theta u + 1 - \theta}{2\theta} u^{n-1} 2\theta du =$$

$$= n \int_0^1 [2\theta u + 1 - \theta] u^{n-1} du = 2\theta n \int_0^1 u^n du + (n - n\theta) \int_0^1 u^{n-1} du =$$

$$2\theta n \frac{u^{n+1}}{n+1} \bigg|_0^1 + n(1-\theta) \frac{u^n}{n} \bigg|_0^1 =$$

$$2\theta \frac{n}{n+1} + \frac{n(1-\theta)}{n} = 1 - \theta + 2\theta \frac{n}{n+1}.$$

Note that $E_\theta(X_{(n)}) \approx 1+\theta$ as you should expect.

By Theorem 4.2b),

$$f_{X_{(1)}}(t) = \frac{n}{2\theta} \left(\frac{\theta - t + 1}{2\theta} \right)^{n-1}$$

for $1-\theta < t < 1+\theta$ and thus

$$E_\theta(X_{(1)}) = \int_{1-\theta}^{1+\theta} x \frac{n}{2\theta} \left(\frac{\theta - x + 1}{2\theta} \right)^{n-1} dx.$$

Use u–substitution with $u = (\theta - x + 1)/2\theta$ and $x = \theta + 1 - 2\theta u$. Hence $x = 1+\theta$ implies $u = 0$, and $x = 1-\theta$ implies $u = 1$ and $dx = -2\theta du$. Thus

$$E_\theta(X_{(1)}) = \int_1^0 \frac{n}{2\theta} (\theta + 1 - 2\theta u) u^{n-1} (-2\theta) du = n \int_0^1 (\theta + 1 - 2\theta u) u^{n-1} du =$$

$$n(\theta+1) \int_0^1 u^{n-1} du - 2\theta n \int_0^1 u^n du = (\theta+1)n/n - 2\theta n/(n+1) = \theta + 1 - 2\theta \frac{n}{n+1}.$$

To show that $\boldsymbol{T}_1$ is not complete try showing $E_\theta(aX_{(1)} + bX_{(n)} + c) = 0$ for some constants a, b, and c. Note that $a = b = 1$ and $c = -2$ works. Hence $E_\theta(X_{(1)} + X_{(n)} - 2) = 0$ for all $\theta > 0$ but $P_\theta(g(\boldsymbol{T}_1) = 0) = P_\theta(X_{(1)} + X_{(n)} - 2 = 0) = 0 < 1$ for all $\theta > 0$. Hence $\boldsymbol{T}_1$ is not complete. (To see that $P_\theta(g(\boldsymbol{T}_1) = 0) = 0$, note that if random variables $Y_1, \ldots, Y_n$ have a joint pdf, then a nonzero linear combination of the random variables is a random variable W with a pdf by Theorem 2.14. So $P(W = 0) = 0$ since the probability that a random variable W is equal to a constant is 0 if W has a pdf. The order statistics have a joint pdf by Theorem 4.2 c). Thus the linear combination $W = g(\boldsymbol{T}_1)$ has a pdf and $P(W = 0) = 0$.)

Definition 4.5. Let $Y_1, \ldots, Y_n$ have pdf or pmf $f(\boldsymbol{y}|\theta)$. A statistic $\boldsymbol{W}(\boldsymbol{Y})$ whose distribution does not depend on θ is called an **ancillary statistic.**

Theorem 4.10, Basu's Theorem. Let $Y_1, \ldots, Y_n$ have pdf or pmf $f(\boldsymbol{y}|\theta)$. If $\boldsymbol{T}(\boldsymbol{Y})$ is a k-dimensional complete sufficient statistic, then $\boldsymbol{T}(\boldsymbol{Y})$ is independent of every ancillary statistic.

Remark 4.4. Basu's Theorem says that if $\boldsymbol{T}$ is minimal sufficient and complete, then $\boldsymbol{T} \perp\!\!\!\perp R$ if R is ancillary. Application: If $\boldsymbol{T}$ is minimal sufficient, R ancillary and R is a function of $\boldsymbol{T}$ (so $R = h(\boldsymbol{T})$ is not independent of $\boldsymbol{T}$), then $\boldsymbol{T}$ is not complete. Since θ is a scalar, usually $\boldsymbol{T}(\boldsymbol{Y})$ is not complete unless $k = 1$: then write the minimal sufficient statistic as $T(\boldsymbol{Y})$.

Example 4.17. Suppose $X_1, \ldots, X_n$ are iid uniform observations on the interval $(\theta, \theta + 1)$, $-\infty < \theta < \infty$. Let $X_{(1)} = \min(X_1, \ldots, X_n)$, $X_{(n)} = \max(X_1, \ldots, X_n)$ and $\boldsymbol{T}(\boldsymbol{X}) = (X_{(1)}, X_{(n)})$ be a minimal sufficient statistic. Then $R = X_{(n)} - X_{(1)}$ is ancillary since $R = \max(X_1 - \theta, \ldots, X_n - \theta) + \theta - [\min(X_1 - \theta, \ldots, X_n - \theta) + \theta] = U_{(n)} - U_{(1)}$ where $U_i = X_i - \theta \sim U(0, 1)$ has a distribution that does not depend on θ. R is not independent of $\boldsymbol{T}$, so $\boldsymbol{T}$ is not complete.

Example 4.18. Let $Y_1, \ldots, Y_n$ be iid from a location family with pdf $f_Y(y|\theta) = f_X(y - \theta)$ where $Y = X + \theta$ and $f_X(y)$ is the standard pdf for the location family (and thus the distribution of X does not depend on θ).
Claim: $\boldsymbol{W} = (Y_1 - \overline{Y}, \ldots, Y_n - \overline{Y})$ is ancillary.

Proof: Since $Y_i = X_i + \theta$,

$$\boldsymbol{W} = \left(X_1 + \theta - \frac{1}{n}\sum_{i=1}^n (X_i + \theta), \ldots, X_n + \theta - \frac{1}{n}\sum_{i=1}^n (X_i + \theta) \right) = (X_1 - \overline{X}, \ldots, X_n - \overline{X})$$

and the distribution of the final vector is free of θ. $\square$

Application: Let $Y_1, \ldots, Y_n$ be iid $N(\mu, \sigma^2)$. For any fixed σ^2, this is a location family with $\theta = \mu$ and complete sufficient statistic $T(\boldsymbol{Y}) = \overline{Y}$. Thus $\overline{Y} \perp\!\!\!\perp \boldsymbol{W}$ by Basu's Theorem. Hence $\overline{Y} \perp\!\!\!\perp S^2$ for any known $\sigma^2 > 0$ since

$$S^2 = \frac{1}{n-1}\sum_{i=1}^n (Y_i - \overline{Y})^2$$

is a function of $\boldsymbol{W}$. Thus $\overline{Y} \perp\!\!\!\perp S^2$ even if $\sigma^2 > 0$ is not known.

4.3 Summary

1. A statistic is a function of the data that does not depend on any unknown parameters.
2. For parametric inference, the data $Y_1,\ldots,Y_n$ comes from a family of parametric distributions $f(\boldsymbol{y}|\boldsymbol{\theta})$ for $\boldsymbol{\theta} \in \Theta$. Often the data are iid and $f(\boldsymbol{y}|\boldsymbol{\theta}) = \prod_{i=1}^n f(y_i|\boldsymbol{\theta})$. The parametric distribution is completely specified by the unknown parameters $\boldsymbol{\theta}$. The statistic is a random vector or random variable and hence also comes from some probability distribution. The distribution of the statistic is called the sampling distribution of the statistic.
3. For iid $N(\mu,\sigma^2)$ data, $\overline{Y} \perp\!\!\!\perp S^2$, $\overline{Y} \sim N(\mu,\sigma^2/n)$ and $\sum_{i=1}^n (Y_i - \overline{Y})^2 \sim \sigma^2\chi^2_{n-1}$.
4. For iid data with cdf F_Y and pdf f_Y, $f_{Y_{(n)}}(t) = n[F_Y(t)]^{n-1} f_Y(t)$ and $f_{Y_{(1)}}(t) = n[1-F_Y(t)]^{n-1} f_Y(t)$.
5. A statistic $\boldsymbol{T}(Y_1,\ldots,Y_n)$ is a *sufficient statistic* for $\boldsymbol{\theta}$ if the conditional distribution of $(Y_1,\ldots,Y_n)$ given $\boldsymbol{T}$ does not depend on $\boldsymbol{\theta}$.
6. A sufficient statistic $\boldsymbol{T}(\boldsymbol{Y})$ is a *minimal sufficient statistic* if for any other sufficient statistic $\boldsymbol{S}(\boldsymbol{Y})$, $\boldsymbol{T}(\boldsymbol{Y})$ is a function of $\boldsymbol{S}(\boldsymbol{Y})$.
7. Suppose that a *statistic* $\boldsymbol{T}(\boldsymbol{Y})$ has a pmf or pdf $f(\boldsymbol{t}|\boldsymbol{\theta})$. Then $\boldsymbol{T}(\boldsymbol{Y})$ is a *complete statistic* if $E_{\boldsymbol{\theta}}[g(\boldsymbol{T}(\boldsymbol{Y}))] = 0$ for all $\boldsymbol{\theta} \in \Theta$ implies that $P_{\boldsymbol{\theta}}[g(\boldsymbol{T}(\boldsymbol{Y})) = 0] = 1$ for all $\boldsymbol{\theta} \in \Theta$.
8. A one-to-one function of a sufficient, minimal sufficient, or complete sufficient statistic is sufficient, minimal sufficient, or complete sufficient, respectively.
9. **Factorization Theorem.** Let $f(\boldsymbol{y}|\boldsymbol{\theta})$ denote the pdf or pmf of a sample $\boldsymbol{Y}$. A statistic $\boldsymbol{T}(\boldsymbol{Y})$ is a sufficient statistic for $\boldsymbol{\theta}$ iff for all sample points $\boldsymbol{y}$ and for all $\boldsymbol{\theta}$ in the parameter space Θ,

$$f(\boldsymbol{y}|\boldsymbol{\theta}) = g(\boldsymbol{T}(\boldsymbol{y})|\boldsymbol{\theta})\, h(\boldsymbol{y})$$

where both g and h are nonnegative functions.
Tips: i) for iid data with marginal support $\mathscr{Y}_i \equiv \mathscr{Y}^*$, $I_{\mathscr{Y}}(\boldsymbol{y}) = I(\text{all } y_i \in \mathscr{Y}^*)$. If $\mathscr{Y}^* = (a,b)$, then $I_{\mathscr{Y}}(\boldsymbol{y}) = I(a < y_{(1)} < y_{(n)} < b) = I(a < y_{(1)})I(y_{(n)} < b)$. Put $I(a < y_{(1)})$ in $g(\boldsymbol{T}(\boldsymbol{y})|\boldsymbol{\theta})$ if a is an unknown parameter but put $I(a < y_{(1)})$ in $h(\boldsymbol{y})$ if a is known. If both a and b are unknown parameters, put $I(a < y_{(1)} < y_{(n)} < b)$ in $g(\boldsymbol{T}(\boldsymbol{y})|\boldsymbol{\theta})$. If $b = \infty$, then $I_{\mathscr{Y}}(\boldsymbol{y}) = I(a < y_{(1)})$. If $\mathscr{Y}^* = [a,b]$, then $I_{\mathscr{Y}}(\boldsymbol{y}) = I(a \le y_{(1)} < y_{(n)} \le b) = I(a \le y_{(1)})I(y_{(n)} \le b)$. ii) Try to make the dimension of $\boldsymbol{T}(\boldsymbol{y})$ as small as possible. Put anything that depends on $\boldsymbol{y}$ but not $\boldsymbol{\theta}$ into $h(\boldsymbol{y})$.
10. **Minimal and complete sufficient statistics for k-parameter exponential families:** Let $Y_1,\ldots,Y_n$ be iid from an exponential family $f(y|\boldsymbol{\theta}) = h(y)c(\boldsymbol{\theta})\exp[\sum_{j=1}^k w_j(\boldsymbol{\theta})t_j(y)]$ with the natural parameterization $f(y|\boldsymbol{\eta}) = h(y)b(\boldsymbol{\eta})\exp[\sum_{j=1}^k \eta_j t_j(y)]$. Then $\boldsymbol{T}(\boldsymbol{Y}) = (\sum_{i=1}^n t_1(Y_i),\ldots,\sum_{i=1}^n t_k(Y_i))$ is

a) a minimal sufficient statistic for $\boldsymbol{\eta}$ if the η_j do not satisfy a linearity constraint and for $\boldsymbol{\theta}$ if the $w_j(\boldsymbol{\theta})$ do not satisfy a linearity constraint.

b) a complete sufficient statistic for $\boldsymbol{\theta}$ and for $\boldsymbol{\eta}$ if $\boldsymbol{\eta}$ is a one-to-one function of $\boldsymbol{\theta}$ and if Ω contains a k-dimensional rectangle.

11. **Completeness of REFs:** Suppose that $Y_1, \ldots, Y_n$ are iid from a kP-REF

$$f(y|\boldsymbol{\theta}) = h(y)c(\boldsymbol{\theta})\exp[w_1(\boldsymbol{\theta})t_1(y) + \cdots + w_k(\boldsymbol{\theta})t_k(y)] \tag{4.10}$$

with $\boldsymbol{\theta} \in \Theta$ and natural parameter $\boldsymbol{\eta} \in \Omega$. Then

$$\boldsymbol{T}(\boldsymbol{Y}) = \left(\sum_{i=1}^n t_1(Y_i), \ldots, \sum_{i=1}^n t_k(Y_i)\right) \text{ is}$$

a) a minimal sufficient statistic for $\boldsymbol{\eta}$ and for $\boldsymbol{\theta}$,

b) a complete sufficient statistic for $\boldsymbol{\theta}$ and for $\boldsymbol{\eta}$ if $\boldsymbol{\eta}$ is a one-to-one function of $\boldsymbol{\theta}$ and if Ω contains a k-dimensional rectangle.

12. For a two-parameter exponential family ($k = 2$), η_1 and η_2 satisfy a linearity constraint if the plotted points fall on a line in a plot of η_1 versus η_2. If the plotted points fall on a nonlinear curve, then $\boldsymbol{T}$ is minimal sufficient but Ω does not contain a two-dimensional rectangle.
13. **LSM Theorem:** Let $f(\boldsymbol{y}|\boldsymbol{\theta})$ be the pmf or pdf of a sample $\boldsymbol{Y}$. Let $c_{\boldsymbol{x},\boldsymbol{y}}$ be a constant. Suppose there exists a function $\boldsymbol{T}(\boldsymbol{y})$ such that for any two sample points $\boldsymbol{x}$ and $\boldsymbol{y}$, the ratio $R_{\boldsymbol{x},\boldsymbol{y}}(\boldsymbol{\theta}) = f(\boldsymbol{x}|\boldsymbol{\theta})/f(\boldsymbol{y}|\boldsymbol{\theta}) = c_{\boldsymbol{x},\boldsymbol{y}}$ for all $\boldsymbol{\theta}$ in Θ iff $\boldsymbol{T}(\boldsymbol{x}) = \boldsymbol{T}(\boldsymbol{y})$. Then $\boldsymbol{T}(\boldsymbol{Y})$ is a minimal sufficient statistic for $\boldsymbol{\theta}$. (Define $0/0 \equiv c_{\boldsymbol{x},\boldsymbol{y}}$.)
14. *Tips for finding sufficient, minimal sufficient and complete sufficient statistics.*

    a) Typically $Y_1, \ldots, Y_n$ are iid so the joint distribution $f(y_1, \ldots, y_n) = \prod_{i=1}^n f(y_i)$ where $f(y_i)$ is the marginal distribution. Use the **factorization theorem** to find the candidate sufficient statistic $\boldsymbol{T}$.

    b) Use factorization to find candidates $\boldsymbol{T}$ that might be minimal sufficient statistics. Try to find $\boldsymbol{T}$ with as small a dimension k as possible. If the support of the random variable depends on θ, often $Y_{(1)}$ or $Y_{(n)}$ will be a component of the minimal sufficient statistic. To prove that $\boldsymbol{T}$ is minimal sufficient, use the **LSM theorem**. **Alternatively prove or recognize that** Y **comes from a regular exponential family.** $\boldsymbol{T}$ will be minimal sufficient for $\boldsymbol{\theta}$ if Y comes from an exponential family as long as the $w_i(\boldsymbol{\theta})$ do not satisfy a linearity constraint.

    c) **To prove that the statistic is complete, prove or recognize that** Y **comes from a regular exponential family**. Check whether $\dim(\Theta) = k$. If $\dim(\Theta) < k$, then the family is usually not a kP-REF and Theorem 4.5 and Corollary 4.6 do not apply. The uniform distribution where one endpoint is known also has a complete sufficient statistic.

    d) Let k be free of the sample size n. Then a k-dimensional complete sufficient statistic is also a minimal sufficient statistic (**Bahadur's Theorem**).

e) To show that a statistic $\boldsymbol{T}$ is not a sufficient statistic, either show that factorization fails or find a minimal sufficient statistic $\boldsymbol{S}$ and show that $\boldsymbol{S}$ is not a function of $\boldsymbol{T}$.

f) To show that $\boldsymbol{T}$ is not minimal sufficient, first try to show that $\boldsymbol{T}$ is not a sufficient statistic. If $\boldsymbol{T}$ is sufficient, find a minimal sufficient statistic $\boldsymbol{S}$ and show that $\boldsymbol{T}$ is not a function of $\boldsymbol{S}$. (Of course $\boldsymbol{S}$ will be a function of $\boldsymbol{T}$.) The **Lehmann–Scheffé (LSM) theorem cannot be used to show that a statistic is not minimal sufficient.**

g) To show that a sufficient statistics $\boldsymbol{T}$ is not complete, find a function $g(\boldsymbol{T})$ such that $E_{\boldsymbol{\theta}}(g(\boldsymbol{T})) = 0$ for all $\boldsymbol{\theta}$ but $g(\boldsymbol{T})$ is not equal to the zero with probability one. Finding such a g is often hard, unless there are clues. For example, if $\boldsymbol{T} = (\overline{X}, \overline{Y}, \ldots.)$ and $\mu_1 = \mu_2$, try $g(\boldsymbol{T}) = \overline{X} - \overline{Y}$. As a **rule of thumb**, a k-dimensional minimal sufficient statistic will generally not be complete if $k > dim(\Theta)$. In particular, if $\boldsymbol{T}$ is k-dimensional and $\boldsymbol{\theta}$ is j-dimensional with $j < k$ (especially $j = 1 < 2 = k$) then $\boldsymbol{T}$ will **generally not be complete.** If you can show that a k-dimensional sufficient statistic $\boldsymbol{T}$ is not minimal sufficient (often hard), then $\boldsymbol{T}$ is not complete by Bahadur's Theorem. Basu's Theorem can sometimes be used to show that a minimal sufficient statistic is not complete. See Remark 4.4 and Example 4.17.

15. A **common** question takes $Y_1, \ldots, Y_n$ iid $U(h_l(\theta), h_u(\theta))$ where the min = $Y_{(1)}$ and the max = $Y_{(n)}$ form the two-dimensional minimal sufficient statistic. Since θ is one dimensional, the minimal sufficient statistic is probably not complete. Find $E_\theta[Y_{(1)}]$ and $E_\theta[Y_{(n)}]$. Then show that $E_\theta[aY_{(1)} + bY_{(n)} + c] \equiv 0$ so that $\boldsymbol{T} = (Y_{(1)}, Y_{(n)})$ is not complete.

4.4 Complements

Some minimal sufficient statistics and complete sufficient statistics are given in Theorem 4.9 for distributions that are not exponential families.

Stigler (1984) presents Kruskal's proof that $\overline{Y} \perp\!\!\!\perp S^2$ when the data are iid $N(\mu, \sigma^2)$, but Zehna (1991) states that there is a flaw in the proof.

The Factorization Theorem was developed with increasing generality by Fisher, Neyman, and Halmos and Savage (1949).

Bahadur's Theorem is due to Bahadur (1958) and Lehmann and Scheffé (1950).

Basu's Theorem is due to Basu (1959). Also see Koehn and Thomas (1975) and Boos and Hughes-Oliver (1998). An interesting alternative method for proving independence between two statistics that works for some important examples is given in Datta and Sarker (2008).

Some techniques for showing whether a statistic is minimal sufficient are illustrated in Sampson and Spencer (1976).

4.5 Problems

PROBLEMS WITH AN ASTERISK * ARE ESPECIALLY USEFUL.

Refer to Chap. 10 for the pdf or pmf of the distributions in the problems below.

4.1. Let $X_1,\ldots,X_n$ be a random sample from a $N(\mu,\sigma^2)$ distribution, which is an exponential family. Show that the sample space of (T_1,T_2) contains an open subset of $\mathbb{R}^2$, if $n \geq 2$ but not if $n = 1$.

Hint: Show that if $n \geq 2$, then $T_1 = \sum_{i=1}^n X_i$ and $T_2 = \sum_{i=1}^n X_i^2$. Then $T_2 = aT_1^2 + b(X_1,\ldots,X_n)$ for some constant a where $b(X_1,\ldots,X_n) = \sum_{i=1}^n (X_i - \overline{X})^2 \in (0,\infty)$. So range$(T_1,T_2) = \{\ (t_1,t_2)|t_2 \geq at_1^2\ \}$. Find a. If $n = 1$ then $b(X_1) \equiv 0$ and the curve cannot contain an open two-dimensional rectangle.

4.2. Let $X_1,\ldots,X_n$ be iid exponential(λ) random variables. Use the Factorization Theorem to show that $T(\boldsymbol{X}) = \sum_{i=1}^n X_i$ is a sufficient statistic for λ.

4.3. Let $X_1,\ldots,X_n$ be iid from a regular exponential family with pdf

$$f(x|\boldsymbol{\eta}) = h(x)b(\boldsymbol{\eta})\exp\left[\sum_{i=1}^k \eta_i t_i(x)\right].$$

Let $\boldsymbol{T}(\boldsymbol{X}) = (T_1(\boldsymbol{X}),\ldots,T_k(\boldsymbol{X}))$ where $T_i(\boldsymbol{X}) = \sum_{j=1}^n t_i(X_j)$.

a) Use the Factorization Theorem to show that $\boldsymbol{T}(\boldsymbol{X})$ is a k-dimensional sufficient statistic for $\boldsymbol{\eta}$.
b) Use the Lehmann Scheffé LSM theorem to show that $\boldsymbol{T}(\boldsymbol{X})$ is a minimal sufficient statistic for $\boldsymbol{\eta}$.
(Hint: in a regular exponential family, if $\sum_{i=1}^k a_i\eta_i = c$ for all $\boldsymbol{\eta}$ in the natural parameter space for some fixed constants $a_1,\ldots,a_k$ and c, then $a_1 = \cdots = a_k = 0$.)

4.4. Let $X_1,\ldots,X_n$ be iid $N(\mu,\gamma_o^2\mu^2)$ random variables where $\gamma_o^2 > 0$ is **known** and $\mu > 0$.

a) Find a sufficient statistic for μ.
b) Show that $(\sum_{i=1}^n X_i, \sum_{i=1}^n X_i^2)$ is a minimal sufficient statistic.
c) Find $E_\mu[\sum_{i=1}^n X_i^2]$.
d) Find $E_\mu[(\sum_{i=1}^n X_i)^2]$.
e) Find

$$E_\mu\left[\frac{n+\gamma_o^2}{1+\gamma_o^2}\sum_{i=1}^n X_i^2 - \left(\sum_{i=1}^n X_i\right)^2\right].$$

(Hint: use c) and d).)
f) Is the minimal sufficient statistic given in b) complete? Explain.

4.5. If $X_1,\ldots,X_n$ are iid with $f(x|\theta) = \exp[-(x-\theta)]$for $x > \theta$, then the joint pdf can be written as

$$f(\boldsymbol{x}|\theta) = e^{n\theta}\exp\left(-\sum x_i\right) I[\theta < x_{(1)}].$$

By the Factorization Theorem, $\boldsymbol{T}(\boldsymbol{X}) = (\sum X_i, X_{(1)})$ is a sufficient statistic. Show that $R(\theta) = f(\boldsymbol{x}|\theta)/f(\boldsymbol{y}|\theta)$ can be constant even though $\boldsymbol{T}(\boldsymbol{x}) \neq \boldsymbol{T}(\boldsymbol{y})$. Hence the Lehmann–Scheffé Theorem does not imply that $\boldsymbol{T}(\boldsymbol{X})$ is a minimal sufficient statistic.

4.6. Find a complete minimal sufficient statistic if $Y_1,\ldots,Y_n$ are iid from the following 1P-REFs.

a) $Y \sim$ binomial (k,ρ) with k known.
b) $Y \sim$ exponential (λ).
c) $Y \sim$ gamma (ν,λ) with ν known.
d) $Y \sim$ geometric (ρ).
e) $Y \sim$ negative binomial (r,ρ) with r known.
f) $Y \sim$ normal (μ,σ^2) with σ^2 known.
g) $Y \sim$ normal (μ,σ^2) with μ known.
h) $Y \sim$ Poisson (θ).

4.7. Find a complete minimal sufficient statistic if $Y_1,\ldots,Y_n$ are iid from the following 1P-REFs.

a) $Y \sim$ Burr Type XII (ϕ,λ) with ϕ known.
b) $Y \sim$ chi(p,σ) with p known
c) $Y \sim$ double exponential (θ,λ) with θ known.
d) $Y \sim$ two-parameter exponential (θ,λ) with θ known.
e) $Y \sim$ generalized negative binomial (μ,κ) with κ known.
f) $Y \sim$ half normal (μ,σ^2) with μ known.
g) $Y \sim$ inverse Gaussian (θ,λ) with λ known.
h) $Y \sim$ inverted gamma (ν,λ) with ν known.
i) $Y \sim$ lognormal (μ,σ^2) with μ known.
j) $Y \sim$ lognormal (μ,σ^2) with σ^2 known.
k) $Y \sim$ Maxwell-Boltzmann (μ,σ) with μ known.
l) $Y \sim$ one-sided stable (σ).
m) $Y \sim$ Pareto (σ,λ) with σ known.
n) $Y \sim$ power (λ).
o) $Y \sim$ Rayleigh (μ,σ) with μ known.
p) $Y \sim$ Topp–Leone (ν).
q) $Y \sim$ truncated extreme value (λ).
r) $Y \sim$ Weibull (ϕ,λ) with ϕ known.

4.8. Find a complete minimal sufficient statistic $\boldsymbol{T}$ if $Y_1,\ldots,Y_n$ are iid from the following 2P-REFs.

a) The beta (δ, ν) distribution.
b) The chi (p, σ) distribution.
c) The gamma (ν, λ) distribution.
d) The lognormal (μ, σ^2) distribution.
e) The normal (μ, σ^2) distribution.

4.9. i) Show that each of the following families is a 1P-REF. ii) Find a complete minimal sufficient statistic if $Y_1, \ldots, Y_n$ are iid from the 1P-REF.

a) Let

$$f(y) = \frac{\log(\theta)}{\theta - 1}\theta^y$$

where $0 < y < 1$ and $\theta > 1$.
Comment:

$$F(y) = \frac{\theta^y - 1}{\theta - 1}$$

for $0 < y < 1$, and the mgf

$$m(t) = \frac{\log(\theta)}{\theta - 1} \frac{e^{(t+\log(\theta))} - 1}{t + \log(\theta)}.$$

b) Y has an inverse Weibull distribution.
c) Y has a Zipf distribution.

4.10. Suppose Y has a log-gamma distribution, $Y \sim \mathrm{LG}(\nu, \lambda)$.

i) Show the Y is a 2P-REF.
ii) If $Y_1, \ldots, Y_n$ are iid $\mathrm{LG}(\nu, \lambda)$, find a complete minimal sufficient statistic.
iii) Show $W = e^Y \sim$ gamma (ν, λ).

Problems from old quizzes and exams. Problems from old qualifying exams are marked with a Q since these problems take longer than quiz and exam problems.

4.11. Suppose that $X_1, \ldots, X_m; Y_1, \ldots, Y_n$ are iid $N(\mu, 1)$ random variables. Find a minimal sufficient statistic for μ.

4.12. Let $X_1, \ldots, X_n$ be iid from a uniform $U(\theta - 1, \theta + 2)$ distribution. Find a sufficient statistic for θ.

4.13. Let $Y_1, \ldots, Y_n$ be iid with a distribution that has pmf $P_\theta(X = x) = \theta(1-\theta)^{x-1}$, $x = 1, 2, \ldots$, where $0 < \theta < 1$. Find a minimal sufficient statistic for θ.

4.14. Let $Y_1, \ldots, Y_n$ be iid Poisson(λ) random variables. Find a minimal sufficient statistic for λ using the fact that the Poisson distribution is a regular exponential family (REF).

4.15. Suppose that $X_1,\ldots,X_n$ are iid from a REF with pdf (with respect to the natural parameterization)

$$f(x) = h(x)c^*(\boldsymbol{\eta})\exp\left[\sum_{i=1}^{4} \eta_i t_i(x)\right].$$

Assume $\dim(\Theta) = 4$. Find a complete minimal sufficient statistic $\boldsymbol{T}(\boldsymbol{X})$ in terms of n, t_1, t_2, t_3, and t_4.

4.16. Let X be a uniform $U(-\theta,\theta)$ random variable (sample size $n=1$). a) Find $E_\theta X$. b) Is $T(X) = X$ a complete sufficient statistic? c) Show that $|X| = \max(-X,X)$ is a minimal sufficient statistic.

4.17. A fact from mathematics is that if the polynomial $P(w) = a_n w^n + a_{n-1} w^{n-1} + \cdots + a_2 w^2 + a_1 w + a_0 \equiv 0$ for all w in a domain that includes an open interval, then $a_n = \cdots = a_1 = a_0 = 0$. Suppose that you are trying to use the Lehmann Scheffé (LSM) theorem to show that $(\sum X_i, \sum X_i^2)$ is a minimal sufficient statistic and that you have managed to show that

$$\frac{f(\boldsymbol{x}|\mu)}{f(\boldsymbol{y}|\mu)} \equiv c$$

iff

$$-\frac{1}{2\gamma_o^2\mu^2}\left[\sum x_i^2 - \sum y_i^2\right] + \frac{1}{\gamma_o^2\mu}\left[\sum x_i - \sum y_i\right] \equiv d \qquad (4.11)$$

for all $\mu > 0$. Parts a) and b) give two different ways to proceed.

a) Let $w = 1/\mu$ and assume that γ_o is known. Identify a_2, a_1 and a_0 and show that $a_i = 0$ implies that $(\sum X_i, \sum X_i^2)$ is a minimal sufficient statistic.
b) Let $\eta_1 = 1/\mu^2$ and $\eta_2 = 1/\mu$. Since (4.11) is a polynomial in $1/\mu$, can η_1 and η_2 satisfy a linearity constraint? If not, why is $(\sum X_i, \sum X_i^2)$ a minimal sufficient statistic?

4.18. Let $X_1,\ldots,X_n$ be iid Exponential(λ) random variables and $Y_1,\ldots,Y_m$ iid Exponential($\lambda/2$) random variables. Assume that the Y_i's and X_j's are independent. Show that the statistic $(\sum_{i=1}^n X_i, \sum_{i=1}^m Y_i)$ is not a complete sufficient statistic.

4.19. Let $X_1,\ldots,X_n$ be iid gamma(ν,λ) random variables. Find a complete, minimal sufficient statistic $(T_1(\boldsymbol{X}), T_2(\boldsymbol{X}))$. (Hint: recall a theorem for exponential families. The gamma pdf is (for x > 0)

$$f(x) = \frac{x^{\nu-1}e^{-x/\lambda}}{\lambda^\nu \Gamma(\nu)}.)$$

4.20. Let $X_1,\ldots,X_n$ be iid uniform($\theta-1,\theta+1$) random variables. The following expectations may be useful:

$$E_\theta \overline{X} = \theta, \ E_\theta X_{(1)} = 1 + \theta - 2\theta \frac{n}{n+1}, \ E_\theta X_{(n)} = 1 - \theta + 2\theta \frac{n}{n+1}.$$

a) Find a minimal sufficient statistic for θ.
b) Show whether the minimal sufficient statistic is complete or not.

4.21. Let $X_1, \ldots, X_n$ be independent identically distributed random variables with pdf

$$f(x) = \sqrt{\frac{\sigma}{2\pi x^3}} \exp\left(-\frac{\sigma}{2x}\right)$$

where x and σ are both positive. Find a sufficient statistic $T(\boldsymbol{X})$ for σ.

4.22. Suppose that $X_1, \ldots, X_n$ are iid beta(δ, ν) random variables. Find a minimal sufficient statistic for (δ, ν). Hint: write as a two-parameter REF.

4.23. Let $X_1, \ldots, X_n$ be iid from a distribution with pdf

$$f(x|\theta) = \theta x^{-2}, \ \ 0 < \theta \le x < \infty.$$

Find a sufficient statistic for θ.

4.24. Let $X_1, \ldots, X_n$ be iid with a distribution that has pdf

$$f(x) = \frac{x}{\sigma^2} \exp\left(\frac{-x}{2\sigma^2}\right)$$

for $x > 0$ and $\sigma^2 > 0$. Find a minimal sufficient statistic for σ^2 using the Lehmann–Scheffé Theorem.

4.25. Let $X_1, \ldots, X_n$ be iid exponential (λ) random variables. Find a minimal sufficient statistic for λ using the fact that the exponential distribution is a 1P-REF.

4.26. Suppose that $X_1, \ldots, X_n$ are iid $N(\mu, \sigma^2)$. Find a complete sufficient statistic for (μ, σ^2).

4.27^Q. Let X_1 and X_2 be iid Poisson (λ) random variables. Show that $T = X_1 + 2X_2$ is not a sufficient statistic for λ. (Hint: the Factorization Theorem uses the word *iff*. Alternatively, find a minimal sufficient statistic S and show that S is not a function of T.)

4.28^Q. Suppose that $X_1, \ldots, X_n$ are iid $N(\sigma, \sigma)$ where $\sigma > 0$.

a) Find a minimal sufficient statistic for σ.
b) Show that $(\overline{X}, S^2)$ is a sufficient statistic but is not a complete sufficient statistic for σ.

4.29. Let $X_1, \ldots, X_n$ be iid binomial$(k = 1, \theta)$ random variables and $Y_1, \ldots, Y_m$ iid binomial$(k = 1, \theta/2)$ random variables. Assume that the Y_i's and X_j's are independent. Show that the statistic $(\sum_{i=1}^n X_i, \sum_{i=1}^m Y_i)$ is not a complete sufficient statistic.

4.30. Suppose that $X_1,\ldots,X_n$ are iid Poisson(λ) where $\lambda > 0$. Show that $(\overline{X}, S^2)$ is not a complete sufficient statistic for λ.

4.31^Q. Let $X_1,\ldots,X_n$ be iid beta(θ,θ). (Hence $\delta = \nu = \theta$.)

a) Find a minimal sufficient statistic for θ.
b) Is the statistic found in a) complete? (prove or disprove)

4.32^Q. Let $X_1,\ldots,X_n$ be independent identically distributed random variables with probability mass function

$$f(x) = P(X = x) = \frac{1}{x^\nu \zeta(\nu)}$$

where $\nu > 1$ and $x = 1,2,3,\ldots$. Here the zeta function

$$\zeta(\nu) = \sum_{x=1}^{\infty} \frac{1}{x^\nu}$$

for $\nu > 1$.

a) Find a minimal sufficient statistic for ν.
b) Is the statistic found in a) complete? (prove or disprove)
c) Give an example of a sufficient statistic that is strictly not minimal.

4.33. Let $X_1,\ldots,X_n$ be a random sample from a half normal distribution with pdf

$$f(x) = \frac{2}{\sqrt{2\pi}\,\sigma} \exp\left(\frac{-(x-\mu)^2}{2\sigma^2}\right)$$

where $\sigma > 0$ and $x > \mu$ and μ is real.
Find a sufficient statistic $\boldsymbol{T} = (T_1, T_2, \ldots, T_k)$ for (μ, σ) with dimension $k \le 3$.

4.34. Let $X_{(1)} = \min_{1 \le i \le n} X_i$. If $X_1,\ldots,X_n$ are iid exponential(1) random variables, find $E(X_{(1)})$.

4.35. Let $X_{(n)} = \max_{1 \le i \le n} X_i$. If $X_1,\ldots,X_n$ are iid uniform(0,1) random variables, find $E(X_{(n)})$.

4.36^Q. Let $X_1,\ldots,X_n$ be iid uniform($\theta, \theta+1$) random variables where θ is real.

a) Find a minimal sufficient statistic for θ.
b) Show whether the minimal sufficient statistic is complete or not.

4.37^Q. Let $Y_1,\ldots,Y_n$ be iid from a distribution with pdf

$$f(y) = 2\,\tau\, y\, e^{-y^2}\,(1 - e^{-y^2})^{\tau - 1}$$

for $y > 0$ and $f(y) = 0$ for $y \le 0$ where $\tau > 0$.

a) Find a minimal sufficient statistic for τ.
b) Is the statistic found in a) complete? Prove or disprove.

Chapter 5
Point Estimation I

A point estimator gives a single value as an estimate of a parameter. For example, $\overline{Y} = 10.54$ is a point estimate of the population mean μ. An interval estimator gives a range (L_n, U_n) of reasonable values for the parameter. Confidence intervals, studied in Chap. 9, are interval estimators. The most widely used point estimators are the maximum likelihood estimators, and method of moments estimators are also widely used. These two methods should be familiar to the reader. Uniformly minimum variance unbiased estimators are discussed in Chap. 6.

5.1 Maximum Likelihood Estimators

Definition 5.1. Let $f(\boldsymbol{y}|\boldsymbol{\theta})$ be the pmf or pdf of a sample $\boldsymbol{Y}$ with parameter space Θ. If $\boldsymbol{Y} = \boldsymbol{y}$ is observed, then the **likelihood function** is $L(\boldsymbol{\theta}) \equiv L(\boldsymbol{\theta}|\boldsymbol{y}) = f(\boldsymbol{y}|\boldsymbol{\theta})$. For each sample point $\boldsymbol{y} = (y_1, \ldots, y_n)$, let $\hat{\boldsymbol{\theta}}(\boldsymbol{y}) \in \Theta$ be a parameter value at which $L(\boldsymbol{\theta}) \equiv L(\boldsymbol{\theta}|\boldsymbol{y})$ attains its maximum as a function of $\boldsymbol{\theta}$ with $\boldsymbol{y}$ held fixed. Then a maximum likelihood estimator (**MLE**) of the parameter $\boldsymbol{\theta}$ based on the sample $\boldsymbol{Y}$ is $\hat{\boldsymbol{\theta}}(\boldsymbol{Y})$.

The following remarks are important.

I) It is crucial to observe that the likelihood function is a function of $\boldsymbol{\theta}$ (and that $y_1, \ldots, y_n$ act as fixed constants). Note that the pdf or pmf $f(\boldsymbol{y}|\boldsymbol{\theta})$ is a function of n variables while $L(\boldsymbol{\theta})$ is a function of k variables if $\boldsymbol{\theta}$ is a $1 \times k$ vector. Often $k = 1$ or $k = 2$ while n could be in the hundreds or thousands.
II) If $Y_1, \ldots, Y_n$ is an independent sample from a population with pdf or pmf $f(y|\boldsymbol{\theta})$, then the likelihood function

$$L(\boldsymbol{\theta}) \equiv L(\boldsymbol{\theta}|y_1, \ldots, y_n) = \prod_{i=1}^{n} f(y_i|\boldsymbol{\theta}). \tag{5.1}$$

D.J. Olive, *Statistical Theory and Inference*, DOI 10.1007/978-3-319-04972-4_5,

$$L(\boldsymbol{\theta}) = \prod_{i=1}^{n} f_i(y_i|\boldsymbol{\theta})$$

if the Y_i are independent but have different pdfs or pmfs.

III) If the MLE $\hat{\boldsymbol{\theta}}$ exists, then $\hat{\boldsymbol{\theta}} \in \Theta$. Hence if $\hat{\boldsymbol{\theta}}$ is not in the parameter space Θ, then $\hat{\boldsymbol{\theta}}$ is not the MLE of $\boldsymbol{\theta}$.

IV) If the MLE is unique, then the MLE is a function of the minimal sufficient statistic. See Levy (1985) and Moore (1971). This fact is useful since exponential families tend to have a tractable log likelihood and an easily found minimal sufficient statistic.

Theorem 5.1: Invariance Principle. If $\hat{\boldsymbol{\theta}}$ is the MLE of $\boldsymbol{\theta}$, then $h(\hat{\boldsymbol{\theta}})$ is the MLE of $h(\boldsymbol{\theta})$ where h is a function with domain Θ.

This theorem will be proved in Sect. 5.4. Really just need $\Theta \in \text{dom}(h)$ so $h(\hat{\boldsymbol{\theta}})$ is well defined: can't have $\log(-7.89)$ or $\sqrt{-1.57}$.

There are **four commonly used techniques** for finding the MLE.

- Potential candidates can be found by differentiating $\log L(\boldsymbol{\theta})$, the log likelihood.
- Potential candidates can be found by differentiating the likelihood $L(\boldsymbol{\theta})$.
- The MLE can sometimes be found by direct maximization of the likelihood $L(\boldsymbol{\theta})$.
- **Invariance Principle**: If $\hat{\boldsymbol{\theta}}$ is the MLE of $\boldsymbol{\theta}$, then $h(\hat{\boldsymbol{\theta}})$ is the MLE of $h(\boldsymbol{\theta})$.

The one-parameter case can often be solved by hand with the following technique. To show that $\hat{\theta}$ is the MLE of θ is equivalent to showing that $\hat{\theta}$ is the global maximizer of $\log L(\theta)$ on Θ where Θ is an interval with endpoints a and b, not necessarily finite. Suppose that $\log L(\theta)$ is continuous on Θ. Show that $\log L(\theta)$ is differentiable on (a,b). Then show that $\hat{\theta}$ is the unique solution to the equation $\frac{d}{d\theta}\log L(\theta) = 0$ and that the second derivative evaluated at $\hat{\theta}$ is negative: $\left.\frac{d^2}{d\theta^2}\log L(\theta)\right|_{\hat{\theta}} < 0$. See Remark 5.1V below.

Remark 5.1. From calculus, recall the following facts.

I) If the function h is continuous on an interval $[a,b]$, then both the max and min of h exist. Suppose that h is continuous on an interval $[a,b]$ and differentiable on (a,b). Solve $h'(\theta) \equiv 0$ and find the places where $h'(\theta)$ does not exist. These values are the **critical points**. Evaluate h at a, b, and the critical points. One of these values will be the min and one the max.

II) Assume h is continuous. Then h has a local max at the critical point θ_o if h is increasing for $\theta < \theta_o$ in a neighborhood of θ_o and if h is decreasing for $\theta > \theta_o$ in a neighborhood of θ_o (and θ_o is a global max if you can remove the phrase "in a neighborhood of θ_o"). The first derivative test is often used: if h is continuous at θ_o and if there exists some $\delta > 0$ such that $h'(\theta) > 0$ for all θ in $(\theta_o - \delta, \theta_o)$ and $h'(\theta) < 0$ for all θ in $(\theta_o, \theta_o + \delta)$, then h has a local max at θ_o.

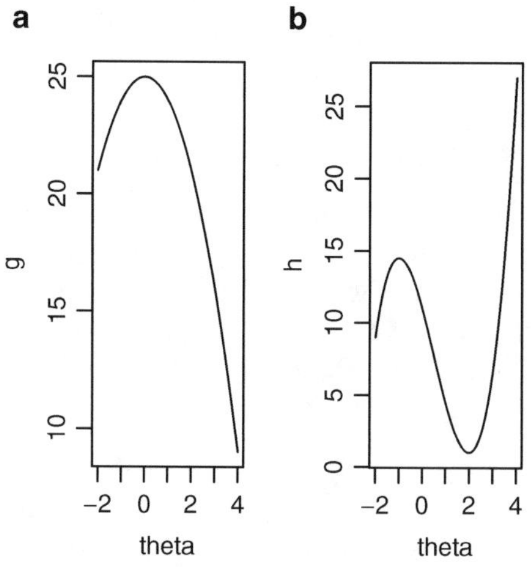

Fig. 5.1 The local max in (**a**) is a global max, but the local max at $\theta = -1$ in (**b**) is not the global max

III) If h is strictly concave ($\frac{d^2}{d\theta^2}h(\theta) < 0$ for all $\theta \in \Theta$), then any local max of h is a global max.

IV) Suppose $h'(\theta_o) = 0$. The second derivative test states that if $\frac{d^2}{d\theta^2}h(\theta_o) < 0$, then h has a local max at θ_o.

V) If $h(\theta)$ is a continuous function on an interval with endpoints $a < b$ (not necessarily finite), differentiable on (a,b) and if the **critical point is unique**, then the critical point is a **global maximum** if it is a local maximum. To see this claim, note that if the critical point is not the global max, then there would be a local minimum and the critical point would not be unique. Also see Casella and Berger (2002, p. 317). Let $a = -2$ and $b = 4$. In Fig. 5.1a, the critical point for $g(\theta) = -\theta^2 + 25$ is at $\theta = 0$, is unique, and is both a local and global maximum. In Fig. 5.1b, $h(\theta) = \theta^3 - 1.5\theta^2 - 6\theta + 11$, the critical point $\theta = -1$ is not unique and is a local max but not a global max.

VI) If h is strictly convex ($\frac{d^2}{d\theta^2}h(\theta) > 0$ for all $\theta \in \Theta$), then any local min of h is a global min. If $h'(\theta_o) = 0$, then the second derivative test states that if $\frac{d^2}{d\theta^2}h(\theta_o) > 0$, then θ_o is a local min.

VII) If $h(\theta)$ is a continuous function on an interval with endpoints $a < b$ (not necessarily finite), differentiable on (a,b) and if the **critical point is unique**, then the critical point is a **global minimum** if it is a local minimum. To see this claim, note that if the critical point is not the global min, then there would be a local maximum and the critical point would not be unique.

Tips: a) $\exp(a) = e^a$ and $\log(y) = \ln(y) = \log_e(y)$ is the **natural logarithm**.
b) $\log(a^b) = b\log(a)$ and $\log(e^b) = b$.
c) $\log(\prod_{i=1}^n a_i) = \sum_{i=1}^n \log(a_i)$.
d) $\log L(\theta) = \log(\prod_{i=1}^n f(y_i|\theta)) = \sum_{i=1}^n \log(f(y_i|\theta))$.

e) If t is a differentiable function and $t(\theta) \neq 0$, then $\frac{d}{d\theta}\log(|t(\theta)|) = \frac{t'(\theta)}{t(\theta)}$ where $t'(\theta) = \frac{d}{d\theta}t(\theta)$. In particular, $\frac{d}{d\theta}\log(\theta) = 1/\theta$.
f) Any additive term that does not depend on θ is treated as a constant with respect to θ and hence has derivative 0 with respect to θ.

Showing that $\hat{\boldsymbol{\theta}}$ is the global maximizer of $\log(L(\boldsymbol{\theta}))$ is much more difficult in the multiparameter case. To show that there is a local max at $\hat{\boldsymbol{\theta}}$ often involves using a Hessian matrix of second derivatives. Calculations involving the Hessian matrix are often very difficult. Often there is no closed form solution for the MLE and a computer needs to be used. For hand calculations, Remark 5.2 and Theorem 5.2 can often be used to avoid using the Hessian matrix.

Definition 5.2. Let the data be $Y_1,\ldots,Y_n$ and suppose that the parameter $\boldsymbol{\theta}$ has components $(\boldsymbol{\theta}_1,\ldots,\boldsymbol{\theta}_k)$. Then $\hat{\boldsymbol{\theta}}_i$ will be called the MLE of $\boldsymbol{\theta}_i$. Without loss of generality, assume that $\boldsymbol{\theta} = (\boldsymbol{\theta}_1,\boldsymbol{\theta}_2)$, that the MLE of $\boldsymbol{\theta}$ is $(\hat{\boldsymbol{\theta}}_1,\hat{\boldsymbol{\theta}}_2)$ and that $\hat{\boldsymbol{\theta}}_2$ is known. The **profile likelihood function** is $L_P(\boldsymbol{\theta}_1) = L(\boldsymbol{\theta}_1,\hat{\boldsymbol{\theta}}_2(\boldsymbol{y}))$ with domain $\{\boldsymbol{\theta}_1 : (\boldsymbol{\theta}_1,\hat{\boldsymbol{\theta}}_2) \in \Theta\}$.

Remark 5.2. Since $L(\boldsymbol{\theta}_1,\boldsymbol{\theta}_2)$ is maximized over Θ by $(\hat{\boldsymbol{\theta}}_1,\hat{\boldsymbol{\theta}}_2)$, the maximizer of the profile likelihood function and of the log profile likelihood function is $\hat{\boldsymbol{\theta}}_1$. The log profile likelihood function can often be maximized using calculus if $\boldsymbol{\theta}_1 = \theta_1$ is a scalar.

Theorem 5.2: Existence of the MLE for a REF (Barndorff–Nielsen 1982): Assume that the natural parameterization of the kP-REF (k-parameter regular exponential family) is used so that Ω is an open k-dimensional convex set (usually an open interval or cross product of open intervals). Then the log likelihood function $\log[L(\boldsymbol{\eta})]$ is a strictly concave function of $\boldsymbol{\eta}$. Hence if $\hat{\boldsymbol{\eta}}$ is a critical point of $\log[\mathrm{L}(\boldsymbol{\eta})]$ and if $\hat{\boldsymbol{\eta}} \in \Omega$, then $\hat{\boldsymbol{\eta}}$ is the unique MLE of $\boldsymbol{\eta}$. Hence the Hessian matrix of second derivatives does not need to be checked! Moreover, the critical point $\hat{\boldsymbol{\eta}}$ is the solution to the equations $T_j(\boldsymbol{y}) = \sum_{m=1}^n t_j(y_m) \stackrel{\text{set}}{=} \sum_{m=1}^n E[t_j(Y_m)] = E[T_j(\boldsymbol{Y})]$ for $j = 1,\ldots,k$.

Proof Sketch. The proof needs several results from nonlinear programming, which is also known as optimization theory. Suppose that $h(\boldsymbol{\eta})$ is a function with continuous first and second derivatives on an open convex set Ω. Suppose the $h(\boldsymbol{\eta})$ has Hessian (or Jacobian) matrix $\boldsymbol{H} \equiv \boldsymbol{H}(\boldsymbol{\eta})$ with ij entry

$$\boldsymbol{H}_{ij}(\boldsymbol{\eta}) = \frac{\partial^2}{\partial\eta_i \partial\eta_j} h(\boldsymbol{\eta}).$$

Let the critical point $\hat{\boldsymbol{\eta}}$ be the solution to

$$\frac{\partial}{\partial\eta_i} h(\boldsymbol{\eta}) \stackrel{\text{set}}{=} 0$$

for $i = 1, \ldots, k$. If $\boldsymbol{H}(\boldsymbol{\eta})$ is positive definite on Ω, then h is strictly convex and $\hat{\boldsymbol{\eta}}$ is a global minimizer of h. Hence $-h$ is strictly concave and $\hat{\boldsymbol{\eta}}$ is a global maximizer of h. See Peressini et al. (1988, pp. 10–13, 54).

Suppose Y comes from a k-parameter exponential family. Then the distribution of $(t_1(Y), \ldots, t_k(Y))$ is a k-parameter exponential family by Theorem 3.5 that has a covariance matrix $\boldsymbol{\Sigma}_t$ by Theorem 3.4. An important fact is that $\boldsymbol{\Sigma}_t$ is positive definite.

Let $Y_1, \ldots, Y_n$ be iid from a kP-REF with pdf or pmf

$$f(y|\boldsymbol{\eta}) = h(y)b(\boldsymbol{\eta})\exp\left[\sum_{i=1}^{k} \eta_i t_i(y)\right].$$

As stated above Definition 3.3, Ω is an open convex set. Now the likelihood function $L(\boldsymbol{\eta}) = \prod_{m=1}^{n} f(y_m|\boldsymbol{\eta}) =$

$$\left[\prod_{m=1}^{n} h(y_m)\right] [b(\boldsymbol{\eta})]^n \exp\left[\eta_1 \sum_{m=1}^{n} t_1(y_m) + \cdots + \eta_k \sum_{m=1}^{n} t_k(y_m)\right].$$

Hence the log likelihood function

$$\log[L(\boldsymbol{\eta})] = d + n\log(b(\boldsymbol{\eta})) + \sum_{j=1}^{k} \eta_j T_j$$

where $T_j \equiv T_j(\boldsymbol{y}) = \sum_{m=1}^{n} t_j(y_m)$.

Now

$$\frac{\partial}{\partial \eta_i} \log[L(\boldsymbol{\eta})] = n\frac{\partial}{\partial \eta_i}\log(b(\boldsymbol{\eta})) + T_i = -nE[t_i(Y)] + T_i =$$

$$-\sum_{m=1}^{n} E[t_i(Y_m)] + \sum_{m=1}^{n} t_i(y_m)$$

for $i = 1, \ldots, k$ by Theorem 3.3a. Since the critical point $\hat{\boldsymbol{\eta}}$ is the solution to the k equations $\frac{\partial}{\partial \eta_i}\log[L(\boldsymbol{\eta})] \stackrel{\text{set}}{=} 0$, the critical point $\hat{\boldsymbol{\eta}}$ is also a solution to $T_i(\boldsymbol{y}) = \sum_{m=1}^{n} t_i(y_m) \stackrel{\text{set}}{=} \sum_{m=1}^{n} E[t_i(Y_m)] = E[T_i(\boldsymbol{Y})]$ for $i = 1, \ldots, k$.

Now

$$\frac{\partial^2}{\partial \eta_i \partial \eta_j} \log[L(\boldsymbol{\eta})] = n\frac{\partial^2}{\partial \eta_i \partial \eta_j}\log(b(\boldsymbol{\eta})) = -n\text{Cov}(t_i(Y), t_j(Y))$$

by Theorem 3.3b, and the covariance matrix $\boldsymbol{\Sigma}_t$ with ij entry $\text{Cov}(t_i(Y), t_j(Y))$ is positive definite. Thus $h(\boldsymbol{\eta}) = -\log[L(\boldsymbol{\eta})]$ has a positive definite Hessian and is strictly convex with global min $\hat{\boldsymbol{\eta}}$ if $\hat{\boldsymbol{\eta}} \in \Omega$. So $\log[L(\boldsymbol{\eta})]$ is strictly concave with global max $\hat{\boldsymbol{\eta}}$ if $\hat{\boldsymbol{\eta}} \in \Omega$.

Remark 5.3. For k-parameter exponential families with $k > 1$, it is usually easier to verify that the family is regular than to calculate the Hessian matrix. For 1P-REFs, check that the critical point is a global maximum using standard calculus techniques such as calculating the second derivative of the log likelihood $\log L(\boldsymbol{\theta})$. For a 1P-REF (one-parameter regular exponential family), verifying that the family is regular is often more difficult than using calculus. Also, often the MLE is desired for a parameter space Θ_U which is not an open set (e.g., for $\Theta_U = [0,1]$ instead of $\Theta = (0,1)$).

Remark 5.4. (Barndorff–Nielsen 1982). The MLE does not exist if $\hat{\boldsymbol{\eta}}$ is not in Ω, an event that occurs with positive probability for discrete distributions. If $\boldsymbol{T}$ is the complete sufficient statistic and C is the closed convex hull of the support of $\boldsymbol{T}$, then the MLE exists iff $\boldsymbol{T} \in \text{int}\, C$ where int C is the interior of C.

Remark 5.5. Suppose $L(\boldsymbol{\theta}) = f(\boldsymbol{y}|\boldsymbol{\theta}) = g(\boldsymbol{y}|\boldsymbol{\theta})I(\boldsymbol{y} \in \mathscr{Y}_{\boldsymbol{\theta}}) = g(\boldsymbol{y}|\boldsymbol{\theta})I(\boldsymbol{\theta} \in A_{\boldsymbol{y}}) = g(\boldsymbol{y}|\boldsymbol{\theta})I(\boldsymbol{\theta} \in A_{\boldsymbol{y}}) + 0I(\boldsymbol{\theta} \in A^c_{\boldsymbol{y}})$ where $A^c_{\boldsymbol{y}}$ is the complement of $A_{\boldsymbol{y}}$. Then $\log(L(\boldsymbol{\theta})) = \log[g(\boldsymbol{y}|\boldsymbol{\theta})]I(\boldsymbol{\theta} \in A_{\boldsymbol{y}}) + (-\infty)I(\boldsymbol{\theta} \in A^c_{\boldsymbol{y}})$. Neither $L(\boldsymbol{\theta})$ nor $\log(L(\boldsymbol{\theta}))$ is maximized for $\boldsymbol{\theta} \in A^c_{\boldsymbol{y}}$, and for $\boldsymbol{\theta} \in A_{\boldsymbol{y}}$, the log likelihood $\log(L(\boldsymbol{\theta})) = \log[g(\boldsymbol{y}|\boldsymbol{\theta})]I(\boldsymbol{y} \in \mathscr{Y}_{\boldsymbol{\theta}}) = \log[g(\boldsymbol{y}|\boldsymbol{\theta})]I(\boldsymbol{\theta} \in A_{\boldsymbol{y}})$. Thus if $L(\boldsymbol{\theta}) = g(\boldsymbol{y}|\boldsymbol{\theta})I(\boldsymbol{y} \in \mathscr{Y}_{\boldsymbol{\theta}})$, **do not do anything to the indicator** when finding $\log(L(\boldsymbol{\theta}))$, but only consider values of $\boldsymbol{\theta}$ for which the indicator is equal to one when maximizing $\log(L(\boldsymbol{\theta}))$.

Remark 5.6. As illustrated in the following examples, the second derivative is evaluated at $\hat{\theta}(\boldsymbol{y})$. The MLE is a statistic and $T_n(\boldsymbol{y}) = \hat{\theta}(\boldsymbol{y})$ is the observed value of the MLE $T_n(\boldsymbol{Y}) = \hat{\theta}(\boldsymbol{Y})$. Often $\boldsymbol{y}$ and $\boldsymbol{Y}$ are suppressed. Hence in the following example, $\hat{\theta} = \overline{y}$ is the observed value of the MLE, while $\hat{\theta} = \overline{Y}$ is the MLE.

Example 5.1. Suppose that $Y_1, \ldots, Y_n$ are iid Poisson (θ). This distribution is a 1P-REF with $\Theta = (0, \infty)$. The likelihood

$$L(\theta) = c\, e^{-n\theta} \exp[\log(\theta) \sum y_i]$$

where the constant c does not depend on θ, and the log likelihood

$$\log(L(\theta)) = d - n\theta + \log(\theta) \sum y_i$$

where $d = \log(c)$ does not depend on θ. Hence

$$\frac{d}{d\theta} \log(L(\theta)) = -n + \frac{1}{\theta} \sum y_i \stackrel{\text{set}}{=} 0,$$

or $\sum y_i = n\theta$, or

$$\hat{\theta} = \overline{y}.$$

Notice that $\hat{\theta}$ is the unique solution and

$$\frac{d^2}{d\theta^2} \log(L(\theta)) = \frac{-\sum y_i}{\theta^2} < 0$$

unless $\sum y_i = 0$. Hence for $\sum y_i > 0$ the log likelihood is strictly concave and $\overline{Y}$ is the MLE of θ. The MLE does not exist if $\sum_{i=1}^n Y_i = 0$ since 0 is not in Θ.

Now suppose that $\Theta = [0, \infty)$. This family is not an exponential family since the same formula for the pmf needs to hold for all values of $\theta \in \Theta$ and 0^0 is not defined. Notice that

$$f(y|\theta) = \frac{e^{-\theta}\theta^y}{y!} I[\theta > 0] + 1\, I[\theta = 0, y = 0].$$

Now

$$I_A(\theta) I_B(\theta) = I_{A \cap B}(\theta)$$

and $I_\emptyset(\theta) = 0$ for all θ. Hence the likelihood

$$L(\theta) = e^{-n\theta} \exp\left[\log(\theta) \sum_{i=1}^n y_i\right] \frac{1}{\prod_{i=1}^n y_i!} I[\theta > 0] + 1\, I\left[\theta = 0, \sum_{i=1}^n y_i = 0\right].$$

If $\sum y_i \neq 0$, then $\overline{y}$ maximizes $L(\theta)$ by the work above. If $\sum y_i = 0$, then $L(\theta) = e^{-n\theta} I(\theta > 0) + I(\theta = 0) = e^{-n\theta} I(\theta \geq 0)$ which is maximized by $\theta = 0 = \overline{y}$. Hence $\overline{Y}$ is the MLE of θ if $\Theta = [0, \infty)$.

By invariance, $t(\overline{Y})$ is the MLE of $t(\theta)$. Hence $(\overline{Y})^2$ is the MLE of θ^2. $\sin(\overline{Y})$ is the MLE of $\sin(\theta)$, etc.

Example 5.2. Suppose that $Y_1, \ldots, Y_n$ are iid $N(\mu, \sigma^2)$ where $\sigma^2 > 0$ and $\mu \in \mathbb{R} = (-\infty, \infty)$. Then

$$L(\mu, \sigma^2) = \left(\frac{1}{\sqrt{2\pi}}\right)^n \frac{1}{(\sigma^2)^{n/2}} \exp\left[\frac{-1}{2\sigma^2} \sum_{i=1}^n (y_i - \mu)^2\right].$$

Notice that

$$\frac{d}{d\mu} \sum_{i=1}^n (y_i - \mu)^2 = \sum_{i=1}^n -2(y_i - \mu) \stackrel{\text{set}}{=} 0$$

or $\sum_{i=1}^n y_i = n\mu$ or $\hat{\mu} = \overline{y}$. Since $\hat{\mu}$ is the unique solution and

$$\frac{d^2}{d\mu^2} \sum_{i=1}^n (y_i - \mu)^2 = 2n > 0,$$

$\hat{\mu} = \overline{y}$ is the minimizer of $h(\mu) = \sum_{i=1}^n (y_i - \mu)^2$. Hence $\overline{y}$ is the maximizer of

$$\exp\left[\frac{-1}{2\sigma^2} \sum_{i=1}^n (y_i - \mu)^2\right]$$

regardless of the value of $\sigma^2 > 0$. Hence $\hat{\mu} = \overline{Y}$ is the MLE of μ and the MLE of σ^2 can be found by maximizing the profile likelihood

$$L_P(\sigma^2) = L(\hat{\mu}(y), \sigma^2) = \left(\frac{1}{\sqrt{2\pi}}\right)^n \frac{1}{(\sigma^2)^{n/2}} \exp\left[\frac{-1}{2\sigma^2} \sum_{i=1}^n (y_i - \overline{y})^2\right].$$

Writing $\tau = \sigma^2$ often helps prevent calculus errors. Then

$$\log(L_p(\tau)) = d - \frac{n}{2}\log(\tau) + \frac{-1}{2\tau}\sum_{i=1}^n (y_i - \overline{y})^2$$

where the constant d does not depend on τ. Hence

$$\frac{d}{d\tau}\log(L_p(\tau)) = \frac{-n}{2}\frac{1}{\tau} + \frac{1}{2\tau^2}\sum_{i=1}^n (y_i - \overline{y})^2 \stackrel{\text{set}}{=} 0,$$

or

$$n\tau = \sum_{i=1}^n (y_i - \overline{y})^2$$

or

$$\hat{\tau} = \frac{1}{n}\sum_{i=1}^n (y_i - \overline{y})^2$$

and the solution $\hat{\tau}$ is the unique critical point. Note that

$$\frac{d^2}{d\tau^2}\log(L_P(\tau)) = \frac{n}{2(\tau)^2} - \left.\frac{\sum(y_i - \overline{y})^2}{(\tau)^3}\right|_{\tau=\hat{\tau}} = \frac{n}{2(\hat{\tau})^2} - \frac{n\hat{\tau}}{(\hat{\tau})^3}\frac{2}{2}$$

$$= \frac{-n}{2(\hat{\tau})^2} < 0.$$

Hence $\hat{\sigma}^2 = \hat{\tau} = \frac{1}{n}\sum_{i=1}^n (Y_i - \overline{Y})^2$ is the MLE of σ^2 by Remark 5.1 V). Thus $(\overline{Y}, \frac{1}{n}\sum_{i=1}^n (Y_i - \overline{Y})^2)$ is the MLE of (μ, σ^2).

Example 5.3. Following Pewsey (2002), suppose that $Y_1, \ldots, Y_n$ are iid HN(μ, σ^2) where μ and σ^2 are both unknown. Let the ith order statistic $Y_{(i)} \equiv Y_{i:n}$. Then the likelihood

$$L(\mu, \sigma^2) = cI[y_{1:n} \geq \mu]\frac{1}{\sigma^n}\exp\left[\left(\frac{-1}{2\sigma^2}\right)\sum(y_i - \mu)^2\right].$$

For any fixed $\sigma^2 > 0$, this likelihood is maximized by making $\sum(y_i - \mu)^2$ as small as possible subject to the constraint $y_{1:n} \geq \mu$. Notice that for any $\mu_o < y_{1:n}$, the terms $(y_i - y_{1:n})^2 < (y_i - \mu_o)^2$. Hence the MLE of μ is

$$\hat{\mu} = Y_{1:n}$$

and the MLE of σ^2 is found by maximizing the log profile likelihood

$$\log(L_P(\sigma^2)) = \log(L(y_{1:n}, \sigma^2)) = d - \frac{n}{2}\log(\sigma^2) - \frac{1}{2\sigma^2}\sum(y_i - y_{1:n})^2,$$

and

$$\frac{d}{d(\sigma^2)}\log(L(y_{1:n}, \sigma^2)) = \frac{-n}{2(\sigma^2)} + \frac{1}{2(\sigma^2)^2}\sum(y_i - y_{1:n})^2 \stackrel{\text{set}}{=} 0.$$

Or $\sum(y_i - y_{1:n})^2 = n\sigma^2$. So

$$\hat{\sigma}^2 \equiv w_n = \frac{1}{n}\sum(y_i - y_{1:n})^2.$$

Since the solution $\hat{\sigma}^2$ is unique and

$$\frac{d^2}{d(\sigma^2)^2}\log(L(y_{1:n},\sigma^2)) =$$

$$\frac{n}{2(\sigma^2)^2} - \frac{\sum(y_i-\mu)^2}{(\sigma^2)^3})\Big|_{\sigma^2=\hat{\sigma}^2} = \frac{n}{2(\hat{\sigma}^2)^2} - \frac{n\hat{\sigma}^2}{(\hat{\sigma^2})^3}\frac{2}{2} = \frac{-n}{2\hat{\sigma}^2} < 0,$$

$(\hat{\mu},\hat{\sigma}^2) = (Y_{1:n}, W_n)$ is MLE of (μ,σ^2).

Example 5.4. Suppose that the random vectors $\boldsymbol{X}_1,\ldots,\boldsymbol{X}_n$ are iid from a multivariate normal $N_p(\boldsymbol{\mu},\boldsymbol{\Sigma})$ distribution where $\boldsymbol{\Sigma}$ is a positive definite matrix. To find the MLE of $(\boldsymbol{\mu},\boldsymbol{\Sigma})$ we will use three results proved in Anderson (1984, p. 62).

$$\text{i) } \sum_{i=1}^n (\boldsymbol{x}_i-\boldsymbol{\mu})^T\boldsymbol{\Sigma}^{-1}(\boldsymbol{x}_i-\boldsymbol{\mu}) = tr(\boldsymbol{\Sigma}^{-1}\boldsymbol{A}) + n(\overline{\boldsymbol{x}}-\boldsymbol{\mu})^T\boldsymbol{\Sigma}^{-1}(\overline{\boldsymbol{x}}-\boldsymbol{\mu})$$

where

$$\boldsymbol{A} = \sum_{i=1}^n (\boldsymbol{x}_i-\overline{\boldsymbol{x}})(\boldsymbol{x}_i-\overline{\boldsymbol{x}})^T.$$

ii) Let $\boldsymbol{C}$ and $\boldsymbol{D}$ be positive definite matrices. Then $\boldsymbol{C} = \frac{1}{n}\boldsymbol{D}$ maximizes

$$h(\boldsymbol{C}) = -n\log(|\boldsymbol{C}|) - tr(\boldsymbol{C}^{-1}\boldsymbol{D})$$

with respect to positive definite matrices.
iii) Since $\boldsymbol{\Sigma}^{-1}$ is positive definite, $(\overline{\boldsymbol{x}}-\boldsymbol{\mu})^T\boldsymbol{\Sigma}^{-1}(\overline{\boldsymbol{x}}-\boldsymbol{\mu}) \geq 0$ as a function of $\boldsymbol{\mu}$ with equality iff $\boldsymbol{\mu} = \overline{\boldsymbol{x}}$.

Since

$$f(\boldsymbol{x}|\boldsymbol{\mu},\boldsymbol{\Sigma}) = \frac{1}{(2\pi)^{p/2}|\boldsymbol{\Sigma}|^{1/2}}\exp\left[-\frac{1}{2}(\boldsymbol{x}-\boldsymbol{\mu})^T\boldsymbol{\Sigma}^{-1}(\boldsymbol{x}-\boldsymbol{\mu})\right],$$

the likelihood function

$$L(\boldsymbol{\mu},\boldsymbol{\Sigma}) = \prod_{i=1}^n f(\boldsymbol{x}_i|\boldsymbol{\mu},\boldsymbol{\Sigma})$$

$$= \frac{1}{(2\pi)^{np/2}|\boldsymbol{\Sigma}|^{n/2}}\exp\left[-\frac{1}{2}\sum_{i=1}^n(\boldsymbol{x}_i-\boldsymbol{\mu})^T\boldsymbol{\Sigma}^{-1}(\boldsymbol{x}_i-\boldsymbol{\mu})\right],$$

and the log likelihood $\log(L(\boldsymbol{\mu}, \boldsymbol{\Sigma})) =$

$$-\frac{np}{2}\log(2\pi) - \frac{n}{2}\log(|\boldsymbol{\Sigma}|) - \frac{1}{2}\sum_{i=1}^{n}(\boldsymbol{x}_i - \boldsymbol{\mu})^T \boldsymbol{\Sigma}^{-1}(\boldsymbol{x}_i - \boldsymbol{\mu})$$

$$= -\frac{np}{2}\log(2\pi) - \frac{n}{2}\log(|\boldsymbol{\Sigma}|) - \frac{1}{2}tr(\boldsymbol{\Sigma}^{-1}\boldsymbol{A}) - \frac{n}{2}(\overline{\boldsymbol{x}} - \boldsymbol{\mu})^T \boldsymbol{\Sigma}^{-1}(\overline{\boldsymbol{x}} - \boldsymbol{\mu})$$

by i). Now the last term is maximized by $\boldsymbol{\mu} = \overline{\boldsymbol{x}}$ by iii) and the middle two terms are maximized by $\frac{1}{n}\boldsymbol{A}$ by ii) since $\boldsymbol{\Sigma}$ and $\boldsymbol{A}$ are both positive definite. Hence the MLE of $(\boldsymbol{\mu}, \boldsymbol{\Sigma})$ is

$$(\hat{\boldsymbol{\mu}}, \hat{\boldsymbol{\Sigma}}) = \left(\overline{\boldsymbol{X}}, \frac{1}{n}\sum_{i=1}^{n}(\boldsymbol{X}_i - \overline{\boldsymbol{X}})(\boldsymbol{X}_i - \overline{\boldsymbol{X}})^T\right).$$

Example 5.5. Let $X_1, \ldots, X_n$ be independent identically distributed random variables from a lognormal (μ, σ^2) distribution with pdf

$$f(x) = \frac{1}{x\sqrt{2\pi\sigma^2}} \exp\left(\frac{-(\log(x) - \mu)^2}{2\sigma^2}\right)$$

where $\sigma > 0$ and $x > 0$ and μ is real. **Assume that σ is known.**

a) Find the maximum likelihood estimator of μ.
b) What is the maximum likelihood estimator of μ^3? Explain.

Solution: a)

$$\hat{\mu} = \frac{\sum \log(X_i)}{n}$$

To see this note that

$$L(\mu) = \left(\prod \frac{1}{x_i\sqrt{2\pi\sigma^2}}\right) \exp\left(\frac{-\sum(\log(x_i) - \mu)^2}{2\sigma^2}\right).$$

So

$$\log(L(\mu)) = \log(c) - \frac{\sum(\log(x_i) - \mu)^2}{2\sigma^2}$$

and the derivative of the log likelihood wrt μ is

$$\frac{\sum 2(\log(x_i) - \mu)}{2\sigma^2}.$$

Setting this quantity equal to 0 gives $n\mu = \sum \log(x_i)$ and the solution $\hat{\mu}$ is unique. The second derivative is $-n/\sigma^2 < 0$, so $\hat{\mu}$ is indeed the global maximum.

b)

$$\left(\frac{\sum \log(X_i)}{n}\right)^3$$

by invariance.

Example 5.6. Suppose that the joint probability distribution function of $X_1, \ldots, X_k$ is

$$f(x_1, x_2, \ldots, x_k|\theta) = \frac{n!}{(n-k)!\theta^k} \exp\left(\frac{-[(\sum_{i=1}^k x_i) + (n-k)x_k]}{\theta}\right)$$

where $0 \le x_1 \le x_2 \le \cdots \le x_k$ and $\theta > 0$.

a) Find the maximum likelihood estimator (MLE) for θ.
b) What is the MLE for θ^2? Explain briefly.

Solution: a) Let $t = [(\sum_{i=1}^k x_i) + (n-k)x_k]$. $L(\theta) = f(\boldsymbol{x}|\theta)$ and $\log(L(\theta)) = \log(f(\boldsymbol{x}|\theta)) =$

$$d - k\log(\theta) - \frac{t}{\theta}.$$

Hence

$$\frac{d}{d\theta}\log(L(\theta)) = \frac{-k}{\theta} + \frac{t}{\theta^2} \stackrel{\text{set}}{=} 0.$$

Hence

$$k\theta = t$$

or

$$\hat{\theta} = \frac{t}{k}.$$

This is a unique solution and

$$\frac{d^2}{d\theta^2}\log(L(\theta)) = \frac{k}{\theta^2} - \frac{2t}{\theta^3}\bigg|_{\theta=\hat{\theta}} = \frac{k}{\hat{\theta}^2} - \frac{2k\hat{\theta}}{\hat{\theta}^3} = -\frac{k}{\hat{\theta}^2} < 0.$$

Hence $\hat{\theta} = T/k$ is the MLE where $T = [(\sum_{i=1}^k X_i) + (n-k)X_k]$.
b) $\hat{\theta}^2$ by the invariance principle.

Example 5.7. Let $X_1, \ldots, X_n$ be independent identically distributed random variables with pdf

$$f(x) = \frac{1}{\lambda} x^{\frac{1}{\lambda}-1},$$

where $\lambda > 0$ and $0 < x \le 1$.

a) Find the maximum likelihood estimator of λ.
b) What is the maximum likelihood estimator of λ^3? Explain.

Solution: a) For $0 < x \le 1$

$$f(x) = \frac{1}{\lambda}\exp\left[\left(\frac{1}{\lambda} - 1\right)\log(x)\right].$$

Hence the likelihood

$$L(\lambda) = \frac{1}{\lambda^n} \exp\left[\left(\frac{1}{\lambda} - 1\right) \sum \log(x_i)\right],$$

and the log likelihood

$$\log(L(\lambda)) = -n\log(\lambda) + \left(\frac{1}{\lambda} - 1\right) \sum \log(x_i).$$

Hence

$$\frac{d}{d\lambda} \log(L(\lambda)) = \frac{-n}{\lambda} - \frac{\sum \log(x_i)}{\lambda^2} \stackrel{\text{set}}{=} 0,$$

or $-\sum \log(x_i) = n\lambda$, or

$$\hat{\lambda} = \frac{-\sum \log(x_i)}{n}.$$

Notice that $\hat{\lambda}$ is the unique solution and that

$$\frac{d^2}{d\lambda^2} \log(L(\lambda)) = \frac{n}{\lambda^2} + \left.\frac{2\sum \log(x_i)}{\lambda^3}\right|_{\lambda=\hat{\lambda}}$$

$$= \frac{n}{\hat{\lambda}^2} - \frac{2n\hat{\lambda}}{\hat{\lambda}^3} = \frac{-n}{\hat{\lambda}^2} < 0.$$

Hence $\hat{\lambda} = -\sum \log(X_i)/n$ is the MLE of λ.

b) By invariance, $\hat{\lambda}^3$ is the MLE of λ.

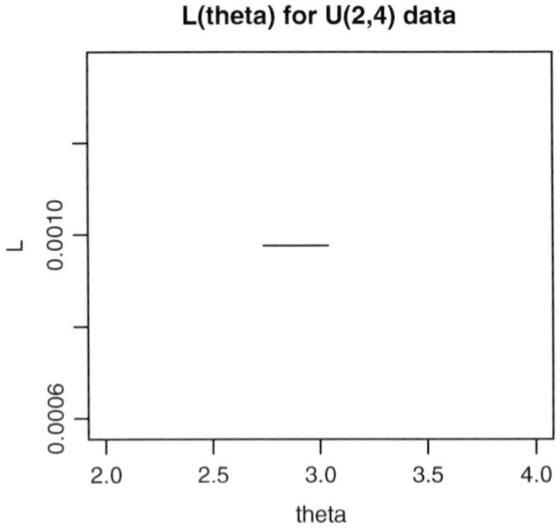

Fig. 5.2 Sample size $n = 10$

Example 5.8. Suppose $Y_1,\ldots,Y_n$ are iid $U(\theta-1,\theta+1)$. Then

$$L(\theta)=\prod_{i=1}^n f(y_i)=\prod_{i=1}^n \frac{1}{2}I(\theta-1\le y_i\le\theta+1)=\frac{1}{2^n}I(\theta-1\le \text{ all } y_i\le\theta+1)$$

$$=\frac{1}{2^n}I(\theta-1\le y_{(1)}\le y_{(n)}\le\theta+1)=\frac{1}{2^n}I(y_{(n)}-1\le\theta\le y_{(1)}+1).$$

Let $0\le c\le 1$. Then any estimator of the form $\hat\theta_c=c[Y_{(n)}-1]+(1-c)[Y_{(1)}+1]$ is an MLE of θ. Figure 5.2 shows $L(\theta)$ for $U(2,4)$ data with $n=10, y_{(1)}=2.0375$ and $y_{(n)}=3.7383$. The height of the plotted line segment is $1/2^{10}\approx 0.00098$.

Remark 5.7. Chapter 10 has many MLE examples.

5.2 Method of Moments Estimators

The method of moments is another useful way for obtaining point estimators. Let $Y_1,\ldots,Y_n$ be an iid sample and let

$$\hat\mu_j=\frac{1}{n}\sum_{i=1}^n Y_i^j \text{ and } \mu_j\equiv\mu_j(\boldsymbol\theta)=E_{\boldsymbol\theta}(Y^j) \tag{5.2}$$

for $j=1,\ldots,k$. So $\hat\mu_j$ is the jth sample moment and μ_j is the jth population moment. Fix k and assume that $\mu_j=\mu_j(\theta_1,\ldots,\theta_k)$. Solve the system

$$\begin{aligned}\hat\mu_1 &\overset{\text{set}}{=} \mu_1(\theta_1,\ldots,\theta_k)\\ &\vdots\\ \hat\mu_k &\overset{\text{set}}{=} \mu_k(\theta_1,\ldots,\theta_k)\end{aligned}$$

for $\tilde{\boldsymbol\theta}$.

Definition 5.3. The solution $\tilde\theta=(\tilde\theta_1,\ldots,\tilde\theta_k)$ is the **method of moments estimator** of $\boldsymbol\theta$. If g is a continuous function of the first k moments and $h(\boldsymbol\theta)=g(\mu_1(\boldsymbol\theta),\ldots,\mu_k(\boldsymbol\theta))$, then the method of moments estimator of $h(\boldsymbol\theta)$ is

$$g(\hat\mu_1,\ldots,\hat\mu_k).$$

Definition 5.3 is similar to the invariance principle for the MLE, but note that g needs to be a continuous function, and the definition only applies to a function of $(\hat\mu_1,\ldots,\hat\mu_k)$ where $k\ge 1$. Hence $\overline{Y}$ is the method of moments estimator of the population mean μ, and $g(\overline{Y})$ is the method of moments estimator of $g(\mu)$ if g is a continuous function. Sometimes the notation $\hat\theta_{MLE}$ and $\hat\theta_{MM}$ will be used to denote the MLE and method of moments estimators of θ, respectively. As with maximum likelihood estimators, not all method of moments estimators exist in closed form, and then numerical techniques must be used.

Example 5.9. Let $Y_1,\ldots,Y_n$ be iid from a distribution with a given pdf or pmf $f(y|\theta)$.

a) If $E(Y)=h(\theta)$, then $\hat{\theta}_{MM}=h^{-1}(\overline{Y})$.
b) The method of moments estimator of $E(Y)=\mu_1$ is $\hat{\mu}_1=\overline{Y}$.
c) The method of moments estimator of $\text{VAR}_\theta(Y)=\mu_2(\theta)-[\mu_1(\theta)]^2$ is

$$\hat{\sigma}^2_{MM}=\hat{\mu}_2-\hat{\mu}_1^2=\frac{1}{n}\sum_{i=1}^n Y_i^2-(\overline{Y})^2=\frac{1}{n}\sum_{i=1}^n (Y_i-\overline{Y})^2\equiv S_M^2.$$

Method of moments estimators need not be unique. For example both $\overline{Y}$ and S_M^2 are method of moment estimators of θ for iid Poisson(θ) data. Generally the method of moments estimators that use small j for $\hat{\mu}_j$ are preferred, so use $\overline{Y}$ for Poisson data.

Proposition 5.3. Let $S_M^2=\frac{1}{n}\sum_{i=1}^n(Y_i-\overline{Y})^2$ and suppose that $E(Y)=h_1(\theta_1,\theta_2)$ and $V(Y)=h_2(\theta_1,\theta_2)$. Then solving

$$\overline{Y}\stackrel{\text{set}}{=}h_1(\theta_1,\theta_2)$$
$$S_M^2\stackrel{\text{set}}{=}h_2(\theta_1,\theta_2)$$

for $\tilde{\boldsymbol{\theta}}$ is a method of moments estimator.

Proof. Notice that $\mu_1=h_1(\theta_1,\theta_2)=\mu_1(\theta_1,\theta_2)$ while $\mu_2-[\mu_1]^2=h_2(\theta_1,\theta_2)$. Hence $\mu_2=h_2(\theta_1,\theta_2)+[h_1(\theta_1,\theta_2)]^2=\mu_2(\theta_1,\theta_2)$. Hence the method of moments estimator is a solution to $\overline{Y}\stackrel{\text{set}}{=}\mu_1(\theta_1,\theta_2)$ and $\frac{1}{n}\sum_{i=1}^n Y_i^2\stackrel{\text{set}}{=}h_2(\theta_1,\theta_2)+[\mu_1(\theta_1,\theta_2)]^2$. Equivalently, solve $\overline{Y}\stackrel{\text{set}}{=}h_1(\theta_1,\theta_2)$ and
$\frac{1}{n}\sum_{i=1}^n Y_i^2-[\overline{Y}]^2=S_M^2\stackrel{\text{set}}{=}h_2(\theta_1,\theta_2)$. □

Example 5.10. Suppose that $Y_1,\ldots,Y_n$ be iid gamma (ν,λ). Then $\hat{\mu}_1\stackrel{\text{set}}{=}E(Y)=\nu\lambda$ and $\hat{\mu}_2\stackrel{\text{set}}{=}E(Y^2)=\text{VAR}(Y)+[E(Y)]^2=\nu\lambda^2+\nu^2\lambda^2=\nu\lambda^2(1+\nu)$. Substitute $\nu=\hat{\mu}_1/\lambda$ into the second equation to obtain

$$\hat{\mu}_2=\frac{\hat{\mu}_1}{\lambda}\lambda^2\left(1+\frac{\hat{\mu}_1}{\lambda}\right)=\lambda\hat{\mu}_1+\hat{\mu}_1^2.$$

Thus

$$\tilde{\lambda}=\frac{\hat{\mu}_2-\hat{\mu}_1^2}{\hat{\mu}_1}=\frac{S_M^2}{\overline{Y}} \text{ and } \tilde{\nu}=\frac{\hat{\mu}_1}{\tilde{\lambda}}=\frac{\hat{\mu}_1^2}{\hat{\mu}_2-\hat{\mu}_1^2}=\frac{[\overline{Y}]^2}{S_M^2}.$$

Alternatively, solve $\overline{Y}\stackrel{\text{set}}{=}\nu\lambda$ and $S_M^2\stackrel{\text{set}}{=}\nu\lambda^2=(\nu\lambda)\lambda$. Hence $\tilde{\lambda}=S_M^2/\overline{Y}$ and

$$\tilde{\nu}=\frac{\overline{Y}}{\tilde{\lambda}}=\frac{[\overline{Y}]^2}{S_M^2}.$$

5.3 Summary

Let $f(\boldsymbol{y}|\boldsymbol{\theta})$ be the pmf or pdf of a sample $\boldsymbol{Y}$. If $\boldsymbol{Y} = \boldsymbol{y}$ is observed, then **the likelihood function** $L(\boldsymbol{\theta}) = f(\boldsymbol{y}|\boldsymbol{\theta})$. For each sample point $\boldsymbol{y} = (y_1, \ldots, y_n)$, let $\hat{\boldsymbol{\theta}}(\boldsymbol{y})$ be a parameter value at which $L(\boldsymbol{\theta}|\boldsymbol{y})$ attains its maximum as a function of $\boldsymbol{\theta}$ with $\boldsymbol{y}$ held fixed. Then a maximum likelihood estimator (**MLE**) of the parameter $\boldsymbol{\theta}$ based on the sample $\boldsymbol{Y}$ is $\hat{\boldsymbol{\theta}}(\boldsymbol{Y})$.

Note: it is crucial to observe that the likelihood function is a function of $\boldsymbol{\theta}$ (and that $y_1, \ldots, y_n$ act as fixed constants).

Note: If $Y_1, \ldots, Y_n$ is an independent sample from a population with pdf or pmf $f(y|\boldsymbol{\theta})$ then the likelihood function

$$L(\boldsymbol{\theta}) = L(\boldsymbol{\theta}|y_1, \ldots, y_n) = \prod_{i=1}^{n} f(y_i|\boldsymbol{\theta}).$$

Note: If the MLE $\hat{\boldsymbol{\theta}}$ exists, then $\hat{\boldsymbol{\theta}} \in \Theta$.

Potential candidates for the MLE can sometimes be found by differentiating $L(\boldsymbol{\theta})$, by direct maximization, or using the profile likelihood. Typically the log likelihood is used as in A) below.

A) Let $Y_1, \ldots, Y_n$ be iid with pdf or pmf $f(y|\theta)$. Then $L(\theta) = \prod_{i=1}^{n} f(y_i|\theta)$. To find the MLE,

  i) find $L(\theta)$ and then find the log likelihood $\log L(\theta)$.
  ii) Find the derivative $\frac{d}{d\theta} \log L(\theta)$, set the derivative equal to zero and solve for θ. The solution is a candidate for the MLE.
  iii) **Invariance Principle:** If $\hat{\theta}$ is the MLE of θ, then $\tau(\hat{\theta})$ is the MLE of $\tau(\theta)$.
  iv) Show that $\hat{\theta}$ is the MLE by showing that $\hat{\theta}$ is the global maximizer of $\log L(\theta)$. Often this is done by noting that $\hat{\theta}$ is the unique solution to the equation $\frac{d}{d\theta} \log L(\theta) = 0$ and that the second derivative evaluated at $\hat{\theta}$ is negative: $\frac{d^2}{d\theta^2} \log L(\theta)|_{\hat{\theta}} < 0$.

B) If $\log L(\theta)$ is strictly concave ($\frac{d^2}{d\theta^2} \log L(\theta) < 0$ for all $\theta \in \Theta$), then any local max of $\log L(\theta)$ is a global max.

C) Know how to find the MLE for the normal distribution (including when μ or σ^2 is known). Memorize the MLEs

$$\overline{Y},\ S_M^2 = \frac{1}{n}\sum_{i=1}^{n}(Y_i - \overline{Y})^2,\ \frac{1}{n}\sum_{i=1}^{n}(Y_i - \mu)^2$$

for the normal and for the uniform distribution. Also $\overline{Y}$ is the MLE for several brand name distributions. Notice that S_M^2 is the method of moments estimator for $V(Y)$ and is the MLE for $V(Y)$ if the data are iid $N(\mu, \sigma^2)$.

D) **On qualifying exams**, the $N(\mu, \mu)$ and $N(\mu, \mu^2)$ distributions are common. See Problems 5.4, 5.30, and 5.35.

E) Indicators are useful. For example, $\prod_{i=1}^n I_A(y_i) = I(\text{all } y_i \in A)$ and $\prod_{j=1}^k I_{A_j}(y) = I_{\cap_{j=1}^k A_j}(y)$. Hence $I(0 \le y \le \theta) = I(0 \le y)I(y \le \theta)$, and $\prod_{i=1}^n I(\theta_1 \le y_i \le \theta_2) = I(\theta_1 \le y_{(1)} \le y_{(n)} \le \theta_2) = I(\theta_1 \le y_{(1)})I(y_{(n)} \le \theta_2)$.

F) Suppose $X_1,\ldots,X_n$ are iid with pdf or pmf $f(x|\lambda)$ and $Y_1,\ldots,Y_n$ are iid with pdf or pmf $g(y|\mu)$. Suppose that the X's are independent of the Y's. Then

$$\sup_{(\lambda,\mu)\in\Theta} L(\lambda,\mu|\boldsymbol{x},\boldsymbol{y}) \le \sup_{\lambda} L_{\boldsymbol{x}}(\lambda) \sup_{\mu} L_{\boldsymbol{y}}(\mu)$$

where $L_{\boldsymbol{x}}(\lambda) = \prod_{i=1}^n f(x_i|\lambda)$. Hence if $\hat{\lambda}$ is the marginal MLE of λ and $\hat{\mu}$ is the marginal MLE of μ, then $(\hat{\lambda},\hat{\mu})$ is the MLE of (λ,μ) provided that $(\hat{\lambda},\hat{\mu})$ is in the parameter space Θ.

G) Let $\hat{\mu}_j = \frac{1}{n}\sum_{i=1}^n Y_i^j$, let $\mu_j = E(Y^j)$ and assume that $\mu_j = \mu_j(\theta_1,\ldots,\theta_k)$. Solve the system

$$\begin{aligned} \hat{\mu}_1 &\stackrel{\text{set}}{=} \mu_1(\theta_1,\ldots,\theta_k) \\ \vdots \quad & \quad\quad \vdots \\ \hat{\mu}_k &\stackrel{\text{set}}{=} \mu_k(\theta_1,\ldots,\theta_k) \end{aligned}$$

for the method of moments estimator $\tilde{\boldsymbol{\theta}}$.

H) If g is a continuous function of the first k moments and $h(\boldsymbol{\theta}) = g(\mu_1(\boldsymbol{\theta}),\ldots,\mu_k(\boldsymbol{\theta}))$, then the method of moments estimator of $h(\boldsymbol{\theta})$ is $g(\hat{\mu}_1,\ldots,\hat{\mu}_k)$.

5.4 Complements

Optimization theory is also known as nonlinear programming and shows how to find the global max and min of a multivariate function. Peressini et al. (1988) is an undergraduate text. Also see Sundaram (1996) and Bertsekas (1999).

Maximum likelihood estimation is widely used in statistical models. See Pawitan (2001) and texts for Categorical Data Analysis, Econometrics, Multiple Linear Regression Generalized Linear Models, Multivariate Analysis, and Survival Analysis.

Suppose that $Y = t(W)$ and $W = t^{-1}(Y)$ where W has a pdf with parameters $\boldsymbol{\theta}$, the transformation t does not depend on any unknown parameters, and the pdf of Y is

$$f_Y(y) = f_W(t^{-1}(y)) \left| \frac{dt^{-1}(y)}{dy} \right|.$$

If $W_1,\ldots,W_n$ are iid with pdf $f_W(w)$, assume that the MLE of $\boldsymbol{\theta}$ is $\hat{\boldsymbol{\theta}}_W(\boldsymbol{w})$ where the w_i are the observed values of W_i and $\boldsymbol{w} = (w_1,\ldots,w_n)$. If $Y_1,\ldots,Y_n$ are iid and the y_i are the observed values of Y_i, then the likelihood is

$$L_Y(\boldsymbol{\theta}) = \left(\prod_{i=1}^n \left| \frac{dt^{-1}(y_i)}{dy} \right| \right) \prod_{i=1}^n f_W(t^{-1}(y_i)|\boldsymbol{\theta}) = c \prod_{i=1}^n f_W(t^{-1}(y_i)|\boldsymbol{\theta}).$$

Hence the log likelihood is $\log(L_Y(\boldsymbol{\theta})) =$

$$d + \sum_{i=1}^{n} \log[f_W(t^{-1}(y_i)|\boldsymbol{\theta})] = d + \sum_{i=1}^{n} \log[f_W(w_i|\boldsymbol{\theta})] = d + \log[L_W(\boldsymbol{\theta})]$$

where $w_i = t^{-1}(y_i)$. Hence maximizing the $\log(L_Y(\boldsymbol{\theta}))$ is equivalent to maximizing $\log(L_W(\boldsymbol{\theta}))$ and

$$\hat{\boldsymbol{\theta}}_Y(\boldsymbol{y}) = \hat{\boldsymbol{\theta}}_W(\boldsymbol{w}) = \hat{\boldsymbol{\theta}}_W(t^{-1}(y_1), \ldots, t^{-1}(y_n)). \tag{5.3}$$

Compare Meeker and Escobar (1998, p. 175).

Example 5.11. Suppose $Y_1, \ldots, Y_n$ are iid lognormal (μ, σ^2). Then $W_i = \log(Y_i) \sim N(\mu, \sigma^2)$ and the MLE $(\hat{\mu}, \hat{\sigma}^2) = (\overline{W}, \frac{1}{n}\sum_{i=1}^n (W_i - \overline{W})^2)$.

One of the most useful properties of the maximum likelihood estimator is the invariance property: if $\hat{\theta}$ is the MLE of θ, then $\tau(\hat{\theta})$ is the MLE of $\tau(\theta)$. Olive (2004) is a good discussion of the MLE invariance principle. Also see Pal and Berry (1992). Many texts either define the MLE of $\tau(\theta)$ to be $\tau(\hat{\theta})$, say that the property is immediate from the definition of the MLE, or quote Zehna (1966). A little known paper, Berk (1967), gives an elegant proof of the invariance property that can be used in introductory statistical courses. The next subsection will show that Berk (1967) answers some questions about the MLE which cannot be answered using Zehna (1966).

5.4.1 Two "Proofs" of the Invariance Principle

"Proof" I) The following argument of Zehna (1966) also appears in Casella and Berger (2002, p. 320). Let $\boldsymbol{\theta} \in \Theta$ and let $h : \Theta \to \Lambda$ be a function. Since the MLE

$$\hat{\boldsymbol{\theta}} \in \Theta, \;\; h(\hat{\boldsymbol{\theta}}) = \hat{\boldsymbol{\lambda}} \in \Lambda.$$

If h is not one to one, then many values of $\boldsymbol{\theta}$ may be mapped to $\boldsymbol{\lambda}$. Let

$$\Theta_{\boldsymbol{\lambda}} = \{\boldsymbol{\theta} : h(\boldsymbol{\theta}) = \boldsymbol{\lambda}\}$$

and define the induced likelihood function $M(\boldsymbol{\lambda})$ by

$$M(\boldsymbol{\lambda}) = \sup_{\boldsymbol{\theta} \in \Theta_{\boldsymbol{\lambda}}} L(\boldsymbol{\theta}). \tag{5.4}$$

Then for any $\boldsymbol{\lambda} \in \Lambda$,

$$M(\boldsymbol{\lambda}) = \sup_{\boldsymbol{\theta} \in \Theta_{\boldsymbol{\lambda}}} L(\boldsymbol{\theta}) \leq \sup_{\boldsymbol{\theta} \in \Theta} L(\boldsymbol{\theta}) = L(\hat{\boldsymbol{\theta}}) = M(\hat{\boldsymbol{\lambda}}). \tag{5.5}$$

Hence $h(\hat{\boldsymbol{\theta}}) = \hat{\boldsymbol{\lambda}}$ maximizes the induced likelihood $M(\boldsymbol{\lambda})$. Zehna (1966) says that since $h(\hat{\boldsymbol{\theta}})$ maximizes the induced likelihood, we should call $h(\hat{\boldsymbol{\theta}})$ the MLE of $h(\boldsymbol{\theta})$, but the definition of MLE says that we should be maximizing a genuine likelihood.

This argument raises two important questions.

- If we call $h(\hat{\boldsymbol{\theta}})$ the MLE of $h(\boldsymbol{\theta})$ and h is not one to one, does $h(\hat{\boldsymbol{\theta}})$ maximize a likelihood or should
- $h(\hat{\boldsymbol{\theta}})$ be called a maximum induced likelihood estimator?
- If $h(\hat{\boldsymbol{\theta}})$ is an MLE, what is the likelihood function $K(h(\boldsymbol{\theta}))$?

Some examples might clarify these questions.

- If the population come from a $N(\mu,\sigma^2)$ distribution, the invariance principle says that the MLE of μ/σ is $\overline{X}/S_M$ where

$$\overline{X} = \frac{1}{n}\sum_{i=1}^{n} X_i \text{ and } S_M^2 = \frac{1}{n}\sum_{i=1}^{n}(X_i - \overline{X})^2$$

are the MLEs of μ and σ^2. Since the function $h(x,y) = x/\sqrt{y}$ is not one to one (e.g., $h(x,y) = 1$ if $x = \sqrt{y}$), what is the likelihood $K(h(\mu,\sigma^2)) = K(\mu/\sigma)$ that is being maximized?
- If X_i comes from a Bernoulli(ρ) population, why is $\overline{X}_n(1-\overline{X}_n)$ the MLE of $\rho(1-\rho)$?

Proof II) Examining the invariance principle for one-to-one functions h is also useful. When h is one to one, let $\boldsymbol{\eta} = h(\boldsymbol{\theta})$. Then the inverse function h^{-1} exists and $\boldsymbol{\theta} = h^{-1}(\boldsymbol{\eta})$. Hence

$$f(\boldsymbol{x}|\boldsymbol{\theta}) = f(\boldsymbol{x}|h^{-1}(\boldsymbol{\eta})) \tag{5.6}$$

is the joint pdf or pmf of $\boldsymbol{x}$. So the likelihood function of $h(\boldsymbol{\theta}) = \boldsymbol{\eta}$ is

$$L^*(\boldsymbol{\eta}) \equiv K(\boldsymbol{\eta}) = L(h^{-1}(\boldsymbol{\eta})). \tag{5.7}$$

Also note that

$$\sup_{\boldsymbol{\eta}} K(\boldsymbol{\eta}|\boldsymbol{x}) = \sup_{\boldsymbol{\eta}} L(h^{-1}(\boldsymbol{\eta})|\boldsymbol{x}) = L(\hat{\boldsymbol{\theta}}|\boldsymbol{x}). \tag{5.8}$$

Thus

$$\hat{\boldsymbol{\eta}} = h(\hat{\boldsymbol{\theta}}) \tag{5.9}$$

is the MLE of $\boldsymbol{\eta} = h(\boldsymbol{\theta})$ when h is one to one.

If h is not one to one, then the new parameters $\boldsymbol{\eta} = h(\boldsymbol{\theta})$ do not give enough information to define $f(\boldsymbol{x}|\boldsymbol{\eta})$. Hence we cannot define the likelihood. That is, a $N(\mu,\sigma^2)$ density cannot be defined by the parameter μ/σ alone. Before concluding that the MLE does not exist if h is not one to one, note that if $X_1,\ldots,X_n$ are iid $N(\mu,\sigma^2)$ then $X_1,\ldots,X_n$ remain iid $N(\mu,\sigma^2)$ even though the investigator did not rename the parameters wisely or is interested in a function $h(\mu,\sigma) = \mu/\sigma$ that is not one to one. Berk (1967) said that if h is not one to one, define

$$w(\boldsymbol{\theta}) = (h(\boldsymbol{\theta}),u(\boldsymbol{\theta})) = (\boldsymbol{\eta},\boldsymbol{\gamma}) = \boldsymbol{\xi} \tag{5.10}$$

such that $w(\boldsymbol{\theta})$ is one to one. Note that the choice

$$w(\boldsymbol{\theta}) = (h(\boldsymbol{\theta}), \boldsymbol{\theta})$$

works. In other words, we can always take u to be the identity function.

The choice of w is not unique, but the inverse function

$$w^{-1}(\boldsymbol{\xi}) = \boldsymbol{\theta}$$

is unique. Hence the likelihood is well defined, and $w(\hat{\boldsymbol{\theta}})$ is the MLE of $\boldsymbol{\xi}$. $\square$

Example 5.12. Following Lehmann (1999, p. 466), let

$$f(x|\sigma) = \frac{1}{\sqrt{2\pi}\,\sigma} \exp\left(\frac{-x^2}{2\sigma^2}\right)$$

where x is real and $\sigma > 0$. Let $\eta = \sigma^k$ so $\sigma = \eta^{1/k} = h^{-1}(\eta)$. Then

$$f^*(x|\eta) = \frac{1}{\sqrt{2\pi}\,\eta^{1/k}} \exp\left(\frac{-x^2}{2\eta^{2/k}}\right) = f(x|\sigma = h^{-1}(\eta)).$$

Notice that calling $h(\hat{\boldsymbol{\theta}})$ the MLE of $h(\boldsymbol{\theta})$ is analogous to calling $\overline{X}_n$ the MLE of μ when the data are from a $N(\mu, \sigma^2)$ population. It is often possible to choose the function u so that if $\boldsymbol{\theta}$ is a $p \times 1$ vector, then so is $\boldsymbol{\xi}$. For the $N(\mu, \sigma^2)$ example with $h(\mu, \sigma^2) = h(\boldsymbol{\theta}) = \mu/\sigma$ we can take $u(\boldsymbol{\theta}) = \mu$ or $u(\boldsymbol{\theta}) = \sigma^2$. For the Ber($\rho$) example, $w(\rho) = (\rho(1-\rho), \rho)$ is a reasonable choice.

To summarize, Berk's proof should be widely used to prove the invariance principle, and

I) changing the names of the parameters does not change the distribution of the sample, e.g., if $X_1, \ldots, X_n$ are iid $N(\mu, \sigma^2)$, then $X_1, \ldots, X_n$ remain iid $N(\mu, \sigma^2)$ regardless of the function $h(\mu, \sigma^2)$ that is of interest to the investigator.

II) The invariance principle holds as long as $h(\hat{\boldsymbol{\theta}})$ is a random variable or random vector: h does not need to be a one-to-one function. If there is interest in $\boldsymbol{\eta} = h(\boldsymbol{\theta})$ where h is not one to one, then additional parameters $\boldsymbol{\gamma} = u(\boldsymbol{\theta})$ need to be specified so that $w(\boldsymbol{\theta}) = \boldsymbol{\xi} = (\boldsymbol{\eta}, \boldsymbol{\gamma}) = (h(\boldsymbol{\theta}), u(\boldsymbol{\theta}))$ has a well-defined likelihood $K(\boldsymbol{\xi}) = L(w^{-1}(\boldsymbol{\xi}))$. Then by Definition 5.2, the MLE is $\hat{\boldsymbol{\xi}} = w(\hat{\boldsymbol{\theta}}) = w(h(\hat{\boldsymbol{\theta}}), u(\hat{\boldsymbol{\theta}}))$ and the MLE of $\boldsymbol{\eta} = h(\boldsymbol{\theta})$ is $\hat{\boldsymbol{\eta}} = h(\hat{\boldsymbol{\theta}})$.

III) Using the identity function $\boldsymbol{\gamma} = u(\boldsymbol{\theta}) = \boldsymbol{\theta}$ always works since $\boldsymbol{\xi} = w(\boldsymbol{\theta}) = (h(\boldsymbol{\theta}), \boldsymbol{\theta})$ is a one-to-one function of $\boldsymbol{\theta}$. However, using $u(\boldsymbol{\theta})$ such that $\boldsymbol{\xi}$ and $\boldsymbol{\theta}$ have the same dimension is often useful.

5.5 Problems

PROBLEMS WITH AN ASTERISK * ARE ESPECIALLY USEFUL.

Refer to Chap. 10 for the pdf or pmf of the distributions in the problems below.

5.1*. Let $Y_1, \ldots, Y_n$ be iid binomial $(k = 1, \rho)$.

a) Assume that $\rho \in \Theta = (0,1)$ and that $0 < \sum_{i=1}^n y_i < n$. Show that the MLE of ρ is $\hat{\rho} = \overline{Y}$.
b) Now assume that $\rho \in \Theta = [0,1]$. Show that $f(y|\rho) = \rho^y(1-\rho)^{1-y} I(0 < \rho < 1) + I(\rho = 0, y = 0) + I(\rho = 1, y = 1)$. Then show that

$$L(\rho) = \rho^{\sum y}(1-\rho)^{n-\sum y} I(0 < \rho < 1) + I(\rho = 0, \sum y = 0) + I(\rho = 1, \sum y = n).$$

If $\sum y = 0$ show that $\hat{\rho} = 0 = \overline{y}$. If $\sum y = n$ show that $\hat{\rho} = 1 = \overline{y}$. Then explain why $\hat{\rho} = \overline{Y}$ if $\Theta = [0,1]$.

5.2. Let (X,Y) have the bivariate density

$$f(x,y) = \frac{1}{2\pi} \exp\left(\frac{-1}{2} [(x - \rho\cos\theta)^2 + (y - \rho\sin\theta)^2] \right).$$

Suppose that there are n independent pairs of observations (X_i, Y_i) from the above density and that ρ is known. Assume that $0 \le \theta \le 2\pi$. Find a candidate for the maximum likelihood estimator $\hat{\theta}$ by differentiating the log likelihood $\log(L(\theta))$. (Do not show that the candidate is the MLE, it is difficult to tell whether the candidate, 0 or 2π is the MLE without the actual data.)

5.3*. Suppose a single observation $X = x$ is observed where X is a random variable with pmf given by the table below. Assume $0 \le \theta \le 1$, and find the MLE $\hat{\theta}_{MLE}(x)$. (Hint: drawing $L(\theta) = L(\theta|x)$ for each of the four values of x may help.)

x	1	2	3	4
$f(x\|\theta)$	1/4	1/4	$\frac{1+\theta}{4}$	$\frac{1-\theta}{4}$

5.4. Let $X_1, \ldots, X_n$ be iid normal $N(\mu, \gamma_o^2 \mu^2)$ random variables where $\gamma_o^2 > 0$ is **known** and $\mu > 0$. Find the log likelihood $\log(\mathrm{L}(\mu|x_1, \ldots, x_n))$ and solve

$$\frac{d}{d\mu} \log(L(\mu|x_1, \ldots, x_n)) = 0$$

for $\hat{\mu}_o$, a potential candidate for the MLE of μ.

5.5. Suppose that $X_1, \ldots, X_n$ are iid uniform $U(0, \theta)$. Use the factorization theorem to write $f(\boldsymbol{x}|\theta) = g(T(\boldsymbol{x})|\theta) I[x_{(1)} \ge 0]$ where $T(\boldsymbol{x})$ is a one-dimensional sufficient statistic. Then plot the likelihood function $L(\theta) = g(T(\boldsymbol{x})|\theta)$ and find the MLE of θ.

5.6. Let $Y_1,\ldots,Y_n$ be iid Burr Type XII(λ,ϕ) with ϕ known. Find the MLE of λ.

5.7. Let $Y_1,\ldots,Y_n$ be iid chi(p,σ) with p known. Find the MLE of σ^2.

5.8. Let $Y_1,\ldots,Y_n$ iid double exponential $DE(\theta,\lambda)$ with θ known. Find the MLE of λ.

5.9. Let $Y_1,\ldots,Y_n$ be iid exponential EXP(λ). Find the MLE of λ.

5.10. If $Y_1,\ldots,Y_n$ are iid gamma $G(\nu,\lambda)$ with ν known, find the MLE of λ.

5.11. If $Y_1,\ldots,Y_n$ are iid geometric geom(ρ), find the MLE of ρ.

5.12. If $Y_1,\ldots,Y_n$ are iid inverse Gaussian $IG(\theta,\lambda)$ with λ known, find the MLE of θ.

5.13. If $Y_1,\ldots,Y_n$ are iid inverse Gaussian $IG(\theta,\lambda)$ with θ known, find the MLE of λ.

5.14. If $Y_1,\ldots,Y_n$ are iid largest extreme value LEV(θ,σ) where σ is known, find the MLE of θ.

5.15. If $Y_1,\ldots,Y_n$ are iid negative binomial $NB(r,\rho)$ with r known, find the MLE of ρ.

5.16. If $Y_1,\ldots,Y_n$ are iid Rayleigh $R(\mu,\sigma)$ with μ known, find the MLE of σ^2.

5.17. If $Y_1,\ldots,Y_n$ are iid Weibull $W(\phi,\lambda)$ with ϕ known, find the MLE of λ.

5.18. If $Y_1,\ldots,Y_n$ are iid binomial $BIN(k,\rho)$ with k known, find the MLE of ρ.

5.19. Suppose $Y_1,\ldots,Y_n$ are iid two-parameter exponential EXP(θ,λ).

a) Show that for any fixed $\lambda > 0$, the log likelihood is maximized by $y_{(1)}$. Hence the MLE $\hat{\theta} = Y_{(1)}$.
b) Find $\hat{\lambda}$ by maximizing the profile likelihood.

5.20. Suppose $Y_1,\ldots,Y_n$ are iid truncated extreme value TEV(λ). Find the MLE of λ.

Problems from old quizzes and exams. Problems from old qualifying exams are marked with a Q since these problems take longer than quiz and exam problems.

Note: Problem 5.21 would be better if it replaced "$\lambda \geq 0$" by "$\lambda > 0$, and assume $\sum x_i > 0$." But problems like 5.21 are extremely common on exams and in texts.

5.21. Suppose that $X_1,\ldots,X_n$ are iid Poisson with pmf

$$f(x|\lambda) = P(X = x|\lambda) = \frac{e^{-\lambda}\lambda^x}{x!}$$

where $x = 0,1,\ldots$ and $\lambda \geq 0$.

a) Find the MLE of λ. (Make sure that you prove that your estimator maximizes the likelihood.)
b) Find the MLE of $(1-\lambda)^2$.

5.22. Suppose that $X_1,\ldots,X_n$ are iid $U(0,\theta)$. Make a plot of $L(\theta|x_1,\ldots,x_n)$.

a) If the uniform density is $f(x) = \frac{1}{\theta}I(0 \le x \le \theta)$, find the MLE of θ if it exists.
b) If the uniform density is $f(x) = \frac{1}{\theta}I(0 < x < \theta)$, find the MLE of θ if it exists.

5.23^Q. Let $X_1,\ldots,X_n$ be a random sample from a normal distribution with **known** mean μ and unknown variance τ.

a) Find the maximum likelihood estimator of the variance τ.
b) Find the maximum likelihood estimator of the standard deviation $\sqrt{\tau}$. Explain how the MLE was obtained.

5.24. Suppose a single observation $X = x$ is observed where X is a random variable with pmf given by the table below. Assume $0 \le \theta \le 1$. and find the MLE $\hat{\theta}_{\text{MLE}}(x)$. (Hint: drawing $L(\theta) = L(\theta|x)$ for each of the values of x may help.)

x	0	1
$f(x\|\theta)$	$\frac{1+\theta}{2}$	$\frac{1-\theta}{2}$

5.25. Suppose that X is a random variable with pdf $f(x|\theta) = (x-\theta)^2/3$ for $\theta - 1 \le x \le 2+\theta$. Hence $L(\theta) = (x-\theta)^2/3$ for $x - 2 \le \theta \le x+1$. Suppose that one observation $X = 7$ was observed. Find the MLE $\hat{\theta}$ for θ. (Hint: evaluate the likelihood at the critical value and the two endpoints. One of these three values has to be the MLE.)

5.26. Let $X_1,\ldots,X_n$ be iid from a distribution with pdf

$$f(x|\theta) = \theta x^{-2}, \;\; 0 < \theta \le x < \infty.$$

a) Find a minimal sufficient statistic for θ.
b) Find the MLE for θ.

5.27. Let $Y_1,\ldots,Y_n$ be iid from a distribution with probability mass function

$$f(y|\theta) = \theta(1-\theta)^y, \;\; \text{where } y = 0,1,\ldots \text{ and } 0 < \theta < 1.$$

Assume $0 < \sum y_i < n$.

a) Find the MLE of θ. (Show that it is the global maximizer.)
b) What is the MLE of $1/\theta^2$? Explain.

5.28^Q. Let $X_1,\ldots,X_n$ be independent identically distributed random variables from a half normal HN(μ,σ^2) distribution with pdf

$$f(x) = \frac{2}{\sqrt{2\pi}\,\sigma}\exp\left(\frac{-(x-\mu)^2}{2\sigma^2}\right)$$

where $\sigma > 0$ and $x > \mu$ and μ is real. **Assume that μ is known.**

a) Find the maximum likelihood estimator of σ^2.
b) What is the maximum likelihood estimator of σ? Explain.

5.29. Let $X_1, \ldots, X_n$ be independent identically distributed random variables from a lognormal (μ, σ^2) distribution with pdf

$$f(x) = \frac{1}{x\sqrt{2\pi\sigma^2}} \exp\left(\frac{-(\log(x) - \mu)^2}{2\sigma^2}\right)$$

where $\sigma > 0$ and $x > 0$ and μ is real. **Assume that σ is known.**

a) Find the maximum likelihood estimator of μ.
b) What is the maximum likelihood estimator of μ^3? Explain.

5.30^Q. Let X be a single observation from a normal distribution with mean θ and with variance θ^2, where $\theta > 0$. Find the maximum likelihood estimator of θ^2.

5.31. Let $X_1, \ldots, X_n$ be independent identically distributed random variables with probability density function

$$f(x) = \frac{\sigma^{1/\lambda}}{\lambda} \exp\left[-\left(1 + \frac{1}{\lambda}\right) \log(x)\right] I[x \geq \sigma]$$

where $x \geq \sigma$, $\sigma > 0$, and $\lambda > 0$. The indicator function $I[x \geq \sigma] = 1$ if $x \geq \sigma$ and 0, otherwise. Find the maximum likelihood estimator (MLE) $(\hat{\sigma}, \hat{\lambda})$ of (σ, λ) with the following steps.

a) Explain why $\hat{\sigma} = X_{(1)} = \min(X_1, \ldots, X_n)$ is the MLE of σ regardless of the value of $\lambda > 0$.
b) Find the MLE $\hat{\lambda}$ of λ if $\sigma = \hat{\sigma}$ (that is, act as if $\sigma = \hat{\sigma}$ is known).

5.32. Let $X_1, \ldots, X_n$ be independent identically distributed random variables with pdf

$$f(x) = \frac{1}{\lambda} \exp\left[-\left(1 + \frac{1}{\lambda}\right) \log(x)\right]$$

where $\lambda > 0$ and $x \geq 1$.

a) Find the maximum likelihood estimator of λ.
b) What is the maximum likelihood estimator of λ^8? Explain.

5.33. Let $X_1, \ldots, X_n$ be independent identically distributed random variables with probability mass function

$$f(x) = e^{-2\theta} \frac{1}{x!} \exp[\log(2\theta)x],$$

for $x = 0, 1, \ldots$, where $\theta > 0$. Assume that at least one $X_i > 0$.

a) Find the maximum likelihood estimator of θ.
b) What is the maximum likelihood estimator of $(\theta)^4$? Explain.

5.34. Let $X_1,\ldots,X_n$ be iid with one of two probability density functions. If $\theta=0$, then

$$f(x|\theta)=\begin{cases}1, & 0\le x\le 1\\ 0, & \text{otherwise.}\end{cases}$$

If $\theta=1$, then

$$f(x|\theta)=\begin{cases}\frac{1}{2\sqrt{x}}, & 0\le x\le 1\\ 0, & \text{otherwise.}\end{cases}$$

Find the maximum likelihood estimator of θ.

Warning: Variants of the following question often appear on qualifying exams.

5.35^Q. Let $Y_1,\ldots,Y_n$ denote a random sample from a $N(a\theta,\theta)$ population.

a) Find the MLE of θ when $a=1$.
b) Find the MLE of θ when a is known but arbitrary.

5.36. Suppose that $X_1,\ldots,X_n$ are iid random variable with pdf

$$f(x|\theta)=(x-\theta)^2/3$$

for $\theta-1\le x\le 2+\theta$.

a) Assume that $n=1$ and that $X=7$ was observed. Sketch the log likelihood function $L(\theta)$ and find the maximum likelihood estimator (MLE) $\hat{\theta}$.
b) Again assume that $n=1$ and that $X=7$ was observed. Find the MLE of

$$h(\theta)=2\theta-\exp(-\theta^2).$$

5.37^Q. Let $X_1,\ldots,X_n$ be independent identically distributed (iid) random variables with probability density function

$$f(x)=\frac{2}{\lambda\sqrt{2\pi}}e^x\exp\left(\frac{-(e^x-1)^2}{2\lambda^2}\right)$$

where $x>0$ and $\lambda>0$.

a) Find the maximum likelihood estimator (MLE) $\hat{\lambda}$ of λ.
b) What is the MLE of λ^2? Explain.

5.38^Q. Let $X_1,\ldots,X_n$ be independent identically distributed random variables from a distribution with pdf

$$f(x) = \frac{2}{\lambda\sqrt{2\pi}} \frac{1}{x} \exp\left[\frac{-(\log(x))^2}{2\lambda^2}\right]$$

where $\lambda > 0$ where and $0 \leq x \leq 1$.

a) Find the maximum likelihood estimator (MLE) of λ.
b) Find the MLE of λ^2.

5.39. Suppose that $X_1, \ldots, X_n$ are iid $U(\theta, \theta+1)$ so that

$$L(\theta) = 1^n I[x_{(1)} \geq \theta] I[x_{(n)} \leq \theta + 1] = I[x_{(n)} - 1 \leq \theta \leq x_{(1)}].$$

a) Sketch $L(\theta)$.
b) An MLE of θ is $\hat{\theta}_{MLE}(\boldsymbol{x}) = t$ for some fixed $t \in [c, d]$. Find $[c, d]$.

5.40. Let $Y_1, \ldots, Y_n$ be independent identically distributed random variables with pdf

$$f_Y(y) = \frac{2\gamma^{3/2}}{\sqrt{\pi}} y^2 \exp(-\gamma y^2)$$

where $\gamma > 0$ and y is real.

a) Find the maximum likelihood estimator of γ. (Make sure that you prove that your answer is the MLE.)
b) What is the maximum likelihood estimator of $1/\gamma$? Explain.

5.41^Q. Suppose that X has probability density function

$$f_X(x) = \frac{\theta}{x^{1+\theta}}, \quad x \geq 1$$

where $\theta > 0$.

a) If $U = X^2$, derive the probability density function $f_U(u)$ of U.
b) Find the method of moments estimator of θ.
c) Find the method of moments estimator of θ^2.

5.42^Q. Suppose that the joint probability distribution function of $X_1, \ldots, X_k$ is

$$f(x_1, x_2, \ldots, x_k | \theta) = \frac{n!}{(n-k)!\theta^k} \exp\left(\frac{-[(\sum_{i=1}^k x_i) + (n-k)x_k]}{\theta}\right)$$

where $0 \leq x_1 \leq x_2 \leq \cdots \leq x_k$ and $\theta > 0$.

a) Find the maximum likelihood estimator (MLE) for θ.
b) What is the MLE for θ^2? Explain briefly.

5.43^Q. Let $X_1, \ldots, X_n$ be iid with pdf

$$f(x) = \frac{\cos(\theta)}{2\cosh(\pi x/2)} \exp(\theta x)$$

where x is real and $|\theta| < \pi/2$.

a) Find the maximum likelihood estimator (MLE) for θ.
b) What is the MLE for $\tan(\theta)$? Explain briefly.

5.44^Q. Let $X_1,\ldots,X_n$ be a random sample from a population with pdf

$$f(x) = \frac{1}{\sigma}\exp\left(-\frac{x-\mu}{\sigma}\right),\ x \geq \mu,$$

where $-\infty < \mu < \infty$, $\sigma > 0$.

a) Find the maximum likelihood estimator of μ and σ.
b) Evaluate $\tau(\mu,\sigma) = P_{\mu,\sigma}[X_1 \geq t]$ where $t > \mu$. Find the maximum likelihood estimator of $\tau(\mu,\sigma)$.

5.45^Q. Let $Y_1,\ldots,Y_n$ be independent identically distributed (iid) random variables from a distribution with probability density function (pdf)

$$f(y) = \frac{1}{2\sqrt{2\pi}}\left(\frac{1}{\theta}\sqrt{\frac{\theta}{y}} + \frac{\theta}{y^2}\sqrt{\frac{y}{\theta}}\right)\frac{1}{\nu}\exp\left[\frac{-1}{2\nu^2}\left(\frac{y}{\theta}+\frac{\theta}{y}-2\right)\right]$$

where $y > 0, \theta > 0$ is **known** and $\nu > 0$.

a) Find the maximum likelihood estimator (MLE) of ν.
b) Find the MLE of ν^2.

5.46^Q. Let $Y_1,\ldots,Y_n$ be independent identically distributed (iid) random variables from a distribution with probability density function (pdf)

$$f(y) = \phi\, y^{-(\phi+1)}\frac{1}{1+y^{-\phi}}\,\frac{1}{\lambda}\exp\left[\frac{-1}{\lambda}\log(1+y^{-\phi})\right]$$

where $y > 0, \phi > 0$ is **known** and $\lambda > 0$.

a) Find the maximum likelihood estimator (MLE) of λ.
b) Find the MLE of λ^2.

5.47^Q. Let $Y_1,\ldots,Y_n$ be independent identically distributed (iid) random variables from an inverse half normal distribution with probability density function (pdf)

$$f(y) = \frac{2}{\sigma\sqrt{2\pi}}\,\frac{1}{y^2}\exp\left(\frac{-1}{2\sigma^2 y^2}\right)$$

where $y > 0$ and $\sigma > 0$.

a) Find the maximum likelihood estimator (MLE) of σ^2.
b) Find the MLE of σ.

5.48^Q. Let $Y_1,\ldots,Y_n$ be independent identically distributed (iid) random variables from a distribution with probability density function (pdf)

$$f(y) = \frac{\theta}{y^2} \exp\left(\frac{-\theta}{y}\right)$$

where $y > 0$ and $\theta > 0$.

a) Find the maximum likelihood estimator (MLE) of θ.
b) Find the MLE of $1/\theta$.

5.49^Q. Let $Y_1, \ldots, Y_n$ be independent identically distributed (iid) random variables from a Lindley distribution with probability density function (pdf)

$$f(y) = \frac{\theta^2}{1+\theta}(1+y)e^{-\theta y}$$

where $y > 0$ and $\theta > 0$.

a) Find the maximum likelihood estimator (MLE) of θ. You may assume that

$$\left.\frac{d^2}{d\theta^2} \log(L(\theta))\right|_{\theta=\hat{\theta}} < 0.$$

b) Find the MLE of $1/\theta$.

5.50. Let $Y_1, \ldots, Y_n$ be iid random variables from a distribution with pdf

$$f(y) = \frac{\theta}{(1+y)^{\theta+1}}$$

where $y > 0$ and $\theta > 0$. Find the MLE of θ.

5.51. Let $Y_1, \ldots, Y_n$ be iid random variables from a distribution with pdf

$$f(y) = \frac{\theta}{2(1+|y|)^{\theta+1}}$$

where $-\infty < y < \infty$ and $\theta > 0$. Find the MLE of θ.

5.52. Let $Y_1, \ldots, Y_n$ be iid random variables from a distribution with pdf

$$f(y) = \frac{2}{\sigma\sqrt{2\pi}} \frac{1}{1+y} \exp\left[\frac{-1}{2\sigma^2}[\log(1+y)]^2\right]$$

where $y > 0$ and $\sigma > 0$. Find the MLE of σ.

Chapter 6
Point Estimation II

Unbiased estimators and mean squared error should be familiar to the reader. A UMVUE is an unbiased point estimator, and complete sufficient statistics are crucial for UMVUE theory. Want point estimators to have small bias and small variance. An estimator with bias that goes to 0 and variance that goes to the FCRLB as the sample size n goes to infinity will often outperform other estimators with bias that goes to zero. Hence the FCRLB will be useful for large sample theory in Chap. 8.

Warning: UMVUE theory is rarely used in practice unless the UMVUE U_n of θ satisfies $U_n = a_n \hat{\theta}_{\text{MLE}}$ where a_n is a constant that could depend on the sample size n. UMVUE theory tends to be useful if the data is iid from a 1P-REF if $U_n = a_n \sum_{i=1}^n t(Y_i)$.

6.1 MSE and Bias

Definition 6.1. Let the sample $\boldsymbol{Y} = (Y_1, \ldots, Y_n)$ where $\boldsymbol{Y}$ has a pdf or pmf $f(\boldsymbol{y}|\boldsymbol{\theta})$ for $\boldsymbol{\theta} \in \Theta$. Assume all relevant expectations exist. Let $\tau(\boldsymbol{\theta})$ be a real valued function of $\boldsymbol{\theta}$, and let $T \equiv T(Y_1, \ldots, Y_n)$ be an estimator of $\tau(\boldsymbol{\theta})$. The **bias** of the estimator T for $\tau(\boldsymbol{\theta})$ is

$$B(T) \equiv B_{\tau(\boldsymbol{\theta})}(T) \equiv \text{Bias}(T) \equiv \text{Bias}_{\tau(\boldsymbol{\theta})}(T) = E_{\boldsymbol{\theta}}(T) - \tau(\boldsymbol{\theta}). \tag{6.1}$$

The *mean squared error* (**MSE**) of an estimator T for $\tau(\boldsymbol{\theta})$ is

$$\text{MSE}(T) \equiv \text{MSE}_{\tau(\boldsymbol{\theta})}(T) = E_{\boldsymbol{\theta}}[(T - \tau(\boldsymbol{\theta}))^2]$$

$$= \text{Var}_{\boldsymbol{\theta}}(T) + [\text{Bias}_{\tau(\boldsymbol{\theta})}(T)]^2. \tag{6.2}$$

T is an *unbiased estimator* of $\tau(\boldsymbol{\theta})$ if

$$E_{\boldsymbol{\theta}}(T) = \tau(\boldsymbol{\theta}) \tag{6.3}$$

D.J. Olive, *Statistical Theory and Inference*, DOI 10.1007/978-3-319-04972-4_6,

for all $\boldsymbol{\theta} \in \Theta$. Notice that $\text{Bias}_{\tau(\boldsymbol{\theta})}(T) = 0$ for all $\boldsymbol{\theta} \in \Theta$ if T is an unbiased estimator of $\tau(\boldsymbol{\theta})$.

Notice that the bias and MSE are functions of $\boldsymbol{\theta}$ for $\boldsymbol{\theta} \in \Theta$. If $\text{MSE}_{\tau(\boldsymbol{\theta})}(T_1) < \text{MSE}_{\tau(\boldsymbol{\theta})}(T_2)$ for all $\boldsymbol{\theta} \in \Theta$, then T_1 is "a better estimator" of $\tau(\boldsymbol{\theta})$ than T_2. So estimators with small MSE are judged to be better than ones with large MSE. Often T_1 has smaller MSE than T_2 for some $\boldsymbol{\theta}$ but larger MSE for other values of $\boldsymbol{\theta}$.

Often θ is real valued. A common problem considers a class of estimators $T_k(\boldsymbol{Y})$ of $\tau(\theta)$ where $k \in \Lambda$. Find the MSE as a function of k and then find the value $k_o \in \Lambda$ that is the global minimizer of $\text{MSE}(k) \equiv \text{MSE}(T_k)$. This type of problem is a lot like the MLE problem except you need to find the global min rather than the global max. This type of problem can often be done if $T_k = kW_1(\boldsymbol{X}) + (1-k)W_2(\boldsymbol{X})$ where both W_1 and W_2 are unbiased estimators of $\tau(\theta)$ and $0 \le k \le 1$.

Example 6.1. If $X_1, \ldots, X_n$ are iid $N(\mu, \sigma^2)$, then $k_o = n+1$ will minimize the MSE for estimators of σ^2 of the form

$$S^2(k) = \frac{1}{k}\sum_{i=1}^{n}(X_i - \overline{X})^2$$

where $k > 0$. See Problem 6.2.

Example 6.2. Find the bias and MSE (as a function of n and c) of an estimator $T = c\sum_{i=1}^n Y_i$ or ($T = b\overline{Y}$) of μ when $Y_1, \ldots, Y_n$ are iid with $E(Y_1) = \mu = \theta$ and $V(Y_i) = \sigma^2$.
Solution: $E(T) = c\sum_{i=1}^n E(Y_i) = nc\mu$, $V(T) = c^2\sum_{i=1}^n V(Y_i) = nc^2\sigma^2$, $B(T) = E(T) - \mu$ and $\text{MSE}(T) = V(T) + [B(T)]^2$. (For $T = b\overline{Y}$, use $c = b/n$.)

Example 6.3. Suppose that $Y_1, \ldots, Y_n$ are independent binomial(m_i, ρ) where the $m_i \ge 1$ are known constants. Let

$$T_1 = \frac{\sum_{i=1}^n Y_i}{\sum_{i=1}^n m_i} \quad \text{and} \quad T_2 = \frac{1}{n}\sum_{i=1}^n \frac{Y_i}{m_i}$$

be estimators of ρ.
a) Find MSE(T_1).
b) Find MSE(T_2).
c) Which estimator is better?
Hint: by the arithmetic–geometric–harmonic mean inequality,

$$\frac{1}{n}\sum_{i=1}^n m_i \ge \frac{n}{\sum_{i=1}^n \frac{1}{m_i}}.$$

Solution: a)

$$E(T_1) = \frac{\sum_{i=1}^n E(Y_i)}{\sum_{i=1}^n m_i} = \frac{\sum_{i=1}^n m_i \rho}{\sum_{i=1}^n m_i} = \rho,$$

so MSE$(T_1) = V(T_1) =$

$$\frac{1}{(\sum_{i=1}^n m_i)^2} V(\sum_{i=1}^n Y_i) = \frac{1}{(\sum_{i=1}^n m_i)^2} \sum_{i=1}^n V(Y_i) = \frac{1}{(\sum_{i=1}^n m_i)^2} \sum_{i=1}^n m_i \rho(1-\rho)$$

$$= \frac{\rho(1-\rho)}{\sum_{i=1}^n m_i}.$$

b)

$$E(T_2) = \frac{1}{n} \sum_{i=1}^n \frac{E(Y_i)}{m_i} = \frac{1}{n} \sum_{i=1}^n \frac{m_i \rho}{m_i} = \frac{1}{n} \sum_{i=1}^n \rho = \rho,$$

so MSE$(T_2) = V(T_2) =$

$$\frac{1}{n^2} V\left(\sum_{i=1}^n \frac{Y_i}{m_i}\right) = \frac{1}{n^2} \sum_{i=1}^n V\left(\frac{Y_i}{m_i}\right) = \frac{1}{n^2} \sum_{i=1}^n \frac{V(Y_i)}{(m_i)^2} = \frac{1}{n^2} \sum_{i=1}^n \frac{m_i \rho(1-\rho)}{(m_i)^2}$$

$$= \frac{\rho(1-\rho)}{n^2} \sum_{i=1}^n \frac{1}{m_i}.$$

c) The hint

$$\frac{1}{n} \sum_{i=1}^n m_i \geq \frac{n}{\sum_{i=1}^n \frac{1}{m_i}}$$

implies that

$$\frac{n}{\sum_{i=1}^n m_i} \leq \frac{\sum_{i=1}^n \frac{1}{m_i}}{n} \quad \text{and} \quad \frac{1}{\sum_{i=1}^n m_i} \leq \frac{\sum_{i=1}^n \frac{1}{m_i}}{n^2}.$$

Hence MSE$(T_1) \leq$ MSE(T_2), and T_1 is better.

6.2 Exponential Families, UMVUEs, and the FCRLB

In the class of unbiased estimators, the UMVUE is best since the UMVUE has the smallest variance, hence the smallest MSE. Often the MLE and method of moments estimator are biased but have a smaller MSE than the UMVUE. MLEs and method of moments estimators are widely used because they often have good statistical properties and are relatively easy to compute. Sometimes the UMVUE, MLE, and method of moments estimators for θ are the same for a 1P-REF when $\hat{\theta} = \frac{1}{n} \sum_{i=1}^n t(Y_i)$ and $\theta = E(\hat{\theta}) = E[t(Y)]$. See Chap. 10 for examples.

Definition 6.2. Let the sample $\boldsymbol{Y} = (Y_1, \ldots, Y_n)$ where $\boldsymbol{Y}$ has a pdf or pmf $f(\boldsymbol{y}|\theta)$ for $\theta \in \Theta$. Assume all relevant expectations exist. Let $\tau(\theta)$ be a real valued function of θ, and let $U \equiv U(Y_1, \ldots, Y_n)$ be an estimator of $\tau(\theta)$. Then U is the *uniformly minimum variance unbiased estimator* (**UMVUE**) of $\tau(\theta)$ if U is an unbiased estimator of $\tau(\theta)$ and if $\text{Var}_\theta(U) \leq \text{Var}_\theta(\text{W})$ for all $\theta \in \Theta$ where W is any other unbiased estimator of $\tau(\theta)$.

The following theorem is the most useful method for finding UMVUEs since if $Y_1, \ldots, Y_n$ are iid from a 1P-REF $f(y|\theta) = h(y)c(\theta)\exp[w(\theta)t(y)]$ where $\eta = w(\theta) \in \Omega = (a,b)$ and $a < b$ are not necessarily finite, then $T(\boldsymbol{Y}) = \sum_{i=1}^n t(Y_i)$ is a complete sufficient statistic. It will turn out that $E_\theta[W(\boldsymbol{Y})|T(\boldsymbol{Y})] \equiv E[W(\boldsymbol{Y})|T(\boldsymbol{Y})]$ does not depend on θ. Hence $U = E[W(\boldsymbol{Y})|T(\boldsymbol{Y})]$ is a statistic.

Theorem 6.1, Lehmann–Scheffé UMVUE (LSU) Theorem: If $T(\boldsymbol{Y})$ is a complete sufficient statistic for θ, then

$$U = g(T(\boldsymbol{Y})) \tag{6.4}$$

is the UMVUE of its expectation $E_\theta(U) = E_\theta[g(T(\boldsymbol{Y}))]$. In particular, if $W(\boldsymbol{Y})$ is *any unbiased estimator* of $\tau(\theta)$, then

$$U \equiv g(T(\boldsymbol{Y})) = E[W(\boldsymbol{Y})|T(\boldsymbol{Y})] \tag{6.5}$$

is the UMVUE of $\tau(\theta)$. If $V_\theta(U) < \infty$ for all $\theta \in \Theta$, then U is the unique UMVUE of $\tau(\theta) = E_\theta[g(T(\boldsymbol{Y}))]$.

The process (6.5) is called Rao–Blackwellization because of the following theorem. The theorem is also called the Rao–Blackwell–Lehmann–Scheffé theorem. Theorem 6.2 shows that if W is an unbiased estimator, then $\phi(T) = E(W|T)$ is a better unbiased estimator than W in that $\text{MSE}_\theta(\phi(T)) \leq \text{MSE}_\theta(W)$ for all $\theta \in \Theta$.

Theorem 6.2, Rao–Blackwell Theorem. Let $W \equiv W(\boldsymbol{Y})$ be an unbiased estimator of $\tau(\theta)$ and let $T \equiv T(\boldsymbol{Y})$ be a sufficient statistic for θ. Then $\phi(T) = E[W|T]$ is an unbiased estimator of $\tau(\theta)$ and $\text{VAR}_\theta[\phi(T)] \leq \text{VAR}_\theta(W)$ for all $\theta \in \Theta$.

Proof. Notice that $\phi(T)$ does not depend on θ by the definition of a sufficient statistic, and that $\phi(T)$ is an unbiased estimator for $\tau(\theta)$ since $\tau(\theta) = E_\theta(W) = E_\theta(E(W|T)) = E_\theta(\phi(T))$ by iterated expectations (Theorem 2.10). By Steiner's formula (Theorem 2.11), $\text{VAR}_\theta(W) =$

$$E_\theta[\text{VAR}(W|T)] + \text{VAR}_\theta[E(W|T)] \geq \text{VAR}_\theta[E(W|T)] = \text{VAR}_\theta[\phi(T)]. \quad \square$$

Tips for finding the UMVUE:

i) From the LSU Theorem, if $T(\boldsymbol{Y})$ is complete sufficient statistic and $g(T(\boldsymbol{Y}))$ is a real valued function, then $U = g(T(\boldsymbol{Y}))$ is **the UMVUE of its expectation** $E_\theta[g(T(\boldsymbol{Y}))]$.

ii) Given a complete sufficient statistic $T(\boldsymbol{Y})$ (e.g., $T(\boldsymbol{Y}) = \sum_{i=1}^n t(Y_i)$ if the data are iid from a 1P-REF), the first method for finding the UMVUE of $\tau(\theta)$ is to guess g and show that $E_\theta[g(T(\boldsymbol{Y}))] = \tau(\theta)$ for all θ.

iii) If $T(\boldsymbol{Y})$ is complete, the second method is to find **any unbiased estimator** $W(\boldsymbol{Y})$ of $\tau(\theta)$. Then $U(\boldsymbol{Y}) = E[W(\boldsymbol{Y})|T(\boldsymbol{Y})]$ is the UMVUE of $\tau(\boldsymbol{\theta})$.

This problem is often very hard because guessing g or finding an unbiased estimator W and computing $E[W(\boldsymbol{Y})|T(\boldsymbol{Y})]$ tend to be difficult. Write down the two methods for finding the UMVUE and simplify $E[W(\boldsymbol{Y})|T(\boldsymbol{Y})]$. If you are asked to find the UMVUE of $\tau(\theta)$, see if an unbiased estimator $W(\boldsymbol{Y})$ is given in the problem. Also check whether you are asked to compute $E[W(\boldsymbol{Y})|T(\boldsymbol{Y}) = t]$ anywhere. Note that $W(\boldsymbol{Y}) = I[Y_1 = k]$ has $E[W(\boldsymbol{Y})] = P(Y_1 = k)$, and the UMVUE of $P(Y_1 = k)$ is $E(I(Y_1 = k)|T(\boldsymbol{Y})] = P[Y_1 = k|T(\boldsymbol{Y})]$ which needs to be simplified. The equality holds since $Z \equiv I(Y_1 = k)|T(\boldsymbol{Y})$ is 1 with probability equal to $P[Y_1 = k|T(\boldsymbol{Y})]$, and $Z = 0$ with probability equal to $1 - P[Y_1 = k|T(\boldsymbol{Y})]$. See a similar calculation in Example 6.6 a).

iv) The following facts can be useful for computing the conditional expectation via Rao–Blackwellization (see Problems 6.7, 6.10, and 6.12). Suppose $Y_1, \ldots, Y_n$ are iid with finite expectation.
a) Then $E[Y_1|\sum_{i=1}^n Y_i = x] = x/n$.
b) If the Y_i are iid Poisson(θ), then $(Y_1|\sum_{i=1}^n Y_i = x) \sim \text{bin}(x, 1/n)$.
c) If the Y_i are iid Bernoulli Ber(ρ), then $(Y_1|\sum_{i=1}^n Y_i = x) \sim \text{Ber}(x/n)$.
d) If the Y_i are iid $N(\mu, \sigma^2)$, then $(Y_1|\sum_{i=1}^n Y_i = x) \sim N[x/n, \sigma^2(1 - 1/n)]$.

Result a) follows since the $Y_i|\sum_{i=1}^n Y_i = x$ have the same distribution. Hence $E[Y_i|\sum_{i=1}^n Y_i = x] = c$ for $i = 1, \ldots, n$ and some constant c. Then $nc =$ $E[\sum_{i=1}^n Y_i|\sum_{i=1}^n Y_i = x] = x$. For b), let $k \in \{0, 1, \ldots, x\}$. Let $W \equiv Y_1|\sum_{i=1}^n Y_i = x$. Then $P(W = k) = P(Y_1 = k|\sum_{i=1}^n Y_i = x) =$

$$\frac{P(Y_1 = k, \sum_{i=1}^n Y_i = x)}{P(\sum_{i=1}^n Y_i = x)} = \frac{P(Y_1 = k, \sum_{i=2}^n Y_i = x - k)}{P(\sum_{i=1}^n Y_i = x)} =$$

$$\frac{P(Y_1 = k)P(\sum_{i=2}^n Y_i = x - k)}{P(\sum_{i=1}^n Y_i = x)}$$

by independence. Now $Y_1 \sim \text{Poisson}(\theta)$, $\sum_{i=2}^n Y_i \sim \text{Poisson}((n-1)\theta)$ and $\sum_{i=1}^n Y_i \sim$ Poisson $(n\theta)$. Algebra will then give result b). For part c), $W \equiv Y_1|\sum_{i=1}^n Y_i = x$ is Bernoulli(π) since $W = 0$ or $W = 1$. Hence $\pi = E(W) = x/n$ by a). For part d), normality follows by Proposition 2.27 and the mean is x/n by a). In Proposition 2.27, $\Sigma_{11} = V(Y_1) = \sigma^2$, $\Sigma_{22} = V(\sum_{i=1}^n Y_i) = n\sigma^2$ and $\Sigma_{12} =$ $\text{Cov}(Y_1, \sum_{i=1}^n Y_i) = \sum_{i=1}^n \text{Cov}(Y_1, Y_i) = \text{Cov}(Y_1, Y_1) = \sigma^2$. Hence the variance is equal to $\Sigma_{11} - \Sigma_{12}\Sigma_{22}^{-1}\Sigma_{21} = \sigma^2 - \sigma^2(n\sigma^2)^{-1}\sigma^2 = \sigma^2(1 - 1/n)$.

Example 6.4. Let $X_1, \ldots, X_n$ be a random sample from a Poisson (λ) distribution. Let $\overline{X}$ and S^2 denote the sample mean and the sample variance, respectively.

a) Show that $\overline{X}$ is uniformly minimum variance unbiased (UMVU) estimator of λ.

b) Show that $E(S^2|\overline{X}) = \overline{X}$.

c) Show that Var$(S^2) \geq$ Var$(\overline{X})$.

d) Show that Var$(S^2) >$ Var$(\overline{X})$.

Solution: a) Since $f(x) = \frac{1}{x!}\exp[\log(\lambda)x]I(x \in \{0,1,\ldots\})$ is a 1P-REF, $\sum_{i=1}^n X_i$ is a complete sufficient statistic and $E(\overline{X}) = \lambda$. Hence $\overline{X} = (\sum_{i=1}^n X_i)/n$ is the UMVUE of λ by the LSU theorem.

b) $E(S^2) = \lambda$ is an unbiased estimator of λ. Hence $E(S^2|\overline{X})$ is the unique UMVUE of λ by the LSU theorem. Thus $E(S^2|\overline{X}) = \overline{X}$ by part a).

c) Note that $\overline{X}$ is the UMVUE and S^2 is an unbiased estimator of λ. Hence $V(\overline{X}) \leq V(S^2)$ by the definition of a UMVUE, and the inequality is strict for at least one value of λ since the UMVUE is unique.

d) By Steiner's formula, $V(S^2) = V(E(S^2|\overline{X})) + E(V(S^2|\overline{X})) = V(\overline{X}) + E(V(S^2|\overline{X})) > V(\overline{X})$.

Often students will be asked to compute a lower bound on the variance of unbiased estimators of $\eta = \tau(\theta)$ when θ is a scalar. Some preliminary results are needed to define the lower bound, known as the FCRLB. The Fisher information, defined below, is also useful for large sample theory in Chap. 8 since often the asymptotic variance of a good estimator of $\tau(\theta)$ is $1/I_n(\tau(\theta))$. Good estimators tend to have a variance $\geq c/n$, so the FCRLB should be c/n for some positive constant c that may depend on the parameters of the distribution. Often $c = [\tau'(\theta)]^2/I_1(\theta)$.

Definition 6.3. Let $\boldsymbol{Y} = (Y_1,\ldots,Y_n)$ have a pdf or pmf $f(\boldsymbol{y}|\theta)$. Then the **information number** or **Fisher Information** is

$$I_{\boldsymbol{Y}}(\theta) \equiv I_n(\theta) = E_\theta\left(\left[\frac{\partial}{\partial\theta}\log(f(\boldsymbol{Y}|\theta))\right]^2\right). \tag{6.6}$$

Let $\eta = \tau(\theta)$ where $\tau'(\theta) \neq 0$. Then

$$I_n(\eta) \equiv I_n(\tau(\theta)) = \frac{I_n(\theta)}{[\tau'(\theta)]^2}. \tag{6.7}$$

Theorem 6.3. a) Equations (6.6) and (6.7) agree if $\tau'(\theta)$ is continuous, $\tau'(\theta) \neq 0$, and $\tau(\theta)$ is one to one and onto so that an inverse function exists such that $\theta = \tau^{-1}(\eta)$.

b) If the $Y_1 \equiv Y$ is from a 1P-REF, then the Fisher information in a sample of size one is

$$I_1(\theta) = -E_\theta\left[\frac{\partial^2}{\partial\theta^2}\log(f(Y|\theta))\right]. \tag{6.8}$$

c) If the $Y_1,\ldots,Y_n$ are iid from a 1P-REF, then

$$I_n(\theta) = nI_1(\theta). \tag{6.9}$$

Hence if $\tau'(\theta)$ exists and is continuous and if $\tau'(\theta) \neq 0$, then

$$I_n(\tau(\theta)) = \frac{nI_1(\theta)}{[\tau'(\theta)]^2}. \tag{6.10}$$

Proof. a) See Lehmann (1999, pp. 467–468).

b) The proof will be for a pdf. For a pmf replace the integrals by sums. By Remark 3.3, the integral and differentiation operators of all orders can be interchanged. Note that

$$0 = E\left[\frac{\partial}{\partial\theta}\log(f(Y|\theta))\right] \tag{6.11}$$

since

$$\frac{\partial}{\partial\theta}1 = 0 = \frac{\partial}{\partial\theta}\int f(y|\theta)dy = \int \frac{\partial}{\partial\theta}f(y|\theta)dy = \int \frac{\frac{\partial}{\partial\theta}f(y|\theta)}{f(y|\theta)}f(y|\theta)dy$$

or

$$0 = \frac{\partial}{\partial\theta}\int f(y|\theta)dy = \int \left[\frac{\partial}{\partial\theta}\log(f(y|\theta))\right] f(y|\theta)dy$$

which is (6.11). Taking second derivatives of the above expression gives

$$0 = \frac{\partial^2}{\partial\theta^2}\int f(y|\theta)dy = \frac{\partial}{\partial\theta}\int \left[\frac{\partial}{\partial\theta}\log(f(y|\theta))\right] f(y|\theta)dy =$$

$$\int \frac{\partial}{\partial\theta}\left(\left[\frac{\partial}{\partial\theta}\log(f(y|\theta))\right] f(y|\theta)\right) dy =$$

$$\int \left[\frac{\partial^2}{\partial\theta^2}\log(f(y|\theta))\right] f(y|\theta)dy + \int \left[\frac{\partial}{\partial\theta}\log(f(y|\theta))\right]\left[\frac{\partial}{\partial\theta}f(y|\theta)\right]\frac{f(y|\theta)}{f(y|\theta)}dy$$

$$= \int \left[\frac{\partial^2}{\partial\theta^2}\log(f(y|\theta))\right] f(y|\theta)dy + \int \left[\frac{\partial}{\partial\theta}\log(f(y|\theta))\right]^2 f(y|\theta)dy$$

or

$$I_1(\theta) = E_\theta\left[\left(\frac{\partial}{\partial\theta}\log f(Y|\theta)\right)^2\right] = -E_\theta\left[\frac{\partial^2}{\partial\theta^2}\log(f(Y|\theta))\right].$$

c) By independence,

$$I_n(\theta) = E_\theta\left[\left(\frac{\partial}{\partial\theta}\log\left(\prod_{i=1}^n f(Y_i|\theta)\right)\right)^2\right] = E_\theta\left[\left(\frac{\partial}{\partial\theta}\sum_{i=1}^n \log(f(Y_i|\theta))\right)^2\right] =$$

$$E_\theta\left[\left(\frac{\partial}{\partial\theta}\sum_{i=1}^n \log(f(Y_i|\theta))\right)\left(\frac{\partial}{\partial\theta}\sum_{j=1}^n \log(f(Y_j|\theta))\right)\right] =$$

$$E_\theta\left[\left(\sum_{i=1}^n \frac{\partial}{\partial\theta}\log(f(Y_i|\theta))\right)\left(\sum_{j=1}^n \frac{\partial}{\partial\theta}\log(f(Y_j|\theta))\right)\right] =$$

$$\sum_{i=1}^n E_\theta\left[\left(\frac{\partial}{\partial\theta}\log(f(Y_i|\theta))\right)^2\right] +$$

$$\sum_{i \neq j}\sum E_\theta\left[\left(\frac{\partial}{\partial\theta}\log(f(Y_i|\theta))\right)\left(\frac{\partial}{\partial\theta}\log(f(Y_j|\theta))\right)\right].$$

Hence

$$I_n(\theta) = nI_1(\theta) + \sum_{i \neq j}\sum E_\theta\left[\left(\frac{\partial}{\partial\theta}\log(f(Y_i|\theta))\right)\right] E_\theta\left[\left(\frac{\partial}{\partial\theta}\log(f(Y_j|\theta))\right)\right]$$

by independence. Hence

$$I_n(\theta) = nI_1(\theta) + n(n-1)\left[E_\theta\left(\frac{\partial}{\partial\theta}\log(f(Y_j|\theta))\right)\right]^2$$

since the Y_i are iid. Thus $I_n(\theta) = nI_1(\theta)$ by Eq. (6.11) which holds since the Y_i are iid from a 1P-REF. □

Definition 6.4. Let $\boldsymbol{Y} = (Y_1, \ldots, Y_n)$ be the data, and consider $\tau(\theta)$ where $\tau'(\theta) \neq 0$. The quantity

$$\text{FCRLB}_n(\tau(\theta)) = \frac{[\tau'(\theta)]^2}{I_n(\theta)}$$

is called the **Fréchet–Cramér–Rao lower bound** (FCRLB) for the variance of unbiased estimators of $\tau(\theta)$. In particular, if $\tau(\theta) = \theta$, then $\text{FCRLB}_n(\theta) = \dfrac{1}{I_n(\theta)}$. The FCRLB is often called the Cramér Rao lower bound (CRLB).

Theorem 6.4, Fréchet–Cramér–Rao Lower Bound or Information Inequality. Let $Y_1, \ldots, Y_n$ be iid from a 1P-REF with pdf or pmf $f(y|\theta)$. Let $W(Y_1, \ldots, Y_n) = W(\boldsymbol{Y})$ be any unbiased estimator of $\tau(\theta) \equiv E_\theta W(\boldsymbol{Y})$. Then

$$\text{VAR}_\theta(W(\boldsymbol{Y})) \geq \text{FCRLB}_n(\tau(\theta)) = \frac{[\tau'(\theta)]^2}{I_n(\theta)} = \frac{[\tau'(\theta)]^2}{nI_1(\theta)} = \frac{1}{I_n(\tau(\theta))}.$$

Proof. By Definition 6.4 and Theorem 6.3c,

$$\text{FCRLB}_n(\tau(\theta)) = \frac{[\tau'(\theta)]^2}{I_n(\theta)} = \frac{[\tau'(\theta)]^2}{nI_1(\theta)} = \frac{1}{I_n(\tau(\theta))}.$$

Since the Y_i are iid from a 1P-REF, by Remark 3.3 the derivative and integral or sum operators can be interchanged when finding the derivative of $E_\theta h(\boldsymbol{Y})$ if

$E_\theta|h(\boldsymbol{Y})| < \infty$. The following argument will be for pdfs. For pmfs, replace the integrals by appropriate sums. Following Casella and Berger (2002, pp. 335–8), the Cauchy Schwarz Inequality is

$$[\mathrm{Cov}(\mathrm{X},\mathrm{Y})]^2 \le \mathrm{V}(\mathrm{X})\mathrm{V}(\mathrm{Y}), \text{ or } \mathrm{V}(\mathrm{X}) \ge \frac{[\mathrm{Cov}(\mathrm{X},\mathrm{Y})]^2}{\mathrm{V}(\mathrm{Y})}.$$

Hence

$$V_\theta(W(\boldsymbol{Y})) \ge \frac{(\mathrm{Cov}_\theta[W(\boldsymbol{Y}), \frac{\partial}{\partial\theta}\log(f(\boldsymbol{Y}|\theta))])^2}{V_\theta[\frac{\partial}{\partial\theta}\log(f(\boldsymbol{Y}|\theta))]}. \tag{6.12}$$

Now

$$E_\theta\left[\frac{\partial}{\partial\theta}\log(f(\boldsymbol{Y}|\theta))\right] = E_\theta\left[\frac{\frac{\partial}{\partial\theta}f(\boldsymbol{Y}|\theta)}{f(\boldsymbol{Y}|\theta)}\right]$$

since the derivative of $\log(h(t))$ is $h'(t)/h(t)$. By the definition of expectation,

$$E_\theta\left[\frac{\partial}{\partial\theta}\log(f(\boldsymbol{Y}|\theta))\right] = \int\cdots\int_{\mathscr{Y}} \frac{\frac{\partial}{\partial\theta}f(\boldsymbol{y}|\theta)}{f(\boldsymbol{y}|\theta)} f(\boldsymbol{y}|\theta)d\boldsymbol{y}$$

$$= \int\cdots\int_{\mathscr{Y}} \frac{\partial}{\partial\theta}f(\boldsymbol{y}|\theta)d\boldsymbol{y} = \frac{d}{d\theta}\int\cdots\int_{\mathscr{Y}} f(\boldsymbol{y}|\theta)d\boldsymbol{y} = \frac{d}{d\theta}1 = 0.$$

Notice that $f(\boldsymbol{y}|\theta) > 0$ on the support $\mathscr{Y}$, that the $f(\boldsymbol{y}|\theta)$ cancelled in the second term, that the derivative was moved outside of the integral by Remark 3.3, and that the integral of $f(\boldsymbol{y}|\theta)$ on the support $\mathscr{Y}$ is equal to 1.

This result implies that

$$\mathrm{Cov}_\theta\left[W(\boldsymbol{Y}), \frac{\partial}{\partial\theta}\log(f(\boldsymbol{Y}|\theta))\right] = E_\theta\left[W(\boldsymbol{Y})\ \frac{\partial}{\partial\theta}\log(f(\boldsymbol{Y}|\theta))\right]$$

$$= E_\theta\left[\frac{W(\boldsymbol{Y})\ \left(\frac{\partial}{\partial\theta}f(\boldsymbol{Y}|\theta)\right)}{f(\boldsymbol{Y}|\theta)}\right]$$

since the derivative of $\log(h(t))$ is $h'(t)/h(t)$. By the definition of expectation, the right-hand side is equal to

$$\int\cdots\int_{\mathscr{Y}} \frac{W(\boldsymbol{y})\frac{\partial}{\partial\theta}f(\boldsymbol{y}|\theta)}{f(\boldsymbol{y}|\theta)} f(\boldsymbol{y}|\theta)d\boldsymbol{y} = \frac{d}{d\theta}\int\cdots\int_{\mathscr{Y}} W(\boldsymbol{y})f(\boldsymbol{y}|\theta)d\boldsymbol{y}$$

$$= \frac{d}{d\theta}E_\theta W(\boldsymbol{Y}) = \tau'(\theta) = \mathrm{Cov}_\theta\left[W(\boldsymbol{Y}), \frac{\partial}{\partial\theta}\log(f(\boldsymbol{Y}|\theta))\right]. \tag{6.13}$$

Since

$$E_\theta\left[\frac{\partial}{\partial\theta}\log\ f(\boldsymbol{Y}|\theta)\right] = 0,$$

$$V_\theta\left[\frac{\partial}{\partial\theta}\log(f(\boldsymbol{Y}|\theta))\right] = E_\theta\left(\left[\frac{\partial}{\partial\theta}\log(f(\boldsymbol{Y}|\theta))\right]^2\right) = I_n(\theta) \tag{6.14}$$

by Definition 6.3. Plugging (6.13) and (6.14) into (6.12) gives the result. □

Theorem 6.4 is not very useful in applications. If the data are iid from a 1P-REF, then $\text{FCRLB}_n(\tau(\theta)) = [\tau'(\theta)]^2/[nI_1(\theta)]$ by Theorem 6.4. Notice that $W(\boldsymbol{Y})$ is an unbiased estimator of $\tau(\theta)$ since $E_\theta W(\boldsymbol{Y}) = \tau(\theta)$. Hence if the data are iid from a 1P-REF and if $\text{VAR}_\theta(W(\boldsymbol{Y})) = \text{FCRLB}_n(\tau(\theta))$ for all $\theta \in \Theta$, then $W(\boldsymbol{Y})$ is the UMVUE of $\tau(\theta)$; however, this technique for finding a UMVUE rarely works since typically equality holds only if
1) the data come from a 1P-REF with complete sufficient statistic T, and
2) $W = a + bT$ is a linear function of T.
The FCRLB inequality will typically be strict for nonlinear functions of T if the data is iid from a 1P-REF. If T is complete, $g(T)$ is the UMVUE of its expectation, and determining that T is the complete sufficient statistic from a 1P-REF is simpler than computing $\text{VAR}_\theta(W)$ and $\text{FCRLB}_n(\tau(\theta))$. If the family is not an exponential family, the FCRLB may **not be a lower bound** on the variance of unbiased estimators of $\tau(\theta)$.

Example 6.5. Let $Y_1, \ldots, Y_n$ be iid random variables with pdf

$$f(y) = \frac{2}{\sqrt{2\pi}\lambda}\frac{1}{y}I_{[0,1]}(y)\exp\left[\frac{-(\log(y))^2}{2\lambda^2}\right]$$

where $\lambda > 0$. Then $[\log(Y_i)]^2 \sim G(1/2, 2\lambda^2) \sim \lambda^2\chi_1^2$.

a) Find the uniformly minimum variance estimator (UMVUE) of λ^2.

b) Find the information number $I_1(\lambda)$.

c) Find the Fréchet–Cramér–Rao lower bound (FCRLB) for estimating $\tau(\lambda) = \lambda^2$.

Solution. a) This is a one-parameter exponential family with complete sufficient statistic $T_n = \sum_{i=1}^n [\log(Y_i)]^2$. Now $E(T_n) = nE([\log(Y_i)]^2) = n\lambda^2$. Hence $E(T_n/n) = \lambda^2$ and T_n/n is the UMVUE of λ^2 by the LSU Theorem.

b) Now

$$\log(f(y|\lambda)) = \log(2/\sqrt{2\pi}) - \log(\lambda) - \log(y) - \frac{[\log(y)]^2}{2\lambda^2}.$$

Hence

$$\frac{d}{d\lambda}\log(f(y|\lambda)) = \frac{-1}{\lambda} + \frac{[\log(y)]^2}{\lambda^3},$$

and

$$\frac{d^2}{d\lambda^2}\log(f(y|\lambda)) = \frac{1}{\lambda^2} - \frac{3[\log(y)]^2}{\lambda^4}.$$

Thus

$$I_1(\lambda) = -E\left[\frac{1}{\lambda^2} - \frac{3[\log(Y)]^2}{\lambda^4}\right] = \frac{-1}{\lambda^2} + \frac{3\lambda^2}{\lambda^4} = \frac{2}{\lambda^2}.$$

c)

$$\text{FCRLB}_n(\tau(\lambda)) = \frac{[\tau'(\lambda)]^2}{nI_1(\lambda)}.$$

Now $\tau(\lambda) = \lambda^2$ and $\tau'(\lambda) = 2\lambda$. So

$$\text{FCRLB}_n(\tau(\lambda)) = \frac{4\lambda^2}{n2/\lambda^2} = \frac{2\lambda^4}{n}.$$

Example 6.6. Suppose that $X_1,\ldots,X_n$ are iid Bernoulli(p) where $n \geq 2$ and $0 < p < 1$ is the unknown parameter.

a) Derive the UMVUE of $\tau(p) = e^2(p(1-p))$.

b) Find the FCRLB for estimating $\tau(p) = e^2(p(1-p))$.

Solution: a) Consider the statistic $W = X_1(1-X_2)$ which is an unbiased estimator of $\psi(p) = p(1-p)$. The statistic $T = \sum_{i=1}^n X_i$ is both complete and sufficient. The possible values of W are 0 or 1. Let $U = \phi(T)$ where

$$\begin{aligned}\phi(t) &= E[X_1(1-X_2)|T=t]\\ &= 0P[X_1(1-X_2)=0|T=t] + 1P[X_1(1-X_2)=1|T=t]\\ &= P[X_1(1-X_2)=1|T=t]\\ &= \frac{P[X_1=1, X_2=0 \text{ and } \sum_{i=1}^n X_i = t]}{P[\sum_{i=1}^n X_i = t]}\\ &= \frac{P[X_1=1]P[X_2=0]P[\sum_{i=3}^n X_i = t-1]}{P[\sum_{i=1}^n X_i = t]}.\end{aligned}$$

Now $\sum_{i=3}^n X_i$ is Bin$(n-2,p)$ and $\sum_{i=1}^n X_i$ is Bin(n,p). Thus

$$\phi(t) = \frac{p(1-p)[\binom{n-2}{t-1}p^{t-1}(1-p)^{n-t-1}]}{\binom{n}{t}p^t(1-p)^{n-t}}$$

$$= \frac{\binom{n-2}{t-1}}{\binom{n}{t}} = \frac{(n-2)!}{(t-1)!(n-2-t+1)!}\;\frac{t(t-1)!(n-t)(n-t-1)!}{n(n-1)(n-2)!} = \frac{t(n-t)}{n(n-1)}$$

$$= \frac{\frac{t}{n}(n-n\frac{t}{n})}{n-1} = \frac{\frac{t}{n}n(1-\frac{t}{n})}{n-1} = \frac{n}{n-1}\overline{x}(1-\overline{x}).$$

Thus $\frac{n}{n-1}\overline{X}(1-\overline{X})$ is the UMVUE of $p(1-p)$ and $e^2U = e^2\frac{n}{n-1}\overline{X}(1-\overline{X})$ is the UMVUE of $\tau(p) = e^2 p(1-p)$.

Alternatively, $\overline{X}$ is a complete sufficient statistic, so try an estimator of the form $U = a(\overline{X})^2 + b\overline{X} + c$. Then U is the UMVUE if $E_p(U) = e^2 p(1-p) = e^2(p-p^2)$. Now $E(\overline{X}) = E(X_1) = p$ and $V(\overline{X}) = V(X_1)/n = p(1-p)/n$ since $\sum X_i \sim \text{Bin}(n,p)$. So $E[(\overline{X})^2] = V(\overline{X}) + [E(\overline{X})]^2 = p(1-p)/n + p^2$. So $E_p(U) = a[p(1-p)/n] + ap^2 + bp + c$

$$= \frac{ap}{n} - \frac{ap^2}{n} + ap^2 + bp + c = \left(\frac{a}{n} + b\right)p + \left(a - \frac{a}{n}\right)p^2 + c.$$

So $c = 0$ and $a - \frac{a}{n} = a\frac{n-1}{n} = -e^2$ or

$$a = \frac{-n}{n-1}e^2.$$

Hence $\frac{a}{n} + b = e^2$ or

$$b = e^2 - \frac{a}{n} = e^2 + \frac{n}{n(n-1)}e^2 = \frac{n}{n-1}e^2.$$

So

$$U = \frac{-n}{n-1}e^2(\overline{X})^2 + \frac{n}{n-1}e^2\overline{X} = \frac{n}{n-1}e^2\overline{X}(1-\overline{X}).$$

b) The FCRLB for $\tau(p)$ is $[\tau'(p)]^2/nI_1(p)$. Now $f(x) = p^x(1-p)^{1-x}$, so $\log f(x) = x\log(p) + (1-x)\log(1-p)$. Hence

$$\frac{\partial \log f}{\partial p} = \frac{x}{p} - \frac{1-x}{1-p}$$

and

$$\frac{\partial^2 \log f}{\partial p^2} = \frac{-x}{p^2} - \frac{1-x}{(1-p)^2}.$$

So

$$I_1(p) = -E\left(\frac{\partial^2 \log f}{\partial p^2}\right) = -\left(\frac{-p}{p^2} - \frac{1-p}{(1-p)^2}\right) = \frac{1}{p(1-p)}.$$

So

$$\text{FCRLB}_n = \frac{[e^2(1-2p)]^2}{\frac{n}{p(1-p)}} = \frac{e^4(1-2p)^2 p(1-p)}{n}.$$

Example 6.7. Let $X_1, \ldots, X_n$ be iid random variables with pdf

$$f(x) = \frac{1}{\lambda}\phi x^{\phi-1}\frac{1}{1+x^\phi}\exp\left[-\frac{1}{\lambda}\log(1+x^\phi)\right]$$

where x, ϕ, and λ are all positive. If ϕ is known, find the uniformly minimum unbiased estimator of λ using the fact that $\log(1+X_i^\phi) \sim \text{Gamma}\ (\nu = 1, \lambda)$.

Solution: This is a regular one-parameter exponential family with complete sufficient statistic $T_n = \sum_{i=1}^n \log(1+X_i^\phi) \sim G(n,\lambda)$. Hence $E(T_n) = n\lambda$ and T_n/n is the UMVUE of λ.

Example 6.8. If $\sum_{i=1}^n Y_i$ is a complete sufficient statistic for θ, then by the LSU theorem, $e^{t\sum_{i=1}^n Y_i}$ is the UMVUE of $E[e^{t\sum_{i=1}^n Y_i}] = m_{\sum_{i=1}^n Y_i}(t)$, the mgf of $\sum_{i=1}^n Y_i$. Refer to Theorems 2.17 and 2.18 for the following special cases.

a) If $Y_1,\ldots,Y_n$ are iid BIN(k,ρ) where k is known, then $\sum_{i=1}^n Y_i \sim$ BIN(nk,ρ), and $e^{t\sum_{i=1}^n Y_i}$ is the UMVUE of

$$[(1-\rho)+\rho e^t]^{nk}.$$

b) If $Y_1,\ldots,Y_n$ are iid EXP(λ), then $\sum_{i=1}^n Y_i \sim G(n,\lambda)$, and $e^{t\sum_{i=1}^n Y_i}$ is the UMVUE of

$$\left(\frac{1}{1-\lambda t}\right)^n$$

for $t < 1/\lambda$.

c) If $Y_1,\ldots,Y_n$ are iid $G(\nu,\lambda)$ where ν is known, then $\sum_{i=1}^n Y_i \sim G(n\nu,\lambda)$, and $e^{t\sum_{i=1}^n Y_i}$ is the UMVUE of

$$\left(\frac{1}{1-\lambda t}\right)^{n\nu}$$

for $t < 1/\lambda$.

d) If $Y_1,\ldots,Y_n$ are iid $N(\mu,\sigma^2)$ where σ^2 is known, then $\sum_{i=1}^n Y_i \sim N(n\mu,n\sigma^2)$, and $e^{t\sum_{i=1}^n Y_i}$ is the UMVUE of

$$\exp(tn\mu + t^2 n\sigma^2/2).$$

e) If $Y_1,\ldots,Y_n$ are iid Poisson(θ), then $\sum_{i=1}^n Y_i \sim$ Poisson($n\theta$), and $e^{t\sum_{i=1}^n Y_i}$ is the UMVUE of

$$\exp(n\theta(e^t-1)).$$

f) If $Y_1,\ldots,Y_n$ are iid NB(r,ρ) where r is known, then $\sum_{i=1}^n Y_i \sim$ NB(nr,ρ), and $e^{t\sum_{i=1}^n Y_i}$ is the UMVUE of

$$\left[\frac{\rho}{1-(1-\rho)e^t}\right]^{nr}$$

for $t < -\log(1-\rho)$.

Example 6.9. Let $X_1,\ldots,X_n$ be a random sample from a Poisson distribution with mean θ.

a) For $a > 0$, find the uniformly minimum variance unbiased estimator (UMVUE) of $g(\theta) = e^{a\theta}$.

b) Prove the identity:

$$E\left[2^{X_1}|T\right] = \left(1+\frac{1}{n}\right)^T.$$

Solution: a) By Example 6.4, $T = \sum_{i=1}^n X_i \sim \text{Poisson}(n\theta)$ is a complete sufficient statistic for θ. Hence the mgf of T is

$$E(e^{tT}) = m_T(t) = \exp[n\theta(e^t - 1)].$$

Thus $n(e^t - 1) = a$, or $e^t = a/n + 1$, or $e^t = (a+n)/n$, or $t = \log[(a+n)/n]$. Thus

$$e^{tT} = (e^t)^T = \left(\frac{a+n}{n}\right)^T = \exp\left[T \log\left(\frac{a+n}{n}\right)\right]$$

is the UMVUE of $e^{a\theta}$ by the LSU theorem.

b) Let $X = X_1$, and note that 2^X is an unbiased estimator of e^θ since

$$2^X = e^{\log(2^X)} = e^{(\log 2)X},$$

$$\text{and } E(2^X) = m_X(\log 2) = \exp[\theta(e^{\log 2} - 1)] = e^\theta.$$

Thus $E[2^X|T]$ is the UMVUE of $E(2^X) = e^\theta$ by the LSU theorem. By part a) with $a = 1$,

$$E[2^X|T] = \left(\frac{1+n}{n}\right)^T.$$

The following theorem compares the UMVUE with the estimator that minimizes the MSE for one-parameter exponential families. Note that the constant c cannot depend on the unknown parameter θ since $cT(\boldsymbol{Y})$ needs to be a statistic. Often $\theta X \sim \theta G(1,1) \sim G(1,\theta)$. Note $c_M/c_U \to 1$ as $n \to \infty$. Hence the UMVUE and the estimator that minimizes the MSE behave similarly in terms of MSE for large n. See Problem 6.35.

Theorem 6.5. Let $Y_1, \ldots, Y_n$ be iid from a one-parameter exponential family with pdf or pmf $f(y|\theta)$ with complete sufficient statistic $T(\boldsymbol{Y}) = \sum_{i=1}^n t(Y_i)$ where $t(Y_i) \sim \theta X$ and X has a known distribution with known mean $E(X)$ and known variance $V(X)$. Let $W_n = c\,T(\boldsymbol{Y})$ be an estimator of θ where c is a constant.

a) The value c that minimizes the MSE is

$$c_M = \frac{E(X)}{V(X) + n[E(X)]^2}.$$

b) The UMVUE of θ is $\dfrac{T(\boldsymbol{Y})}{nE(X)}$ which uses $c_U = \dfrac{1}{nE(X)}$.

Proof. a) $E(W_n) = c\sum_{i=1}^n E(t(Y_i)) = cn\theta E(X)$, and $V(W_n) = c^2 \sum_{i=1}^n V(t(Y_i)) = c^2 n\theta^2 V(X)$. Hence $\text{MSE}(c) \equiv \text{MSE}(W_n) = V(W_n) + [E(W_n) - \theta]^2 = c^2 n\theta^2 V(X) + (cn\theta E(X) - \theta)^2$. Thus

$$\frac{d\,\text{MSE}(c)}{dc} = 2cn\theta^2 V(X) + 2(cn\theta E(X) - \theta)n\theta E(X) \stackrel{\text{set}}{=} 0,$$

or

$$c(n\theta^2 V(X) + n^2\theta^2[E(X)]^2) = n\theta^2 E(X),$$

or

$$c = \frac{E(X)}{V(X) + n[E(X)]^2},$$

which is unique. Now

$$\frac{d^2 \text{ MSE}(c)}{dc^2} = 2[n\theta^2 V(X) + n^2\theta^2[E(X)]^2] > 0.$$

So MSE(c) is convex and $c = c_M$ is the minimizer.

b) $E[c_U T(\boldsymbol{Y})] = \theta$, hence $c_U T(\boldsymbol{Y})$ is the UMVUE of θ by the LSU theorem. □

Remark 6.1. Chapter 10 has several UMVUE examples.

6.3 Summary

1) The **bias** of the estimator T for $\tau(\boldsymbol{\theta})$ is

$$B(T) \equiv B_{\tau(\boldsymbol{\theta})}(T) \equiv \text{Bias}_{\tau(\boldsymbol{\theta})}(T) = E_{\boldsymbol{\theta}} T - \tau(\boldsymbol{\theta})$$

and the MSE is

$$\text{MSE}_{\tau(\boldsymbol{\theta})}(T) = E_{\boldsymbol{\theta}}[(T - \tau(\boldsymbol{\theta}))^2] = V_{\boldsymbol{\theta}}(T) + [\text{Bias}_{\tau(\boldsymbol{\theta})}(T)]^2.$$

2) T is an *unbiased estimator* of $\tau(\boldsymbol{\theta})$ if $E_{\boldsymbol{\theta}}[T] = \tau(\boldsymbol{\theta})$ for all $\boldsymbol{\theta} \in \Theta$.

3) Let $U \equiv U(Y_1, \ldots, Y_n)$ be an estimator of $\tau(\theta)$. Then U is the **UMVUE** of $\tau(\theta)$ if U is an unbiased estimator of $\tau(\theta)$ and if $\text{VAR}_\theta(U) \le \text{VAR}_\theta(W)$ for all $\theta \in \Theta$ where W is any other unbiased estimator of $\tau(\theta)$.

4) If $Y_1, \ldots, Y_n$ are iid from a 1P-REF $f(y|\theta) = h(y)c(\theta)\exp[w(\theta)t(y)]$ where $\eta = w(\theta) \in \Omega = (a,b)$, and if $T \equiv T(\boldsymbol{Y}) = \sum_{i=1}^n t(Y_i)$, then by the LSU Theorem, $g(T)$ is the UMVUE of its expectation $\tau(\theta) = E_\theta(g(T))$.

5) Given a complete sufficient statistic $T(\boldsymbol{Y})$ and any unbiased estimator $W(\boldsymbol{Y})$ of $\tau(\theta)$, then $U(\boldsymbol{Y}) = E[W(\boldsymbol{Y})|T(\boldsymbol{Y})]$ is the UMVUE of $\tau(\theta)$.

7) $I_n(\theta) = E_\theta[(\frac{\partial}{\partial\theta} \log f(\boldsymbol{Y}|\theta))^2]$.

8) $\text{FCRLB}_n(\tau(\theta)) = \dfrac{[\tau'(\theta)]^2}{I_n(\theta)}$.

9) If $Y_1, \ldots, Y_n$ are iid from a 1P-REF $f(y|\theta) = h(y)c(\theta)\exp[w(\theta)t(y)]$, then a)

$$I_1(\theta) = -E_\theta\left[\frac{\partial^2}{\partial\theta^2} \log(f(Y|\theta))\right].$$

b)

$$I_n(\tau(\theta)) = \frac{nI_1(\theta)}{[\tau'(\theta)]^2}.$$

c)

$$\text{FCRLB}_n(\tau(\theta)) = \frac{[\tau'(\theta)]^2}{nI_1(\theta)}.$$

d) Information inequality: Let $Y_1,\ldots,Y_n$ be iid from a 1P-REF and let $W(\boldsymbol{Y})$ be any unbiased estimator of $\tau(\theta) \equiv E_\theta[W(\boldsymbol{Y})]$. Then

$$\text{VAR}_\theta(W(\boldsymbol{Y})) \geq \text{FCRLB}_n(\tau(\theta)) = \frac{[\tau'(\theta)]^2}{nI_1(\theta)}.$$

e) Rule of thumb for a 1P-REF: Let $T(\boldsymbol{Y}) = \sum_{i=1}^n t(Y_i)$ and $\tau(\theta) = E_\theta(g(T(\boldsymbol{Y}))$. Then $g(T(\boldsymbol{Y}))$ is the UMVUE of $\tau(\theta)$ by LSU, but the information inequality is strict for nonlinear functions $g(T(\boldsymbol{Y}))$. Expect the equality

$$\text{VAR}_\theta(g(T(\boldsymbol{Y})) = \frac{[\tau'(\theta)]^2}{nI_1(\theta)}$$

only if g is a linear function, i.e., $g(T) = a + bT$ for some fixed constants a and b.

10) If the family is not an exponential family, the FCRLB may **not be a lower bound** on the variance of unbiased estimators of $\tau(\theta)$.

6.4 Complements

For a more precise statement of when the FCRLB is achieved and for some counterexamples, see Wijsman (1973) and Joshi (1976). Although the FCRLB is not very useful for finding UMVUEs, similar ideas are useful for finding the asymptotic variances of UMVUEs and MLEs. See Chap. 8 and Portnoy (1977).

Karakostas (1985) has useful references for UMVUEs. Also see Guenther (1978) and Hudson (1978).

6.5 Problems

PROBLEMS WITH AN ASTERISK * ARE ESPECIALLY USEFUL.

Refer to Chap. 10 for the pdf or pmf of the distributions in the problems below.

6.1*. Let W be an estimator of $\tau(\theta)$. Show that

$$\text{MSE}_{\tau(\theta)}(W) = \text{Var}_\theta(W) + [\text{Bias}_{\tau(\theta)}(W)]^2.$$

6.2^Q. Let $X_1,\ldots,X_n$ be independent identically distributed random variable from a $N(\mu,\sigma^2)$ distribution. Hence $E(X_1)=\mu$ and $\text{VAR}(X_1)=\sigma^2$. Consider estimators of σ^2 of the form

$$S^2(k)=\frac{1}{k}\sum_{i=1}^{n}(X_i-\overline{X})^2$$

where $k>0$ is a constant to be chosen. Determine the value of k which gives the smallest mean square error. (Hint: Find the MSE as a function of k, then take derivatives with respect to k. Also, use Theorem 4.1c and Remark 5.1 VII.)

6.3. Let $X_1,\ldots,X_n$ be iid $N(\mu,1)$ random variables. Find $\tau(\mu)$ such that $T(X_1,\ldots,X_n)=(\sum_{i=1}^n X_i)^2$ is the UMVUE of $\tau(\mu)$.

6.4. Let $X\sim N(\mu,\sigma^2)$ where σ^2 is known. Find the Fisher information $I_1(\mu)$.

6.5. Let $X\sim N(\mu,\sigma^2)$ where μ is known. Find the Fisher information $I_1(\sigma^2)$.

6.6. Let $X_1,\ldots,X_n$ be iid $N(\mu,\sigma^2)$ random variables where μ is **known** and $\sigma^2>0$. Then $W=\sum_{i=1}^n(X_i-\mu)^2$ is a complete sufficient statistic and $W\sim\sigma^2\chi^2_n$. From Chap. 10,

$$EY^k=\frac{2^k\Gamma(k+n/2)}{\Gamma(n/2)}$$

if $Y\sim\chi^2_n$. Hence

$$T_k(X_1,\ldots,X_n)\equiv\frac{\Gamma(n/2)W^k}{2^k\Gamma(k+n/2)}$$

is the UMVUE of $\tau_k(\sigma^2)=\sigma^{2k}$ for $k>0$. Note that $\tau_k(\theta)=(\theta)^k$ and $\theta=\sigma^2$.

a) Show that

$$\text{Var}_\theta T_k(X_1,\ldots,X_n)=\sigma^{4k}\left[\frac{\Gamma(n/2)\Gamma(2k+n/2)}{\Gamma(k+n/2)\Gamma(k+n/2)}-1\right]\equiv c_k\sigma^{4k}.$$

b) Let $k=2$ and show that $\text{Var}_\theta[T_2]-\text{FCRLB}(\tau_2(\theta))>0$ where $\text{FCRLB}(\tau_2(\theta))$ is for estimating $\tau_2(\sigma^2)=\sigma^4$ and $\theta=\sigma^2$.

6.7^Q. Let $X_1,\ldots,X_n$ be independent, identically distributed $N(\mu,1)$ random variables where μ is unknown and $n\geq 2$. Let t be a fixed real number. Then the expectation

$$E_\mu(I_{(-\infty,t]}(X_1))=P_\mu(X_1\leq t)=\Phi(t-\mu)$$

for all μ where $\Phi(x)$ is the cumulative distribution function of a $N(0,1)$ random variable.

a) Show that the sample mean $\overline{X}$ is a sufficient statistic for μ.

b) Explain why (or show that) $\overline{X}$ is a complete sufficient statistic for μ.

c) Using the fact that the conditional distribution of X_1 given $\overline{X} = \overline{x}$ is the $N(\overline{x}, 1 - 1/n)$ distribution where the second parameter $1 - 1/n$ is the variance of conditional distribution, find

$$E_\mu(I_{(-\infty,t]}(X_1)|\overline{X} = \overline{x}) = E_\mu[I_{(-\infty,t]}(W)]$$

where $W \sim N(\overline{x}, 1 - 1/n)$. (Hint: your answer should be $\Phi(g(\overline{x}))$ for some function g.)

d) What is the uniformly minimum variance unbiased estimator for $\Phi(t - \mu)$?

Problems from old quizzes and exams. Problems from old qualifying exams are marked with a Q.

6.8. Suppose that X is Poisson with pmf

$$f(x|\lambda) = P(X = x|\lambda) = \frac{e^{-\lambda}\lambda^x}{x!}$$

where $x = 0, 1, \ldots$ and $\lambda > 0$. Find the Fisher information $I_1(\lambda)$.

6.9. Let $X_1, \ldots, X_n$ be iid Exponential(β) random variables and $Y_1, \ldots, Y_m$ iid Exponential($\beta/2$) random variables. Assume that the Y_i's and X_j's are independent.

a) Find the joint pdf $f(x_1, \ldots, x_n, y_1, \ldots, y_m)$ and show that this pdf is a regular exponential family with complete sufficient statistic $T = \sum_{i=1}^n X_i + 2\sum_{i=1}^m Y_i$.

b) Find the function $\tau(\beta)$ such that T is the UMVUE of $\tau(\beta)$. (Hint: find $E_\beta[T]$. The theorems of this chapter apply since $X_1, \ldots, X_n, 2Y_1, \ldots, 2Y_m$ are iid.)

6.10. Let $X_1, \ldots, X_n$ be independent, identically distributed $N(\mu, 1)$ random variables where μ is unknown.

a) Find $E_\mu[X_1^2]$.

b) Using the fact that the conditional distribution of X_1 given $\overline{X} = \overline{x}$ is the $N(\overline{x}, 1 - 1/n)$ distribution where the second parameter $1 - 1/n$ is the variance of conditional distribution, find

$$E_\mu(X_1^2|\overline{X} = \overline{x}).$$

[Hint: this expected value is equal to $E(W^2)$ where $W \sim N(\overline{x}, 1 - 1/n)$.]

c) What is the MLE for $\mu^2 + 1$? (Hint: you may use the fact that the MLE for μ is $\overline{X}$.)

d) What is the uniformly minimum variance unbiased estimator for $\mu^2 + 1$? Explain.

6.11. Let $X_1, \ldots, X_n$ be a random sample from a Poisson(λ) population.

a) Find the Fréchet–Cramér–Rao lower bound $\text{FCRLB}_n(\lambda^2)$ for the variance of an unbiased estimator of $\tau(\lambda) = \lambda^2$.

b) The UMVUE for λ^2 is $T(X_1,\ldots,X_n) = (\overline{X})^2 - \overline{X}/n$. Will $\text{Var}_\lambda[T] = \text{FCRLB}_n(\lambda^2)$ or will $\text{Var}_\lambda[T] > \text{FCRLB}_n(\lambda^2)$? Explain. (Hint: use the rule of thumb 9e from Sect. 6.3.)

6.12. Let $X_1,\ldots,X_n$ be independent, identically distributed Poisson(λ) random variables where $\lambda > 0$ is unknown.

a) Find $E_\lambda[X_1^2]$.

b) Using the fact that the conditional distribution of X_1 given $\sum_{i=1}^n X_i = y$ is the Binomial($y, 1/n$) distribution, find

$$E_\lambda\left(X_1^2 | \sum_{i=1}^n X_i = y\right).$$

c) Find $\tau(\lambda)$ such that $E_\lambda(X_1^2|\sum_{i=1}^n X_i)$ is the uniformly minimum variance unbiased estimator for $\tau(\lambda)$.

6.13. Let $X_1,\ldots,X_n$ be iid Bernoulli(ρ) random variables.

a) Find the Fisher information $I_1(\rho)$.

b) Find the Fréchet–Cramér–Rao lower bound for unbiased estimators of $\tau(\rho) = \rho$.

c) The MLE for ρ is $\overline{X}$. Find Var($\overline{X}$).

d) Does the MLE achieve the FCRLB? Is this surprising? Explain.

6.14^Q. Let $X_1,\ldots,X_n$ be independent, identically distributed exponential(θ) random variables where $\theta > 0$ is unknown. Consider the class of estimators of θ

$$\left\{T_n(c) = c\sum_{i=1}^n X_i \mid c > 0\right\}.$$

Determine the value of c that minimizes the mean square error MSE. Show work and prove that your value of c is indeed the global minimizer.

6.15. Let $X_1,\ldots,X_n$ be iid from a distribution with pdf

$$f(x|\theta) = \theta x^{\theta-1} I(0 < x < 1), \quad \theta > 0.$$

a) Find the MLE of θ.

b) What is the MLE of $1/\theta^2$? Explain.

c) Find the Fisher information $I_1(\theta)$. You may use the fact that $-\log(X) \sim$ exponential($1/\theta$).

d) Find the Fréchet–Cramér–Rao lower bound for unbiased estimators of $\tau(\theta) = 1/\theta^2$.

6.16. Let $X_1,\ldots,X_n$ be iid random variables with $E(X) = \mu$ and $\text{Var}(X) = 1$. Suppose that $T = \sum_{i=1}^n X_i$ is a complete sufficient statistic. Find the UMVUE of μ^2.

6.17. Let $X_1,\ldots,X_n$ be iid exponential(λ) random variables.

a) Find $I_1(\lambda)$.

b) Find the FCRLB for estimating $\tau(\lambda) = \lambda^2$.

c) If $T = \sum_{i=1}^n X_i$, it can be shown that the UMVUE of λ^2 is

$$W = \frac{\Gamma(n)}{\Gamma(2+n)} T^2.$$

Do you think that $\text{Var}_\lambda(W)$ is equal to the FCRLB in part b)? Explain briefly.

6.18. Let $X_1,\ldots,X_n$ be iid $N(\mu,\sigma^2)$ where μ is known and $n > 1$. Suppose interest is in estimating $\theta = \sigma^2$. You should have memorized the fact that

$$\frac{(n-1)S^2}{\sigma^2} \sim \chi^2_{n-1}.$$

a) Find the MSE of S^2 for estimating σ^2.

b) Find the MSE of T for estimating σ^2 where

$$T = \frac{1}{n} \sum_{i=1}^n (x_i - \mu)^2.$$

6.19^Q**.** Let $X_1,\ldots,X_n$ be independent identically distributed random variables from a $N(\mu,\sigma^2)$ distribution. Hence $E(X_1) = \mu$ and $\text{VAR}(X_1) = \sigma^2$. Suppose that μ is known and consider estimates of σ^2 of the form

$$S^2(k) = \frac{1}{k} \sum_{i=1}^n (X_i - \mu)^2$$

where k is a constant to be chosen. Note: $E(\chi^2_m) = m$ and $\text{VAR}(\chi^2_m) = 2m$. Determine the value of k which gives the smallest mean square error. (Hint: Find the MSE as a function of k, then take derivatives with respect to k.)

6.20^Q**.** Let $X_1,\ldots,X_n$ be independent identically distributed random variables with pdf

$$f(x|\theta) = \frac{2x}{\theta} e^{-x^2/\theta}, \quad x > 0$$

and $f(x|\theta) = 0$ for $x \leq 0$.

a) Show that X_1^2 is an unbiased estimator of θ. (Hint: use the substitution $W = X^2$ and find the pdf of W or use u-substitution with $u = x^2/\theta$.)

b) Find the Cramér–Rao lower bound for the variance of an unbiased estimator of θ.

c) Find the uniformly minimum variance unbiased estimator (UMVUE) of θ.

6.21^Q**.** See Mukhopadhyay (2000, p. 377). Let $X_1, \ldots, X_n$ be iid $N(\theta, \theta^2)$ normal random variables with mean θ and variance θ^2. Let

$$T_1 = \overline{X} = \frac{1}{n}\sum_{i=1}^{n} X_i$$

and let

$$T_2 = c_n S = c_n \sqrt{\frac{\sum_{i=1}^{n}(X_i - \overline{X})^2}{n-1}}$$

where the constant c_n is such that $E_\theta[c_n S] = \theta$. You do not need to find the constant c_n. Consider estimators $W(\alpha)$ of θ of the form.

$$W(\alpha) = \alpha T_1 + (1-\alpha) T_2$$

where $0 \le \alpha \le 1$.

a) Find the variance

$$\mathrm{Var}_\theta[W(\alpha)] = \mathrm{Var}_\theta(\alpha T_1 + (1-\alpha) T_2).$$

b) Find the mean square error of $W(\alpha)$ in terms of $\mathrm{Var}_\theta(T_1), \mathrm{Var}_\theta(T_2)$ and α.

c) Assume that

$$\mathrm{Var}_\theta(T_2) \approx \frac{\theta^2}{2n}.$$

Determine the value of α that gives the smallest mean square error. (Hint: Find the MSE as a function of α, then take the derivative with respect to α. Set the derivative equal to zero and use the above approximation for $\mathrm{Var}_\theta(T_2)$. Show that your value of α is indeed the global minimizer.)

6.22^Q**.** Suppose that $X_1, \ldots, X_n$ are iid normal distribution with mean 0 and variance σ^2. Consider the following estimators: $T_1 = \frac{1}{2}|X_1 - X_2|$ and $T_2 = \sqrt{\frac{1}{n}\sum_{i=1}^{n} X_i^2}$.

a) Is T_1 unbiased for σ? Evaluate the mean square error (MSE) of T_1.

b) Is T_2 unbiased for σ? If not, find a suitable multiple of T_2 which is unbiased for σ.

6.23^Q. Let $X_1, \ldots, X_n$ be independent identically distributed random variables with pdf (probability density function)

$$f(x) = \frac{1}{\lambda} \exp\left(-\frac{x}{\lambda}\right)$$

where x and λ are both positive. Find the uniformly minimum variance unbiased estimator (UMVUE) of λ^2.

6.24^Q. Let $X_1, \ldots, X_n$ be independent identically distributed random variables with pdf (probability density function)

$$f(x) = \sqrt{\frac{\sigma}{2\pi x^3}} \exp\left(-\frac{\sigma}{2x}\right)$$

where x and σ are both positive. Then $X_i = \dfrac{\sigma}{W_i}$ where $W_i \sim \chi^2_1$. Find the uniformly minimum variance unbiased estimator (UMVUE) of $\dfrac{1}{\sigma}$.

6.25^Q. Let $X_1, \ldots, X_n$ be a random sample from the distribution with density

$$f(x) = \begin{cases} \frac{2x}{\theta^2}, & 0 < x < \theta \\ 0 & \text{elsewhere.} \end{cases}$$

Let $T = max(X_1, \ldots, X_n)$. To estimate θ consider estimators of the form CT. Determine the value of C which gives the smallest mean square error.

6.26^Q. Let $X_1, \ldots, X_n$ be a random sample from a distribution with pdf

$$f(x) = \frac{2x}{\theta^2}, \; 0 < x < \theta.$$

Let $T = c\overline{X}$ be an estimator of θ where c is a constant.

a) Find the mean square error (MSE) of T as a function of c (and of θ and n).

b) Find the value c that minimizes the MSE. Prove that your value is the minimizer.

6.27^Q. Suppose that $X_1, \ldots, X_n$ are iid Bernoulli(p) where $n \geq 2$ and $0 < p < 1$ is the unknown parameter.

a) Derive the UMVUE of $\nu(p)$, where $\nu(p) = e^2(p(1-p))$.

b) Find the Cramér–Rao lower bound for estimating $\nu(p) = e^2(p(1-p))$.

6.28. Let $X_1, \ldots, X_n$ be independent identically distributed Poisson(λ) random variables. Find the UMVUE of

$$\frac{\lambda}{n} + \lambda^2.$$

6.29. Let $Y_1,\ldots,Y_n$ be iid Poisson(θ) random variables.

a) Find the UMVUE for θ.

b) Find the Fisher information $I_1(\theta)$.

c) Find the FCRLB for unbiased estimators of $\tau(\theta) = \theta$.

d) The MLE for θ is $\overline{Y}$. Find Var($\overline{Y}$).

e) Does the MLE achieve the FCRLB? Is this surprising? Explain.

6.30^Q. Suppose that $Y_1,\ldots,Y_n$ are independent binomial(m_i,ρ) where the $m_i \geq 1$ are known constants. Let

$$T_1 = \frac{\sum_{i=1}^n Y_i}{\sum_{i=1}^n m_i} \quad \text{and} \quad T_2 = \frac{1}{n}\sum_{i=1}^n \frac{Y_i}{m_i}$$

be estimators of ρ.

a) Find MSE(T_1).

b) Find MSE(T_2).

c) Which estimator is better?

Hint: by the arithmetic–geometric–harmonic mean inequality,

$$\frac{1}{n}\sum_{i=1}^n m_i \geq \frac{n}{\sum_{i=1}^n \frac{1}{m_i}}.$$

6.31^Q. Let $Y_1,\ldots,Y_n$ be iid gamma($\alpha = 10, \beta$) random variables. Let $T = c\overline{Y}$ be an estimator of β where c is a constant.

a) Find the mean square error (MSE) of T as a function of c (and of β and n).

b) Find the value c that minimizes the MSE. Prove that your value is the minimizer.

6.32^Q. Let $Y_1,\ldots,Y_n$ be independent identically distributed random variables with pdf (probability density function)

$$f(y) = (2-2y)I_{(0,1)}(y)\ \nu\ \exp[(1-\nu)(-\log(2y-y^2))]$$

where $\nu > 0$ and $n > 1$. The indicator $I_{(0,1)}(y) = 1$ if $0 < y < 1$ and $I_{(0,1)}(y) = 0$, otherwise.

a) Find a complete sufficient statistic.

b) Find the Fisher information $I_1(\nu)$ if $n = 1$.

c) Find the Cramér–Rao lower bound (CRLB) for estimating $1/\nu$.

d) Find the uniformly minimum unbiased estimator (UMVUE) of ν.

Hint: You may use the fact that $T_n = -\sum_{i=1}^n \log(2Y_i - Y_i^2) \sim G(n, 1/\nu)$, and

$$E(T_n^r) = \frac{1}{\nu^r}\frac{\Gamma(r+n)}{\Gamma(n)}$$

for $r > -n$. Also $\Gamma(1+x) = x\Gamma(x)$ for $x > 0$.

6.33^Q**.** Let $Y_1,\ldots,Y_n$ be iid random variables from a distribution with pdf

$$f(y) = \frac{\theta}{2(1+|y|)^{\theta+1}}$$

where $\theta > 0$ and y is real. Then $W = \log(1+|Y|)$ has pdf $f(w) = \theta e^{-w\theta}$ for $w > 0$.

a) Find a complete sufficient statistic.

b) Find the (Fisher) information number $I_1(\theta)$.

c) Find the uniformly minimum variance unbiased estimator (UMVUE) for θ.

6.34^Q**.** Suppose that $X_1, X_2, \ldots, X_n$ are independent identically distributed random variables from normal distribution with unknown mean μ and known variance σ^2. Consider the parametric function $g(\mu) = e^{2\mu}$.

a) Derive the uniformly minimum variance unbiased estimator (UMVUE) of $g(\mu)$.

b) Find the Cramér–Rao lower bound (CRLB) for the variance of an unbiased estimator of $g(\mu)$.

c) Is the CRLB attained by the variance of the UMVUE of $g(\mu)$?

6.35. Let $Y_1,\ldots,Y_n$ be iid from a one-parameter exponential family with pdf or pmf $f(y|\theta)$ with complete sufficient statistic $T(\boldsymbol{Y}) = \sum_{i=1}^n t(Y_i)$ where $t(Y_i) \sim \theta X$ and X has a known distribution with known mean $E(X)$ and known variance $V(X)$. Let $W_n = cT(\boldsymbol{Y})$ be an estimator of θ where c is a constant. For parts a)-x) complete i)-iv).

i) Find the mean square error (MSE) of W_n as a function of c (and of n, $E(X)$, and $V(X)$).

ii) Find the value of c that minimizes the MSE. Prove that your value is the minimizer using the first and second derivative of MSE(c).

iii) Find the value of c that minimizes the MSE using Theorem 6.5.

iv) Find the uniformly minimum variance unbiased estimator (UMVUE) of θ.

a) $Y_1,\ldots,Y_n$ are iid beta($\delta = 1, \nu$), $t(Y) = -\log(1-Y) \sim \dfrac{1}{\nu}\text{EXP}(1)$, $\theta = 1/\nu$.

b) $Y_1,\ldots,Y_n$ are iid beta($\delta, \nu = 1$), $t(Y) = -\log(Y) \sim \dfrac{1}{\delta}\text{EXP}(1)$, $\theta = 1/\delta$.

c) $Y_1,\ldots,Y_n$ are iid Burr type III (ϕ,λ) with ϕ known, $t(Y) = \log(1+Y^{-\phi}) \sim \lambda\text{EXP}(1)$, $\theta = \lambda$.

d) $Y_1,\ldots,Y_n$ are iid Burr type X (τ), $t(Y) = -\log(1-e^{-y^2}) \sim \frac{1}{\tau}\text{EXP}(1)$, $\theta = 1/\tau$.

e) $Y_1,\ldots,Y_n$ are iid Burr type XII (ϕ,λ) with ϕ known, $t(Y) = \log(1+Y^{\phi}) \sim \lambda\text{EXP}(1)$, $\theta = \lambda$.

f) $Y_1,\ldots,Y_n$ are iid chi(p,σ) with p known, $t(Y) = Y^2 \sim \sigma^2 G(p/2,2)$, $\theta = \sigma^2$.

g) $Y_1,\ldots,Y_n$ are iid double exponential (μ,λ) with μ known, $t(Y) = |Y-\mu| \sim \lambda\text{EXP}(1)$, $\theta = \lambda$.

h) $Y_1,\ldots,Y_n$ are iid EXP(λ), $t(Y) = Y \sim \lambda\text{EXP}(1)$, $\theta = \lambda$.

i) $Y_1,\ldots,Y_n$ are iid two-parameter exponential (μ,λ) with μ known, $t(Y) = Y_i - \mu \sim \lambda\text{EXP}(1)$, $\theta = \lambda$.

j) $Y_1,\ldots,Y_n$ are iid gamma (ν,λ) with ν known, $t(Y)=Y\sim\lambda G(\nu,1)$, $\theta=\lambda$.

k) $Y_1,\ldots,Y_n$ are iid Gomp(μ,ν) with μ known, $t(Y)=e^{\mu Y}-1\sim\dfrac{1}{\nu}\text{EXP}(\mu)$, $\theta=1/\nu$.

l) $Y_1,\ldots,Y_n$ are iid half normal (μ,σ^2) with μ known, $t(Y)=(Y-\mu)^2\sim\sigma^2G(1/2,2)$, $\theta=\sigma^2$.

m) $Y_1,\ldots,Y_n$ are iid IEXP(μ), $t(Y)=1/Y\sim\frac{1}{\mu}\text{EXP}(1)$, $\theta=1/\mu$.

n) $Y_1,\ldots,Y_n$ are iid IW(ϕ,λ) with ϕ known, $t(Y)=\frac{1}{Y^\phi}\sim\lambda\text{EXP}(1)$, $\theta=\lambda$.

o) $Y_1,\ldots,Y_n$ are iid inverted gamma (ν,λ) with ν known, $t(Y)=1/Y\sim\frac{1}{\lambda}G(\nu,1)$, $\theta=1/\lambda$.

p) $Y_1,\ldots,Y_n$ are iid LG(ν,λ) with ν known, $t(Y)=e^Y\sim\lambda G(\nu,1)$, $\theta=\lambda$.

q) $Y_1,\ldots,Y_n$ are iid Maxwell–Boltzmann (μ,σ) with μ known, $t(Y)=(Y-\mu)^2\sim\sigma^2G(3/2,2)$, $\theta=\sigma^2$.

r) $Y_1,\ldots,Y_n$ are iid MDL(μ,ϕ) with μ known, $t(Y)=\log(\frac{\mu}{\mu-Y})\sim\frac{1}{\phi}\text{EXP}(1)$, $\theta=1/\phi$.

s) $Y_1,\ldots,Y_n$ are iid $N(\mu,\sigma^2)$ with μ known, $t(Y)=(Y-\mu)^2\sim\sigma^2G(1/2,2)$, $\theta=\sigma^2$.

t) $Y_1,\ldots,Y_n$ are iid one-sided stable (σ), $t(Y)=1/Y\sim\frac{1}{\sigma}G(1/2,2)$, $\theta=1/\sigma$.

u) $Y_1,\ldots,Y_n$ are iid power (λ) distribution, $t(Y)=-\log(Y)\sim\lambda\text{EXP}(1)$, $\theta=\lambda$.

v) $Y_1,\ldots,Y_n$ are iid Rayleigh (μ,σ) with μ known, $t(Y)=(Y-\mu)^2\sim\sigma^2\text{EXP}(2)$, $\theta=\sigma^2$.

w) $Y_1,\ldots,Y_n$ are iid Topp–Leone (ν), $t(Y)=-\log(2Y-Y^2)\sim\frac{1}{\nu}\text{EXP}(1)$, $\theta=1/\nu$.

x) $Y_1,\ldots,Y_n$ are iid truncated extreme value (λ), $t(Y)=e^Y-1\sim\lambda\text{EXP}(1)$, $\theta=\lambda$.

6.36^Q. Let $Y_1,\ldots,Y_n$ be iid from a one-parameter exponential family with pdf or pmf $f(y|\theta)$ with complete sufficient statistic $T(\boldsymbol{Y})=\sum_{i=1}^n t(Y_i)$ where $t(Y_i)\sim\theta X$ and X has a known distribution with known mean $E(X)$ and known variance $V(X)$. Let $W_n=cT(\boldsymbol{Y})$ be an estimator of θ where c is a constant.

a) Find the mean square error (MSE) of W_n as a function of c (and of n, $E(X)$, and $V(X)$).

b) Find the value of c that minimizes the MSE. Prove that your value is the minimizer.

c) Find the uniformly minimum variance unbiased estimator (UMVUE) of θ.

6.37^Q. Let $X_1,\ldots,X_n$ be a random sample from a Poisson (λ) distribution. Let $\overline{X}$ and S^2 denote the sample mean and the sample variance, respectively.

a) Show that $\overline{X}$ is uniformly minimum variance unbiased (UMVU) estimator of λ

b) Show that $E(S^2|\overline{X})=\overline{X}$.

c) Show that Var$(S^2)>$ Var$(\overline{X})$.

6.38^Q. Let $X_1,\ldots,X_n$ be a random sample from a Poisson distribution with mean θ.

a) For $a>0$, find the uniformly minimum variance unbiased estimator (UMVUE) of $g(\theta)=e^{a\theta}$.

b) Prove the identity:

$$E\left[2^{X_1}|T\right] = \left(1 + \frac{1}{n}\right)^T.$$

6.39^Q**.** Let $X_1,\ldots,X_n$ be independent identically distributed from a $N(\mu,\sigma^2)$ population, where σ^2 is known. Let $\overline{X}$ be the sample mean.

a) Find $E(\overline{X}-\mu)^2$.

b) Using a), find the UMVUE of μ^2.

c) Find $E(\overline{X}-\mu)^3$. [Hint: Show that if Y is a $N(0,\sigma^2)$ random variable, then $E(Y^3)=0$].

d) Using c), find the UMVUE of μ^3.

6.40^Q**.** Let $Y_1,\ldots,Y_n$ be iid from a uniform $U(0,\theta)$ distribution where $\theta > 0$. Then $T = \max(Y_1,\ldots,Y_n)$ is a complete sufficient statistic.

a) Find $E(T^k)$ for $k > 0$.

b) Find the UMVUE of θ^k for $k > 0$.

6.41. Let $Y_1,\ldots,Y_n$ be iid from a distribution with probability distribution function (pdf)

$$f(y) = \frac{\theta}{(1+y)^{\theta+1}}$$

where $y > 0$ and $\theta > 0$.

a) Find a minimal sufficient statistic for θ.

b) Is the statistic found in a) complete? (prove or disprove)

c) Find the Fisher information $I_1(\theta)$ if $n = 1$.

d) Find the Cramér–Rao lower bound (CRLB) for estimating θ^2.

Chapter 7
Testing Statistical Hypotheses

A hypothesis is a statement about a population parameter $\boldsymbol{\theta}$, and in hypothesis testing there are two competing hypotheses called the null hypothesis $Ho \equiv H_0$ and the alternative hypothesis $H_1 \equiv H_A$. Let Θ_1 and Θ_0 be disjoint sets with $\Theta_i \subset \Theta$ where Θ is the parameter space. Then $Ho : \boldsymbol{\theta} \in \Theta_0$ and $H_1 : \boldsymbol{\theta} \in \Theta_1$.

When a researcher wants strong evidence about a hypothesis, usually this hypothesis is H_1. For example, if Ford claims that their latest car gets 30 mpg on average, then $Ho : \mu = 30$ and $H_1 : \mu > 30$ are reasonable hypotheses where $\theta = \mu$ is the population mean mpg of the car.

The power of a test, $\beta(\boldsymbol{\theta}) = P_{\boldsymbol{\theta}}(Ho \text{ is rejected})$, equals the probability that the test rejects Ho. For a level α test, the probability of rejecting Ho when $\boldsymbol{\theta} \in \Theta_0 =$ $P_{\boldsymbol{\theta}}(\text{type I error})$, and this probability is bounded above by α. Given the bound α on the type I error, want the power to be high when $\boldsymbol{\theta} \in \Theta_1$. UMP tests have good power, and likelihood ratio tests often perform well when UMP tests do not exist. The Neyman–Pearson lemma and a theorem for exponential families are useful for finding UMP tests.

7.1 Hypothesis Tests and Power

Definition 7.1. Assume that the data $\boldsymbol{Y} = (Y_1, \ldots, Y_n)$ has pdf or pmf $f(\boldsymbol{y}|\boldsymbol{\theta})$ for $\boldsymbol{\theta} \in \Theta$. A **hypothesis test** is a rule for rejecting Ho.

Definition 7.2. A **type I error** is rejecting Ho when Ho is true. A **type II error** is failing to reject Ho when Ho is false. $P_{\boldsymbol{\theta}}(\text{reject } Ho) = P_{\boldsymbol{\theta}}(\text{type I error})$ if $\boldsymbol{\theta} \in \Theta_0$ while $P_{\boldsymbol{\theta}}(\text{reject } Ho) = 1 - P_{\boldsymbol{\theta}}(\text{type II error})$ if $\boldsymbol{\theta} \in \Theta_1$.

Definition 7.3. The **power function** of a hypothesis test is

$$\beta(\boldsymbol{\theta}) = P_{\boldsymbol{\theta}}(Ho \text{ is rejected})$$

for $\boldsymbol{\theta} \in \Theta$.

D.J. Olive, *Statistical Theory and Inference*, DOI 10.1007/978-3-319-04972-4_7,

Often there is a rejection region R and an acceptance region. **Reject Ho** if the observed statistic $T(\boldsymbol{y}) \in R$, otherwise **fail to reject Ho.** Then
$\beta(\boldsymbol{\theta}) = P_{\boldsymbol{\theta}}(T(\boldsymbol{Y}) \in R) = P_{\boldsymbol{\theta}}(\text{reject } Ho)$.

Definition 7.4. For $0 \le \alpha \le 1$, a test with power function $\beta(\boldsymbol{\theta})$ is a **size** α **test** if

$$\sup_{\boldsymbol{\theta} \in \Theta_0} \beta(\boldsymbol{\theta}) = \alpha$$

and a **level** α **test** if

$$\sup_{\boldsymbol{\theta} \in \Theta_0} \beta(\boldsymbol{\theta}) \le \alpha.$$

Notice that for $\theta \in \Theta_0$, $\beta(\theta) = P_\theta(\text{type I error})$ and for $\theta \in \Theta_1$, $\beta(\theta) = 1 - P_\theta(\text{type II error})$. We would like $\beta(\theta) \approx 0$ for $\theta \in \Theta_0$ and $\beta(\theta) \approx 1$ for $\theta \in \Theta_1$, but this may not be possible even if the sample size n is large. The tradeoff is that decreasing the probability of a type I error increases the probability of a type II error while decreasing the probability of a type II error increases the probability of a type I error. The size or level of the test gives an upper bound α on the probability of the type I error. Typically the level is fixed, e.g., $\alpha = 0.05$, and then we attempt to find tests that have a small probability of type II error. The following example is a level 0.07 and size 0.0668 test.

Example 7.1. Suppose that $Y \sim N(\mu, 1/9)$ where $\mu \in \{0, 1\}$. Let $Ho: \mu = 0$ and $H_1: \mu = 1$. Let $T(Y) = Y$ and suppose that we reject Ho if $Y \ge 0.5$. Let $Z \sim N(0,1)$ and $\sigma = 1/3$. Then

$$\beta(0) = P_0(Y \ge 0.5) = P_0\left(\frac{Y-0}{1/3} \ge \frac{0.5}{1/3}\right) = P(Z \ge 1.5) \approx 0.0668.$$

$$\beta(1) = P_1(Y \ge 0.5) = P_1\left(\frac{Y-1}{1/3} \ge \frac{0.5-1}{1/3}\right) = P(Z \ge -1.5) \approx 0.9332.$$

Definition 7.5. Suppose the null hypothesis is $H_0: \boldsymbol{\theta} = \boldsymbol{\theta}_0$, and suppose that a test statistic $T_n(\boldsymbol{y})$ is observed. The **p-value** is the probability, assuming H_0 is true, of getting a test statistic $T_n(\boldsymbol{Y})$ at least as extreme as the test statistic $T_n(\boldsymbol{y})$ actually observed where "as extreme" depends on the alternative hypothesis. For an α level test, reject H_0 if p-value $\le \alpha$ while if p-value $> \alpha$, fail to reject H_0.

Suppose $T_n(\boldsymbol{Y}) \sim N(0,1)$ if $H_0: \mu = 0$ is true. If $H_1: \mu > 0$, then this right tailed test has p-value = $P_{H_0}(T_n(\boldsymbol{Y}) \ge T_n(\boldsymbol{y})) = P(Z \ge T_n(\boldsymbol{y}))$. If $H_1: \mu < 0$, then this left tailed test has p-value = $P_{H_0}(T_n(\boldsymbol{Y}) \le T_n(\boldsymbol{y})) = P(Z \le T_n(\boldsymbol{y}))$. If $H_1: \mu \ne 0$, then this two tailed test has p-value = $P_{H_0}(T_n(\boldsymbol{Y}) \ge |T_n(\boldsymbol{y})|) = P_{H_0}(T_n(\boldsymbol{Y}) \le -T_n(\boldsymbol{y})) + P_{H_0}(T_n(\boldsymbol{Y}) \ge T_n(\boldsymbol{y})) = P(Z \ge |T_n(\boldsymbol{y})|)$.

Typically α is small, so H_0 is rejected if the p-value is small. If the p-value = 0, then it is impossible that the test statistic $T_n(\boldsymbol{y})$ would have occurred if H_0 was true. If the p-value = 1, it is impossible that the test statistic $T_n(\boldsymbol{y})$ would have occurred if H_1 was true. If the p-value $= \delta$, then δ is the smallest value of α that would lead to rejecting H_0 when $T_n(\boldsymbol{Y}) = T_n(\boldsymbol{y})$.

If the distribution of $T_n(\boldsymbol{Y})$ is discrete, then the p-value may only take on a countable number of values in [0,1], so p-value = α is impossible for some values of α. For example, suppose the test is a left tailed test so H_0 is rejected if $T_n(\boldsymbol{y}) \leq c$ for some constant c. If $T_n(\boldsymbol{Y}) \sim$ discrete uniform $(1, \ldots, 100)$ when H_0 is true, then p-value = $k/100$ when $T_n(\boldsymbol{y}) = k \in \{1, \ldots, 100\}$. For a left tailed test, if $\alpha = 0.05$, reject H_0 when p-value ≤ 0.05 or when $T_n(\boldsymbol{y}) \leq 5$. If $\alpha = 0.049$, reject H_0 when p-value ≤ 0.049 or when p-value ≤ 0.04 or when $T_n(\boldsymbol{y}) \leq 4$.

7.2 Exponential Families, the Neyman–Pearson Lemma, and UMP Tests

Definition 7.6. Consider all level α tests of $Ho : \theta \in \Theta_0$ vs. $H_1 : \theta \in \Theta_1$. A **uniformly most powerful (UMP) level α test** is a level α test with power function $\beta_{UMP}(\theta)$ such that $\beta_{UMP}(\theta) \geq \beta(\theta)$ for every $\theta \in \Theta_1$ where β is the power function for any level α test of Ho vs. H_1.

The following three theorems can be used to find a UMP test that is both a **level** α and a **size** α test.

Theorem 7.1, The Neyman–Pearson Lemma (NPL). Consider testing $H_0 : \theta = \theta_0$ vs. $H_1 : \theta = \theta_1$ where the pdf or pmf corresponding to θ_i is $f(\boldsymbol{y}|\theta_i)$ for $i = 0, 1$. Suppose the test rejects H_0 if $f(\boldsymbol{y}|\theta_1) > kf(\boldsymbol{y}|\theta_0)$, and rejects H_0 with probability γ if $f(\boldsymbol{y}|\theta_1) = kf(\boldsymbol{y}|\theta_0)$ for some $k \geq 0$. If

$$\alpha = \beta(\theta_0) = P_{\theta_0}[f(\boldsymbol{Y}|\theta_1) > kf(\boldsymbol{Y}|\theta_0)] + \gamma P_{\theta_0}[f(\boldsymbol{Y}|\theta_1) = kf(\boldsymbol{Y}|\theta_0)],$$

then this test is a UMP level α test.

Proof. The proof is for pdfs. Replace the integrals by sums for pmfs. Following Ferguson (1967, p. 202), a test can be written as a test function $\psi(\boldsymbol{y}) \in [0, 1]$ where $\psi(\boldsymbol{y})$ is the probability that the test rejects H_0 when $\boldsymbol{Y} = \boldsymbol{y}$. The Neyman–Pearson (NP) test function is

$$\phi(\boldsymbol{y}) = \begin{cases} 1, \ f(\boldsymbol{y}|\theta_1) > kf(\boldsymbol{y}|\theta_0) \\ \gamma, \ f(\boldsymbol{y}|\theta_1) = kf(\boldsymbol{y}|\theta_0) \\ 0, \ f(\boldsymbol{y}|\theta_1) < kf(\boldsymbol{y}|\theta_0) \end{cases}$$

and $\alpha = E_{\theta_0}[\phi(\boldsymbol{Y})]$. Consider any level α test $\psi(\boldsymbol{y})$. Since $\psi(\boldsymbol{y})$ is a level α test,

$$E_{\theta_0}[\psi(\boldsymbol{Y})] \leq E_{\theta_0}[\phi(\boldsymbol{Y})] = \alpha. \tag{7.1}$$

Then the NP test is UMP if the power

$$\beta_\psi(\theta_1) = E_{\theta_1}[\psi(\boldsymbol{Y})] \le \beta_\phi(\theta_1) = E_{\theta_1}[\phi(\boldsymbol{Y})].$$

Let $f_i(\boldsymbol{y}) = f(\boldsymbol{y}|\theta_i)$ for $i = 0, 1$. Notice that $\phi(\boldsymbol{y}) = 1 \ge \psi(\boldsymbol{y})$ if $f_1(\boldsymbol{y}) > kf_0(\boldsymbol{y})$ and $\phi(\boldsymbol{y}) = 0 \le \psi(\boldsymbol{y})$ if $f_1(\boldsymbol{y}) < kf_0(\boldsymbol{y})$. Hence

$$\int [\phi(\boldsymbol{y}) - \psi(\boldsymbol{y})][f_1(\boldsymbol{y}) - kf_0(\boldsymbol{y})]d\boldsymbol{y} \ge 0 \tag{7.2}$$

since the integrand is nonnegative. Hence the difference in powers is

$$\beta_\phi(\theta_1) - \beta_\psi(\theta_1) = E_{\theta_1}[\phi(\boldsymbol{Y})] - E_{\theta_1}[\psi(\boldsymbol{Y})] \ge k(E_{\theta_0}[\phi(\boldsymbol{Y})] - E_{\theta_0}[\psi(\boldsymbol{Y})]) \ge 0$$

where the first inequality follows from (7.2) and the second inequality from Eq. (7.1). □

Remark 7.1. A test of hypotheses of the form $H_0 : f(\boldsymbol{y}) = f_0(\boldsymbol{y})$ vs. $H_1 : f(\boldsymbol{Y}) = f_1(\boldsymbol{y})$ can be done using the Neyman–Pearson lemma since $H_i : \theta = \theta_i$ indicates that $f(\boldsymbol{y}) = f_{\theta_i}(\boldsymbol{y}) = f(\boldsymbol{y}|\theta_i)$ where $\theta_i = i$ for $i = 0, 1$.

Theorem 7.2, One-Sided UMP Tests via the Neyman–Pearson Lemma. Suppose that the hypotheses are of the form $H_0 : \theta \le \theta_0$ vs. $H_1 : \theta > \theta_0$ or $H_0 : \theta \ge \theta_0$ vs. $H_1 : \theta < \theta_0$, or that the inequality in H_0 is replaced by equality. Also assume that

$$\sup_{\theta \in \Theta_0} \beta(\theta) = \beta(\theta_0).$$

Pick $\theta_1 \in \Theta_1$ and use the Neyman–Pearson lemma to find the UMP test for $H_0^* : \theta = \theta_0$ vs. $H_A^* : \theta = \theta_1$. Then the UMP test rejects H_0^* if $f(\boldsymbol{y}|\theta_1) > kf(\boldsymbol{y}|\theta_0)$, and rejects H_0^* with probability γ if $f(\boldsymbol{y}|\theta_1) = kf(\boldsymbol{y}|\theta_0)$ for some $k \ge 0$ where $\alpha = \beta(\theta_0)$. This test is also the UMP level α test for $H_0 : \theta \in \Theta_0$ vs. $H_1 : \theta \in \Theta_1$ if k *does not depend on the value of* $\theta_1 \in \Theta_1$. If $R = f(\boldsymbol{Y}|\theta_1)/f(\boldsymbol{Y}|\theta_0)$, then $\alpha = P_{\theta_0}(R > k) + \gamma P_{\theta_0}(R = k)$.

Theorem 7.3, One-Sided UMP Tests for Exponential Families. Let $Y_1, \ldots, Y_n$ be a sample with a joint pdf or pmf from a one-parameter exponential family where $w(\theta)$ is increasing and $T(\boldsymbol{y})$ is the complete sufficient statistic. Alternatively, let $Y_1, \ldots, Y_n$ be iid with pdf or pmf

$$f(y|\theta) = h(y)c(\theta)\exp[w(\theta)t(y)]$$

from a one-parameter exponential family where θ is real and $w(\theta)$ is increasing. Here $T(\boldsymbol{y}) = \sum_{i=1}^n t(y_i)$. I) Let $\theta_1 > \theta_0$. Consider the test that rejects H_o if $T(\boldsymbol{y}) > k$ and rejects H_0 with probability γ if $T(\boldsymbol{y}) = k$ where
$\alpha = P_{\theta_0}(T(\boldsymbol{Y}) > k) + \gamma P_{\theta_0}(T(\boldsymbol{Y}) = k)$. This test is the UMP level α test for
a) $H_0 : \theta = \theta_0$ vs. $H_1 : \theta = \theta_1$,
b) $H_0 : \theta = \theta_0$ vs. $H_1 : \theta > \theta_0$, and
c) $H_0 : \theta \le \theta_0$ vs. $H_1 : \theta > \theta_0$.

II) Let $\theta_1 < \theta_0$. Consider the test that rejects H_0 if $T(\boldsymbol{y}) < k$ and rejects H_0 with probability γ if $T(\boldsymbol{y}) = k$ where $\alpha = P_{\theta_0}(T(\boldsymbol{Y}) < k) + \gamma P_{\theta_0}(T(\boldsymbol{Y}) = k)$. This test is the UMP level α test for
d) $H_0 : \theta = \theta_0$ vs. $H_1 : \theta = \theta_1$
e) $H_0 : \theta = \theta_0$ vs. $H_1 : \theta < \theta_0$, and
f) $H_0 : \theta \geq \theta_0$ vs. $H_1 : \theta < \theta_0$.

Proof. I) Let $\theta_1 > \theta_0$. a) Then

$$\frac{f(\boldsymbol{y}|\theta_1)}{f(\boldsymbol{y}|\theta_0)} = \left[\frac{c(\theta_1)}{c(\theta_0)}\right]^n \frac{\exp[w(\theta_1)\sum_{i=1}^n t(y_i)]}{\exp[w(\theta_0)\sum_{i=1}^n t(y_i)]} > c$$

iff

$$[w(\theta_1) - w(\theta_0)] \sum_{i=1}^n t(y_i) > d$$

iff $\sum_{i=1}^n t(y_i) > k$ since $w(\theta)$ is increasing. Hence the result holds by the NP lemma. b) The test in a) did not depend on $\theta_1 > \theta_0$, so the test is UMP by Theorem 7.2. c) In a), $\theta_0 < \theta_1$ were arbitrary, so $\sup_{\theta \in \Theta_0} \beta(\theta) = \beta(\theta_0)$ where $\Theta_0 = \{\theta \in \Theta | \theta \leq \theta_0\}$. So the test is UMP by Theorem 7.2. The proof of II) is similar, but $\theta_1 < \theta_0$ so $[w(\theta_1) - w(\theta_0)] < 0$, and there is a sign change. □

Remark 7.2. a) The UMP level α tests in Theorems 7.1–7.3 are also UMP size α tests. b) As a mnemonic, note that the *inequality used in the rejection region is the same as the inequality in the alternative hypothesis.* Usually $\gamma = 0$ if f is a pdf. Suppose that the parameterization is

$$f(y|\theta) = h(y)c(\theta)\exp[\tilde{w}(\theta)\tilde{t}(y)]$$

where $\tilde{w}(\theta)$ is decreasing. Then set $w(\theta) = -\tilde{w}(\theta)$ and $t(y) = -\tilde{t}(y)$. In this text, $w(\theta)$ is an increasing function if $w(\theta_0) < w(\theta_1)$ for $\theta_0 < \theta_1$ and nondecreasing if $w(\theta_0) \leq w(\theta_1)$. Some texts use "strictly increasing" for "increasing" and use "increasing" for "nondecreasing." c) A **simple hypothesis** consists of exactly one distribution for the sample. A **composite hypothesis** consists of more than one distribution for the sample.

If the data are iid from a one-parameter exponential family, then Theorem 7.3 is simpler to use than the Neyman–Pearson lemma since the test statistic T will have a distribution from an exponential family by Theorem 3.5. This result makes finding the cutoff value k easier. To find a UMP test via the Neyman–Pearson lemma, you need to check that the cutoff value k does not depend on $\theta_1 \in \Theta_1$ and usually need to transform the NP test statistic to put the test in *useful form.* With exponential families, the transformed test statistic is often T.

Example 7.2. Suppose that $X_1, \ldots, X_{10}$ are iid Poisson with unknown mean λ. Derive the most powerful level $\alpha = 0.10$ test for $H_0 : \lambda = 0.30$ versus $H_1 : \lambda = 0.40$.

Solution: Since

$$f(x|\lambda) = \frac{1}{x!} e^{-\lambda} \exp[\log(\lambda)x]$$

and $\log(\lambda)$ is an increasing function of λ, by Theorem 7.3 the UMP test rejects Ho if $\sum x_i > k$ and rejects Ho with probability γ if $\sum x_i = k$ where $\alpha = 0.1 = P_{\lambda=0.3}(\sum X_i > k) + \gamma P_{\lambda=0.3}(\sum X_i = k)$. Notice that

$$\gamma = \frac{\alpha - P_{Ho}(\sum X_i > k)}{P_{Ho}(\sum X_i = k)}. \tag{7.3}$$

Alternatively use the Neyman–Pearson lemma. Let

$$r = f(\boldsymbol{x}|0.4)/f(\boldsymbol{x}|0.3) = \frac{e^{-n\lambda_1} \lambda_1^{\sum x_i}}{\prod x_i!} \frac{\prod x_i!}{e^{-n\lambda_0} \lambda_0^{\sum x_i}} = e^{-n(\lambda_1 - \lambda_0)} \left(\frac{\lambda_1}{\lambda_0}\right)^{\sum x_i}.$$

Since $\lambda_1 = 0.4 > 0.3 = \lambda_0$, $r > c$ is equivalent to $\sum x_i > k$ and the NP UMP test has the same form as the UMP test found using the much simpler Theorem 7.3.

k	0	1	2	3	4	5
$P(T=k)$	0.0498	0.1494	0.2240	0.2240	0.1680	0.1008
$F(k)$	0.0498	0.1992	0.4232	0.6472	0.8152	0.9160

If Ho is true, then $T = \sum_{i=1}^{10} X_i \sim \text{Pois}(3)$ since $3 = 10\lambda_0 = 10(0.3)$. The above table gives the probability that $T = k$ and $F(k) = P(T \leq k)$. First find the smallest integer k such that $P_{\lambda=0.30}(\sum X_i > k) = P(T > k) < \alpha = 0.1$. Since $P(T > k) = 1 - F(k)$, find the smallest value of k such that $F(k) > 0.9$. This happens with $k = 5$. Next use (7.3) to find γ.

$$\gamma = \frac{0.1 - (1 - 0.9160)}{0.1008} = \frac{0.1 - 0.084}{0.1008} = \frac{0.016}{0.1008} \approx 0.1587.$$

Hence the $\alpha = 0.1$ UMP test rejects Ho if $T \equiv \sum_{i=1}^{10} X_i > 5$ and rejects Ho with probability 0.1587 if $\sum_{i=1}^{10} X_i = 5$. Equivalently, the test function $\phi(T)$ gives the probability of rejecting Ho for a given value of T where

$$\phi(T) = \begin{cases} 1, & T > 5 \\ 0.1587, & T = 5 \\ 0, & T < 5. \end{cases}$$

Example 7.3. Let $X_1, \ldots, X_n$ be independent identically distributed random variables from a distribution with pdf

$$f(x) = \frac{2}{\lambda\sqrt{2\pi}} \frac{1}{x} \exp\left[\frac{-(\log(x))^2}{2\lambda^2}\right]$$

where $\lambda > 0$ and $0 \le x \le 1$.

a) What is the UMP (uniformly most powerful) level α test for $H_0 : \lambda = 1$ vs. $H_1 : \lambda = 2$?

b) If possible, find the UMP level α test for $H_0 : \lambda = 1$ vs. $H_1 : \lambda > 1$.

Solution. a) By the NP lemma reject Ho if

$$\frac{f(\boldsymbol{x}|\lambda = 2)}{f(\boldsymbol{x}|\lambda = 1)} > k'.$$

The (left-hand side) LHS =

$$\frac{\frac{1}{2^n} \exp[\frac{-1}{8} \sum [\log(x_i)]^2]}{\exp[\frac{-1}{2} \sum [\log(x_i)]^2]}.$$

So reject Ho if

$$\frac{1}{2^n} \exp\left[\sum [\log(x_i)]^2 \left(\frac{1}{2} - \frac{1}{8}\right)\right] > k'$$

or if $\sum [\log(X_i)]^2 > k$ where $P_{\lambda=1}(\sum [\log(X_i)]^2 > k) = \alpha$.

b) In the above argument, with any $\lambda_1 > 1$, get

$$\sum [\log(x_i)]^2 \left(\frac{1}{2} - \frac{1}{2\lambda_1^2}\right)$$

and

$$\frac{1}{2} - \frac{1}{2\lambda_1^2} > 0$$

for any $\lambda_1^2 > 1$. Hence the UMP test is the same as in a).

Theorem 7.3 gives the same UMP test as a) for both a) and b) since the pdf is a 1P-REF and $w(\lambda^2) = -1/(2\lambda^2)$ is an increasing function of λ^2. Also, it can be shown that $\sum [\log(X_i)]^2 \sim \lambda^2 \chi^2_n$, so $k = \chi^2_{n,1-\alpha}$ where $P(W > \chi^2_{n,1-\alpha}) = \alpha$ if $W \sim \chi^2_n$.

Example 7.4. Let $X_1, \ldots, X_n$ be independent identically distributed (iid) random variables with probability density function

$$f(x) = \frac{2}{\lambda\sqrt{2\pi}} e^x \exp\left(\frac{-(e^x - 1)^2}{2\lambda^2}\right)$$

where $x > 0$ and $\lambda > 0$.

a) What is the UMP (uniformly most powerful) level α test for $H_0 : \lambda = 1$ vs. $H_1 : \lambda = 2$?

b) If possible, find the UMP level α test for $H_0 : \lambda = 1$ vs. $H_1 : \lambda > 1$.

a) By the NP lemma reject Ho if

$$\frac{f(\boldsymbol{x}|\lambda=2)}{f(\boldsymbol{x}|\lambda=1)} > k'.$$

The LHS =

$$\frac{\frac{1}{2^n}\exp[\frac{-1}{8}\sum(e^{x_i}-1)^2]}{\exp[\frac{-1}{2}\sum(e^{x_i}-1)^2]}.$$

So reject Ho if

$$\frac{1}{2^n}\exp\left[\sum(e^{x_i}-1)^2\left(\frac{1}{2}-\frac{1}{8}\right)\right] > k'$$

or if $\sum(e^{x_i}-1)^2 > k$ where $P_1(\sum(e^{X_i}-1)^2 > k) = \alpha$.

b) In the above argument, with any $\lambda_1 > 1$, get

$$\sum(e^{x_i}-1)^2\left(\frac{1}{2}-\frac{1}{2\lambda_1^2}\right)$$

and

$$\frac{1}{2}-\frac{1}{2\lambda_1^2} > 0$$

for any $\lambda_1^2 > 1$. Hence the UMP test is the same as in a).

Alternatively, use the fact that this is an exponential family where $w(\lambda^2) = -1/(2\lambda^2)$ is an increasing function of λ^2 with $T(X_i) = (e^{X_i}-1)^2$. Hence the same test in a) is UMP for both a) and b) by Theorem 7.3.

Example 7.5. Let $X_1,\ldots,X_n$ be independent identically distributed random variables from a half normal HN(μ,σ^2) distribution with pdf

$$f(x) = \frac{2}{\sigma\sqrt{2\pi}}\exp\left(\frac{-(x-\mu)^2}{2\sigma^2}\right)$$

where $\sigma > 0$ and $x > \mu$ and μ is real. **Assume that μ is known.**

a) What is the UMP (uniformly most powerful) level α test for $H_0: \sigma^2 = 1$ vs. $H_1: \sigma^2 = 4$?

b) If possible, find the UMP level α test for $H_0: \sigma^2 = 1$ vs. $H_1: \sigma^2 > 1$.

Solution: a) By the NP lemma reject Ho if

$$\frac{f(\boldsymbol{x}|\sigma^2=4)}{f(\boldsymbol{x}|\sigma^2=1)} > k'.$$

The LHS =

$$\frac{\frac{1}{2^n}\exp\left[\left(\frac{-\sum(x_i-\mu)^2}{2(4)}\right)\right]}{\exp\left[\left(\frac{-\sum(x_i-\mu)^2}{2}\right)\right]}.$$

So reject Ho if

$$\frac{1}{2^n}\exp\left[\sum(x_i-\mu)^2\left(\frac{-1}{8}+\frac{1}{2}\right)\right] > k'$$

or if $\sum(x_i-\mu)^2 > k$ where $P_{\sigma^2=1}(\sum(X_i-\mu)^2 > k) = \alpha$.

Under Ho, $\sum(X_i-\mu)^2 \sim \chi^2_n$ so $k = \chi^2_n(1-\alpha)$ where $P(\chi^2_n > \chi^2_n(1-\alpha)) = \alpha$.

b) In the above argument,

$$\frac{-1}{2(4)}+0.5 = \frac{-1}{8}+0.5 > 0$$

but

$$\frac{-1}{2\sigma_1^2}+0.5 > 0$$

for any $\sigma_1^2 > 1$. Hence the UMP test is the same as in a).

Alternatively, use the fact that this is an exponential family where $w(\sigma^2) = -1/(2\sigma^2)$ is an increasing function of σ^2 with $T(X_i) = (X_i-\mu)^2$. Hence the test in a) is UMP for a) and b) by Theorem 7.3.

Example 7.6. Let $Y_1, \ldots., Y_n$ be iid with pdf

$$f(y) = \frac{1}{2\sqrt{2\pi}\ \theta y^2}\ \frac{y^2-\theta^2}{\sqrt{\frac{y}{\theta}}-\sqrt{\frac{\theta}{y}}}\ I(y>0)\ \frac{1}{\nu}\ \exp\left[\frac{-1}{2\nu^2}\ t(y)\right]$$

where $\nu > 0$, θ is **known** and $t(y)$ is a function such that $t(Y) \sim \nu^2\chi^2_1$.

a) Find the UMP level α test for $H_0: \nu = 1$ versus $H_1: \nu = 1.19$.

b) Suppose $n = 12$ and $\alpha = 0.05$. Find the power $\beta(1.19)$ when $\nu = 1.19$.

Solution: a) This is an exponential family. Note that $2\nu^2$ is increasing, so $1/(2\nu^2)$ is decreasing and $w(\lambda) = -1/(2\nu^2)$ is increasing. Thus the UMP test rejects H_0 if $\sum_{i=1}^n t(y_i) > k$ where $\alpha = P_1(\sum_{i=1}^n t(Y_i) > k)$.

b) Use α to find k and then find the power. If H_0 is true so $\nu = 1$, then $\sum_{i=1}^n t(Y_i) \sim \chi^2_{12}$. Thus $k = \chi^2_{12}(0.95) = 21.03$ using a chi-square table. If $\nu = 1.19$, then $\sum_{i=1}^n t(Y_i) \sim (1.19)^2\chi^2_{12}$. So $\beta(1.19) = P_{1.19}(\sum_{i=1}^n t(Y_i) > 21.03) =$ $P(X > 21.03/(1.19)^2) = P(X > 14.8506) = 0.25$ using a chi-square table where $X \sim \chi^2_{12}$.

Example 7.7. Let $Y_1, \ldots, Y_n$ be independent identically distributed random variables with pdf

$$f(y) = \frac{\sqrt{2} y^2 e^{\frac{-1}{2\sigma^2} y^2}}{\sigma^3 \sqrt{\pi}}$$

where $y \geq 0$ and $\sigma > 0$. You may use the fact that $W = t(Y) = Y^2 \sim \sigma^2 \chi^2_3$.

a) What is the UMP (uniformly most powerful) level α test for $H_0 : \sigma = 1$ versus $H_1 : \sigma > 1$?

b) If $n = 20$ and $\alpha = 0.05$, then find the power $\beta(\sqrt{1.8311})$ of the above UMP test if $\sigma = \sqrt{1.8311}$. Let $P(\chi^2_d \leq \chi^2_{d,\delta}) = \delta$. The tabled values below give $\chi^2_{d,\delta}$.

d	δ							
	0.01	0.05	0.1	0.25	0.75	0.9	0.95	0.99
20	8.260	10.851	12.443	15.452	23.828	28.412	31.410	37.566
40	22.164	26.509	29.051	33.660	45.616	51.805	55.758	63.691
60	37.485	43.188	46.459	52.294	66.981	74.397	79.082	88.379

Solution. a) This family is a regular one-parameter exponential family where $w(\sigma) = -1/(2\sigma^2)$ is increasing. Hence the level α UMP test rejects H_0 when $\sum_{i=1}^n y_i^2 > k$ where $\alpha = P_1(\sum_{i=1}^n Y_i^2 > k) = P_1(T(\boldsymbol{Y}) > k)$.

b) Since $T(\boldsymbol{Y}) \sim \sigma^2 \chi^2_{3n}$, $\dfrac{T(\boldsymbol{Y})}{\sigma^2} \sim \chi^2_{3n}$. Hence

$$\alpha = 0.05 = P_1(T(\boldsymbol{Y}) > k) = P(\chi^2_{60} > \chi^2_{60,1-\alpha}),$$

and $k = \chi^2_{60,1-\alpha} = 79.082$. Hence the power

$$\beta(\sigma) = P_\sigma(T(\boldsymbol{Y}) > 79.082) = P\left(\frac{T(\boldsymbol{Y})}{\sigma^2} > \frac{79.082}{\sigma^2}\right) = P\left(\chi^2_{60} > \frac{79.082}{\sigma^2}\right)$$

$$= P\left(\chi^2_{60} > \frac{79.082}{1.8311}\right) = P(\chi^2_{60} > 43.188) = 1 - 0.05 = 0.95.$$

7.3 Likelihood Ratio Tests

Definition 7.7. Let $(Y_1, \ldots, Y_n)$ be the data with joint pdf or joint pmf $f(\boldsymbol{y}|\boldsymbol{\theta})$ where $\boldsymbol{\theta}$ is a vector of unknown parameters with parameter space Θ. Let $\hat{\boldsymbol{\theta}}$ be the MLE of $\boldsymbol{\theta}$ and let $\hat{\boldsymbol{\theta}}_0$ be the MLE of $\boldsymbol{\theta}$ if the parameter space is Θ_0 (where $\Theta_0 \subset \Theta$). A *likelihood ratio test (LRT) statistic* for testing $H_0 : \boldsymbol{\theta} \in \Theta_0$ versus $H_1 : \boldsymbol{\theta} \in \Theta_0^c$ is

$$\lambda(\boldsymbol{y}) = \frac{L(\hat{\boldsymbol{\theta}}_0|\boldsymbol{y})}{L(\hat{\boldsymbol{\theta}}|\boldsymbol{y})} = \frac{\sup_{\Theta_0} L(\boldsymbol{\theta}|\boldsymbol{y})}{\sup_{\Theta} L(\boldsymbol{\theta}|\boldsymbol{y})}. \tag{7.4}$$

The **likelihood ratio test** (LRT) has a rejection region of the form

$$R = \{\boldsymbol{y} | \lambda(\boldsymbol{y}) \leq c\}$$

where $0 \le c \le 1$, and $\alpha = \sup_{\boldsymbol{\theta} \in \Theta_0} P_{\boldsymbol{\theta}}(\lambda(\boldsymbol{Y}) \le c)$. Suppose $\boldsymbol{\theta}_0 \in \Theta_0$ and $\sup_{\boldsymbol{\theta} \in \Theta_0} P_{\boldsymbol{\theta}}(\lambda(\boldsymbol{Y}) \le c) = P_{\boldsymbol{\theta}_0}(\lambda(\boldsymbol{Y}) \le c)$. Then $\alpha = P_{\boldsymbol{\theta}_0}(\lambda(\boldsymbol{Y}) \le c)$.

Rule of Thumb 7.1: Asymptotic Distribution of the LRT. Let $Y_1, \ldots, Y_n$ be iid. Then under strong regularity conditions, $-2\log\lambda(\boldsymbol{Y}) \approx \chi^2_j$ for large n where $j = r - q$, r is the number of free parameters specified by $\boldsymbol{\theta} \in \Theta$, and q is the number of free parameters specified by $\boldsymbol{\theta} \in \Theta_0$. Hence the approximate LRT rejects H_0 if $-2\log\lambda(\boldsymbol{y}) > c$ where $P(\chi^2_j > c) = \alpha$. Thus $c = \chi^2_{j,1-\alpha}$ where $P(\chi^2_j > \chi^2_{j,1-\alpha}) = \alpha$.

Often $\hat{\boldsymbol{\theta}}$ is called the *unrestricted MLE* of $\boldsymbol{\theta}$, and $\hat{\boldsymbol{\theta}}_0$ is called the *restricted MLE* of $\boldsymbol{\theta}$. Often $\boldsymbol{\theta} = \theta$ is a scalar parameter, $\Theta_0 = (a, \theta_0]$ and $\Theta_1 = \Theta_0^c = (\theta_0, b)$ or $\Theta_0 = [\theta_0, b)$ and $\Theta_1 = (a, \theta_0)$.

Remark 7.3. Suppose the problem wants the rejection region in useful form. Find the two MLEs and write $L(\boldsymbol{\theta}|\boldsymbol{y})$ in terms of a sufficient statistic. Then you should either I) simplify the LRT test statistic $\lambda(\boldsymbol{y})$ and try to find an equivalent test that uses test statistic $T(\boldsymbol{y})$ where the distribution of $T(\boldsymbol{Y})$ is known (i.e., put the LRT in useful form). Often the LRT rejects H_0 if $T > k$ (or $T < k$). Getting the test into useful form can be very difficult. Monotone transformations such as log or power transformations can be useful. II) If you cannot find a statistic T with a simple distribution, state that the Rule of Thumb 7.1 suggests that the LRT test rejects H_o if $-2\log\lambda(\boldsymbol{y}) > \chi^2_{j,1-\alpha}$ where $\alpha = P(-2\log\lambda(\boldsymbol{Y}) > \chi^2_{j,1-\alpha})$. Using II) is dangerous because for many data sets the asymptotic result will not be valid.

Example 7.8. Let $X_1, \ldots, X_n$ be independent identically distributed random variables from a $N(\mu, \sigma^2)$ distribution where the variance σ^2 is known. We want to test $H_0 : \mu = \mu_0$ against $H_1 : \mu \ne \mu_0$.

a) Derive the likelihood ratio test.

b) Let λ be the likelihood ratio. Show that $-2\log\lambda$ is a function of $(\overline{X} - \mu_0)$.

c) Assuming that H_0 is true, find $P(-2\log\lambda > 3.84)$.

Solution: a) The likelihood function

$$L(\mu) = (2\pi\sigma^2)^{-n/2} \exp\left[\frac{-1}{2\sigma^2} \sum (x_i - \mu)^2\right]$$

and the MLE for μ is $\hat{\mu} = \overline{x}$. Thus the numerator of the likelihood ratio test statistic is $L(\mu_0)$ and the denominator is $L(\overline{x})$. So the test is reject H_0 if $\lambda(\boldsymbol{x}) = L(\mu_0)/L(\overline{x}) \le c$ where $\alpha = P_{\mu_0}(\lambda(\boldsymbol{X}) \le c)$.

b) As a statistic, $\log\lambda = \log L(\mu_0) - \log L(\overline{X}) =$ $-\frac{1}{2\sigma^2}[\sum(X_i - \mu_0)^2 - \sum(X_i - \overline{X})^2] = \frac{-n}{2\sigma^2}[\overline{X} - \mu_0]^2$ since $\sum(X_i - \mu_0)^2 =$ $\sum(X_i - \overline{X} + \overline{X} - \mu_0)^2 = \sum(X_i - \overline{X})^2 + n(\overline{X} - \mu_0)^2$. So $-2\log\lambda = \frac{n}{\sigma^2}[\overline{X} - \mu_0]^2$.

c) $-2\log\lambda \sim \chi^2_1$ and from a chi-square table, $P(-2\log\lambda > 3.84) = 0.05$.

Example 7.9. Let $Y_1, \ldots, Y_n$ be iid $N(\mu, \sigma^2)$ random variables where μ and σ^2 are unknown. Set up the likelihood ratio test for $Ho: \mu = \mu_0$ versus $H_A: \mu \neq \mu_0$.

Solution: Under Ho, $\mu = \mu_0$ is known and the MLE

$$(\hat{\mu}_0, \hat{\sigma}_0^2) = \left(\mu_0, \frac{1}{n}\sum_{i=1}^{n}(Y_i - \mu_0)^2\right).$$

Recall that

$$(\hat{\mu}, \hat{\sigma}^2) = \left(\overline{Y}, \frac{1}{n}\sum_{i=1}^{n}(Y_i - \overline{Y})^2\right).$$

Now

$$L(\mu, \sigma^2) = \prod_{i=1}^{n} \frac{1}{\sigma\sqrt{2\pi}} \exp\left[\frac{1}{2\sigma^2}(y_i - \mu)^2\right].$$

Thus

$$\lambda(\boldsymbol{y}) = \frac{L(\hat{\mu}_0, \hat{\sigma}_0^2|\boldsymbol{y})}{L(\hat{\mu}, \hat{\sigma}^2|\boldsymbol{y})} = \frac{\frac{1}{(\hat{\sigma}_0^2)^{n/2}} \exp\left[\frac{1}{2\hat{\sigma}_0^2}\sum_{i=1}^{n}(y_i - \mu_0)^2\right]}{\frac{1}{(\hat{\sigma}^2)^{n/2}} \exp\left[\frac{1}{2\hat{\sigma}^2}\sum_{i=1}^{n}(y_i - \overline{y})^2\right]} =$$

$$\left(\frac{\hat{\sigma}^2}{\hat{\sigma}_0^2}\right)^{n/2} \frac{\exp(n/2)}{\exp(n/2)} = \left(\frac{\hat{\sigma}^2}{\hat{\sigma}_0^2}\right)^{n/2}.$$

The LRT rejects Ho iff $\lambda(\boldsymbol{y}) \leq c$ where $\sup_{\sigma^2} P_{\mu_0, \sigma^2}(\lambda(\boldsymbol{Y}) \leq c) = \alpha$.

On an exam the above work may be sufficient, but to implement the LRT, more work is needed. Notice that the LRT rejects Ho iff $\hat{\sigma}^2/\hat{\sigma}_0^2 \leq c'$ iff $\hat{\sigma}_0^2/\hat{\sigma}^2 \geq k'$. Using

$$\sum_{i=1}^{n}(y_i - \mu_0)^2 = \sum_{i=1}^{n}(y_i - \overline{y})^2 + n(\overline{y} - \mu_0)^2,$$

the LRT rejects Ho iff

$$\left[1 + \frac{n(\overline{y} - \mu_0)^2}{\sum_{i=1}^{n}(y_i - \overline{y})^2}\right] \geq k\text{''}$$

iff

$$\frac{\sqrt{n}\,|\overline{y} - \mu_0|}{\left[\frac{\sum_{i=1}^{n}(y_i - \overline{y})^2}{n-1}\right]^{1/2}} = \sqrt{n}\frac{|\overline{y} - \mu_0|}{s} \geq k$$

where s is the observed sample standard deviation. Hence the LRT is equivalent to the usual t test with test statistic

$$T_0 = \frac{\overline{Y} - \mu_0}{S/\sqrt{n}}$$

that rejects Ho iff $|T_0| \geq k$ with $k = t_{n-1,1-\alpha/2}$ where $P(T \leq t_{n-1,1-\alpha/2}) = 1 - \alpha/2$ when $T \sim t_{n-1}$.

Example 7.10. Suppose that $X_1, \ldots, X_n$ are iid $N(0, \sigma^2)$ where $\sigma > 0$ is the unknown parameter. With preassigned $\alpha \in (0,1)$, derive a level α likelihood ratio test for the null hypothesis $H_0 : \sigma^2 = \sigma_0^2$ against an alternative hypothesis $H_A : \sigma^2 \neq \sigma_0^2$.

Solution: The likelihood function is given by

$$L(\sigma^2) = (2\pi\sigma^2)^{-\frac{n}{2}} \exp\left(-\frac{1}{2\sigma^2}\sum_{i=1}^{n} x_i^2\right)$$

for all $\sigma^2 > 0$, and $\hat{\sigma}^2(\boldsymbol{x}) = \sum_{i=1}^n x_i^2/n$ is the MLE for σ^2. Under Ho, $\hat{\sigma}_0^2 = \sigma_o^2$ since σ_0^2 is the only value in the parameter space $\Theta_0 = \{\sigma_0^2\}$. Thus

$$\lambda(\boldsymbol{x}) = \frac{L(\hat{\sigma}_0^2|\boldsymbol{x})}{L(\hat{\sigma}^2|\boldsymbol{x})} = \frac{\sup_{\Theta_0} L(\sigma^2|\boldsymbol{x})}{\sup_{\sigma^2} L(\sigma^2|\boldsymbol{x})} = \frac{(2\pi\sigma_0^2)^{\frac{-n}{2}} \exp\left(-\frac{1}{2\sigma_0^2}\sum_{i=1}^n x_i^2\right)}{(2\pi\hat{\sigma}^2)^{\frac{-n}{2}} \exp\left(\frac{-n}{2}\right)}.$$

So

$$\lambda(\boldsymbol{x}) = \left(\frac{\hat{\sigma}^2}{\sigma_0^2}\right)^{n/2} \exp\left(\frac{-n\hat{\sigma}^2}{2\sigma_0^2}\right) e^{n/2} = \left[\frac{\hat{\sigma}^2}{\sigma_0^2}\exp\left(1 - \frac{\hat{\sigma}^2}{\sigma_0^2}\right)\right]^{n/2}.$$

The LRT rejects H_0 if $\lambda(\boldsymbol{x}) \leq c$ where $P_{\sigma_0^2}(\lambda(\boldsymbol{X}) \leq c) = \alpha$.

The function $g(u) = ue^{1-u}I(u > 0)$ monotonically increases for $0 < u < d$, monotonically decreases for $d < u < \infty$, and attains its maximum at $u = d$, for some $d > 0$. So $\lambda(\boldsymbol{x})$ will be small in the two tail areas.

Under H_0, $T = \sum_{i=1} X_i^2/\sigma_0^2 \sim \chi_n^2$. Hence the LR test will reject Ho if $T < a$ or $T > b$ where $0 < a < b$. The a and b correspond to horizontal line drawn on the χ_n^2 pdf such that the tail area is α. Hence a and b need to be found numerically. An approximation that should be good for large n rejects Ho if $T < \chi^2_{n,\frac{\alpha}{2}}$ or $T > \chi^2_{n,1-\frac{\alpha}{2}}$ where $P(\chi_n^2 < \chi_{n,\alpha}^2) = \alpha$.

Example 7.11. Consider independent random variables $X_1, \ldots, X_n$, where $X_i \sim$ $\mathrm{N}(\theta_i, \sigma^2)$, $1 \leq i \leq n$, and σ^2 is known.

a) Find the most powerful test of

$$H_0 : \theta_i = 0, \forall i, \text{ versus } H_1 : \theta_i = \theta_{i0}, \forall i,$$

where θ_{i0} are known. Derive (and simplify) the exact critical region for a level α test.

b) Find the likelihood ratio test of

$$H_0 : \theta_i = 0, \forall i, \text{ versus } H_1 : \theta_i \neq 0, \text{ for some } i.$$

Derive (and simplify) the exact critical region for a level α test.

c) Find the power of the test in (a), when $\theta_{i0} = n^{-1/3}, \forall i$. What happens to this power expression as $n \to \infty$?

Solution: a) In Neyman–Pearson's lemma, let $\theta = 0$ if H_0 is true and $\theta = 1$ if H_1 is true. Then want to find $f(\boldsymbol{x}|\theta = 1)/f(\boldsymbol{x}|\theta = 0) \equiv f_1(\boldsymbol{x})/f_0(\boldsymbol{x})$. Since

$$f(\boldsymbol{x}) = \frac{1}{(\sqrt{2\pi}\,\sigma)^n} \exp\left[\frac{-1}{2\sigma^2}\sum_{i=1}^n (x_i - \theta_i)^2\right],$$

$$\frac{f_1(\boldsymbol{x})}{f_0(\boldsymbol{x})} = \frac{\exp[\frac{-1}{2\sigma^2}\sum_{i=1}^n (x_i - \theta_{i0})^2]}{\exp[\frac{-1}{2\sigma^2}\sum_{i=1}^n x_i^2]} = \exp\left(\frac{-1}{2\sigma^2}\left[\sum_{i=1}^n (x_i - \theta_{i0})^2 - \sum_{i=1}^n x_i^2\right]\right) =$$

$$\exp\left(\frac{-1}{2\sigma^2}\left[-2\sum_{i=1}^n x_i\theta_{i0} + \sum_{i=1}^n \theta_{i0}^2\right]\right) > k'$$

if $\frac{-1}{2\sigma^2}[-2\sum_{i=1}^n x_i\theta_{i0} + \sum_{i=1}^n \theta_{i0}^2]) > k$" or if $\sum_{i=1}^n x_i\theta_{i0} > k$. Under Ho, $\sum_{i=1}^n X_i\theta_{i0} \sim N(0, \sigma^2\sum_{i=1}^n \theta_{i0}^2)$. Thus

$$\frac{\sum_{i=1}^n X_i\theta_{i0}}{\sigma\sqrt{\sum_{i=1}^n \theta_{i0}^2}} \sim N(0,1).$$

By Neyman–Pearson's lemma, reject Ho if

$$\frac{\sum_{i=1}^n X_i\theta_{i0}}{\sigma\sqrt{\sum_{i=1}^n \theta_{i0}^2}} > z_{1-\alpha}$$

where $P(Z < z_{1-\alpha}) = 1 - \alpha$ when $Z \sim N(0,1)$.

b) The MLE under Ho is $\hat{\theta}_i = 0$ for $i = 1, \ldots, n$, while the unrestricted MLE is $\hat{\theta}_i = x_i$ for $i = 1, \ldots, n$ since $\overline{x_i} = x_i$ when the sample size is 1. Hence

$$\lambda(\boldsymbol{x}) = \frac{L(\hat{\theta}_i = 0)}{L(\hat{\theta}_i = x_i)} = \frac{\exp[\frac{-1}{2\sigma^2}\sum_{i=1}^n x_i^2]}{\exp[\frac{-1}{2\sigma^2}\sum_{i=1}^n (x_i - x_i)^2]} = \exp\left[\frac{-1}{2\sigma^2}\sum_{i=1}^n x_i^2\right] \le c'$$

if $\frac{-1}{2\sigma^2}\sum_{i=1}^n x_i^2 \le c$", or if $\sum_{i=1}^n x_i^2 \ge c$. Under Ho, $X_i \sim N(0, \sigma^2)$, $X_i/\sigma \sim N(0,1)$, and $\sum_{i=1}^n X_i^2/\sigma^2 \sim \chi_n^2$. So the LRT is reject Ho if $\sum_{i=1}^n X_i^2/\sigma^2 \ge \chi^2_{n,1-\alpha}$ where $P(W \ge \chi^2_{n,1-\alpha}) = 1 - \alpha$ if $W \sim \chi_n^2$.

c) Power = P(reject Ho) =

$$P\left(\frac{n^{-1/3}\sum_{i=1}^n X_i}{\sigma\sqrt{n\,n^{-2/3}}} > z_{1-\alpha}\right) = P\left(\frac{n^{-1/3}\sum_{i=1}^n X_i}{\sigma\, n^{1/6}} > z_{1-\alpha}\right) =$$

$$P\left(\frac{n^{-1/2}\sum_{i=1}^n X_i}{\sigma} > z_{1-\alpha}\right) = P\left(\sum_{i=1}^n X_i > \sigma\, z_{1-\alpha}\, n^{1/2}\right)$$

where $\sum_{i=1}^n X_i \sim N(n\, n^{-1/3}, n\, \sigma^2) \sim N(n^{2/3}, n\, \sigma^2)$. So

$$\frac{\sum_{i=1}^n X_i - n^{2/3}}{\sqrt{n}\,\sigma} \sim N(0,1), \text{ and power } = P\left(\frac{\sum_{i=1}^n X_i}{\sqrt{n}\,\sigma} > z_{1-\alpha}\right) =$$

$$P\left(\frac{\sum_{i=1}^n X_i - n^{2/3}}{\sqrt{n}\,\sigma} > z_{1-\alpha} - \frac{n^{2/3}}{\sqrt{n}\,\sigma}\right) = 1 - \Phi\left(z_{1-\alpha} - \frac{n^{2/3}}{\sqrt{n}\,\sigma}\right) =$$

$$1 - \Phi\left(z_{1-\alpha} - \frac{n^{1/6}}{\sigma}\right) \to 1 - \Phi(-\infty) = 1$$

as $n \to \infty$.

Example 7.12. Let $X_1, \ldots, X_m$ be iid from a distribution with pdf

$$f(x) = \mu x^{\mu - 1},$$

for $0 < x < 1$ where $\mu > 0$. Let $Y_1, \ldots, Y_n$ be iid from a distribution with pdf

$$g(y) = \theta y^{\theta - 1},$$

for $0 < y < 1$ where $\theta > 0$. Let

$$T_1 = \sum_{i=1}^m \log(X_i) \text{ and } T_2 = \sum_{j=1}^n \log(Y_i).$$

Find the likelihood ratio test statistic for $H_0 : \mu = \theta$ versus $H_1 : \mu \neq \theta$ in terms of T_1, T_2, and the MLEs. Simplify.

Solution: $L(\mu) = \mu^m \exp[(\mu - 1)\sum \log(x_i)]$, and $\log(L(\mu)) = m \log(\mu) + (\mu - 1)\sum \log(x_i)$. Hence

$$\frac{d \log(L(\mu))}{d\mu} = \frac{m}{\mu} + \sum \log(x_i) \stackrel{\text{set}}{=} 0.$$

Or $\mu \sum \log(x_i) = -m$ or $\hat{\mu} = -m/T_1$, unique. Now

$$\frac{d^2 \log(L(\mu))}{d\mu^2} = \frac{-m}{\mu^2} < 0.$$

Hence $\hat{\mu}$ is the MLE of μ. Similarly $\hat{\theta} = \dfrac{-n}{\sum_{j=1}^n \log(Y_j)} = \dfrac{-n}{T_2}$. Under H_0 combine the two samples into one iid sample of size $m + n$ with MLE

$$\hat{\mu}_0 = \frac{-(m+n)}{T_1 + T_2}.$$

Now the likelihood ratio statistic

$$\lambda = \frac{L(\hat{\mu}_0)}{L(\hat{\mu}, \hat{\theta})} = \frac{\hat{\mu}_0^{m+n} \exp[(\hat{\mu}_0 - 1)(\sum \log(X_i) + \sum \log(Y_i))]}{\hat{\mu}^m \hat{\theta}^n \exp[(\hat{\mu} - 1) \sum \log(X_i) + (\hat{\theta} - 1) \sum \log(Y_i)]}$$

$$= \frac{\hat{\mu}_0^{m+n} \exp[(\hat{\mu}_0 - 1)(T_1 + T_2)]}{\hat{\mu}^m \hat{\theta}^n \exp[(\hat{\mu} - 1)T_1 + (\hat{\theta} - 1)T_2]} = \frac{\hat{\mu}_0^{m+n} \exp[-(m+n)] \exp[-(T_1 + T_2)]}{\hat{\mu}^m \hat{\theta}^n \exp(-m) \exp(-n) \exp[-(T_1 + T_2)]}$$

$$= \frac{\hat{\mu}_0^{m+n}}{\hat{\mu}^m \hat{\theta}^n} = \frac{\left(\frac{-(m+n)}{T_1+T_2}\right)^{m+n}}{\left(\frac{-m}{T_1}\right)^m \left(\frac{-n}{T_2}\right)^n}.$$

7.4 Summary

For hypothesis testing there is a null hypothesis Ho and an alternative hypothesis $H_1 \equiv H_A$. A **hypothesis test** is a rule for rejecting Ho. Either reject Ho or fail to reject Ho. A **simple hypothesis** consists of exactly one distribution for the sample. A **composite hypothesis** consists of more than one distribution for the sample.

The **power** $\beta(\boldsymbol{\theta}) = P_{\boldsymbol{\theta}}(\text{reject } H_0)$ is the probability of rejecting H_0 when $\boldsymbol{\theta}$ is the true value of the parameter. Often the power function cannot be calculated, but you should be prepared to calculate the power for a sample of size one for a test of the form $H_0 : f(y) = f_0(y)$ versus $H_1 : f(y) = f_1(y)$ or if the test is of the form $\sum t(Y_i) > k$ or $\sum t(Y_i) < k$ when $\sum t(Y_i)$ has an easily handled distribution under H_1, e.g., binomial, normal, Poisson, or χ^2_p. To compute the power, you need to find k and γ for the given value of α.

For a left tailed test, ***p*-value** = $P_{H_0}(T_n(\boldsymbol{Y}) \le T_n(\boldsymbol{y}))$. For a right tailed test, ***p*-value** = $P_{H_0}(T_n(\boldsymbol{Y}) \ge T_n(\boldsymbol{y}))$. If the test statistic $T_n(\boldsymbol{Y})$ has a sampling distribution that is symmetric about 0, then for a two tailed test, ***p*-value** = $P_{H_0}(T_n(\boldsymbol{Y}) \ge |T_n(\boldsymbol{y})|)$. Reject Ho if p-value $\le \alpha$.

Consider all level α tests of $H_0 : \theta \in \Theta_0$ vs. $H_1 : \theta_1 \in \Theta_1$. A **uniformly most powerful** (UMP) level α test is a level α test with power function $\beta_{UMP}(\theta)$ such that $\beta_{UMP}(\theta) \ge \beta(\theta)$ for every $\theta \in \Theta_1$ where β is the power function for any level α test of H_0 vs. H_1.

One-Sided UMP Tests for Exponential Families. Let $Y_1, \ldots, Y_n$ be iid with pdf or pmf

$$f(y|\theta) = h(y)c(\theta)\exp[w(\theta)t(y)]$$

from a one-parameter exponential family where θ is real and $w(\theta)$ is increasing. Let $T(\mathbf{y}) = \sum_{i=1}^n t(y_i)$. Then the UMP test for $H_0 : \theta \leq \theta_0$ vs. $H_A : \theta > \theta_0$ rejects H_0 if $T(\mathbf{y}) > k$ and rejects H_0 with probability γ if $T(\mathbf{y}) = k$ where
$\alpha = P_{\theta_0}(T(\mathbf{Y}) > k) + \gamma P_{\theta_0}(T(\mathbf{Y}) = k)$. The UMP test for $H_0 : \theta \geq \theta_0$ vs. $H_A : \theta < \theta_0$ rejects H_0 if $T(\mathbf{y}) < k$ and rejects H_0 with probability γ if $T(\mathbf{y}) = k$ where
$\alpha = P_{\theta_0}(T(\mathbf{Y}) < k) + \gamma P_{\theta_0}(T(\mathbf{Y}) = k)$.

Fact: if f is a pdf, then usually $\gamma = 0$. For a pmf and $H_A : \theta > \theta_0$,

$$\gamma = \frac{\alpha - P_{\theta_0}[T(\mathbf{Y}) > k]}{P_{\theta_0}[T(\mathbf{Y}) = k]}.$$

For a pmf and $H_A : \theta < \theta_0$,

$$\gamma = \frac{\alpha - P_{\theta_0}[T(\mathbf{Y}) < k]}{P_{\theta_0}[T(\mathbf{Y}) = k]}.$$

As a mnemonic, note that the *inequality used in the rejection region is the same as the inequality in the alternative hypothesis.* Suppose that the parameterization is

$$f(y|\theta) = h(y)c(\theta)\exp[\tilde{w}(\theta)\tilde{t}(y)]$$

where $\tilde{w}(\theta)$ is decreasing. Then set $w(\theta) = -\tilde{w}(\theta)$ and $t(y) = -\tilde{t}(y)$.

Recall that $w(\theta)$ is increasing on Θ if $w'(\theta) > 0$ for $\theta \in \Theta$, and $w(\theta)$ is decreasing on Θ if $w'(\theta) < 0$ for $\theta \in \Theta$. Also $w(\theta)$ is nondecreasing on Θ if $w'(\theta) \geq 0$ for $\theta \in \Theta$, and $w(\theta)$ is nonincreasing on Θ if $w'(\theta) \leq 0$ for $\theta \in \Theta$.

The Neyman–Pearson Lemma: Consider testing $H_0 : \theta = \theta_0$ vs. $H_1 : \theta = \theta_1$ where the pdf or pmf corresponding to θ_i is $f(\mathbf{y}|\theta_i)$ for $i = 0, 1$. Suppose the test rejects H_0 if $f(\mathbf{y}|\theta_1) > kf(\mathbf{y}|\theta_0)$, and rejects H_0 with probability γ if $f(\mathbf{y}|\theta_1) = kf(\mathbf{y}|\theta_0)$ for some $k \geq 0$. If

$$\alpha = \beta(\theta_0) = P_{\theta_0}[f(\mathbf{Y}|\theta_1) > kf(\mathbf{Y}|\theta_0)] + \gamma P_{\theta_0}[f(\mathbf{Y}|\theta_1) = kf(\mathbf{Y}|\theta_0)],$$

then this test is a UMP level α test and a UMP size α test.

One-Sided UMP Tests via the Neyman–Pearson Lemma: Suppose that the hypotheses are of the form $H_0 : \theta \leq \theta_0$ vs. $H_1 : \theta > \theta_0$ or $H_0 : \theta \geq \theta_0$ vs. $H_1 : \theta < \theta_0$, or that the inequality in H_0 is replaced by equality. Also assume that

$$\sup_{\theta \in \Theta_0} \beta(\theta) = \beta(\theta_0).$$

Pick $\theta_1 \in \Theta_1$ and use the Neyman–Pearson lemma to find the UMP test for $H_0^* : \theta = \theta_0$ vs. $H_A^* : \theta = \theta_1$. Then the UMP test rejects H_0^* if $f(\mathbf{y}|\theta_1) > kf(\mathbf{y}|\theta_0)$, and rejects H_0^* with probability γ if $f(\mathbf{y}|\theta_1) = kf(\mathbf{y}|\theta_0)$ for some $k \geq 0$ where

$\alpha = \beta(\theta_0)$. This test is also the UMP level α test for $H_0 : \theta \in \Theta_0$ vs. $H_1 : \theta \in \Theta_1$ if k *does not depend on the value of* $\theta_1 \in \Theta_1$. If $R = f(\boldsymbol{Y}|\theta_1)/f(\boldsymbol{Y}|\theta_0)$, then $\alpha = P_{\theta_0}(R > k) + \gamma P_{\theta_0}(R = k)$.

Fact: if f is a pdf, then usually $\gamma = 0$ and $\alpha = P_{\theta_0}[f(\boldsymbol{Y}|\theta_1) > kf(\boldsymbol{Y}|\theta_0)]$. So γ is important when f is a pmf. For a pmf,

$$\gamma = \frac{\alpha - P_{\theta_0}[f(\boldsymbol{Y}|\theta_1) > kf(\boldsymbol{Y}|\theta_0)]}{P_{\theta_0}[f(\boldsymbol{Y}|\theta_1) = kf(\boldsymbol{Y}|\theta_0)]}.$$

Often it is too hard to give the UMP test in useful form. Then simply specify when the test rejects H_0 and specify α in terms of k (e.g., $\alpha = P_{\theta_0}(T > k) + \gamma P_{\theta_0}(T = k)$).

The problem will be harder if you are asked to put the test in useful form. To find an UMP test with the NP lemma, often the ratio $\dfrac{f(\boldsymbol{y}|\theta_1)}{f(\boldsymbol{y}|\theta_0)}$ is computed. The test will certainly reject H_0 if the ratio is large, but usually the distribution of the ratio is not easy to use. Hence try to get an equivalent test by simplifying and transforming the ratio. Ideally, the ratio can be transformed into a statistic T whose distribution is tabled.

If the test rejects H_0 if $T > k$ (or if $T > k$ and with probability γ if $T = k$, or if $T < k$, or if $T < k$ and with probability γ if $T = k$) the test is in **useful form** if for a given α, you find k and γ. If you are asked to find the power (perhaps with a table), put the test in useful form.

Let $(Y_1, \ldots, Y_n)$ be the data with joint pdf or pmf $f(\boldsymbol{y}|\boldsymbol{\theta})$ where $\boldsymbol{\theta}$ is a vector of unknown parameters with parameter space Θ. Let $\hat{\boldsymbol{\theta}}$ be the MLE of $\boldsymbol{\theta}$ and let $\hat{\boldsymbol{\theta}}_0$ be the MLE of $\boldsymbol{\theta}$ if the parameter space is Θ_0 (where $\Theta_0 \subset \Theta$). A LRT statistic for testing $H_0 : \boldsymbol{\theta} \in \Theta_0$ versus $H_1 : \boldsymbol{\theta} \in \Theta_0^c$ is

$$\lambda(\boldsymbol{y}) = \frac{L(\hat{\boldsymbol{\theta}}_0|\boldsymbol{y})}{L(\hat{\boldsymbol{\theta}}|\boldsymbol{y})}.$$

The **LRT** has a rejection region of the form

$$R = \{\boldsymbol{y}|\lambda(\boldsymbol{y}) \le c\}$$

where $0 \le c \le 1$ and $\alpha = \sup_{\boldsymbol{\theta} \in \Theta_0} P_{\boldsymbol{\theta}}(\lambda(\boldsymbol{Y}) \le c)$.

Fact: Often $\Theta_0 = (a, \theta_0]$ and $\Theta_1 = (\theta_0, b)$ or $\Theta_0 = [\theta_0, b)$ and $\Theta_1 = (a, \theta_0)$.

If you are not asked to find the power or to put the LRT into useful form, it is often enough to find the two MLEs and write $L(\boldsymbol{\theta}|\boldsymbol{y})$ in terms of a sufficient statistic. Simplify the statistic $\lambda(\boldsymbol{y})$ and state that the LRT test rejects H_o if $\lambda(\boldsymbol{y}) \le c$ where $\alpha = \sup_{\boldsymbol{\theta} \in \Theta_0} P_{\boldsymbol{\theta}}(\lambda(\boldsymbol{Y}) \le c)$. If the sup is achieved at $\boldsymbol{\theta}_0 \in \Theta_0$, then $\alpha = P_{\boldsymbol{\theta}_0}(\lambda(\boldsymbol{Y}) \le c)$.

Put the LRT into useful form if asked to find the power. Try to simplify λ or transform λ so that the test rejects H_0 if some statistic $T > k$ (or $T < k$). Getting the test into useful form can be very difficult. Monotone transformations such as log or power transformations can be useful. If you cannot find a statistic T with a simple distribution, use the large sample approximation to the LRT that rejects H_o if $-2\log\lambda(\boldsymbol{y}) > \chi^2_{j,1-\alpha}$ where $P(\chi^2_j > \chi^2_{j,1-\alpha}) = \alpha$. Here $j = r - q$ where r is the number of free parameters specified by $\theta \in \Theta$, and q is the number of free parameters specified by $\theta \in \Theta_0$.

A common LRT problem is $X_1,\ldots,X_n$ are iid with pdf $f(x|\theta)$ while $Y_1,\ldots,Y_m$ are iid with pdf $f(y|\mu)$. $H_0: \mu = \theta$ and $H_1: \mu \neq \theta$. Then under H_0, $X_1,\ldots,X_n,Y_1,\ldots,Y_m$ are an iid sample of size $n+m$ with pdf $f(y|\theta)$. Hence if $f(y|\theta)$ is the $N(\mu,1)$ pdf, then $\hat{\mu}_0 (= \hat{\theta}_0) = \dfrac{\sum_{i=1}^n X_i + \sum_{j=1}^m Y_j}{n+m}$, the sample mean of the combined sample.

Some distribution facts useful for testing.

Memorize Theorems 2.17 and 4.1.

Suppose $Y_1,\ldots,Y_n$ are iid $N(\mu,\sigma^2)$. Then $Z = \frac{Y-\mu}{\sigma} \sim N(0,1)$.

$Z = \frac{\overline{Y}-\mu}{\sigma/\sqrt{n}} \sim N(0,1)$ while $a + cY_i \sim N(a + c\mu, c^2\sigma^2)$.

Suppose $Z, Z_1,\ldots,Z_n$ are iid N(0,1). Then $Z^2 \sim \chi^2_1$.

Also $a + cZ_i \sim N(a,c^2)$ while $\sum_{i=1}^n Z_i^2 \sim \chi^2_n$.

If X_i are independent $\chi^2_{k_i} \equiv \chi^2(k_i)$ for $i = 1,\ldots,n$, then $\sum_{i=1}^n X_i \sim \chi^2(\sum_{i=1}^n k_i)$.

Let $W \sim \text{EXP}(\lambda)$ and let $c > 0$. Then $cW \sim \text{EXP}(c\lambda)$.

Let $W \sim \text{gamma}(\nu,\lambda)$ and let $c > 0$. Then $cW \sim \text{gamma}(\nu,c\lambda)$.

If $W \sim \text{EXP}(\lambda) \sim \text{gamma}(1,\lambda)$, then $2W/\lambda \sim \text{EXP}(2) \sim \text{gamma}(1,2) \sim \chi^2(2)$.

Let $k \geq 0.5$ and let $2k$ be an integer. If $W \sim \text{gamma}(k,\lambda)$, then $2W/\lambda \sim \text{gamma}(k,2) \sim \chi^2(2k)$.

Let $W_1,\ldots,W_n$ be independent $\text{gamma}(\nu_i,\lambda)$. Then $\sum_{i=1}^n W_i \sim \text{gamma}(\sum_{i=1}^n \nu_i, \lambda)$.

7.5 Complements

Example 7.13. As an application of this example, see Problem 7.25. Let $Y_1,\ldots,Y_n$ be iid from a one-parameter exponential family with pdf $f(y|\theta)$ and complete sufficient statistic $T(\boldsymbol{Y}) = \sum_{i=1}^n t(Y_i)$ where $\theta > 0$, $t(Y_i) \sim c\theta\chi^2_j$ and j is some positive integer, often 1 or 2. Usually the constant $c = -1, -1/2, 1/2$ or 1. Suppose

$w(\theta)$ is increasing and let $P(\chi^2_d \leq \chi^2_{d,\delta}) = \delta$. Let a be a positive integer and $b > 0$ some constant. Then $t(Y) \sim G(\frac{a}{2}, b\theta) \sim b\theta\frac{2}{2}G(\frac{a}{2}, 1) \sim \frac{b}{2}\theta G(\frac{a}{2}, 2) \sim \frac{b}{2}\theta\chi^2_a$. Hence $c = b/2$ and $j = a$. If $-t(Y) \sim G(\frac{a}{2}, b\theta)$, then $t(Y) \sim \frac{-b}{2}\theta\chi^2_a$. Hence $c = -b/2$ and $j = a$. Note that $T(\boldsymbol{Y}) \sim c\theta\chi^2_{nj}$, so $\frac{T(\boldsymbol{Y})}{c\theta} \sim \chi^2_{nj}$. Consider the UMP test for $H_0: \theta = \theta_0$ versus $H_1: \theta = \theta_1$, or $H_1: \theta > \theta_0$ or $H_1: \theta < \theta_0$.

i) Let $c > 0$ and $\theta_1 > \theta_0$. Then the UMP test rejects H_0 if $T(\boldsymbol{Y}) > k$ where $\alpha = P_{\theta_0}(T(\boldsymbol{Y}) > k) =$

$$P\left(\frac{T(\boldsymbol{Y})}{c\theta_0} > \frac{k}{c\theta_0}\right) = P(\chi^2_{nj} > \chi^2_{nj,1-\alpha}),$$

and $k = c\theta_0\chi^2_{nj,1-\alpha}$. Hence the power $\beta(\theta) = P_\theta(T(\boldsymbol{Y}) > k) =$

$$P_\theta(T(\boldsymbol{Y}) > c\theta_0\chi^2_{nj,1-\alpha}) = P\left(\frac{T(\boldsymbol{Y})}{c\theta} > \frac{\theta_0}{\theta}\chi^2_{nj,1-\alpha}\right) = P\left(\chi^2_{nj} > \frac{\theta_0}{\theta}\chi^2_{nj,1-\alpha}\right).$$

ii) Let $c < 0$ and $\theta_1 > \theta_0$. Then the UMP test rejects H_0 if $T(\boldsymbol{Y}) > k$ where $\alpha = P_{\theta_0}(T(\boldsymbol{Y}) > k) =$

$$P\left(\frac{T(\boldsymbol{Y})}{c\theta_0} < \frac{k}{c\theta_0}\right) = P(\chi^2_{nj} < \chi^2_{nj,\alpha}),$$

and $k = c\theta_0\chi^2_{nj,\alpha}$. Hence the power $\beta(\theta) = P_\theta(T(\boldsymbol{Y}) > k) =$

$$P_\theta(T(\boldsymbol{Y}) > c\theta_0\chi^2_{nj,\alpha}) = P\left(\frac{T(\boldsymbol{Y})}{c\theta} < \frac{\theta_0}{\theta}\chi^2_{nj,\alpha}\right) = P(\chi^2_{nj} < \frac{\theta_0}{\theta}\chi^2_{nj,\alpha}).$$

iii) Let $c > 0$ and $\theta_1 < \theta_0$. Then the UMP test rejects H_0 if $T(\boldsymbol{Y}) < k$ where $\alpha = P_{\theta_0}(T(\boldsymbol{Y}) < k) =$

$$P\left(\frac{T(\boldsymbol{Y})}{c\theta_0} < \frac{k}{c\theta_0}\right) = P(\chi^2_{nj} < \chi^2_{nj,\alpha}),$$

and $k = c\theta_0\chi^2_{nj,\alpha}$. Hence the power $\beta(\theta) = P_\theta(T(\boldsymbol{Y}) < k) =$

$$P_\theta(T(\boldsymbol{Y}) < c\theta_0\chi^2_{nj,\alpha}) = P\left(\frac{T(\boldsymbol{Y})}{c\theta} < \frac{\theta_0}{\theta}\chi^2_{nj,\alpha}\right) = P\left(\chi^2_{nj} < \frac{\theta_0}{\theta}\chi^2_{nj,\alpha}\right).$$

iv) Let $c < 0$ and $\theta_1 < \theta_0$. Then the UMP test rejects H_0 if $T(\boldsymbol{Y}) < k$ where $\alpha = P_{\theta_0}(T(\boldsymbol{Y}) < k) =$

$$P\left(\frac{T(\boldsymbol{Y})}{c\theta_0} > \frac{k}{c\theta_0}\right) = P(\chi^2_{nj} > \chi^2_{nj,1-\alpha}),$$

and $k = c\theta_0\chi^2_{nj,1-\alpha}$. Hence the power $\beta(\theta) = P_\theta(T(\boldsymbol{Y}) < k) =$

$$P_\theta(T(\boldsymbol{Y}) < c\theta_0\chi^2_{nj,1-\alpha}) = P\left(\frac{T(\boldsymbol{Y})}{c\theta} > \frac{\theta_0}{\theta}\chi^2_{nj,1-\alpha}\right) = P\left(\chi^2_{nj} > \frac{\theta_0}{\theta}\chi^2_{nj,1-\alpha}\right).$$

Definition 7.8. Let $Y_1,\ldots,Y_n$ have pdf or pmf $f(\boldsymbol{y}|\theta)$ for $\theta \in \Theta$. Let $T(\boldsymbol{Y})$ be a statistic. Then $f(\boldsymbol{y}|\theta)$ has a **monotone likelihood ratio** (MLR) in statistic T if for any two values $\theta_0, \theta_1 \in \Theta$ with $\theta_0 < \theta_1$, the ratio $f(\boldsymbol{y}|\theta_1)/f(\boldsymbol{y}|\theta_0)$ depends on the vector $\boldsymbol{y}$ only through $T(\boldsymbol{y})$, and this ratio is an increasing function of $T(\boldsymbol{y})$ over the possible values of $T(\boldsymbol{y})$.

Remark 7.4. Theorem 7.3 is a corollary of the following theorem, because under the conditions of Theorem 7.3, $f(\boldsymbol{y}|\theta)$ has MLR in $T(\boldsymbol{y}) = \sum_{i=1}^n t(y_i)$.

Theorem 7.4, MLR UMP Tests. Let $Y_1,\ldots,Y_n$ be a sample with a joint pdf or pmf $f(y|\theta)$ that has MLR in statistic $T(\boldsymbol{y})$. I) The UMP test for $H_0 : \theta \leq \theta_0$ vs. $H_1 : \theta > \theta_0$ rejects H_0 if $T(\boldsymbol{y}) > k$ and rejects H_0 with probability γ if $T(\boldsymbol{y}) = k$ where $\alpha = P_{\theta_0}(T(\boldsymbol{Y}) > k) + \gamma P_{\theta_0}(T(\boldsymbol{Y}) = k)$. II) The UMP test for $H_0 : \theta \geq \theta_0$ vs. $H_1 : \theta < \theta_0$ rejects H_0 if $T(\boldsymbol{y}) < k$ and rejects H_0 with probability γ if $T(\boldsymbol{y}) = k$ where $\alpha = P_{\theta_0}(T(\boldsymbol{Y}) < k) + \gamma P_{\theta_0}(T(\boldsymbol{Y}) = k)$.

Proof. I) Pick $\theta_1 > \theta_0$ and consider $H_0^* : \theta = \theta_0$ versus $H_1^* : \theta = \theta_1$. Let h be the increasing function such that

$$\frac{f(\boldsymbol{y}|\theta_1)}{f(\boldsymbol{y}|\theta_0)} = h(T(\boldsymbol{y})) > c$$

iff $T(\boldsymbol{y}) > k$. So the NP UMP test is equivalent to the test that rejects H_0^* if $T(\boldsymbol{y}) > k$ and rejects H_0^* with probability γ if $T(\boldsymbol{y}) = k$ where $\alpha =$ $P_{\theta_0}(T(\boldsymbol{Y}) > k) + \gamma P_{\theta_0}(T(\boldsymbol{Y}) = k)$. Since this test does not depend on $\theta_1 > \theta_0$, it is also the UMP test for $H_0^* : \theta = \theta_0$ versus $H_1 : \theta > \theta_0$ by Theorem 7.2. Since $\theta_0 < \theta_1$ was arbitrary, $\sup_{\theta \in \Theta_0} \beta(\theta) = \beta(\theta_0)$ if $\Theta_0 = \{\theta \in \Theta | \theta \leq \theta_0\}$, and the result follows. The proof of II) is similar, but $\theta_1 < \theta_0$. Thus h is an increasing function with respect to $f(\boldsymbol{y}|\theta_0)/f(\boldsymbol{y}|\theta_1)$, but a decreasing function of $T(\boldsymbol{y})$ with respect to $f(\boldsymbol{y}|\theta_1)/f(\boldsymbol{y}|\theta_0)$. Thus

$$\frac{f(\boldsymbol{y}|\theta_1)}{f(\boldsymbol{y}|\theta_0)} = h(T(\boldsymbol{y})) > c$$

iff $T(\boldsymbol{y}) < k$. □

Lehmann and Romano (2005) is an authoritative Ph.D. level text on testing statistical hypotheses. Many of the most used statistical tests of hypotheses are likelihood ratio tests, and several examples are given in DeGroot and Schervish (2012). Scott (2007) discusses the asymptotic distribution of the LRT test.

Birkes (1990) and Solomen (1975) compare the LRT and UMP tests. Rohatgi (1984, p. 725) claims that if the Neyman–Pearson and likelihood ratio tests exist for a given size α, then the two tests are equivalent, but this claim seems to contradict Solomen (1975). Exponential families have the MLR property, and Pfanzagl (1968) gives a partial converse.

7.6 Problems

PROBLEMS WITH AN ASTERISK * ARE ESPECIALLY USEFUL.

Refer to Chap. 10 for the pdf or pmf of the distributions in the problems below.

7.1. Let $X_1,\ldots,X_n$ be iid $N(\mu,\sigma^2)$, $\sigma^2>0$. Let $\Theta_0=\{(\mu_0,\sigma^2):\mu_0 \text{ fixed}, \sigma^2>0\}$ and let $\Theta=\{(\mu,\sigma^2):\mu\in\mathbb{R},\sigma^2>0\}$. Consider testing $H_0:\theta=(\mu,\sigma^2)\in\Theta_0$ vs. H_1: not H_0. The MLE $\hat{\theta}=(\hat{\mu},\hat{\sigma}^2)=(\overline{X},\frac{1}{n}\sum_{i=1}^n(X_i-\overline{X})^2)$ while the restricted MLE is $\hat{\theta}_0=(\hat{\mu}_0,\hat{\sigma}_0^2)=(\mu_0,\dfrac{1}{n}\sum_{i=1}^n(X_i-\mu_0)^2)$.

a) Show that the likelihood ratio statistic

$$\lambda(\boldsymbol{x})=(\hat{\sigma}^2/\hat{\sigma}_0^2)^{n/2}=\left[1+\frac{n(\overline{x}-\mu_0)^2}{\sum_{i=1}^n(x_i-\overline{x})^2}\right]^{-n/2}.$$

b) Show that H_0 is rejected iff $|\sqrt{n}(\overline{X}-\mu_o)/S|\ge k$ and find k if $n=11$ and $\alpha=0.05$. (Hint: show that H_0 is rejected iff $n(\overline{X}-\mu_0)^2/\sum_{i=1}^n(X_i-\overline{X})^2\ge c$, then multiply both sides by a constant such that the left-hand side has a $(t_{n-1})^2$ distribution. Use the fact that

$$\frac{\overline{X}-\mu_0}{S/\sqrt{n}}\sim t_{n-1}$$

under H_0. Use a t-table to find k.)

7.2. Let $X_1,\ldots,X_n$ be a random sample from the distribution with pdf

$$f(x|\theta)=\frac{x^{\theta-1}e^{-x}}{\Gamma(\theta)},\ x>0,\ \theta>0.$$

For a) and b) do not put the rejection region into useful form.

a) Use the Neyman–Pearson lemma to find the UMP size α test for testing $H_0:\theta=1$ vs. $H_1:\theta=\theta_1$ where θ_1 is a fixed number greater than 1.

b) Find the uniformly most powerful level α test of

$$H_0\colon\theta=1 \text{ versus } H_1\colon\theta>1.$$

Justify your steps. Hint: Use the statistic in part a).

7.3. Let $H_0:X_1,\ldots,X_n$ are iid $U(0,10)$ and $H_1:X_1,\ldots,X_n$ are iid $U(4,7)$. Suppose you had a sample of size $n=1000$. How would you decide which hypothesis is true?

Problems from old quizzes and exams. Problems from old qualifying exams are marked with a Q.

7.4. Let $X_1, \ldots, X_{10}$ be iid Bernoulli(p). The most powerful level $\alpha = 0.0547$ test of $H_o : p = 1/2$ vs. $H_1 : p = 1/4$ rejects H_0 if $\sum_{i=1}^{10} x_i \le 2$. H_0 is not rejected if $\sum_{i=1}^{10} x_i > 2$. Find the power of this test if $p = 1/4$.

7.5. Let $X_1, \ldots, X_n$ be iid exponential(β). Hence the pdf is

$$f(x|\beta) = \frac{1}{\beta} \exp(-x/\beta)$$

where $0 \le x$ and $0 < \beta$.

a) Find the MLE of β.

b) Find the level α likelihood ratio test for the hypotheses $H_0 : \beta = \beta_o$ vs. $H_1 : \beta \ne \beta_o$.

7.6^Q. Let $X_1, \ldots, X_n$ be independent, identically distributed random variables from a distribution with a beta(θ, θ) pdf

$$f(x|\theta) = \frac{\Gamma(2\theta)}{\Gamma(\theta)\Gamma(\theta)} [x(1-x)]^{\theta-1}$$

where $0 < x < 1$ and $\theta > 0$.

a) Find the UMP (uniformly most powerful) level α test for $H_0 : \theta = 1$ vs. $H_1 : \theta = 2$.

b) If possible, find the UMP level α test for $H_0 : \theta = 1$ vs. $H_1 : \theta > 1$.

7.7. Let $X_1, \ldots, X_n$ be iid $N(\mu_1, 1)$ random variables and let $Y_1, \ldots, Y_n$ be iid $N(\mu_2, 1)$ random variables that are independent of the X's.

a) Find the α level likelihood ratio test for $H_0 : \mu_1 = \mu_2$ vs. $H_1 : \mu_1 \ne \mu_2$. You may assume that $(\overline{X}, \overline{Y})$ is the MLE of (μ_1, μ_2) and that under the restriction $\mu_1 = \mu_2 = \mu$, say, then the restricted MLE

$$\hat{\mu} = \frac{\sum_{i=1}^n X_i + \sum_{i=1}^n Y_i}{2n}.$$

b) If λ is the LRT test statistic of the above test, use the approximation

$$-2 \log \lambda \approx \chi^2_d$$

for the appropriate degrees of freedom d to find the rejection region of the test **in useful form** if $\alpha = 0.05$.

7.8. Let $X_1, \ldots, X_n$ be independent identically distributed random variables from a distribution with pdf

$$f(x) = \frac{2}{\sigma\sqrt{2\pi}} \frac{1}{x} \exp\left(\frac{-[\log(x)]^2}{2\sigma^2} \right)$$

where $\sigma > 0$ and $x \ge 1$.

If possible, find the UMP level α test for $H_0 : \sigma = 1$ vs. $H_1 : \sigma > 1$.

7.9. Let $X_1,\ldots,X_n$ be independent identically distributed random variables from a distribution with pdf

$$f(x) = \frac{2}{\sigma\sqrt{2\pi}} \exp\left(\frac{-(x-\mu)^2}{2\sigma^2}\right)$$

where $\sigma > 0$ and $x > \mu$ and μ is real. **Assume that μ is known.**

a) What is the UMP (uniformly most powerful) level α test for $H_0 : \sigma^2 = 1$ vs. $H_1 : \sigma^2 = 4$?

b) If possible, find the UMP level α test for $H_0 : \sigma^2 = 1$ vs. $H_1 : \sigma^2 > 1$.

7.10^Q. Let $X_1,\ldots,X_n$ be a random sample from the distribution with pdf

$$f(x,\theta) = \frac{x^{\theta-1}e^{-x}}{\Gamma(\theta)}, \; x > 0, \; \theta > 0.$$

Find the uniformly most powerful level α test of

H: $\theta = 1$ versus K: $\theta > 1$.

7.11^Q. Let $X_1,\ldots,X_n$ be independent identically distributed random variables from a $N(\mu,\sigma^2)$ distribution where the variance σ^2 is known. We want to test $H_0 : \mu = \mu_0$ against $H_1 : \mu \neq \mu_0$.

a) Derive the likelihood ratio test.

b) Let λ be the likelihood ratio. Show that $-2\log\lambda$ is a function of $(\overline{X} - \mu_0)$.

c) Assuming that H_0 is true, find $P(-2\log\lambda > 3.84)$.

7.12^Q. Let $X_1,\ldots,X_n$ be iid from a distribution with pdf

$$f(x) = \frac{2x}{\lambda}\exp(-x^2/\lambda)$$

where λ and x are both positive. Find the level α UMP test for $H_0 : \lambda = 1$ vs. $H_1 : \lambda > 1$.

7.13^Q. Let $X_1,\ldots,X_n$ be iid from a distribution with pdf

$$f(x|\theta) = \frac{(\log\theta)\theta^x}{\theta - 1}$$

where $0 < x < 1$ and $\theta > 1$. Find the UMP (uniformly most powerful) level α test of $H_0 : \theta = 2$ vs. $H_1 : \theta = 4$.

7.14^Q**.** Let $X_1,\ldots,X_n$ be independent identically distributed random variables from a distribution with pdf

$$f(x)=\frac{x^2\exp\left(\frac{-x^2}{2\sigma^2}\right)}{\sigma^3\sqrt{2}\,\Gamma(3/2)}$$

where $\sigma>0$ and $x\geq 0$.

a) What is the UMP (uniformly most powerful) level α test for $H_0:\sigma=1$ vs. $H_1:\sigma=2$?

b) If possible, find the UMP level α test for $H_0:\sigma=1$ vs. $H_1:\sigma>1$.

7.15^Q**.** Let $X_1,\ldots,X_n$ be independent identically distributed random variables from a distribution with pdf

$$f(x)=\frac{2}{\sigma\sqrt{2\pi}}\frac{1}{x}\exp\left(\frac{-[\log(x)]^2}{2\sigma^2}\right)$$

where $\sigma>0$ and $x\geq 1$.

a) What is the UMP (uniformly most powerful) level α test for $H_0:\sigma=1$ vs. $H_1:\sigma=2$?

b) If possible, find the UMP level α test for $H_0:\sigma=1$ vs. $H_1:\sigma>1$.

7.16^Q**.** Suppose X is an observable random variable with its pdf given by $f(x)$, $x\in\mathbb{R}$. Consider two functions defined as follows:

$$f_0(x)=\begin{cases}\frac{3}{64}x^2 & 0\leq x\leq 4\\ 0 & \text{elsewhere}\end{cases}$$

$$f_1(x)=\begin{cases}\frac{3}{16}\sqrt{x} & 0\leq x\leq 4\\ 0 & \text{elsewhere.}\end{cases}$$

Determine the most powerful level α test for $H_0:f(x)=f_0(x)$ versus $H_A:f(x)=f_1(x)$ in the simplest implementable form. Also, find the power of the test when $\alpha=0.01$

7.17^Q**.** Let X be one observation from the probability density function

$$f(x)=\theta x^{\theta-1},\ 0<x<1,\ \theta>0.$$

a) Find the most powerful level α test of $H_0:\theta=1$ versus $H_1:\theta=2$.

b) For testing $H_0:\theta\leq 1$ versus $H_1:\theta>1$, find the size and the power function of the test which rejects H_0 if $X>\dfrac{5}{8}$.

c) Is there a UMP test of $H_0 : \theta \le 1$ versus $H_1 : \theta > 1$? If so, find it. If not, prove so.

7.18. Let $X_1, \ldots, X_n$ be iid $N(\mu, \sigma^2)$ random variables where $\sigma^2 > 0$ **is known**.

a) Find the UMVUE of μ^2.

(Hint: try estimators of the form $aT^2 + b$ where T is the complete sufficient statistic for μ.)

b) Suppose $\sigma^2 = 1$ and $n = 4$. Then the (uniformly) most powerful level $\alpha = 0.05$ test for $H_0 : \mu = 0$ vs. $H_1 : \mu = 2$ rejects H_0 if and only if

$$\sum_{i=1}^{4} x_i > 3.29.$$

Find the power $\beta(2)$ of this test if $\mu = 2$.

7.19^Q. Let $X_1, \ldots, X_n$ be independent identically distributed random variables from a half normal $HN(\mu, \sigma^2)$ distribution with pdf

$$f(x) = \frac{2}{\sigma\sqrt{2\pi}} \exp\left(\frac{-(x-\mu)^2}{2\sigma^2}\right)$$

where $\sigma > 0$ and $x > \mu$ and μ is real. **Assume that μ is known.**

a) What is the UMP (uniformly most powerful) level α test for $H_0 : \sigma^2 = 1$ vs. $H_1 : \sigma^2 = 4$?

b) If possible, find the UMP level α test for $H_0 : \sigma^2 = 1$ vs. $H_1 : \sigma^2 > 1$.

7.20^Q. Suppose that the test statistic $T(X)$ for testing $H_0 : \lambda = 1$ versus $H_1 : \lambda > 1$ has an exponential$(1/\lambda_1)$ distribution if $\lambda = \lambda_1$. The test rejects H_0 if $T(X) < \log(100/95)$.

a) Find the power of the test if $\lambda_1 = 1$.

b) Find the power of the test if $\lambda_1 = 50$.

c) Find the p-value of this test.

7.21^Q. Let $X_1, \ldots, X_n$ be independent identically distributed random variables from a Burr type X distribution with pdf

$$f(x) = 2\,\tau\, x\, e^{-x^2}\,(1 - e^{-x^2})^{\tau-1}$$

where $\tau > 0$ and $x > 0$.

a) What is the UMP (uniformly most powerful) level α test for $H_0 : \tau = 2$ versus $H_1 : \tau = 4$?

b) If possible, find the UMP level α test for $H_0 : \tau = 2$ versus $H_1 : \tau > 2$.

7.22^Q. Let $X_1, \ldots, X_n$ be independent identically distributed random variables from an inverse exponential distribution with pdf

$$f(x) = \frac{\theta}{x^2} \exp\left(\frac{-\theta}{x}\right)$$

where $\theta > 0$ and $x > 0$.

a) What is the UMP (uniformly most powerful) level α test for $H_0 : \theta = 1$ versus $H_1 : \theta = 2$?

b) If possible, find the UMP level α test for $H_0 : \theta = 1$ versus $H_1 : \theta > 1$.

7.23^Q. Suppose that X is an observable random variable with its pdf given by $f(x)$. Consider the two functions defined as follows: $f_0(x)$ is the probability density function of a Beta distribution with $\alpha = 1$ and $\beta = 2$ and and $f_1(x)$ is the pdf of a Beta distribution with $\alpha = 2$ and $\beta = 1$.

a) Determine the UMP level $\alpha = 0.10$ test for $H_0 : f(x) = f_0(x)$ versus $H_1 : f(x) = f_1(x)$. (Find the constant.)

b) Find the power of the test in a).

7.24^Q. The pdf of a bivariate normal distribution is $f(x,y) =$

$$\frac{1}{2\pi\sigma_1\sigma_2(1-\rho^2)^{1/2}} \exp\left(\frac{-1}{2(1-\rho^2)}\left[\left(\frac{x-\mu_1}{\sigma_1}\right)^2 -2\rho\left(\frac{x-\mu_1}{\sigma_1}\right)\left(\frac{y-\mu_2}{\sigma_2}\right) + \left(\frac{y-\mu_2}{\sigma_2}\right)^2\right]\right)$$

where $-1 < \rho < 1$, $\sigma_1 > 0$, $\sigma_2 > 0$, while x, y, μ_1, and μ_2 are all real. Let $(X_1, Y_1), \ldots, (X_n, Y_n)$ be a random sample from a bivariate normal distribution. Let $\hat{\theta}(\boldsymbol{x}, \boldsymbol{y})$ be the observed value of the MLE of θ, and let $\hat{\theta}(\boldsymbol{X}, \boldsymbol{Y})$ be the MLE as a random variable. Let the (unrestricted) MLEs be $\hat{\mu}_1$, $\hat{\mu}_2$, $\hat{\sigma}_1$, $\hat{\sigma}_2$, and $\hat{\rho}$. Then

$$T_1 = \sum_{i=1}^{n} \left(\frac{x_i - \hat{\mu}_1}{\hat{\sigma}_1}\right)^2 = \frac{n\hat{\sigma}_1^2}{\hat{\sigma}_1^2} = n, \text{ and } T_3 = \sum_{i=1}^{n} \left(\frac{y_i - \hat{\mu}_2}{\hat{\sigma}_2}\right)^2 = \frac{n\hat{\sigma}_2^2}{\hat{\sigma}_2^2} = n,$$

$$\text{and } T_2 = \sum_{i=1}^{n} \left(\frac{x_i - \hat{\mu}_1}{\hat{\sigma}_1}\right)\left(\frac{y_i - \hat{\mu}_2}{\hat{\sigma}_2}\right) = n\hat{\rho}.$$

Consider testing $H_0 : \rho = 0$ vs. $H_A : \rho \neq 0$. The (restricted) MLEs for μ_1, μ_2, σ_1, and σ_2 do not change under H_0, and hence are still equal to $\hat{\mu}_1$, $\hat{\mu}_2$, $\hat{\sigma}_1$, and $\hat{\sigma}_2$.

a) Using the above information, find the likelihood ratio test for $H_0 : \rho = 0$ vs. $H_A : \rho \neq 0$. Denote the likelihood ratio test statistic by $\lambda(\boldsymbol{x}, \boldsymbol{y})$.

b) Find the large sample (asymptotic) likelihood ratio test that uses test statistic $-2\log(\lambda(\boldsymbol{x}, \boldsymbol{y}))$.

7.25. Refer to Example 7.13. Let $Y_1,\ldots,Y_n$ be iid from a one-parameter exponential family with pdf $f(y|\theta)$ with complete sufficient statistic $T(\boldsymbol{Y}) = \sum_{i=1}^n t(Y_i)$ where $\theta > 0$, $t(Y_i) \sim c\ \theta\ \chi^2_j$ and j is some positive integer, often 1 or 2. For the following exponential families, assume $w(\theta)$ is increasing, $n = 20$, $\alpha = 0.05$, and find i) the UMP level α test and ii) the power $\beta(\theta_1)$. Let $P(\chi^2_d \le \chi^2_{d,\delta}) = \delta$. The tabled values below give $\chi^2_{d,\delta}$.

a) $Y_1,\ldots,Y_n$ are iid Burr type III ($\phi = 1,\lambda$), $t(Y) = \log(1+Y^{-1}) \sim \frac{\lambda}{2}\chi^2_2$, test $H_0 : \lambda = 2$ versus $H_1 : \lambda = 3.8386$.

b) $Y_1,\ldots,Y_n$ are iid Burr type XII ($\phi = 1,\lambda$), $t(Y) = \log(1+Y) \sim \frac{\lambda}{2}\chi^2_2$, test $H_0 : \lambda = 2$ versus $H_1 : \lambda = 3.8386$.

c) $Y_1,\ldots,Y_n$ are iid BS($\nu,\mu = 1$), $t(Y) \sim \nu^2\chi^2_1$, test $H_0 : \nu = 1$ versus $H_1 : \nu = \sqrt{3.8027}$.

d) $Y_1,\ldots,Y_n$ are iid chi($p = 1,\sigma$) with p known, $t(Y) = Y^2 \sim \sigma^2\chi^2_1$, test $H_0 : \sigma = 1$ versus $H_1 : \sigma = \sqrt{3.8027}$.

e) $Y_1,\ldots,Y_n$ are iid double exponential ($\mu = 0,\lambda$), $t(Y) = |Y| \sim \frac{\lambda}{2}\chi^2_2$, test $H_0 : \lambda = 2$ versus $H_1 : \lambda = 3.8386$.

f) $Y_1,\ldots,Y_n$ are iid $EXP(\lambda)$, $t(Y) = Y \sim \frac{\lambda}{2}\chi^2_2$, test $H_0 : \lambda = 2$ versus $H_1 : \lambda = 3.8386$.

g) $Y_1,\ldots,Y_n$ are iid gamma ($\nu = 1,\lambda$), $t(Y) = Y \sim \frac{\lambda}{2}\chi^2_2$, test $H_0 : \lambda = 2$ versus $H_1 : \lambda = 3.8386$.

h) $Y_1,\ldots,Y_n$ are iid half normal ($\mu = 0,\sigma^2$), $t(Y) = Y^2 \sim \sigma^2\chi^2_1$, test $H_0 : \sigma = 1$ versus $H_1 : \sigma = \sqrt{3.8027}$.

i) $Y_1,\ldots,Y_n$ are iid inverse Weibull IW($\phi = 1,\lambda$), $t(Y) = 1/Y \sim \frac{\lambda}{2}\chi^2_2$, test $H_0 : \lambda = 2$ versus $H_1 : \lambda = 3.8386$.

j) $Y_1,\ldots,Y_n$ are iid inverted gamma ($\nu = 1,\lambda$), $t(Y) = 1/Y \sim \frac{\lambda}{2}\chi^2_2$, test $H_0 : \lambda = 2$ versus $H_1 : \lambda = 3.8386$.

k) $Y_1,\ldots,Y_n$ are iid LG($\nu = 1,\lambda$), $t(Y) = e^Y \sim \dfrac{\lambda}{2}\chi^2_2$, test $H_0 : \lambda = 2$ versus $H_1 : \lambda = 3.8386$.

l) $Y_1,\ldots,Y_n$ are iid Maxwell–Boltzmann ($\mu = 0,\sigma$), $t(Y) = Y^2 \sim \sigma^2\chi^2_3$, test $H_0 : \sigma = 1$ versus $H_1 : \sigma = \sqrt{1.8311}$.

m) $Y_1,\ldots,Y_n$ are iid N($\mu = 0,\sigma^2$), $t(Y) = Y^2 \sim \sigma^2\chi^2_1$, test $H_0 : \sigma = 1$ versus $H_1 : \sigma = \sqrt{3.8027}$.

n) $Y_1,\ldots,Y_n$ are iid Pareto ($\sigma = 1,\lambda$), $t(Y) = \log(Y) \sim \frac{\lambda}{2}\chi^2_2$, test $H_0 : \lambda = 2$ versus $H_1 : \lambda = 3.8386$.

o) $Y_1,\ldots,Y_n$ are iid power (λ) distribution, $t(Y) = -\log(Y) \sim \frac{\lambda}{2}\chi^2_2$, test $H_0 : \lambda = 2$ versus $H_1 : \lambda = 3.8386$.

p) $Y_1,\ldots,Y_n$ are iid Rayleigh ($\mu = 0,\sigma$), $t(Y) = Y^2 \sim \sigma^2\chi^2_2$, test $H_0 : \sigma = 1$ versus $H_1 : \sigma = \sqrt{1.9193}$.

q) $Y_1,\ldots,Y_n$ are iid truncated extreme value (λ), $t(Y) = e^Y - 1 \sim \frac{\lambda}{2}\chi^2_2$, test $H_0 : \lambda = 2$ versus $H_1 : \lambda = 3.8386$.

d	δ							
	0.01	0.05	0.1	0.25	0.75	0.9	0.95	0.99
20	8.260	10.851	12.443	15.452	23.828	28.412	31.410	37.566
40	22.164	26.509	29.051	33.660	45.616	51.805	55.758	63.691
60	37.485	43.188	46.459	52.294	66.981	74.397	79.082	88.379

7.26^Q. Let $Y_1, \ldots, Y_n$ be independent identically distributed random variables with pdf

$$f(y) = e^y I(y \geq 0) \frac{1}{\lambda} \exp\left[\frac{-1}{\lambda}(e^y - 1)\right]$$

where $y > 0$ and $\lambda > 0$.

a) Show that $W = e^Y - 1 \sim \frac{\lambda}{2}\chi^2_2$.

b) What is the UMP (uniformly most powerful) level α test for $H_0 : \lambda = 2$ versus $H_1 : \lambda > 2$?

c) If $n = 20$ and $\alpha = 0.05$, then find the power $\beta(3.8386)$ of the above UMP test if $\lambda = 3.8386$. Let $P(\chi^2_d \leq \chi^2_{d,\delta}) = \delta$. The above tabled values give $\chi^2_{d,\delta}$.

7.27^Q. Let $Y_1, \ldots, Y_n$ be independent identically distributed $N(\mu = 0, \sigma^2)$ random variables with pdf

$$f(y) = \frac{1}{\sqrt{2\pi\sigma^2}} \exp\left(\frac{-y^2}{2\sigma^2}\right)$$

where y is real and $\sigma^2 > 0$.

a) Show $W = Y^2 \sim \sigma^2 \chi^2_1$.

b) What is the UMP (uniformly most powerful) level α test for $H_0 : \sigma^2 = 1$ versus $H_1 : \sigma^2 > 1$?

c) If $n = 20$ and $\alpha = 0.05$, then find the power $\beta(3.8027)$ of the above UMP test if $\sigma^2 = 3.8027$. Let $P(\chi^2_d \leq \chi^2_{d,\delta}) = \delta$. The tabled values below give $\chi^2_{d,\delta}$.

d	δ							
	0.01	0.05	0.1	0.25	0.75	0.9	0.95	0.99
20	8.260	10.851	12.443	15.452	23.828	28.412	31.410	37.566
30	14.953	18.493	20.599	24.478	34.800	40.256	43.773	50.892
40	22.164	26.509	29.051	33.660	45.616	51.805	55.758	63.691

7.28^Q. Let $Y_1, \ldots, Y_n$ be independent identically distributed random variables with pdf

$$f(y) = \frac{y}{\sigma^2} \exp\left[-\frac{1}{2}\left(\frac{y}{\sigma}\right)^2\right]$$

where $\sigma > 0$, μ is real, and $y \geq 0$.

a) Show $W = Y^2 \sim \sigma^2 \chi_2^2$. Equivalently, show $Y^2/\sigma^2 \sim \chi_2^2$.

b) What is the UMP (uniformly most powerful) level α test for $H_0 : \sigma = 1$ versus $H_1 : \sigma > 1$?

c) If $n = 20$ and $\alpha = 0.05$, then find the power $\beta(\sqrt{1.9193})$ of the above UMP test if $\sigma = \sqrt{1.9193}$. Let $P(\chi_d^2 \leq \chi_{d,\delta}^2) = \delta$. The above tabled values give $\chi_{d,\delta}^2$.

7.29^Q. Consider independent random variables $X_1, \ldots, X_n$, where $X_i \sim N(\theta_i, \sigma^2)$, $1 \leq i \leq n$, and σ^2 is known.

a) Find the most powerful test of

$$H_0 : \theta_i = 0, \forall i, \text{ versus } H_1 : \theta_i = \theta_{i0}, \forall i,$$

where θ_{i0} are known. Derive (and simplify) the exact critical region for a level α test.

b) Find the likelihood ratio test of

$$H_0 : \theta_i = 0, \forall i, \text{ versus } H_1 : \theta_i \neq 0, \text{ for some } i.$$

Derive (and simplify) the exact critical region for a level α test.

c) Find the power of the test in (a), when $\theta_{i0} = n^{-1/3}, \forall i$. What happens to this power expression as $n \to \infty$?

7.30. Consider a population with three kinds of individuals labeled 1, 2, and 3 occurring in the proportions $p_1 = \theta^2$, $p_2 = 2\theta(1-\theta)$, $p_3 = (1-\theta)^2$, where $0 < \theta < 1$. For a sample $X_1, \ldots, X_n$ from this population, let N_1, N_2, and N_3 denote the number of X_j equal to 1, 2, and 3, respectively. Consider testing $H_0 : \theta = \theta_0$ versus $H_1 : \theta = \theta_1$, where $0 < \theta_0 < \theta_1 < 1$. Let $\theta_k \in \{\theta_0, \theta_1\}$ for $k = i, j$.

a) Show that the ratio $\dfrac{f(\boldsymbol{x}|\theta_i)}{f(\boldsymbol{x}|\theta_j)}$ used in the Neyman–Pearson Lemma is an increasing function of $2N_1 + N_2$. [Hint: $N_i \equiv N_i(\boldsymbol{X})$. Let $n_i \equiv n_i(\boldsymbol{x})$ be the observed value of N_i. Then $f(\boldsymbol{x}|\theta) = d_{\boldsymbol{x}} p_1^{n_1} p_2^{n_2} p_3^{n_3}$ where the constant $d_{\boldsymbol{x}}$ does not depend on θ and $n_1 + n_2 + n_3 = n$. Write $f(\boldsymbol{x}|\theta)$ as a function of $\theta, d_{\boldsymbol{x}}, n, n_2$, and $2n_1 + n_2$ and simplify. Then simplify the ratio.]

b) Suppose that $c > 0$ and $0 < \alpha < 1$. Show that the test that rejects H_0 if and only if $2N_1 + N_2 \geq c$ is a most powerful test.

7.31^Q. Let $X_1, \ldots, X_m$ be iid from a distribution with pdf

$$f(x) = \mu x^{\mu - 1},$$

for $0 < x < 1$ where $\mu > 0$. Let $Y_1, \ldots, Y_n$ be iid from a distribution with pdf

$$g(y) = \theta y^{\theta - 1},$$

for $0 < y < 1$ where $\theta > 0$. Let

$$T_1 = \sum_{i=1}^{m} \log(X_i) \text{ and } T_2 = \sum_{j=1}^{n} \log(Y_i).$$

Find the likelihood ratio test statistic for $H_0 : \mu = \theta$ versus $H_1 : \mu \neq \theta$ in terms of T_1, T_2, and the MLEs. Simplify.

Chapter 8
Large Sample Theory

Large sample theory, also called asymptotic theory, is used to approximate the distribution of an estimator when the sample size n is large. This theory is extremely useful if the exact sampling distribution of the estimator is complicated or unknown. To use this theory, one must determine what the estimator is estimating, the rate of convergence, the asymptotic distribution, and how large n must be for the approximation to be useful. Moreover, the (asymptotic) standard error (SE), an estimator of the asymptotic standard deviation, must be computable if the estimator is to be useful for inference.

This chapter discusses the central limit theorem, the delta method, asymptotically efficient estimators, convergence in distribution, and convergence in probability. Results on multivariate limit theorems in Sects. 8.6 and 8.7 may be omitted when first reading this chapter. Chapter 9 uses large sample theory to create large sample confidence intervals and large sample tests of hypotheses.

8.1 The CLT, Delta Method, and an Exponential Family Limit Theorem

Theorem 8.1: The Central Limit Theorem (CLT). Let $Y_1, \ldots, Y_n$ be iid with $E(Y) = \mu$ and $\text{VAR}(Y) = \sigma^2$. Let the sample mean $\overline{Y}_n = \frac{1}{n}\sum_{i=1}^n Y_i$. Then

$$\sqrt{n}(\overline{Y}_n - \mu) \xrightarrow{D} N(0, \sigma^2).$$

Hence

$$\sqrt{n}\left(\frac{\overline{Y}_n - \mu}{\sigma}\right) = \sqrt{n}\left(\frac{\sum_{i=1}^n Y_i - n\mu}{n\sigma}\right) \xrightarrow{D} N(0,1).$$

Note that the sample mean is estimating the *population mean* μ with a $\sqrt{n}$ convergence rate, the asymptotic distribution is normal, and the SE $= S/\sqrt{n}$ where S is the *sample standard deviation*. For many distributions the central limit theorem

D.J. Olive, *Statistical Theory and Inference*, DOI 10.1007/978-3-319-04972-4_8,

provides a good approximation if the sample size $n > 30$. A special case of the CLT is proven at the end of Sect. 8.4.

Notation. The notation $X \sim Y$ and $X \stackrel{D}{=} Y$ both mean that the random variables X and Y have the same distribution. See Definition 1.24. The notation $Y_n \stackrel{D}{\to} X$ means that for large n we can approximate the cdf of Y_n by the cdf of X. The distribution of X is the limiting distribution or asymptotic distribution of Y_n, and the limiting distribution does not depend on n. For the CLT, notice that

$$Z_n = \sqrt{n}\left(\frac{\overline{Y}_n - \mu}{\sigma}\right) = \left(\frac{\overline{Y}_n - \mu}{\sigma/\sqrt{n}}\right)$$

is the z-score of $\overline{Y}$. If $Z_n \stackrel{D}{\to} N(0,1)$, then the notation $Z_n \approx N(0,1)$, also written as $Z_n \sim AN(0,1)$, means approximate the cdf of Z_n by the standard normal cdf. Similarly, the notation

$$\overline{Y}_n \approx N(\mu, \sigma^2/n),$$

also written as $\overline{Y}_n \sim AN(\mu, \sigma^2/n)$, means approximate the cdf of $\overline{Y}_n$ as if $\overline{Y}_n \sim N(\mu, \sigma^2/n)$. Note that the approximate distribution, unlike the limiting distribution, does depend on n. The standard error $S/\sqrt{n}$ approximates the asymptotic standard deviation $\sqrt{\sigma^2/n}$ of $\overline{Y}$.

The two main applications of the CLT are to give the limiting distribution of $\sqrt{n}(\overline{Y}_n - \mu)$ and the limiting distribution of $\sqrt{n}(Y_n/n - \mu_X)$ for a random variable Y_n such that $Y_n = \sum_{i=1}^n X_i$ where the X_i are iid with $E(X) = \mu_X$ and $\text{VAR}(X) = \sigma_X^2$. Several of the random variables in Theorems 2.17 and 2.18 can be approximated in this way. The CLT says that $\overline{Y}_n \sim AN(\mu, \sigma^2/n)$. The delta method says that if $T_n \sim AN(\theta, \sigma^2/n)$, and if $g'(\theta) \neq 0$, then $g(T_n) \sim AN(g(\theta), \sigma^2[g'(\theta)]^2/n)$. Hence a smooth function $g(T_n)$ of a well-behaved statistic T_n tends to be well behaved (asymptotically normal with a $\sqrt{n}$ convergence rate).

Example 8.1. a) Let $Y_1, \ldots, Y_n$ be iid Ber(ρ). Then $E(Y) = \rho$ and $\text{VAR}(Y) = \rho(1-\rho)$. Hence

$$\sqrt{n}(\overline{Y}_n - \rho) \stackrel{D}{\to} N(0, \rho(1-\rho))$$

by the CLT.

b) Now suppose that $Y_n \sim \text{BIN}(n, \rho)$. Then $Y_n \stackrel{D}{=} \sum_{i=1}^n X_i$ where $X_1, \ldots, X_n$ are iid Ber(ρ). Hence

$$\sqrt{n}\left(\frac{Y_n}{n} - \rho\right) \stackrel{D}{\to} N(0, \rho(1-\rho))$$

since

$$\sqrt{n}\left(\frac{Y_n}{n} - \rho\right) \stackrel{D}{=} \sqrt{n}(\overline{X}_n - \rho) \stackrel{D}{\to} N(0, \rho(1-\rho))$$

by a).

c) Now suppose that $Y_n \sim \text{BIN}(k_n, \rho)$ where $k_n \to \infty$ as $n \to \infty$. Then

$$\sqrt{k_n}\left(\frac{Y_n}{k_n} - \rho\right) \approx N(0, \rho(1-\rho))$$

or

$$\frac{Y_n}{k_n} \approx N\left(\rho, \frac{\rho(1-\rho)}{k_n}\right) \quad \text{or} \quad Y_n \approx N\left(k_n\rho, k_n\rho(1-\rho)\right).$$

Theorem 8.2: The Delta Method. If g does not depend on n, $g'(\theta) \neq 0$, and

$$\sqrt{n}(T_n - \theta) \xrightarrow{D} N(0, \sigma^2),$$

then

$$\sqrt{n}(g(T_n) - g(\theta)) \xrightarrow{D} N(0, \sigma^2[g'(\theta)]^2).$$

Example 8.2. Let $Y_1, \ldots, Y_n$ be iid with $E(Y) = \mu$ and $\text{VAR}(Y) = \sigma^2$. Then by the CLT,

$$\sqrt{n}(\overline{Y}_n - \mu) \xrightarrow{D} N(0, \sigma^2).$$

Let $g(\mu) = \mu^2$. Then $g'(\mu) = 2\mu \neq 0$ for $\mu \neq 0$. Hence

$$\sqrt{n}((\overline{Y}_n)^2 - \mu^2) \xrightarrow{D} N(0, 4\sigma^2\mu^2)$$

for $\mu \neq 0$ by the delta method.

Example 8.3. Let $X \sim$ Binomial(n, p) where the positive integer n is large and $0 < p < 1$. Find the limiting distribution of $\sqrt{n}\left[g\left(\frac{X}{n}\right)^2 - p^2\right]$.

Solution. Example 8.1b gives the limiting distribution of $\sqrt{n}(\frac{X}{n} - p)$. Let $g(p) = p^2$. Then $g'(p) = 2p$ and by the delta method,

$$\sqrt{n}\left[\left(\frac{X}{n}\right)^2 - p^2\right] = \sqrt{n}\left(g\left(\frac{X}{n}\right) - g(p)\right) \xrightarrow{D}$$

$$N(0, p(1-p)(g'(p))^2) = N(0, p(1-p)4p^2) = N(0, 4p^3(1-p)).$$

Example 8.4. Let $X_n \sim$ Poisson$(n\lambda)$ where the positive integer n is large and $0 < \lambda$.

a) Find the limiting distribution of $\sqrt{n}\left(\frac{X_n}{n} - \lambda\right)$.

b) Find the limiting distribution of $\sqrt{n}\left[\sqrt{\frac{X_n}{n}} - \sqrt{\lambda}\right]$.

Solution. a) $X_n \stackrel{D}{=} \sum_{i=1}^n Y_i$ where the Y_i are iid Poisson(λ). Hence $E(Y) = \lambda = Var(Y)$. Thus by the CLT,

$$\sqrt{n}\left(\frac{X_n}{n} - \lambda\right) \stackrel{D}{=} \sqrt{n}\left(\frac{\sum_{i=1}^n Y_i}{n} - \lambda\right) \stackrel{D}{\to} N(0, \lambda).$$

b) Let $g(\lambda) = \sqrt{\lambda}$. Then $g'(\lambda) = \frac{1}{2\sqrt{\lambda}}$ and by the delta method,

$$\sqrt{n}\left[\sqrt{\frac{X_n}{n}} - \sqrt{\lambda}\right] = \sqrt{n}\left(g\left(\frac{X_n}{n}\right) - g(\lambda)\right) \stackrel{D}{\to}$$

$$N(0, \lambda\,(g'(\lambda))^2) = N\left(0, \lambda\frac{1}{4\lambda}\right) = N\left(0, \frac{1}{4}\right).$$

Example 8.5. Let $Y_1, \ldots, Y_n$ be independent and identically distributed (iid) from a Gamma(α, β) distribution.

a) Find the limiting distribution of $\sqrt{n}\ (\overline{Y} - \alpha\beta\ )$.

b) Find the limiting distribution of $\sqrt{n}\ (\ (\overline{Y})^2 - c\ )$ for appropriate constant c.

Solution: a) Since $E(Y) = \alpha\beta$ and $V(Y) = \alpha\beta^2$, by the CLT $\sqrt{n}\ (\overline{Y} - \alpha\beta\ ) \stackrel{D}{\to} N(0, \alpha\beta^2)$.

b) Let $\mu = \alpha\beta$ and $\sigma^2 = \alpha\beta^2$. Let $g(\mu) = \mu^2$ so $g'(\mu) = 2\mu$ and $[g'(\mu)]^2 = 4\mu^2 = 4\alpha^2\beta^2$. Then by the delta method, $\sqrt{n}\ (\ (\overline{Y})^2 - c\ ) \stackrel{D}{\to} N(0, \sigma^2[g'(\mu)]^2) = N(0, 4\alpha^3\beta^4)$ where $c = \mu^2 = \alpha^2\beta^2$.

Remark 8.1. Note that if $\sqrt{n}(T_n - k) \stackrel{D}{\to} N(0, \sigma^2)$, then evaluate the derivative at k. Thus use $g'(k)$ where $k = \alpha\beta$ in the above example. A common error occurs when k is a simple function of θ, for example $k = \theta/2$ with $g(\mu) = \mu^2$. Thus $g'(\mu) = 2\mu$ so $g'(\theta/2) = 2\theta/2 = \theta$. Then the common delta method error is to plug in $g'(\theta) = 2\theta$ instead of $g'(k) = \theta$. See Problems 8.3, 8.34, 8.36, 8.37, and 8.38.

Barndorff–Nielsen (1982), Casella and Berger (2002, pp. 472, 515), Cox and Hinkley (1974, p. 286), Lehmann and Casella (1998, Section 6.3), Schervish (1995, p. 418), and many others suggest that under regularity conditions if $Y_1, \ldots, Y_n$ are iid from a one-parameter regular exponential family, and if $\hat{\theta}$ is the MLE of θ, then

$$\sqrt{n}(\tau(\hat{\theta}) - \tau(\theta)) \stackrel{D}{\to} N\left(0, \frac{[\tau'(\theta)]^2}{I_1(\theta)}\right) = N[0, FCRLB_1(\tau(\theta))] \tag{8.1}$$

where the Fréchet–Cramér–Rao lower bound for $\tau(\theta)$ is

$$\text{FCRLB}_1(\tau(\theta)) = \frac{[\tau'(\theta)]^2}{I_1(\theta)}$$

and the Fisher information based on a sample of size one is

$$I_1(\theta) = -E_\theta\left[\frac{\partial^2}{\partial\theta^2}\log(f(X|\theta))\right].$$

Hence $\tau(\hat{\theta}) \sim AN[\tau(\theta), \text{FCRLB}_n(\tau(\theta))]$ where $\text{FCRLB}_n(\tau(\theta)) = \text{FCRLB}_1(\tau(\theta))/n$. Notice that if

$$\sqrt{n}(\hat{\theta}-\theta) \xrightarrow{D} N\left(0, \frac{1}{I_1(\theta)}\right),$$

then (8.1) follows by the delta method. Also recall that $\tau(\hat{\theta})$ is the MLE of $\tau(\theta)$ by the invariance principle and that

$$I_1(\tau(\theta)) = \frac{I_1(\theta)}{[\tau'(\theta)]^2}$$

if $\tau'(\theta) \neq 0$ by Definition 6.3.

For a 1P-REF, $\overline{T}_n = \frac{1}{n}\sum_{i=1}^n t(Y_i)$ is the UMVUE and generally the MLE of its expectation $\mu_t \equiv \mu_T = E_\theta(\overline{T}_n) = E_\theta[t(Y)]$. Let $\sigma_t^2 = \text{VAR}_\theta[t(Y)]$. These values can be found by using the distribution of $t(Y)$ (see Theorems 3.6 and 3.7) or by the following result.

Proposition 8.3. Suppose Y is a 1P-REF with pdf or pmf

$$f(y|\theta) = h(y)c(\theta)\exp[w(\theta)t(y)]$$

and natural parameterization

$$f(y|\eta) = h(y)b(\eta)\exp[\eta t(y)].$$

Then a)

$$\mu_t = E[t(Y)] = \frac{-c'(\theta)}{c(\theta)w'(\theta)} = \frac{-\partial}{\partial\eta}\log(b(\eta)), \tag{8.2}$$

and b)

$$\sigma_t^2 = V[t(Y)] = \frac{\frac{-\partial^2}{\partial\theta^2}\log(c(\theta)) - [w''(\theta)]\mu_t}{[w'(\theta)]^2} = \frac{-\partial^2}{\partial\eta^2}\log(b(\eta)). \tag{8.3}$$

Proof. The proof will be for pdfs. For pmfs replace the integrals by sums. By Theorem 3.3, only the middle equalities need to be shown. By Remark 3.3 the derivative and integral operators can be interchanged for a 1P-REF. a) Since $1 = \int f(y|\theta)dy$,

$$0 = \frac{\partial}{\partial\theta}1 = \frac{\partial}{\partial\theta}\int h(y)\exp[w(\theta)t(y) + \log(c(\theta))]dy$$

$$= \int h(y) \frac{\partial}{\partial \theta} \exp[w(\theta)t(y) \ + \ \log(c(\theta))] dy$$

$$= \int h(y) \exp[w(\theta)t(y) \ + \ \log(c(\theta))] \left(w'(\theta)t(y) + \frac{c'(\theta)}{c(\theta)} \right) dy$$

or

$$E[w'(\theta)t(Y)] = \frac{-c'(\theta)}{c(\theta)}$$

or

$$E[t(Y)] = \frac{-c'(\theta)}{c(\theta)w'(\theta)}.$$

b) Similarly,

$$0 = \int h(y) \frac{\partial^2}{\partial \theta^2} \exp[w(\theta)t(y) \ + \ \log(c(\theta))] dy.$$

From the proof of a) and since $\frac{\partial}{\partial \theta} \log(c(\theta)) = c'(\theta)/c(\theta)$,

$$0 = \int h(y) \frac{\partial}{\partial \theta} \left[\exp[w(\theta)t(y) \ + \ \log(c(\theta))] \left(w'(\theta)t(y) + \frac{\partial}{\partial \theta} \log(c(\theta)) \right) \right] dy$$

$$= \int h(y) \exp[w(\theta)t(y) \ + \ \log(c(\theta))] \left(w'(\theta)t(y) + \frac{\partial}{\partial \theta} \log(c(\theta)) \right)^2 dy$$

$$+ \int h(y) \exp[w(\theta)t(y) \ + \ \log(c(\theta))] \left(w''(\theta)t(y) + \frac{\partial^2}{\partial \theta^2} \log(c(\theta)) \right) dy.$$

So

$$E \left(w'(\theta)t(Y) + \frac{\partial}{\partial \theta} \log(c(\theta)) \right)^2 = -E \left(w''(\theta)t(Y) + \frac{\partial^2}{\partial \theta^2} \log(c(\theta)) \right). \quad (8.4)$$

Using a) shows that the left-hand side of (8.4) equals

$$E \left(w'(\theta) \left(t(Y) + \frac{c'(\theta)}{c(\theta)w'(\theta)} \right) \right)^2 = [w'(\theta)]^2 \ \mathrm{VAR}(t(Y))$$

while the right-hand side of (8.4) equals

$$- \left(w''(\theta)\mu_t + \frac{\partial^2}{\partial \theta^2} \log(c(\theta)) \right)$$

and the result follows. □

The simplicity of the following result is rather surprising. When (as is usually the case) $\frac{1}{n} \sum_{i=1}^n t(Y_i)$ is the MLE of μ_t, $\hat{\eta} = g^{-1}(\frac{1}{n} \sum_{i=1}^n t(Y_i))$ is the MLE of η by the invariance principle.

Theorem 8.4. Let $Y_1, \ldots, Y_n$ be iid from a 1P-REF with pdf or pmf

$$f(y|\theta) = h(y)c(\theta)\exp[w(\theta)t(y)]$$

and natural parameterization

$$f(y|\eta) = h(y)b(\eta)\exp[\eta t(y)].$$

Let

$$E(t(Y)) = \mu_t \equiv g(\eta)$$

and $\text{VAR}(t(Y)) = \sigma_t^2$.

a) Then

$$\sqrt{n}\left[\frac{1}{n}\sum_{i=1}^{n} t(Y_i) - \mu_t\right] \xrightarrow{D} N(0, I_1(\eta))$$

where

$$I_1(\eta) = \sigma_t^2 = g'(\eta) = \frac{[g'(\eta)]^2}{I_1(\eta)}.$$

b) If $\eta = g^{-1}(\mu_t)$, $\hat{\eta} = g^{-1}(\frac{1}{n}\sum_{i=1}^{n} t(Y_i))$, and $g^{-1'}(\mu_t) \neq 0$ exists, then

$$\sqrt{n}[\hat{\eta} - \eta] \xrightarrow{D} N\left(0, \frac{1}{I_1(\eta)}\right).$$

c) Suppose the conditions in b) hold. If $\theta = w^{-1}(\eta)$, $\hat{\theta} = w^{-1}(\hat{\eta})$, $w^{-1'}$ exists and is continuous, and $w^{-1'}(\eta) \neq 0$, then

$$\sqrt{n}[\hat{\theta} - \theta] \xrightarrow{D} N\left(0, \frac{1}{I_1(\theta)}\right).$$

d) If the conditions in c) hold, if τ' is continuous and if $\tau'(\theta) \neq 0$, then

$$\sqrt{n}[\tau(\hat{\theta}) - \tau(\theta)] \xrightarrow{D} N\left(0, \frac{[\tau'(\theta)]^2}{I_1(\theta)}\right).$$

Proof. a) The result follows by the central limit theorem if $V(t(Y)) = \sigma_t^2 = I_1(\eta) = g'(\eta)$. Since $\log(f(y|\eta)) = \log(h(y)) + \log(b(\eta)) + \eta t(y)$,

$$\frac{\partial}{\partial \eta}\log(f(y|\eta)) = \frac{\partial}{\partial \eta}\log(b(\eta)) + t(y) = -\mu_t + t(y) = -g(\eta) + t(y)$$

by Proposition 8.3 a). Hence

$$\frac{\partial^2}{\partial \eta^2}\log(f(y|\eta)) = \frac{\partial^2}{\partial \eta^2}\log(b(\eta)) = -g'(\eta),$$

and thus by Proposition 8.3 b)

$$I_1(\eta) = \frac{-\partial^2}{\partial \eta^2} \log(b(\eta)) = \sigma_t^2 = g'(\eta).$$

b) By the delta method,

$$\sqrt{n}(\hat{\eta} - \eta) \xrightarrow{D} N(0, \sigma_t^2 [g^{-1'}(\mu_t)]^2),$$

but

$$g^{-1'}(\mu_t) = \frac{1}{g'(g^{-1}(\mu_t))} = \frac{1}{g'(\eta)}.$$

Since $\sigma_t^2 = I_1(\eta) = g'(\eta)$, it follows that $\sigma_t^2 = [g'(\eta)]^2 / I_1(\eta)$, and

$$\sigma_t^2 [g^{-1'}(\mu_t)]^2 = \frac{[g'(\eta)]^2}{I_1(\eta)} \frac{1}{[g'(\eta)]^2} = \frac{1}{I_1(\eta)}.$$

So

$$\sqrt{n}(\hat{\eta} - \eta) \xrightarrow{D} N\left(0, \frac{1}{I_1(\eta)}\right).$$

c) By the delta method,

$$\sqrt{n}(\hat{\theta} - \theta) \xrightarrow{D} N\left(0, \frac{[w^{-1'}(\eta)]^2}{I_1(\eta)}\right),$$

but

$$\frac{[w^{-1'}(\eta)]^2}{I_1(\eta)} = \frac{1}{I_1(\theta)}.$$

The last equality holds since by Theorem 6.3c, if $\theta = g(\eta)$, if g' exists and is continuous, and if $g'(\theta) \neq 0$, then $I_1(\theta) = I_1(\eta)/[g'(\eta)]^2$. Use $\eta = w(\theta)$ so $\theta = g(\eta) = w^{-1}(\eta)$.

d) The result follows by the delta method. □

Remark 8.2. Following DasGupta (2008, pp. 241–242), let $\psi(\eta) = -\log(b(\eta))$. Then $E_\eta[t(Y_1)] = \mu_t = \psi'(\eta) = g(\eta)$ by Proposition 8.3a, and the MLE $\hat{\eta}$ is the solution of $\frac{1}{n}\sum_{i=1}^n t(y_i) \stackrel{\text{set}}{=} E_\eta[t(Y_1)] = g(\eta)$ by Theorem 5.2, if the MLE exists. Now $g(\eta) = E_\eta[t(Y_1)]$ is an increasing function of η since $g'(\eta) = \psi''(\eta) = V_\eta(t(Y)) > 0$ (1P-REFs do not contain degenerate distributions). So for large n, with probability tending to one, the MLE $\hat{\eta}$ exists and $\hat{\eta} = g^{-1}(\frac{1}{n}\sum_{i=1}^n t(Y_i))$. Since $g'(\eta)$ exists, $g(\eta)$ and $g^{-1}(\eta)$ are continuous and the delta method can be applied to $\hat{\eta}$ as in Theorem 8.4b. By the proof of Theorem 8.4a), $\psi''(\eta) = I_1(\eta)$. Notice that if $\hat{\eta}$ is the MLE of η, then $\frac{1}{n}\sum_{i=1}^n t(Y_i)$ is the MLE of $\mu_t = E[t(Y_1)]$ by invariance. Hence if n is large enough, Theorem 8.4a, b is for the MLE of $E[t(Y_1)]$ and the MLE of η.

8.2 Asymptotically Efficient Estimators

Definition 8.1. Let $Y_1, \ldots, Y_n$ be iid random variables. Let $T_n \equiv T_n(Y_1, \ldots, Y_n)$ be an estimator of a parameter μ_T such that

$$\sqrt{n}(T_n - \mu_T) \xrightarrow{D} N(0, \sigma_A^2).$$

Then the *asymptotic variance* of $\sqrt{n}(T_n - \mu_T)$ is σ_A^2 and the *asymptotic variance* (AV) of T_n is σ_A^2/n. If S_A^2 is a consistent estimator of σ_A^2, then the (asymptotic) *standard error* (SE) of T_n is $S_A/\sqrt{n}$. If $Y_1, \ldots, Y_n$ are iid with cdf F, then $\sigma_A^2 \equiv \sigma_A^2(F)$ depends on F.

Remark 8.3. Consistent estimators are defined in the following section. The parameter σ_A^2 is a function of both the estimator T_n and the underlying distribution F of Y_1. Frequently $n\text{VAR}(T_n)$ converges in distribution to σ_A^2, but not always. See Staudte and Sheather (1990, p. 51) and Lehmann (1999, p. 232).

Example 8.6. If $Y_1, \ldots, Y_n$ are iid from a distribution with mean μ and variance σ^2, then by the central limit theorem,

$$\sqrt{n}(\overline{Y}_n - \mu) \xrightarrow{D} N(0, \sigma^2).$$

Recall that $\text{VAR}(\overline{Y}_n) = \sigma^2/n = AV(\overline{Y}_n)$ and that the standard error $\text{SE}(\overline{Y}_n) = S_n/\sqrt{n}$ where S_n^2 is the sample variance. Note that $\sigma_A^2(F) = \sigma^2$. If F is a $N(\mu, 1)$ cdf, then $\sigma_A^2(F) = 1$; but if F is the $G(\nu = 7, \lambda = 1)$ cdf, then $\sigma_A^2(F) = 7$.

Definition 8.2. Let $T_{1,n}$ and $T_{2,n}$ be two estimators of a parameter θ such that

$$n^{\delta}(T_{1,n} - \theta) \xrightarrow{D} N(0, \sigma_1^2(F))$$

and

$$n^{\delta}(T_{2,n} - \theta) \xrightarrow{D} N(0, \sigma_2^2(F)),$$

then the **asymptotic relative efficiency** of $T_{1,n}$ with respect to $T_{2,n}$ is

$$\text{ARE}(T_{1,n}, T_{2,n}) = \frac{\sigma_2^2(F)}{\sigma_1^2(F)}.$$

This definition brings up several issues. First, both estimators must have the same convergence rate n^{δ}. Usually $\delta = 0.5$. If $T_{i,n}$ has convergence rate n^{δ_i}, then estimator $T_{1,n}$ is judged to be "better" than $T_{2,n}$ if $\delta_1 > \delta_2$. Secondly, the two estimators need to estimate the same parameter θ. This condition will often not hold unless the distribution is symmetric about μ. Then $\theta = \mu$ is a natural choice. Thirdly, estimators are often judged by their Gaussian efficiency with respect to the sample mean (thus F is the normal distribution). Since the normal distribution is a location–scale family, it is often enough to compute the ARE for the standard normal distribution. If the data come from a distribution F and the ARE can be computed, then $T_{1,n}$ is judged to be a "better" estimator (for the data distribution F) than $T_{2,n}$ if the $ARE > 1$. Similarly, $T_{1,n}$ is judged to be a "worse" estimator than $T_{2,n}$ if the ARE < 1. *Notice that the "better" estimator has the smaller asymptotic variance.*

The *population median* is any value MED(Y) such that

$$P(Y \leq \text{MED}(Y)) \geq 0.5 \text{ and } P(Y \geq \text{MED}(Y)) \geq 0.5. \tag{8.5}$$

In simulation studies, typically the underlying distribution F belongs to a symmetric location–scale family. There are at least two reasons for using such distributions. First, if the distribution is symmetric, then the population median MED(Y) is the point of symmetry and the natural parameter to estimate. Under the symmetry assumption, there are many estimators of MED(Y) that can be compared via their ARE with respect to the sample mean or the maximum likelihood estimator (MLE). Secondly, once the ARE is obtained for one member of the family, it is typically obtained for *all members of the location–scale family.* That is, suppose that $Y_1, \ldots, Y_n$ are iid from a location–scale family with parameters μ and σ. Then $Y_i = \mu + \sigma Z_i$ where the Z_i are iid from the same family with $\mu = 0$ and $\sigma = 1$. Typically

$$\text{AV}[T_{i,n}(\boldsymbol{Y})] = \sigma^2 \text{AV}[T_{i,n}(\boldsymbol{Z})],$$

so

$$\text{ARE}[T_{1,n}(\boldsymbol{Y}), T_{2,n}(\boldsymbol{Y})] = \text{ARE}[T_{1,n}(\boldsymbol{Z}), T_{2,n}(\boldsymbol{Z})].$$

Theorem 8.5. Let $Y_1, \ldots, Y_n$ be iid with a pdf f that is positive at the population median: $f(\text{MED}(Y)) > 0$. Then

$$\sqrt{n}(\text{MED}(n) - \text{MED}(Y)) \xrightarrow{D} N\left(0, \frac{1}{4[f(\text{MED}(Y))]^2}\right).$$

Example 8.7. Let $Y_1, \ldots, Y_n$ be iid $N(\mu, \sigma^2)$, $T_{1,n} = \overline{Y}$ and let $T_{2,n} = \text{MED}(n)$ be the sample median. Let $\theta = \mu = E(Y) = \text{MED}(Y)$. Find ARE$(T_{1,n}, T_{2,n})$.

Solution: By the CLT, $\sigma_1^2(F) = \sigma^2$ when F is the $N(\mu, \sigma^2)$ distribution. By Theorem 8.5,

$$\sigma_2^2(F) = \frac{1}{4[f(\text{MED}(Y))]^2} = \frac{1}{4\left[\frac{1}{\sqrt{2\pi\sigma^2}} \exp\left(\frac{-0}{2\sigma^2}\right)\right]^2} = \frac{\pi\sigma^2}{2}.$$

Hence

$$\text{ARE}(T_{1,n}, T_{2,n}) = \frac{\pi\sigma^2/2}{\sigma^2} = \frac{\pi}{2} \approx 1.571$$

and the sample mean $\overline{Y}$ is a "better" estimator of μ than the sample median MED(n) for the family of normal distributions.

Recall from Definition 6.3 that $I_1(\theta)$ is the information number for θ based on a sample of size 1. Also recall that $I_1(\tau(\theta)) = I_1(\theta)/[\tau'(\theta)]^2 = 1/\text{FCRLB}_1[\tau(\theta)]$. See Definition 6.4. The following definition says that if T_n is an asymptotically efficient estimator of $\tau(\theta)$, then
$T_n \sim AN[\tau(\theta), \text{FCRLB}_n(\tau(\theta))]$.

Definition 8.3. Assume $\tau'(\theta) \neq 0$. Then an estimator T_n of $\tau(\theta)$ is **asymptotically efficient** if

$$\sqrt{n}(T_n - \tau(\theta)) \xrightarrow{D} N\left(0, \frac{[\tau'(\theta)]^2}{I_1(\theta)}\right) \sim N(0, \text{FCRLB}_1[\tau(\theta)]). \tag{8.6}$$

In particular, the estimator T_n of θ is asymptotically efficient if

$$\sqrt{n}(T_n - \theta) \xrightarrow{D} N\left(0, \frac{1}{I_1(\theta)}\right) \sim N(0, \text{FCRLB}_1[\theta]). \tag{8.7}$$

Following Lehmann (1999, p. 486), if $T_{2,n}$ is an asymptotically efficient estimator of θ, if $I_1(\theta)$ and $v(\theta)$ are continuous functions, and if $T_{1,n}$ is an estimator such that

$$\sqrt{n}(T_{1,n} - \theta) \xrightarrow{D} N(0, v(\theta)),$$

then under regularity conditions, $v(\theta) \geq 1/I_1(\theta)$ and

$$\text{ARE}(T_{1,n}, T_{2,n}) = \frac{\frac{1}{I_1(\theta)}}{v(\theta)} = \frac{1}{I_1(\theta)v(\theta)} \leq 1.$$

Hence asymptotically efficient estimators are "better" than estimators of the form $T_{1,n}$. When $T_{2,n}$ is asymptotically efficient,

$$\text{AE}(T_{1,n}) = \text{ARE}(T_{1,n}, T_{2,n}) = \frac{1}{I_1(\theta)v(\theta)}$$

is sometimes called the asymptotic efficiency of $T_{1,n}$.

Notice that for a 1P-REF, $\overline{T}_n = \frac{1}{n}\sum_{i=1}^n t(Y_i)$ is an asymptotically efficient estimator of $g(\eta) = E(t(Y))$ by Theorem 8.4. $\overline{T}_n$ is the UMVUE of $E(t(Y))$ by the LSU theorem.

The following rule of thumb suggests that MLEs and UMVUEs are often asymptotically efficient. The rule often holds for location families where the support does not depend on θ. The rule does not hold for the uniform $(0, \theta)$ family.

Rule of Thumb 8.1: A "Standard Limit Theorem": Let $\hat{\theta}_n$ be the MLE or UMVUE of θ. If $\tau'(\theta) \neq 0$, then under strong regularity conditions,

$$\sqrt{n}[\tau(\hat{\theta}_n) - \tau(\theta)] \xrightarrow{D} N\left(0, \frac{[\tau'(\theta)]^2}{I_1(\theta)}\right).$$

8.3 Modes of Convergence and Consistency

Definition 8.4. Let $\{Z_n, n = 1, 2, \ldots\}$ be a sequence of random variables with cdfs F_n, and let X be a random variable with cdf F. Then Z_n **converges in distribution to** X, written

$$Z_n \xrightarrow{D} X,$$

or Z_n *converges in law to* X, written $Z_n \xrightarrow{L} X$, if

$$\lim_{n\to\infty} F_n(t) = F(t)$$

at each continuity point t of F. The distribution of X is called the **limiting distribution** or the **asymptotic distribution** of Z_n.

An important fact is that **the limiting distribution does not depend on the sample size** n. Notice that the CLT, delta method, and Theorem 8.4 give the limiting distributions of $Z_n = \sqrt{n}(\overline{Y}_n - \mu)$, $Z_n = \sqrt{n}(g(T_n) - g(\theta))$, and $Z_n = \sqrt{n}[\frac{1}{n}\sum_{i=1}^n t(Y_i) - E(t(Y))]$, respectively.

Convergence in distribution is useful because if the distribution of X_n is unknown or complicated and the distribution of X is easy to use, then for large n we can approximate the probability that X_n is in an interval by the probability that X is in the interval. To see this, notice that if $X_n \xrightarrow{D} X$, then $P(a < X_n \leq b) = F_n(b) - F_n(a) \to F(b) - F(a) = P(a < X \leq b)$ if F is continuous at a and b. Convergence in distribution is useful for constructing large sample confidence intervals and tests of hypotheses. See Chap. 9.

Warning: Convergence in distribution says that the cdf $F_n(t)$ of X_n gets close to the cdf of F(t) of X as $n \to \infty$ provided that t is a continuity point of F. Hence for any $\varepsilon > 0$ there exists N_t such that if $n > N_t$, then $|F_n(t) - F(t)| < \varepsilon$. Notice that N_t depends on the value of t. Convergence in distribution does not imply that the random variables $X_n \equiv X_n(\omega)$ converge to the random variable $X \equiv X(\omega)$ for all ω.

Example 8.8. Suppose that $X_n \sim U(-1/n, 1/n)$. Then the cdf $F_n(x)$ of X_n is

$$F_n(x) = \begin{cases} 0, & x \leq \frac{-1}{n} \\ \frac{nx}{2} + \frac{1}{2}, & \frac{-1}{n} \leq x \leq \frac{1}{n} \\ 1, & x \geq \frac{1}{n}. \end{cases}$$

Sketching $F_n(x)$ shows that it has a line segment rising from 0 at $x = -1/n$ to 1 at $x = 1/n$ and that $F_n(0) = 0.5$ for all $n \geq 1$. Examining the cases $x < 0$, $x = 0$, and $x > 0$ shows that as $n \to \infty$,

$$F_n(x) \to \begin{cases} 0, & x < 0 \\ \frac{1}{2} & x = 0 \\ 1, & x > 0. \end{cases}$$

Notice that if X is a random variable such that $P(X = 0) = 1$, then X has cdf

$$F_X(x) = \begin{cases} 0, & x < 0 \\ 1, & x \geq 0. \end{cases}$$

Since $x = 0$ is the only discontinuity point of $F_X(x)$ and since $F_n(x) \to F_X(x)$ for all continuity points of $F_X(x)$ (i.e., for $x \neq 0$),

$$X_n \xrightarrow{D} X.$$

Example 8.9. Suppose $Y_n \sim U(0,n)$. Then $F_n(t) = t/n$ for $0 < t \leq n$ and $F_n(t) = 0$ for $t \leq 0$. Hence $\lim_{n\to\infty} F_n(t) = 0$ for $t \leq 0$. If $t > 0$ and $n > t$, then $F_n(t) = t/n \to 0$ as $n \to \infty$. Thus $\lim_{n\to\infty} F_n(t) = 0$ for all t and Y_n does not converge in distribution to any random variable Y since $H(t) \equiv 0$ is not a cdf.

Definition 8.5. A sequence of random variables X_n *converges in distribution to a constant* $\tau(\theta)$, written

$$X_n \xrightarrow{D} \tau(\theta), \text{ if } X_n \xrightarrow{D} X$$

where $P(X = \tau(\theta)) = 1$. The distribution of the random variable X is said to be *degenerate at* $\tau(\theta)$ or to be a *point mass at* $\tau(\theta)$.

Definition 8.6. A sequence of random variables X_n *converges in probability to a constant* $\tau(\theta)$, written

$$X_n \xrightarrow{P} \tau(\theta),$$

if for every $\varepsilon > 0$,

$$\lim_{n\to\infty} P(|X_n - \tau(\theta)| < \varepsilon) = 1 \text{ or, equivalently, } \lim_{n\to\infty} P(|X_n - \tau(\theta)| \geq \varepsilon) = 0.$$

The sequence X_n **converges in probability to** X, written

$$X_n \xrightarrow{P} X,$$

if $X_n - X \xrightarrow{P} 0$.

Notice that $X_n \xrightarrow{P} X$ if for every $\varepsilon > 0$,

$$\lim_{n\to\infty} P(|X_n - X| < \varepsilon) = 1, \text{ or, equivalently, } \lim_{n\to\infty} P(|X_n - X| \geq \varepsilon) = 0.$$

Definition 8.7. A sequence of estimators T_n of $\tau(\theta)$ is **consistent** for $\tau(\theta)$ if

$$T_n \xrightarrow{P} \tau(\theta)$$

for every $\theta \in \Theta$. If T_n is consistent for $\tau(\theta)$, then T_n is a **consistent estimator** of $\tau(\theta)$.

Consistency is a weak property that is usually satisfied by good estimators. T_n is a consistent estimator for $\tau(\theta)$ if the probability that T_n falls in any neighborhood of $\tau(\theta)$ goes to one, regardless of the value of $\theta \in \Theta$. The probability $P \equiv P_\theta$ is the "true" probability distribution or underlying probability that depends on θ.

Definition 8.8. For a real number $r > 0$, Y_n **converges in rth mean** to a random variable Y, written $Y_n \xrightarrow{r} Y$, if

$$E(|Y_n - Y|^r) \to 0$$

as $n \to \infty$. In particular, if $r = 2$, Y_n **converges in quadratic mean** to Y, written

$$Y_n \xrightarrow{2} Y \text{ or } Y_n \xrightarrow{qm} Y,$$

if $E[(Y_n - Y)^2] \to 0$ as $n \to \infty$.

Lemma 8.6: Generalized Chebyshev's Inequality. Let $u : \mathbb{R} \to [0, \infty)$ be a non-negative function. If $E[u(Y)]$ exists then for any $c > 0$,

$$P[u(Y) \geq c] \leq \frac{E[u(Y)]}{c}.$$

If $\mu = E(Y)$ exists, then taking $u(y) = |y - \mu|^r$ and $\tilde{c} = c^r$ gives
Markov's Inequality: for $r > 0$ and any $c > 0$,

$$P(|Y - \mu| \geq c] = P(|Y - \mu|^r \geq c^r] \leq \frac{E[|Y - \mu|^r]}{c^r}.$$

If $r = 2$ and $\sigma^2 = \text{VAR}(Y)$ exists, then we obtain
Chebyshev's Inequality:

$$P(|Y - \mu| \geq c] \leq \frac{\text{VAR}(Y)}{c^2}.$$

Proof. The proof is given for pdfs. For pmfs, replace the integrals by sums. Now

$$E[u(Y)] = \int_{\mathbb{R}} u(y)f(y)dy = \int_{\{y: u(y) \geq c\}} u(y)f(y)dy + \int_{\{y: u(y) < c\}} u(y)f(y)dy$$

$$\geq \int_{\{y: u(y) \geq c\}} u(y)f(y)dy$$

since the integrand $u(y)f(y) \geq 0$. Hence

$$E[u(Y)] \geq c \int_{\{y: u(y) \geq c\}} f(y)dy = cP[u(Y) \geq c]. \ \square$$

The following proposition gives sufficient conditions for T_n to be a consistent estimator of $\tau(\theta)$. Notice that $\text{MSE}_{\tau(\theta)}(T_n) \to 0$ for all $\theta \in \Theta$ is equivalent to $T_n \stackrel{qm}{\to} \tau(\theta)$ for all $\theta \in \Theta$.

Proposition 8.7. a) If

$$\lim_{n \to \infty} \text{MSE}_{\tau(\theta)}(T_n) = 0$$

for all $\theta \in \Theta$, then T_n is a consistent estimator of $\tau(\theta)$.

b) If

$$\lim_{n \to \infty} \text{VAR}_\theta(T_n) = 0 \text{ and } \lim_{n \to \infty} E_\theta(T_n) = \tau(\theta)$$

for all $\theta \in \Theta$, then T_n is a consistent estimator of $\tau(\theta)$.

Proof. a) Using Lemma 8.6 with $Y = T_n$, $u(T_n) = (T_n - \tau(\theta))^2$, and $c = \varepsilon^2$ shows that for any $\varepsilon > 0$,

$$P_\theta(|T_n - \tau(\theta)| \geq \varepsilon) = P_\theta[(T_n - \tau(\theta))^2 \geq \varepsilon^2] \leq \frac{E_\theta[(T_n - \tau(\theta))^2]}{\varepsilon^2}.$$

Hence

$$\lim_{n \to \infty} E_\theta[(T_n - \tau(\theta))^2] = \lim_{n \to \infty} \text{MSE}_{\tau(\theta)}(T_n) \to 0$$

is a sufficient condition for T_n to be a consistent estimator of $\tau(\theta)$.

b) Referring to Definition 6.1,

$$\text{MSE}_{\tau(\theta)}(T_n) = \text{VAR}_\theta(T_n) + [\text{Bias}_{\tau(\theta)}(T_n)]^2$$

where $\text{Bias}_{\tau(\theta)}(T_n) = E_\theta(T_n) - \tau(\theta)$. Since $\text{MSE}_{\tau(\theta)}(T_n) \to 0$ if both $\text{VAR}_\theta(T_n) \to 0$ and $\text{Bias}_{\tau(\theta)}(T_n) = E_\theta(T_n) - \tau(\theta) \to 0$, the result follows from a). $\square$

The following result shows estimators that converge at a $\sqrt{n}$ rate are consistent. Use this result and the delta method to show that $g(T_n)$ is a consistent estimator of $g(\theta)$. Note that b) follows from a) with $X_\theta \sim N(0, v(\theta))$. The WLLN shows that $\overline{Y}$ is a consistent estimator of $E(Y) = \mu$ if $E(Y)$ exists.

Proposition 8.8. a) Let X_θ be a random variable with a distribution depending on θ, and $0 < \delta \leq 1$. If

$$n^\delta (T_n - \tau(\theta)) \stackrel{D}{\to} X_\theta$$

for all $\theta \in \Theta$, then $T_n \stackrel{P}{\to} \tau(\theta)$.

b) If

$$\sqrt{n}(T_n - \tau(\theta)) \xrightarrow{D} N(0, v(\theta))$$

for all $\theta \in \Theta$, then T_n is a consistent estimator of $\tau(\theta)$.

Proposition 8.9. A sequence of random variables X_n *converges almost everywhere* (or *almost surely*, or *with probability 1*) to X if

$$P(\lim_{n\to\infty} X_n = X) = 1.$$

This type of convergence will be denoted by

$$X_n \xrightarrow{ae} X.$$

Notation such as "X_n converges to X ae" will also be used. Sometimes "ae" will be replaced with "as" or "wp1." We say that X_n *converges almost everywhere* to $\tau(\theta)$, written

$$X_n \xrightarrow{ae} \tau(\theta),$$

if $P(\lim_{n\to\infty} X_n = \tau(\theta)) = 1$.

Theorem 8.9. Let Y_n be a sequence of iid random variables with $E(Y_i) = \mu$. Then
a) **Strong Law of Large Numbers (SLLN):** $\overline{Y}_n \xrightarrow{ae} \mu$, and
b) **Weak Law of Large Numbers (WLLN):** $\overline{Y}_n \xrightarrow{P} \mu$.

Proof of WLLN when $V(Y_i) = \sigma^2$: By Chebyshev's inequality, for every $\varepsilon > 0$,

$$P(|\overline{Y}_n - \mu| \geq \varepsilon) \leq \frac{V(\overline{Y}_n)}{\varepsilon^2} = \frac{\sigma^2}{n\varepsilon^2} \to 0$$

as $n \to \infty$. $\square$

8.4 Slutsky's Theorem and Related Results

Theorem 8.10: Slutsky's Theorem. Suppose $Y_n \xrightarrow{D} Y$ and $W_n \xrightarrow{P} w$ for some constant w. Then
a) $Y_n + W_n \xrightarrow{D} Y + w$,
b) $Y_n W_n \xrightarrow{D} wY$, and
c) $Y_n / W_n \xrightarrow{D} Y/w$ if $w \neq 0$.

Theorem 8.11. a) If $X_n \xrightarrow{P} X$ then $X_n \xrightarrow{D} X$.
b) If $X_n \xrightarrow{ae} X$ then $X_n \xrightarrow{P} X$ and $X_n \xrightarrow{D} X$.
c) If $X_n \xrightarrow{r} X$ then $X_n \xrightarrow{P} X$ and $X_n \xrightarrow{D} X$.

d) $X_n \xrightarrow{P} \tau(\theta)$ iff $X_n \xrightarrow{D} \tau(\theta)$.
e) If $X_n \xrightarrow{P} \theta$ and τ is continuous at θ, then $\tau(X_n) \xrightarrow{P} \tau(\theta)$.
f) If $X_n \xrightarrow{D} \theta$ and τ is continuous at θ, then $\tau(X_n) \xrightarrow{D} \tau(\theta)$.

Suppose that for all $\theta \in \Theta$, $T_n \xrightarrow{D} \tau(\theta)$, $T_n \xrightarrow{r} \tau(\theta)$ or $T_n \xrightarrow{ae} \tau(\theta)$. Then T_n is a consistent estimator of $\tau(\theta)$ by Theorem 8.11. We are assuming that the function τ does not depend on n.

Example 8.10. Let $Y_1,\ldots,Y_n$ be iid with mean $E(Y_i) = \mu$ and variance $V(Y_i) = \sigma^2$. Then the sample mean $\overline{Y}_n$ is a consistent estimator of μ since i) the SLLN holds (use Theorem 8.9 and 8.11), ii) the WLLN holds, and iii) the CLT holds (use Proposition 8.8). Since

$$\lim_{n\to\infty} \text{VAR}_\mu(\overline{Y}_n) = \lim_{n\to\infty} \sigma^2/n = 0 \text{ and } \lim_{n\to\infty} E_\mu(\overline{Y}_n) = \mu,$$

$\overline{Y}_n$ is also a consistent estimator of μ by Proposition 8.7b. By the delta method and Proposition 8.8b, $T_n = g(\overline{Y}_n)$ is a consistent estimator of $g(\mu)$ if $g'(\mu) \neq 0$ for all $\mu \in \Theta$. By Theorem 8.11e, $g(\overline{Y}_n)$ is a consistent estimator of $g(\mu)$ if g is continuous at μ for all $\mu \in \Theta$.

Theorem 8.12. Assume that the function g does not depend on n. a) **Generalized Continuous Mapping Theorem:** If $X_n \xrightarrow{D} X$ and the function g is such that $P[X \in C(g)] = 1$ where $C(g)$ is the set of points where g is continuous, then $g(X_n) \xrightarrow{D} g(X)$.
b) **Continuous Mapping Theorem:** If $X_n \xrightarrow{D} X$ and the function g is continuous, then $g(X_n) \xrightarrow{D} g(X)$.

Remark 8.4. For Theorem 8.11, a) follows from Slutsky's Theorem by taking $Y_n \equiv X = Y$ and $W_n = X_n - X$. Then $Y_n \xrightarrow{D} Y = X$ and $W_n \xrightarrow{P} 0$. Hence $X_n = Y_n + W_n \xrightarrow{D} Y + 0 = X$. The convergence in distribution parts of b) and c) follow from a). Part f) follows from d) and e). Part e) implies that if T_n is a consistent estimator of θ and τ is a continuous function, then $\tau(T_n)$ is a consistent estimator of $\tau(\theta)$. Theorem 8.12 says that convergence in distribution is preserved by continuous functions, and even some discontinuities are allowed as long as the set of continuity points is assigned probability 1 by the asymptotic distribution. Equivalently, the set of discontinuity points is assigned probability 0.

Example 8.11 (Ferguson 1996, p. 40). If $X_n \xrightarrow{D} X$ then $1/X_n \xrightarrow{D} 1/X$ if X is a continuous random variable since $P(X = 0) = 0$ and $x = 0$ is the only discontinuity point of $g(x) = 1/x$.

Example 8.12. Show that if $Y_n \sim t_n$, a t distribution with n degrees of freedom, then $Y_n \xrightarrow{D} Z$ where $Z \sim N(0,1)$.

Solution: $Y_n \stackrel{D}{=} Z/\sqrt{V_n/n}$ where $Z \perp\!\!\!\perp V_n \sim \chi^2_n$. If $W_n = \sqrt{V_n/n} \stackrel{P}{\to} 1$, then the result follows by Slutsky's Theorem. But $V_n \stackrel{D}{=} \sum_{i=1}^n X_i$ where the iid $X_i \sim \chi^2_1$. Hence $V_n/n \stackrel{P}{\to} 1$ by the WLLN and $\sqrt{V_n/n} \stackrel{P}{\to} 1$ by Theorem 8.11e.

Theorem 8.13: Continuity Theorem. Let Y_n be sequence of random variables with characteristic functions $\phi_n(t)$. Let Y be a random variable with cf $\phi(t)$.

a)

$$Y_n \stackrel{D}{\to} Y \text{ iff } \phi_n(t) \to \phi(t) \; \forall t \in \mathbb{R}.$$

b) Also assume that Y_n has mgf m_n and Y has mgf m. Assume that all of the mgfs m_n and m are defined on $|t| \le d$ for some $d > 0$. Then if $m_n(t) \to m(t)$ as $n \to \infty$ for all $|t| < c$ where $0 < c < d$, then $Y_n \stackrel{D}{\to} Y$.

Application: Proof of a Special Case of the CLT. Following Rohatgi (1984, pp. 569–569), let $Y_1, \ldots, Y_n$ be iid with mean μ, variance σ^2 and mgf $m_Y(t)$ for $|t| < t_o$. Then

$$Z_i = \frac{Y_i - \mu}{\sigma}$$

has mean 0, variance 1, and mgf $m_Z(t) = \exp(-t\mu/\sigma) m_Y(t/\sigma)$ for $|t| < \sigma t_o$. Want to show that

$$W_n = \sqrt{n}\left(\frac{\overline{Y}_n - \mu}{\sigma}\right) \stackrel{D}{\to} N(0,1).$$

Notice that $W_n =$

$$n^{-1/2} \sum_{i=1}^n Z_i = n^{-1/2} \sum_{i=1}^n \left(\frac{Y_i - \mu}{\sigma}\right) = n^{-1/2} \frac{\sum_{i=1}^n Y_i - n\mu}{\sigma} = \frac{n^{-1/2}}{\frac{1}{n}} \frac{\overline{Y}_n - \mu}{\sigma}.$$

Thus

$$m_{W_n}(t) = E(e^{tW_n}) = E\left[\exp(tn^{-1/2} \sum_{i=1}^n Z_i)\right] = E\left[\exp(\sum_{i=1}^n tZ_i/\sqrt{n})\right]$$

$$= \prod_{i=1}^n E[e^{tZ_i/\sqrt{n}}] = \prod_{i=1}^n m_Z(t/\sqrt{n}) = [m_Z(t/\sqrt{n})]^n.$$

Set $\psi(x) = \log(m_Z(x))$. Then

$$\log[m_{W_n}(t)] = n \log[m_Z(t/\sqrt{n})] = n\psi(t/\sqrt{n}) = \frac{\psi(t/\sqrt{n})}{\frac{1}{n}}.$$

Now $\psi(0) = \log[m_Z(0)] = \log(1) = 0$. Thus by L'Hôpital's rule (where the derivative is with respect to n), $\lim_{n\to\infty} \log[m_{W_n}(t)] =$

$$\lim_{n\to\infty} \frac{\psi(t/\sqrt{n}\,)}{\frac{1}{n}} = \lim_{n\to\infty} \frac{\psi'(t/\sqrt{n}\,)[\frac{-t/2}{n^{3/2}}]}{(\frac{-1}{n^2})} = \frac{t}{2} \lim_{n\to\infty} \frac{\psi'(t/\sqrt{n}\,)}{\frac{1}{\sqrt{n}}}.$$

Now

$$\psi'(0) = \frac{m_Z'(0)}{m_Z(0)} = E(Z_i)/1 = 0,$$

so L'Hôpital's rule can be applied again, giving $\lim_{n\to\infty} \log[m_{W_n}(t)] =$

$$\frac{t}{2} \lim_{n\to\infty} \frac{\psi''(t/\sqrt{n}\,)[\frac{-t}{2n^{3/2}}]}{\left(\frac{-1}{2n^{3/2}}\right)} = \frac{t^2}{2} \lim_{n\to\infty} \psi''(t/\sqrt{n}\,) = \frac{t^2}{2}\psi''(0).$$

Now

$$\psi''(t) = \frac{d}{dt}\frac{m_Z'(t)}{m_Z(t)} = \frac{m_Z''(t) m_Z(t) - (m_Z'(t))^2}{[m_Z(t)]^2}.$$

So

$$\psi''(0) = m_Z''(0) - [m_Z'(0)]^2 = E(Z_i^2) - [E(Z_i)]^2 = 1.$$

Hence $\lim_{n\to\infty} \log[m_{W_n}(t)] = t^2/2$ and

$$\lim_{n\to\infty} m_{W_n}(t) = \exp(t^2/2)$$

which is the N(0,1) mgf. Thus by the continuity theorem,

$$W_n = \sqrt{n}\left(\frac{\overline{Y}_n - \mu}{\sigma}\right) \xrightarrow{D} N(0,1).$$

8.5 Order Relations and Convergence Rates

Definition 8.10 (Lehmann 1999, pp. 53–54). a) A sequence of random variables W_n is *tight* or *bounded in probability*, written $W_n = O_P(1)$, if for every $\varepsilon > 0$ there exist positive constants D_ε and N_ε such that

$$P(|W_n| \leq D_\varepsilon) \geq 1 - \varepsilon$$

for all $n \geq N_\varepsilon$. Also $W_n = O_P(X_n)$ if $|W_n/X_n| = O_P(1)$.

b) The sequence $W_n = o_P(n^{-\delta})$ if $n^\delta W_n = o_P(1)$ which means that

$$n^\delta W_n \xrightarrow{P} 0.$$

c) W_n has the *same order as* X_n *in probability*, written $W_n \asymp_P X_n$, if for every $\varepsilon > 0$ there exist positive constants N_ε and $0 < d_\varepsilon < D_\varepsilon$ such that

$$P(d_\varepsilon \leq \left|\frac{W_n}{X_n}\right| \leq D_\varepsilon) \geq 1 - \varepsilon$$

for all $n \geq N_\varepsilon$.

d) Similar notation is used for a $k \times r$ matrix $\boldsymbol{A}_n = [a_{i,j}(n)]$ if each element $a_{i,j}(n)$ has the desired property. For example, $\boldsymbol{A}_n = O_P(n^{-1/2})$ if each $a_{i,j}(n) = O_P(n^{-1/2})$.

Definition 8.11. Let $\hat{\boldsymbol{\beta}}_n$ be an estimator of a $p \times 1$ vector $\boldsymbol{\beta}$, and let $W_n = \|\hat{\boldsymbol{\beta}}_n - \boldsymbol{\beta}\|$.
a) If $W_n \asymp_P n^{-\delta}$ for some $\delta > 0$, then both W_n and $\hat{\beta}_n$ have (tightness) **rate** n^{δ}.
b) If there exists a constant κ such that

$$n^{\delta}(W_n - \kappa) \xrightarrow{D} X$$

for some nondegenerate random variable X, then both W_n and $\hat{\boldsymbol{\beta}}_n$ have *convergence rate* n^{δ}.

Proposition 8.14. Suppose there exists a constant κ such that

$$n^{\delta}(W_n - \kappa) \xrightarrow{D} X.$$

a) Then $W_n = O_P(n^{-\delta})$.
b) If X is not degenerate, then $W_n \asymp_P n^{-\delta}$.

The above result implies that if W_n has convergence rate n^{δ}, then W_n has tightness rate n^{δ}, and the term "tightness" will often be omitted. Part a) is proved, for example, in Lehmann (1999, p. 67).

The following result shows that if $W_n \asymp_P X_n$, then $X_n \asymp_P W_n$, $W_n = O_P(X_n)$, and $X_n = O_P(W_n)$. Notice that if $W_n = O_P(n^{-\delta})$, then n^{δ} is a lower bound on the rate of W_n. As an example, if the CLT holds then $\overline{Y}_n = O_P(n^{-1/3})$, but $\overline{Y}_n \asymp_P n^{-1/2}$.

Proposition 8.15. a) If $W_n \asymp_P X_n$, then $X_n \asymp_P W_n$.
b) If $W_n \asymp_P X_n$, then $W_n = O_P(X_n)$.
c) If $W_n \asymp_P X_n$, then $X_n = O_P(W_n)$.
d) $W_n \asymp_P X_n$ iff $W_n = O_P(X_n)$ and $X_n = O_P(W_n)$.

Proof. a) Since $W_n \asymp_P X_n$,

$$P\left(d_\varepsilon \le \left|\frac{W_n}{X_n}\right| \le D_\varepsilon\right) = P\left(\frac{1}{D_\varepsilon} \le \left|\frac{X_n}{W_n}\right| \le \frac{1}{d_\varepsilon}\right) \ge 1 - \varepsilon$$

for all $n \ge N_\varepsilon$. Hence $X_n \asymp_P W_n$.
b) Since $W_n \asymp_P X_n$,

$$P(|W_n| \le |X_n D_\varepsilon|) \ge P\left(d_\varepsilon \le \left|\frac{W_n}{X_n}\right| \le D_\varepsilon\right) \ge 1 - \varepsilon$$

for all $n \ge N_\varepsilon$. Hence $W_n = O_P(X_n)$.
c) Follows by a) and b).

d) If $W_n \asymp_P X_n$, then $W_n = O_P(X_n)$ and $X_n = O_P(W_n)$ by b) and c). Now suppose $W_n = O_P(X_n)$ and $X_n = O_P(W_n)$. Then

$$P(|W_n| \le |X_n| D_{\varepsilon/2}) \ge 1 - \varepsilon/2$$

for all $n \ge N_1$, and

$$P(|X_n| \le |W_n| 1/d_{\varepsilon/2}) \ge 1 - \varepsilon/2$$

for all $n \ge N_2$. Hence

$$P(A) \equiv P\left(\left| \frac{W_n}{X_n} \right| \le D_{\varepsilon/2} \right) \ge 1 - \varepsilon/2$$

and

$$P(B) \equiv P\left(d_{\varepsilon/2} \le \left| \frac{W_n}{X_n} \right| \right) \ge 1 - \varepsilon/2$$

for all $n \ge N = \max(N_1, N_2)$. Since $P(A \cap B) = P(A) + P(B) - P(A \cup B) \ge P(A) + P(B) - 1$,

$$P(A \cap B) = P\left(d_{\varepsilon/2} \le \left| \frac{W_n}{X_n} \right| \le D_{\varepsilon/2} \right) \ge 1 - \varepsilon/2 + 1 - \varepsilon/2 - 1 = 1 - \varepsilon$$

for all $n \ge N$. Hence $W_n \asymp_P X_n$. $\square$

The following result is used to prove the following Theorem 8.17 which says that if there are K estimators $T_{j,n}$ of a parameter $\boldsymbol{\beta}$, such that $\|T_{j,n} - \boldsymbol{\beta}\| = O_P(n^{-\delta})$ where $0 < \delta \le 1$, and if T_n^* picks one of these estimators, then $\|T_n^* - \boldsymbol{\beta}\| = O_P(n^{-\delta})$.

Proposition 8.16: Pratt (1959). Let $X_{1,n}, \ldots, X_{K,n}$ each be $O_P(1)$ where K is fixed. Suppose $W_n = X_{i_n,n}$ for some $i_n \in \{1, \ldots, K\}$. Then

$$W_n = O_P(1). \tag{8.8}$$

Proof.

$$P(\max\{X_{1,n}, \ldots, X_{K,n}\} \le x) = P(X_{1,n} \le x, \ldots, X_{K,n} \le x) \le$$

$$F_{W_n}(x) \le P(\min\{X_{1,n}, \ldots, X_{K,n}\} \le x) = 1 - P(X_{1,n} > x, \ldots, X_{K,n} > x).$$

Since K is finite, there exists $B > 0$ and N such that $P(X_{i,n} \le B) > 1 - \varepsilon/2K$ and $P(X_{i,n} > -B) > 1 - \varepsilon/2K$ for all $n > N$ and $i = 1, \ldots, K$. Bonferroni's inequality states that $P(\cap_{i=1}^K A_i) \ge \sum_{i=1}^K P(A_i) - (K-1)$. Thus

$$F_{W_n}(B) \ge P(X_{1,n} \le B, \ldots, X_{K,n} \le B) \ge$$

$$K(1 - \varepsilon/2K) - (K-1) = K - \varepsilon/2 - K + 1 = 1 - \varepsilon/2$$

and

$$-F_{W_n}(-B) \geq -1 + P(X_{1,n} > -B, \ldots, X_{K,n} > -B) \geq$$
$$-1 + K(1 - \varepsilon/2K) - (K-1) = -1 + K - \varepsilon/2 - K + 1 = -\varepsilon/2.$$

Hence

$$F_{W_n}(B) - F_{W_n}(-B) \geq 1 - \varepsilon \text{ for } n > N. \ \square$$

Theorem 8.17. Suppose $\|T_{j,n} - \boldsymbol{\beta}\| = O_P(n^{-\delta})$ for $j = 1, \ldots, K$ where $0 < \delta \leq 1$. Let $T_n^* = T_{i_n,n}$ for some $i_n \in \{1, \ldots, K\}$ where, for example, $T_{i_n,n}$ is the $T_{j,n}$ that minimized some criterion function. Then

$$\|T_n^* - \boldsymbol{\beta}\| = O_P(n^{-\delta}). \tag{8.9}$$

Proof. Let $X_{j,n} = n^{\delta}\|T_{j,n} - \boldsymbol{\beta}\|$. Then $X_{j,n} = O_P(1)$ so by Proposition 8.16, $n^{\delta}\|T_n^* - \boldsymbol{\beta}\| = O_P(1)$. Hence $\|T_n^* - \boldsymbol{\beta}\| = O_P(n^{-\delta})$. $\square$

8.6 Multivariate Limit Theorems

Many of the univariate results of the previous five sections can be extended to random vectors. As stated in Sect. 2.7, the notation for random vectors is rather awkward. For the limit theorems, the vector $\boldsymbol{X}$ is typically a $k \times 1$ column vector and $\boldsymbol{X}^T$ is a row vector. Let $\|\boldsymbol{x}\| = \sqrt{x_1^2 + \cdots + x_k^2}$ be the Euclidean norm of $\boldsymbol{x}$.

Definition 8.12. Let $\boldsymbol{X}_n$ be a sequence of random vectors with joint cdfs $F_n(\boldsymbol{x})$ and let $\boldsymbol{X}$ be a random vector with joint cdf $F(\boldsymbol{x})$.

a) $\boldsymbol{X}_n$ **converges in distribution** to $\boldsymbol{X}$, written $\boldsymbol{X}_n \xrightarrow{D} \boldsymbol{X}$, if $F_n(\boldsymbol{x}) \to F(\boldsymbol{x})$ as $n \to \infty$ for all points $\boldsymbol{x}$ at which $F(\boldsymbol{x})$ is continuous. The distribution of $\boldsymbol{X}$ is the **limiting distribution** or **asymptotic distribution** of $\boldsymbol{X}_n$.

b) $\boldsymbol{X}_n$ **converges in probability** to $\boldsymbol{X}$, written $\boldsymbol{X}_n \xrightarrow{P} \boldsymbol{X}$, if for every $\varepsilon > 0$, $P(\|\boldsymbol{X}_n - \boldsymbol{X}\| > \varepsilon) \to 0$ as $n \to \infty$.

c) Let $r > 0$ be a real number. Then $\boldsymbol{X}_n$ **converges in** r**th mean** to $\boldsymbol{X}$, written $\boldsymbol{X}_n \xrightarrow{r} \boldsymbol{X}$, if $E(\|\boldsymbol{X}_n - \boldsymbol{X}\|^r) \to 0$ as $n \to \infty$.

d) $\boldsymbol{X}_n$ **converges almost everywhere** to $\boldsymbol{X}$, written $\boldsymbol{X}_n \xrightarrow{ae} \boldsymbol{X}$, if $P(\lim_{n\to\infty} \boldsymbol{X}_n = \boldsymbol{X}) = 1$.

Theorems 8.18, 8.19, and 8.21 below are the multivariate extensions of the limit theorems in Sect. 8.1. When the limiting distribution of $\boldsymbol{Z}_n = \sqrt{n}(\boldsymbol{g}(\boldsymbol{T}_n) - \boldsymbol{g}(\boldsymbol{\theta}))$ is multivariate normal $N_k(\boldsymbol{0}, \boldsymbol{\Sigma})$, approximate the joint cdf of $\boldsymbol{Z}_n$ with the joint cdf of the $N_k(\boldsymbol{0}, \boldsymbol{\Sigma})$ distribution. Thus to find probabilities, manipulate $\boldsymbol{Z}_n$ as if $\boldsymbol{Z}_n \approx N_k(\boldsymbol{0}, \boldsymbol{\Sigma})$. To see that the CLT is a special case of the MCLT below, let $k = 1$, $E(X) = \mu$ and $V(X) = \boldsymbol{\Sigma} = \sigma^2$.

Theorem 8.18: the Multivariate Central Limit Theorem (MCLT). If $\boldsymbol{X}_1, \ldots, \boldsymbol{X}_n$ are iid $k \times 1$ random vectors with $E(\boldsymbol{X}) = \boldsymbol{\mu}$ and $\text{Cov}(\boldsymbol{X}) = \boldsymbol{\Sigma}$, then

$$\sqrt{n}(\overline{\boldsymbol{X}}_n - \boldsymbol{\mu}) \xrightarrow{D} N_k(\boldsymbol{0}, \boldsymbol{\Sigma})$$

where the sample mean

$$\overline{\boldsymbol{X}}_n = \frac{1}{n} \sum_{i=1}^{n} \boldsymbol{X}_i.$$

The MCLT is proven after Theorem 8.25. To see that the delta method is a special case of the multivariate delta method, note that if T_n and parameter θ are real valued, then $\boldsymbol{D}_{\boldsymbol{g}(\boldsymbol{\theta})} = g'(\theta)$.

Theorem 8.19: The Multivariate Delta Method. If $\boldsymbol{g}$ does not depend on n and

$$\sqrt{n}(\boldsymbol{T}_n - \boldsymbol{\theta}) \xrightarrow{D} N_k(\boldsymbol{0}, \boldsymbol{\Sigma}),$$

then

$$\sqrt{n}(\boldsymbol{g}(\boldsymbol{T}_n) - \boldsymbol{g}(\boldsymbol{\theta})) \xrightarrow{D} N_d(\boldsymbol{0}, \boldsymbol{D}_{\boldsymbol{g}(\boldsymbol{\theta})} \boldsymbol{\Sigma} \boldsymbol{D}^T_{\boldsymbol{g}(\boldsymbol{\theta})})$$

where the $d \times k$ Jacobian matrix of partial derivatives

$$\boldsymbol{D}_{\boldsymbol{g}(\boldsymbol{\theta})} = \begin{bmatrix} \frac{\partial}{\partial \theta_1} g_1(\boldsymbol{\theta}) & \ldots & \frac{\partial}{\partial \theta_k} g_1(\boldsymbol{\theta}) \\ \vdots & & \vdots \\ \frac{\partial}{\partial \theta_1} g_d(\boldsymbol{\theta}) & \ldots & \frac{\partial}{\partial \theta_k} g_d(\boldsymbol{\theta}) \end{bmatrix}.$$

Here the mapping $\boldsymbol{g} : \mathbb{R}^k \to \mathbb{R}^d$ needs to be differentiable in a neighborhood of $\boldsymbol{\theta} \in \mathbb{R}^k$.

Example 8.13. If Y has a Weibull distribution, $Y \sim W(\phi, \lambda)$, then the pdf of Y is

$$f(y) = \frac{\phi}{\lambda} y^{\phi - 1} e^{-\frac{y^\phi}{\lambda}}$$

where $\lambda, y,$ and ϕ are all positive. If $\mu = \lambda^{1/\phi}$ so $\mu^\phi = \lambda$, then the Weibull pdf

$$f(y) = \frac{\phi}{\mu} \left(\frac{y}{\mu}\right)^{\phi - 1} \exp\left[-\left(\frac{y}{\mu}\right)^\phi\right].$$

Let $(\hat{\mu}, \hat{\phi})$ be the MLE of (μ, ϕ). According to Bain (1978, p. 215),

$$\sqrt{n}\left(\begin{pmatrix} \hat{\mu} \\ \hat{\phi} \end{pmatrix} - \begin{pmatrix} \mu \\ \phi \end{pmatrix}\right) \xrightarrow{D} N\left(\begin{pmatrix} 0 \\ 0 \end{pmatrix}, \begin{pmatrix} 1.109\frac{\mu^2}{\phi^2} & 0.257\mu \\ 0.257\mu & 0.608\phi^2 \end{pmatrix}\right)$$

$= N_2(\boldsymbol{0}, \boldsymbol{I}^{-1}(\boldsymbol{\theta}))$ where $\boldsymbol{I}(\boldsymbol{\theta})$ is given in Definition 8.13.

Let column vectors $\boldsymbol{\theta} = (\mu \;\; \phi)^T$ and $\boldsymbol{\eta} = (\lambda \;\; \phi)^T$. Then

$$\boldsymbol{\eta} = \boldsymbol{g}(\boldsymbol{\theta}) = \begin{pmatrix} \lambda \\ \phi \end{pmatrix} = \begin{pmatrix} \mu^{\phi} \\ \phi \end{pmatrix} = \begin{pmatrix} g_1(\boldsymbol{\theta}) \\ g_2(\boldsymbol{\theta}) \end{pmatrix}.$$

So

$$\boldsymbol{D}_{\boldsymbol{g}(\boldsymbol{\theta})} = \begin{bmatrix} \frac{\partial}{\partial \theta_1} g_1(\boldsymbol{\theta}) & \frac{\partial}{\partial \theta_2} g_1(\boldsymbol{\theta}) \\ \frac{\partial}{\partial \theta_1} g_2(\boldsymbol{\theta}) & \frac{\partial}{\partial \theta_2} g_2(\boldsymbol{\theta}) \end{bmatrix} = \begin{bmatrix} \frac{\partial}{\partial \mu} \mu^{\phi} & \frac{\partial}{\partial \phi} \mu^{\phi} \\ \frac{\partial}{\partial \mu} \phi & \frac{\partial}{\partial \phi} \phi \end{bmatrix} = \begin{bmatrix} \phi \mu^{\phi-1} & \mu^{\phi} \log(\mu) \\ 0 & 1 \end{bmatrix}.$$

Thus by the multivariate delta method,

$$\sqrt{n}\left(\begin{pmatrix} \hat{\lambda} \\ \hat{\phi} \end{pmatrix} - \begin{pmatrix} \lambda \\ \phi \end{pmatrix} \right) \xrightarrow{D} N_2(\boldsymbol{0}, \boldsymbol{\Sigma})$$

where (see Definition 8.15 below)

$$\boldsymbol{\Sigma} = \boldsymbol{I}(\boldsymbol{\eta})^{-1} = [\boldsymbol{I}(\boldsymbol{g}(\boldsymbol{\theta}))]^{-1} = \boldsymbol{D}_{\boldsymbol{g}(\boldsymbol{\theta})} \boldsymbol{I}^{-1}(\boldsymbol{\theta}) \boldsymbol{D}^T_{\boldsymbol{g}(\boldsymbol{\theta})} =$$

$$\begin{bmatrix} 1.109\lambda^2(1 + 0.4635\log(\lambda) + 0.5482(\log(\lambda))^2) & 0.257\phi\lambda + 0.608\lambda\phi\log(\lambda) \\ 0.257\phi\lambda + 0.608\lambda\phi\log(\lambda) & 0.608\phi^2 \end{bmatrix}.$$

Definition 8.13. Let X be a random variable with pdf or pmf $f(x|\boldsymbol{\theta})$. Then the **information matrix**

$$\boldsymbol{I}(\boldsymbol{\theta}) = [\boldsymbol{I}_{i,j}]$$

where

$$\boldsymbol{I}_{i,j} = E\left[\frac{\partial}{\partial \theta_i} \log(f(X|\boldsymbol{\theta})) \frac{\partial}{\partial \theta_j} \log(f(X|\boldsymbol{\theta})) \right].$$

Definition 8.14. An estimator $\boldsymbol{T}_n$ of $\boldsymbol{\theta}$ is **asymptotically efficient** if

$$\sqrt{n}(\boldsymbol{T}_n - \boldsymbol{\theta}) \xrightarrow{D} N_k(\boldsymbol{0}, \boldsymbol{I}^{-1}(\boldsymbol{\theta})).$$

Following Lehmann (1999, p. 511), if $\boldsymbol{T}_n$ is asymptotically efficient and if the estimator $\boldsymbol{W}_n$ satisfies

$$\sqrt{n}(\boldsymbol{W}_n - \boldsymbol{\theta}) \xrightarrow{D} N_k(\boldsymbol{0}, \boldsymbol{J}(\boldsymbol{\theta}))$$

where $\boldsymbol{J}(\boldsymbol{\theta})$ and $\boldsymbol{I}^{-1}(\boldsymbol{\theta})$ are continuous functions of $\boldsymbol{\theta}$, then under regularity conditions, $\boldsymbol{J}(\boldsymbol{\theta}) - \boldsymbol{I}^{-1}(\boldsymbol{\theta})$ is a positive semi-definite matrix, and $\boldsymbol{T}_n$ is "better" than $\boldsymbol{W}_n$.

Definition 8.15. Assume that $\boldsymbol{\eta} = \boldsymbol{g}(\boldsymbol{\theta})$. Then

$$\boldsymbol{I}(\boldsymbol{\eta}) = \boldsymbol{I}(\boldsymbol{g}(\boldsymbol{\theta})) = [\boldsymbol{D}_{\boldsymbol{g}(\boldsymbol{\theta})}\boldsymbol{I}^{-1}(\boldsymbol{\theta})\boldsymbol{D}^T_{\boldsymbol{g}(\boldsymbol{\theta})}]^{-1}.$$

Notice that this definition agrees with the multivariate delta method if

$$\sqrt{n}(\boldsymbol{T}_n - \boldsymbol{\theta}) \xrightarrow{D} N_k(\boldsymbol{0}, \boldsymbol{\Sigma})$$

where $\boldsymbol{\Sigma} = \boldsymbol{I}^{-1}(\boldsymbol{\theta})$.

Now suppose that $X_1, \ldots, X_n$ are iid random variables from a k-parameter REF

$$f(x|\boldsymbol{\theta}) = h(x)c(\boldsymbol{\theta})\exp\left[\sum_{i=1}^k w_i(\boldsymbol{\theta})t_i(x)\right] \tag{8.10}$$

with natural parameterization

$$f(x|\boldsymbol{\eta}) = h(x)b(\boldsymbol{\eta})\exp\left[\sum_{i=1}^k \eta_i t_i(x)\right]. \tag{8.11}$$

Then the complete minimal sufficient statistic is

$$\overline{\boldsymbol{T}}_n = \frac{1}{n}\left(\sum_{i=1}^n t_1(X_i), \ldots, \sum_{i=1}^n t_k(X_i)\right)^T.$$

Let $\boldsymbol{\mu}_T = (E(t_1(X), \ldots, E(t_k(X)))^T$. From Theorem 3.3, for $\boldsymbol{\eta} \in \Omega$,

$$E(t_i(X)) = \frac{-\partial}{\partial \eta_i}\log(b(\boldsymbol{\eta})),$$

and

$$\text{Cov}(t_i(X), t_j(X)) \equiv \sigma_{i,j} = \frac{-\partial^2}{\partial \eta_i \partial \eta_j}\log(b(\boldsymbol{\eta})).$$

Proposition 8.20. If the random variable X is a kP–REF with pmf or pdf (8.11), then the information matrix

$$\boldsymbol{I}(\boldsymbol{\eta}) = [\boldsymbol{I}_{i,j}]$$

where

$$\boldsymbol{I}_{i,j} = E\left[\frac{\partial}{\partial \eta_i}\log(f(X|\boldsymbol{\eta}))\frac{\partial}{\partial \eta_j}\log(f(X|\boldsymbol{\eta}))\right] = -E\left[\frac{\partial^2}{\partial \eta_i \partial \eta_j}\log(f(X|\boldsymbol{\eta}))\right].$$

Several authors, including Barndorff–Nielsen (1982), have noted that the multivariate CLT can be used to show that $\sqrt{n}(\overline{\boldsymbol{T}}_n - \boldsymbol{\mu}_T) \xrightarrow{D} N_k(\boldsymbol{0}, \boldsymbol{\Sigma})$. The fact that $\boldsymbol{\Sigma} = \boldsymbol{I}(\boldsymbol{\eta})$ appears in Lehmann (1983, p. 127).

Theorem 8.21. If $X_1,\ldots,X_n$ are iid from a k-parameter regular exponential family, then

$$\sqrt{n}(\overline{\boldsymbol{T}}_n - \boldsymbol{\mu}_T) \xrightarrow{D} N_k(\boldsymbol{0}, \boldsymbol{I}(\boldsymbol{\eta})).$$

Proof. By the multivariate central limit theorem,

$$\sqrt{n}(\overline{\boldsymbol{T}}_n - \boldsymbol{\mu}_T) \xrightarrow{D} N_k(\boldsymbol{0}, \boldsymbol{\Sigma})$$

where $\boldsymbol{\Sigma} = [\sigma_{i,j}]$. Hence the result follows if $\sigma_{i,j} = \boldsymbol{I}_{i,j}$. Since

$$\log(f(x|\boldsymbol{\eta})) = \log(h(x)) + \log(b(\boldsymbol{\eta})) + \sum_{l=1}^{k} \eta_l t_l(x),$$

$$\frac{\partial}{\partial \eta_i} \log(f(x|\boldsymbol{\eta})) = \frac{\partial}{\partial \eta_i} \log(b(\boldsymbol{\eta})) + t_i(X).$$

Hence

$$-\boldsymbol{I}_{i,j} = E\left[\frac{\partial^2}{\partial \eta_i \partial \eta_j} \log(f(X|\boldsymbol{\eta}))\right] = \frac{\partial^2}{\partial \eta_i \partial \eta_j} \log(b(\boldsymbol{\eta})) = -\sigma_{i,j}. \quad \square$$

To obtain standard results, use the multivariate delta method, assume that both $\boldsymbol{\theta}$ and $\boldsymbol{\eta}$ are $k \times 1$ vectors, and assume that $\boldsymbol{\eta} = \boldsymbol{g}(\boldsymbol{\theta})$ is a one-to-one mapping so that the inverse mapping is $\boldsymbol{\theta} = \boldsymbol{g}^{-1}(\boldsymbol{\eta})$. If $\boldsymbol{D}_{\boldsymbol{g}(\boldsymbol{\theta})}$ is nonsingular, then

$$\boldsymbol{D}^{-1}_{\boldsymbol{g}(\boldsymbol{\theta})} = \boldsymbol{D}_{\boldsymbol{g}^{-1}(\boldsymbol{\eta})} \tag{8.12}$$

(see Searle 1982, p. 339), and

$$\boldsymbol{I}(\boldsymbol{\eta}) = [\boldsymbol{D}_{\boldsymbol{g}(\boldsymbol{\theta})} \boldsymbol{I}^{-1}(\boldsymbol{\theta}) \boldsymbol{D}^T_{\boldsymbol{g}(\boldsymbol{\theta})}]^{-1} = [\boldsymbol{D}^{-1}_{\boldsymbol{g}(\boldsymbol{\theta})}]^T \boldsymbol{I}(\boldsymbol{\theta}) \boldsymbol{D}^{-1}_{\boldsymbol{g}(\boldsymbol{\theta})} = \boldsymbol{D}^T_{\boldsymbol{g}^{-1}(\boldsymbol{\eta})} \boldsymbol{I}(\boldsymbol{\theta}) \boldsymbol{D}_{\boldsymbol{g}^{-1}(\boldsymbol{\eta})}. \tag{8.13}$$

Compare Lehmann (1999, p. 500) and Lehmann (1983, p. 127).

For example, suppose that $\boldsymbol{\mu}_T$ and $\boldsymbol{\eta}$ are $k \times 1$ vectors, and

$$\sqrt{n}(\hat{\boldsymbol{\eta}} - \boldsymbol{\eta}) \xrightarrow{D} N_k(\boldsymbol{0}, \boldsymbol{I}^{-1}(\boldsymbol{\eta}))$$

where $\boldsymbol{\mu}_T = \boldsymbol{g}(\boldsymbol{\eta})$ and $\boldsymbol{\eta} = \boldsymbol{g}^{-1}(\boldsymbol{\mu}_T)$. Also assume that $\overline{\boldsymbol{T}}_n = \boldsymbol{g}(\hat{\boldsymbol{\eta}})$ and $\hat{\boldsymbol{\eta}} = \boldsymbol{g}^{-1}(\overline{\boldsymbol{T}}_n)$. Then by the multivariate delta method and Theorem 8.21,

$$\sqrt{n}(\overline{\boldsymbol{T}}_n - \boldsymbol{\mu}_T) = \sqrt{n}(\boldsymbol{g}(\hat{\boldsymbol{\eta}}) - \boldsymbol{g}(\boldsymbol{\eta})) \xrightarrow{D} N_k[\boldsymbol{0}, \boldsymbol{I}(\boldsymbol{\eta})] = N_k[\boldsymbol{0}, \boldsymbol{D}_{\boldsymbol{g}(\boldsymbol{\eta})} \boldsymbol{I}^{-1}(\boldsymbol{\eta}) \boldsymbol{D}^T_{\boldsymbol{g}(\boldsymbol{\eta})}].$$

Hence

$$\boldsymbol{I}(\boldsymbol{\eta}) = \boldsymbol{D}_{\boldsymbol{g}(\boldsymbol{\eta})} \boldsymbol{I}^{-1}(\boldsymbol{\eta}) \boldsymbol{D}^T_{\boldsymbol{g}(\boldsymbol{\eta})}.$$

Similarly,

$$\sqrt{n}(\boldsymbol{g}^{-1}(\overline{\boldsymbol{T}}_n) - \boldsymbol{g}^{-1}(\boldsymbol{\mu}_T)) = \sqrt{n}(\hat{\boldsymbol{\eta}} - \boldsymbol{\eta}) \xrightarrow{D} N_k[\boldsymbol{0}, \boldsymbol{I}^{-1}(\boldsymbol{\eta})] =$$

$$N_k[\boldsymbol{0}, \boldsymbol{D}_{\boldsymbol{g}^{-1}(\boldsymbol{\mu}_T)} \boldsymbol{I}(\boldsymbol{\eta}) \boldsymbol{D}^T_{\boldsymbol{g}^{-1}(\boldsymbol{\mu}_T)}].$$

Thus

$$\boldsymbol{I}^{-1}(\boldsymbol{\eta}) = \boldsymbol{D}_{\boldsymbol{g}^{-1}(\boldsymbol{\mu}_T)} \boldsymbol{I}(\boldsymbol{\eta}) \boldsymbol{D}^T_{\boldsymbol{g}^{-1}(\boldsymbol{\mu}_T)} = \boldsymbol{D}_{\boldsymbol{g}^{-1}(\boldsymbol{\mu}_T)} \boldsymbol{D}_{\boldsymbol{g}(\boldsymbol{\eta})} \boldsymbol{I}^{-1}(\boldsymbol{\eta}) \boldsymbol{D}^T_{\boldsymbol{g}(\boldsymbol{\eta})} \boldsymbol{D}^T_{\boldsymbol{g}^{-1}(\boldsymbol{\mu}_T)}$$

as expected by Eq. (8.13). Typically $\hat{\boldsymbol{\theta}}$ is a function of the sufficient statistic $\boldsymbol{T}_n$ and is the unique MLE of $\boldsymbol{\theta}$. Replacing $\boldsymbol{\eta}$ by $\boldsymbol{\theta}$ in the above discussion shows that $\sqrt{n}(\hat{\boldsymbol{\theta}} - \boldsymbol{\theta}) \xrightarrow{D} N_k(\boldsymbol{0}, \boldsymbol{I}^{-1}(\boldsymbol{\theta}))$ is equivalent to $\sqrt{n}(\boldsymbol{T}_n - \boldsymbol{\mu}_T) \xrightarrow{D} N_k(\boldsymbol{0}, \boldsymbol{I}(\boldsymbol{\theta}))$ provided that $\boldsymbol{D}_{\boldsymbol{g}(\boldsymbol{\theta})}$ is nonsingular.

8.7 More Multivariate Results

Definition 8.16. If the estimator $\boldsymbol{g}(\boldsymbol{T}_n) \xrightarrow{P} \boldsymbol{g}(\boldsymbol{\theta})$ for all $\boldsymbol{\theta} \in \Theta$, then $\boldsymbol{g}(\boldsymbol{T}_n)$ is a **consistent estimator** of $\boldsymbol{g}(\boldsymbol{\theta})$.

Proposition 8.22. If $0 < \delta \leq 1$, $\boldsymbol{X}$ is a random vector, and

$$n^{\delta}(\boldsymbol{g}(\boldsymbol{T}_n) - \boldsymbol{g}(\boldsymbol{\theta})) \xrightarrow{D} \boldsymbol{X},$$

then $\boldsymbol{g}(\boldsymbol{T}_n) \xrightarrow{P} \boldsymbol{g}(\boldsymbol{\theta})$.

Theorem 8.23. If $\boldsymbol{X}_1, \ldots, \boldsymbol{X}_n$ are iid, $E(\|\boldsymbol{X}\|) < \infty$ and $E(\boldsymbol{X}) = \boldsymbol{\mu}$, then

a) WLLN: $\overline{\boldsymbol{X}}_n \xrightarrow{P} \boldsymbol{\mu}$ and

b) SLLN: $\overline{\boldsymbol{X}}_n \xrightarrow{ae} \boldsymbol{\mu}$.

Theorem 8.24: Continuity Theorem. Let $\boldsymbol{X}_n$ be a sequence of $k \times 1$ random vectors with characteristic function $\phi_n(\boldsymbol{t})$ and let $\boldsymbol{X}$ be a $k \times 1$ random vector with cf $\phi(\boldsymbol{t})$. Then

$$\boldsymbol{X}_n \xrightarrow{D} \boldsymbol{X} \text{ iff } \phi_n(\boldsymbol{t}) \to \phi(\boldsymbol{t})$$

for all $\boldsymbol{t} \in \mathbb{R}^k$.

Theorem 8.25: Cramér Wold Device. Let $\boldsymbol{X}_n$ be a sequence of $k \times 1$ random vectors and let $\boldsymbol{X}$ be a $k \times 1$ random vector. Then

$$\boldsymbol{X}_n \xrightarrow{D} \boldsymbol{X} \text{ iff } \boldsymbol{t}^T \boldsymbol{X}_n \xrightarrow{D} \boldsymbol{t}^T \boldsymbol{X}$$

for all $\boldsymbol{t} \in \mathbb{R}^k$.

Application: Proof of the MCLT Theorem 8.18. Note that for fixed $\boldsymbol{t}$, the $\boldsymbol{t}^T\boldsymbol{X}_i$ are iid random variables with mean $\boldsymbol{t}^T\boldsymbol{\mu}$ and variance $\boldsymbol{t}^T\boldsymbol{\Sigma}\boldsymbol{t}$. Hence by the CLT, $\boldsymbol{t}^T\sqrt{n}(\overline{\boldsymbol{X}}_n-\boldsymbol{\mu})\xrightarrow{D}N(\boldsymbol{0},\boldsymbol{t}^T\boldsymbol{\Sigma}\boldsymbol{t})$. The right-hand side has distribution $\boldsymbol{t}^T\boldsymbol{X}$ where $\boldsymbol{X}\sim N_k(\boldsymbol{0},\boldsymbol{\Sigma})$. Hence by the Cramér Wold Device, $\sqrt{n}(\overline{\boldsymbol{X}}_n-\boldsymbol{\mu})\xrightarrow{D}N_k(\boldsymbol{0},\boldsymbol{\Sigma})$. □

Theorem 8.26. a) If $\boldsymbol{X}_n\xrightarrow{P}\boldsymbol{X}$, then $\boldsymbol{X}_n\xrightarrow{D}\boldsymbol{X}$.
b)

$$\boldsymbol{X}_n\xrightarrow{P}\boldsymbol{g}(\boldsymbol{\theta})\ \text{ iff }\ \boldsymbol{X}_n\xrightarrow{D}\boldsymbol{g}(\boldsymbol{\theta}).$$

Let $g(n)\geq 1$ be an increasing function of the sample size n: $g(n)\uparrow\infty$, e.g., $g(n)=\sqrt{n}$. See White (1984, p. 15). If a $k\times 1$ random vector $\boldsymbol{T}_n-\boldsymbol{\mu}$ converges to a nondegenerate multivariate normal distribution with convergence rate $\sqrt{n}$, then $\boldsymbol{T}_n$ has (tightness) rate $\sqrt{n}$.

Definition 8.17. Let $\boldsymbol{A}_n=[a_{i,j}(n)]$ be an $r\times c$ random matrix.
a) $\boldsymbol{A}_n=O_P(X_n)$ if $a_{i,j}(n)=O_P(X_n)$ for $1\leq i\leq r$ and $1\leq j\leq c$.
b) $\boldsymbol{A}_n=o_p(X_n)$ if $a_{i,j}(n)=o_p(X_n)$ for $1\leq i\leq r$ and $1\leq j\leq c$.
c) $\boldsymbol{A}_n\asymp_P(1/(g(n))$ if $a_{i,j}(n)\asymp_P(1/(g(n))$ for $1\leq i\leq r$ and $1\leq j\leq c$.
d) Let $\boldsymbol{A}_{1,n}=\boldsymbol{T}_n-\boldsymbol{\mu}$ and $\boldsymbol{A}_{2,n}=\boldsymbol{C}_n-c\boldsymbol{\Sigma}$ for some constant $c>0$. If $\boldsymbol{A}_{1,n}\asymp_P(1/(g(n))$ and $\boldsymbol{A}_{2,n}\asymp_P(1/(g(n))$, then $(\boldsymbol{T}_n,\boldsymbol{C}_n)$ has (tightness) rate $g(n)$.

Remark 8.5. Following Severini (2005, p. 354), let W_n, X_n, Y_n and Z_n be sequences of random variables such that $Y_n>0$ and $Z_n>0$. (Often Y_n and Z_n are deterministic, e.g., $Y_n=n^{-1/2}$.)

a) If $W_n=O_P(1)$ and $X_n=O_P(1)$, then $W_n+X_n=O_P(1)$ and $W_nX_n=O_P(1)$, thus $O_P(1)+O_P(1)=O_P(1)$ and $O_P(1)O_P(1)=O_P(1)$.

b) If $W_n=O_P(1)$ and $X_n=o_P(1)$, then $W_n+X_n=O_P(1)$ and $W_nX_n=o_P(1)$, thus $O_P(1)+o_P(1)=O_P(1)$ and $O_P(1)o_P(1)=o_P(1)$.

c) If $W_n=O_P(Y_n)$ and $X_n=O_P(Z_n)$, then $W_n+X_n=O_P(\max(Y_n,Z_n))$ and $W_nX_n=O_P(Y_nZ_n)$, thus $O_P(Y_n)+O_P(Z_n)=O_P(\max(Y_n,Z_n))$ and $O_P(Y_n)O_P(Z_n)=O_P(Y_nZ_n)$.

Recall that the smallest integer function $\lceil x\rceil$ rounds up, e.g., $\lceil 7.7\rceil=8$.

Definition 8.18. The *sample* ρ *quantile* $\hat{\xi}_{n,\rho}=Y_{(\lceil n\rho\rceil)}$. The *population quantile* $\xi_\rho=Q(\rho)=\inf\{y:F(y)\geq\rho\}$.

Theorem 8.27 (Serfling 1980, p. 80): Let $0<\rho_1<\rho_2<\cdots<\rho_k<1$. Suppose that F has a density f that is positive and continuous in neighborhoods of $\xi_{\rho_1},\ldots,\xi_{\rho_k}$. Then

$$\sqrt{n}[(\hat{\xi}_{n,\rho_1},\ldots,\hat{\xi}_{n,\rho_k})^T-(\xi_{\rho_1},\ldots,\xi_{\rho_k})^T]\xrightarrow{D}N_k(\boldsymbol{0},\boldsymbol{\Sigma})$$

where $\boldsymbol{\Sigma} = (\sigma_{ij})$ and

$$\sigma_{ij} = \frac{\rho_i(1-\rho_j)}{f(\xi_{\rho_i})f(\xi_{\rho_j})}$$

for $i \leq j$ and $\sigma_{ij} = \sigma_{ji}$ for $i > j$.

Theorem 8.28: Continuous Mapping Theorem. Let $\boldsymbol{X}_n \in \mathbb{R}^k$. If $\boldsymbol{X}_n \xrightarrow{D} \boldsymbol{X}$ and if the function $\boldsymbol{g} : \mathbb{R}^k \to \mathbb{R}^j$ is continuous and does not depend on n, then $\boldsymbol{g}(\boldsymbol{X}_n) \xrightarrow{D} \boldsymbol{g}(\boldsymbol{X})$.

The following theorem is taken from Severini (2005, pp. 345–349).

Theorem 8.29: Let $\boldsymbol{X}_n = (X_{1n}, \ldots, X_{kn})^T$ be a sequence of $k \times 1$ random vectors, let $\boldsymbol{Y}_n$ be a sequence of $k \times 1$ random vectors and let $\boldsymbol{X} = (X_1, \ldots, X_k)^T$ be a $k \times 1$ random vector. Let $\boldsymbol{W}_n$ be a sequence of $k \times k$ nonsingular random matrices and let $\boldsymbol{C}$ be a $k \times k$ constant nonsingular matrix.

a) $\boldsymbol{X}_n \xrightarrow{P} \boldsymbol{X}$ iff $X_{in} \xrightarrow{P} X_i$ for $i = 1, \ldots, k$.

b) **Slutsky's Theorem**: If $\boldsymbol{X}_n \xrightarrow{D} \boldsymbol{X}$, if $\boldsymbol{Y}_n \xrightarrow{P} \boldsymbol{c}$ for some constant $k \times 1$ vector $\boldsymbol{c}$, and if $\boldsymbol{W}_n \xrightarrow{D} \boldsymbol{C}$, then i) $\boldsymbol{X}_n + \boldsymbol{Y}_n \xrightarrow{D} \boldsymbol{X} + \boldsymbol{c}$.

ii) $\boldsymbol{Y}_n^T \boldsymbol{X}_n \xrightarrow{D} \boldsymbol{c}^T \boldsymbol{X}$.

iii) $\boldsymbol{W}_n \boldsymbol{X}_n \xrightarrow{D} \boldsymbol{C}\boldsymbol{X}$, $\boldsymbol{X}_n^T \boldsymbol{W}_n \xrightarrow{D} \boldsymbol{X}^T \boldsymbol{C}$, $\boldsymbol{W}_n^{-1} \boldsymbol{X}_n \xrightarrow{D} \boldsymbol{C}^{-1} \boldsymbol{X}$ and $\boldsymbol{X}_n^T \boldsymbol{W}_n^{-1} \xrightarrow{D} \boldsymbol{X}^T \boldsymbol{C}^{-1}$.

8.8 Summary

1) **CLT**: Let $Y_1, \ldots, Y_n$ be iid with $E(Y) = \mu$ and $V(Y) = \sigma^2$. Then $\sqrt{n}(\overline{Y}_n - \mu) \xrightarrow{D} N(0, \sigma^2)$.

2) **Delta Method**: If $g'(\theta) \neq 0$ and $\sqrt{n}(T_n - \theta) \xrightarrow{D} N(0, \sigma^2)$, then $\sqrt{n}(g(T_n) - g(\theta)) \xrightarrow{D} N(0, \sigma^2[g'(\theta)]^2)$.

3) **1P-REF Limit Theorem**: Let $Y_1, \ldots, Y_n$ be iid from a 1P-REF with pdf or pmf $f(y|\theta) = h(y)c(\theta)\exp[w(\theta)t(y)]$ and natural parameterization $f(y|\eta) = h(y)b(\eta)\exp[\eta t(y)]$. Let $E(t(Y)) = \mu_t \equiv g(\eta)$ and $V(t(Y)) = \sigma_t^2$. Then $\sqrt{n}[\frac{1}{n}\sum_{i=1}^n t(Y_i) - \mu_t] \xrightarrow{D} N(0, I_1(\eta))$ where $I_1(\eta) = \sigma_t^2 = g'(\eta)$.

4) **Limit theorem for the Sample Median:**
$\sqrt{n}(\text{MED}(n) - \text{MED}(Y)) \xrightarrow{D} N\left(0, \frac{1}{4f^2(\text{MED}(Y))}\right)$.

5) If $n^\delta(T_{1,n} - \theta) \xrightarrow{D} N(0, \sigma_1^2(F))$ and $n^\delta(T_{2,n} - \theta) \xrightarrow{D} N(0, \sigma_2^2(F))$, then the **asymptotic relative efficiency** of $T_{1,n}$ with respect to $T_{2,n}$ is

$$\text{ARE}(T_{1,n}, T_{2,n}) = \frac{\sigma_2^2(F)}{\sigma_1^2(F)}.$$

The "better" estimator has the smaller asymptotic variance or $\sigma_i^2(F)$.

6) An estimator T_n of $\tau(\theta)$ is **asymptotically efficient** if

$$\sqrt{n}(T_n - \tau(\theta)) \xrightarrow{D} N\left(0, \frac{[\tau'(\theta)]^2}{I_1(\theta)}\right).$$

7) For a 1P-REF, $\frac{1}{n}\sum_{i=1}^n t(Y_i)$ is an asymptotically efficient estimator of $g(\eta) = E(t(Y))$.

8) Rule of thumb: If $\hat{\theta}_n$ is the MLE or UMVUE of θ, then $T_n = \tau(\hat{\theta}_n)$ is an asymptotically efficient estimator of $\tau(\theta)$. Hence if $\tau'(\theta) \neq 0$, then

$$\sqrt{n}[\tau(\hat{\theta}_n) - \tau(\theta)] \xrightarrow{D} N\left(0, \frac{[\tau'(\theta)]^2}{I_1(\theta)}\right).$$

9) $X_n \xrightarrow{D} X$ if

$$\lim_{n\to\infty} F_n(t) = F(t)$$

at each continuity point t of F.

10) $X_n \xrightarrow{P} \tau(\theta)$ if for every $\varepsilon > 0$,

$$\lim_{n\to\infty} P(|X_n - \tau(\theta)| < \varepsilon) = 1 \text{ or, equivalently, } \lim_{n\to\infty} P(|X_n - \tau(\theta)| \geq \varepsilon) = 0.$$

11) T_n is a **consistent estimator** of $\tau(\theta)$ if $T_n \xrightarrow{P} \tau(\theta)$ for every $\theta \in \Theta$.

12) T_n is a **consistent estimator** of $\tau(\theta)$ if any of the following three conditions holds:

i) $\lim_{n\to\infty} \text{VAR}_\theta(T_n) = 0$ and $\lim_{n\to\infty} E_\theta(T_n) = \tau(\theta)$ for all $\theta \in \Theta$.

ii) $\text{MSE}_{\tau(\theta)}(T_n) \to 0$ for all $\theta \in \Theta$.

iii) $E[(T_n - \tau(\theta))^2] \to 0$ for all $\theta \in \Theta$.

13) If

$$\sqrt{n}(T_n - \tau(\theta)) \xrightarrow{D} N(0, v(\theta))$$

for all $\theta \in \Theta$, then T_n is a consistent estimator of $\tau(\theta)$.

14) **WLLN:** Let $Y_1, \ldots, Y_n, \ldots$ be a sequence of iid random variables with $E(Y_i) = \mu$. Then $\overline{Y}_n \xrightarrow{P} \mu$. Hence $\overline{Y}_n$ is a consistent estimator of μ.

15) i) If $X_n \xrightarrow{P} X$, then $X_n \xrightarrow{D} X$.

ii) $T_n \xrightarrow{P} \tau(\theta)$ iff $T_n \xrightarrow{D} \tau(\theta)$.

iii) If $T_n \xrightarrow{P} \theta$ and τ is continuous at θ, then $\tau(T_n) \xrightarrow{P} \tau(\theta)$. Hence if T_n is a consistent estimator of θ, then $\tau(T_n)$is a consistent estimator of $\tau(\theta)$ if τ is a continuous function on Θ.

16) **Slutsky's Theorem**: If $Y_n \xrightarrow{D} Y$ and $W_n \xrightarrow{P} w$ for some constant w, then $Y_n W_n \xrightarrow{D} wY$, $Y_n + W_n \xrightarrow{D} Y + w$ and $Y_n/W_n \xrightarrow{D} Y/w$ for $w \neq 0$.

8.9 Complements

Some authors state that if $X_n \xrightarrow{P} X$ and g is continuous, then $g(X_n) \xrightarrow{P} g(X)$, but Sen and Singer (1993, p. 59) says that g needs to be uniformly continuous.

The following extension of the delta method is sometimes useful.

Theorem 8.30. Suppose that g does not depend on n, $g'(\theta) = 0$, $g''(\theta) \neq 0$ and

$$\sqrt{n}(T_n - \theta) \xrightarrow{D} N(0, \tau^2(\theta)).$$

Then

$$n[g(T_n) - g(\theta)] \xrightarrow{D} \frac{1}{2}\tau^2(\theta)g''(\theta)\chi^2_1.$$

Example 8.14. Let $X_n \sim$ Binomial(n,p) where the positive integer n is large and $0 < p < 1$. Let $g(\theta) = \theta^3 - \theta$. Find the limiting distribution of $n\left[g\left(\frac{X_n}{n}\right) - c\right]$ for appropriate constant c when $p = \frac{1}{\sqrt{3}}$.

Solution: Since $X_n \stackrel{D}{=} \sum_{i=1}^n Y_i$ where $Y_i \sim \text{BIN}(1,p)$,

$$\sqrt{n}\left(\frac{X_n}{n} - p\right) \xrightarrow{D} N(0, p(1-p))$$

by the CLT. Let $\theta = p$. Then $g'(\theta) = 3\theta^2 - 1$ and $g''(\theta) = 6\theta$. Notice that

$$g(1/\sqrt{3}) = (1/\sqrt{3})^3 - 1/\sqrt{3} = (1/\sqrt{3})\left(\frac{1}{3} - 1\right) = \frac{-2}{3\sqrt{3}} = c.$$

Also $g'(1/\sqrt{3}) = 0$ and $g''(1/\sqrt{3}) = 6/\sqrt{3}$. Since $\tau^2(p) = p(1-p)$,

$$\tau^2(1/\sqrt{3}) = \frac{1}{\sqrt{3}}\left(1 - \frac{1}{\sqrt{3}}\right).$$

Hence

$$n\left[g\left(\frac{X_n}{n}\right) - \left(\frac{-2}{3\sqrt{3}}\right)\right] \xrightarrow{D} \frac{1}{2}\frac{1}{\sqrt{3}}\left(1 - \frac{1}{\sqrt{3}}\right)\frac{6}{\sqrt{3}}\chi^2_1 = \left(1 - \frac{1}{\sqrt{3}}\right)\chi^2_1.$$

A nice review of large sample theory is Chernoff (1956), and there are several Ph.D. level texts on large sample theory including, in roughly increasing order of difficulty, Lehmann (1999), Ferguson (1996), Sen and Singer (1993), and Serfling (1980). Cramér (1946) is also an important reference, and White (1984) considers asymptotic theory for econometric applications. Lecture notes are available from (www.stat.psu.edu/~dhunter/asymp/lectures/). Also see DasGupta (2008), Davidson (1994), Jiang (2010), Polansky (2011), Sen et al. (2010) van der Vaart (1998).

In analysis, convergence in probability is a special case of convergence in measure and convergence in distribution is a special case of weak convergence. See Ash (1972, p. 322) and Sen and Singer (1993, p. 39). Almost sure convergence is also known as strong convergence. See Sen and Singer (1993, p. 34). Since $\overline{Y} \xrightarrow{P} \mu$ iff $\overline{Y} \xrightarrow{D} \mu$, the WLLN refers to weak convergence. Technically X_n and X need to share a common probability space for convergence in probability and almost sure convergence.

Perlman (1972) and Wald (1949) give general results on the consistency of the MLE while Berk (1972), Lehmann (1980), and Schervish (1995, p. 418) discuss the asymptotic normality of the MLE in exponential families. Theorems 8.4 and 8.21 appear in Olive (2007). Also see Cox (1984) and McCulloch (1988). A similar result to Theorem 8.21 for linear exponential families where $t_i(\boldsymbol{x}) = x_i$, is given by (Brown, 1986, p. 172). Portnoy (1977) gives large sample theory for unbiased estimators in exponential families. Although $\overline{T}_n$ is the UMVUE of $E(t(Y)) = \mu_t$, asymptotic efficiency of UMVUEs is not simple in general. See Pfanzagl (1993).

Casella and Berger (2002, pp. 112, 133) give results similar to Proposition 8.3. Some of the order relations of Sect. 8.5 are discussed in Mann and Wald (1943). The multivariate delta method appears, for example, in Ferguson (1996, p. 45), Lehmann (1999, p. 315), Mardia et al. (1979, p. 52), Sen and Singer (1993, p. 136) or Serfling (1980, p. 122).

Suppose $\boldsymbol{\theta} = \boldsymbol{g}^{-1}(\boldsymbol{\eta})$. In analysis, the fact that

$$\boldsymbol{D}_{\boldsymbol{g}(\boldsymbol{\theta})}^{-1} = \boldsymbol{D}_{\boldsymbol{g}^{-1}(\boldsymbol{\eta})}$$

is a corollary of the inverse mapping theorem (or of the inverse function theorem). See Apostol (1957, p. 146), Bickel and Doksum (2007, p. 517), Marsden and Hoffman (1993, p. 393) and Wade (2000, p. 353).

According to Rohatgi (1984, p. 616), if $Y_1, \ldots, Y_n$ are iid with pdf $f(y)$, if $Y_{r_n:n}$ is the r_nth order statistic, $r_n/n \to \rho$, $F(\xi_\rho) = \rho$ and if $f(\xi_\rho) > 0$, then

$$\sqrt{n}(Y_{r_n:n} - \xi_\rho) \xrightarrow{D} N\left(0, \frac{\rho(1-\rho)}{[f(\xi_\rho)]^2}\right).$$

So there are many asymptotically equivalent ways of defining the sample ρ quantile.

8.10 Problems

PROBLEMS WITH AN ASTERISK * ARE ESPECIALLY USEFUL.

Refer to Chap. 10 for the pdf or pmf of the distributions in the problems below.

8.1*. a) Enter the following *R* function that is used to illustrate the central limit theorem when the data $Y_1, \ldots, Y_n$ are iid from an exponential distribution. The function generates a data set of size n and computes $\overline{Y}_1$ from the data set. This step is repeated $nruns = 100$ times. The output is a vector $(\overline{Y}_1, \overline{Y}_2, \ldots, \overline{Y}_{100})$. A histogram of these means should resemble a symmetric normal density once n is large enough.

```
cltsim <- function(n=100, nruns=100){
ybar <- 1:nruns
for(i in 1:nruns){
  ybar[i] <- mean(rexp(n))}
list(ybar=ybar)}
```

b) The following commands will plot four histograms with $n = 1, 5, 25$, and 200. Save the plot in *Word*.

```
> z1 <- cltsim(n=1)
> z5 <- cltsim(n=5)
> z25 <- cltsim(n=25)
> z200 <- cltsim(n=200)
> par(mfrow=c(2,2))
> hist(z1$ybar)
> hist(z5$ybar)
> hist(z25$ybar)
> hist(z200$ybar)
```

c) Explain how your plot illustrates the central limit theorem.

d) Repeat parts a)–c), but in part a), change *rexp(n)* to *rnorm(n)*. Then $Y_1, \ldots, Y_n$ are iid N(0,1) and $\overline{Y} \sim N(0, 1/n)$.

8.2*. Let $X_1, \ldots, X_n$ be iid from a normal distribution with unknown mean μ and known variance σ^2. Let

$$\overline{X} = \frac{\sum_{i=1}^n X_i}{n}$$

Find the limiting distribution of $\sqrt{n}((\overline{X})^3 - c)$ for an appropriate constant c.

8.3^Q. Let $X_1, \ldots, X_n$ be a random sample from a population with pdf

$$f(x) = \begin{cases} \frac{\theta x^{\theta-1}}{3^\theta} & 0 < x < 3 \\ 0 & \text{elsewhere} \end{cases}$$

The method of moments estimator for θ is $T_n = \dfrac{\overline{X}}{3 - \overline{X}}$.

a) Find the limiting distribution of $\sqrt{n}(T_n - \theta)$ as $n \to \infty$.

b) Is T_n asymptotically efficient? Why?

c) Find a consistent estimator for θ and show that it is consistent.

8.4^Q. From Theorems 2.17 and 2.18, if $Y_n = \sum_{i=1}^n X_i$ where X_i is iid from a nice distribution, then Y_n also has a nice distribution. If $E(X) = \mu$ and VAR$(X) = \sigma^2$, then by the CLT

$$\sqrt{n}(\overline{X}_n - \mu) \xrightarrow{D} N(0, \sigma^2).$$

Hence

$$\sqrt{n}\left(\frac{Y_n}{n} - \mu\right) \xrightarrow{D} N(0, \sigma^2).$$

Find μ, σ^2 and the distribution of X_i if

i) $Y_n \sim$ BIN(n, ρ) where BIN stands for binomial.

ii) $Y_n \sim \chi^2_n$.

iii) $Y_n \sim G(n\nu, \lambda)$ where G stands for gamma.

iv) $Y_n \sim$ NB(n, ρ) where NB stands for negative binomial.

v) $Y_n \sim$ POIS$(n\theta)$ where POIS stands for Poisson.

vi) $Y_n \sim N(n\mu, n\sigma^2)$.

8.5*. Suppose that $X_n \sim U(-1/n, 1/n)$.

a) What is the cdf $F_n(x)$ of X_n?

b) What does $F_n(x)$ converge to?

(Hint: consider $x < 0$, $x = 0$ and $x > 0$.)

c) $X_n \xrightarrow{D} X$. What is X?

8.6. Continuity Theorem problem: Let X_n be sequence of random variables with cdfs F_n and mgfs m_n. Let X be a random variable with cdf F and mgf m. Assume that all of the mgfs m_n and m are defined if $|t| \le d$ for some $d > 0$. Thus if $m_n(t) \to m(t)$ as $n \to \infty$ for all $|t| < c$ where $0 < c < d$, then $X_n \xrightarrow{D} X$.

Let

$$m_n(t) = \frac{1}{[1 - (\lambda + \frac{1}{n})t]}$$

for $t < 1/(\lambda + 1/n)$. Then what is $m(t)$ and what is X?

8.7. Let $Y_1, \ldots, Y_n$ be iid, $T_{1,n} = \overline{Y}$ and let $T_{2,n} =$ MED(n) be the sample median. Let $\theta = \mu$.

Then

$$\sqrt{n}(\text{MED}(n) - \text{MED}(Y)) \xrightarrow{D} N\left(0, \frac{1}{4f^2(\text{MED}(Y))}\right)$$

where the population median is MED(Y) (and MED(Y) $= \mu = \theta$ for a) and b) below).

a) Find ARE$(T_{1,n}, T_{2,n})$ if F is the cdf of the normal $N(\mu, \sigma^2)$ distribution.

b) Find ARE$(T_{1,n}, T_{2,n})$ if F is the cdf of the double exponential $DE(\theta, \lambda)$ distribution.

8.8^Q**.** Let $X_1,\ldots,X_n$ be independent identically distributed random variables with probability density function

$$f(x) = \theta x^{\theta-1}, \; 0 < x < 1, \; \theta > 0.$$

a) Find the MLE of $\frac{1}{\theta}$. Is it unbiased? Does it achieve the information inequality lower bound?

b) Find the asymptotic distribution of the MLE of $\frac{1}{\theta}$.

c) Show that $\overline{X}_n$ is unbiased for $\frac{\theta}{\theta+1}$. Does $\overline{X}_n$ achieve the information inequality lower bound?

d) Find an estimator of $\frac{1}{\theta}$ from part (c) above using $\overline{X}_n$ which is different from the MLE in (a). Find the asymptotic distribution of your estimator using the delta method.

e) Find the asymptotic relative efficiency of your estimator in (d) with respect to the MLE in (b).

8.9. Many multiple linear regression estimators $\hat{\boldsymbol{\beta}}$ satisfy

$$\sqrt{n}(\hat{\boldsymbol{\beta}} - \boldsymbol{\beta}) \xrightarrow{D} N_p(0, V(\hat{\boldsymbol{\beta}}, F)\; \boldsymbol{W}) \tag{8.14}$$

when

$$\frac{\boldsymbol{X}^T\boldsymbol{X}}{n} \xrightarrow{P} \boldsymbol{W}^{-1}, \tag{8.15}$$

and when the errors e_i are iid with a cdf F and a unimodal pdf f that is symmetric with a unique maximum at 0. When the variance $V(e_i)$ exists,

$$V(\text{OLS}, F) = V(e_i) = \sigma^2 \text{ while } V(L_1, F) = \frac{1}{4[f(0)]^2}.$$

In the multiple linear regression model,

$$Y_i = x_{i,1}\beta_1 + x_{i,2}\beta_2 + \cdots + x_{i,p}\beta_p + e_i = \boldsymbol{x}_i^T\boldsymbol{\beta} + e_i \tag{8.16}$$

for $i = 1,\ldots,n$. In matrix notation, these n equations become

$$\boldsymbol{Y} = \boldsymbol{X}\boldsymbol{\beta} + \boldsymbol{e}, \tag{8.17}$$

where $\boldsymbol{Y}$ is an $n \times 1$ vector of dependent variables, $\boldsymbol{X}$ is an $n \times p$ matrix of predictors, $\boldsymbol{\beta}$ is a $p \times 1$ vector of unknown coefficients, and $\boldsymbol{e}$ is an $n \times 1$ vector of unknown errors.

a) What is the ijth element of the matrix

$$\frac{\boldsymbol{X}^T\boldsymbol{X}}{n}?$$

b) Suppose $x_{k,1} = 1$ and that $x_{k,j} \sim X_j$ are iid with $E(X_j) = 0$ and $V(X_j) = 1$ for $k = 1,\ldots,n$ and $j = 2,\ldots,p$. Assume that X_i and X_j are independent for $i \neq j$, $i > 1$ and $j > 1$. (Often $x_{k,j} \sim N(0,1)$ in simulations.) Then what is $\boldsymbol{W}^{-1}$ for model (8.16)?

c) Suppose $p = 2$ and $Y_i = \alpha + \beta X_i + e_i$. Show

$$(\boldsymbol{X}^T\boldsymbol{X})^{-1} = \begin{bmatrix} \frac{\sum X_i^2}{n\sum(X_i-\overline{X})^2} & \frac{-\sum X_i}{n\sum(X_i-\overline{X})^2} \\ \\ \frac{-\sum X_i}{n\sum(X_i-\overline{X})^2} & \frac{n}{n\sum(X_i-\overline{X})^2} \end{bmatrix}.$$

d) Under the conditions of c), let $S_x^2 = \sum(X_i - \overline{X})^2/n$. Show that

$$n(\boldsymbol{X}^T\boldsymbol{X})^{-1} = \left(\frac{\boldsymbol{X}^T\boldsymbol{X}}{n}\right)^{-1} = \begin{bmatrix} \frac{\frac{1}{n}\sum X_i^2}{S_x^2} & \frac{-\overline{X}}{S_x^2} \\ \\ \frac{-\overline{X}}{S_x^2} & \frac{1}{S_x^2} \end{bmatrix}.$$

e) If X_i is iid with variance $V(X)$, then $n(\boldsymbol{X}^T\boldsymbol{X})^{-1} \xrightarrow{P} \boldsymbol{W}$. What is $\boldsymbol{W}$?

f) Now suppose that n is divisible by 5 and the $n/5$ of X_i are at 0.1, $n/5$ at 0.3, $n/5$ at 0.5, $n/5$ at 0.7 and $n/5$ at 0.9. (Hence if $n = 100$, 20 of the X_i are at 0.1, 0.3, 0.5, 0.7 and 0.9.)

Find $\sum X_i^2/n$, $\overline{X}$, and S_x^2. (Your answers should not depend on n.)

g) Under the conditions of f), estimate $V(\hat{\alpha})$ and $V(\hat{\beta})$ if L_1 is used and if the e_i are iid $N(0, 0.01)$.

Hint: Estimate $\boldsymbol{W}$ with $n(\boldsymbol{X}^T\boldsymbol{X})^{-1}$ and $V(\hat{\boldsymbol{\beta}}, F) = V(L_1, F) = \frac{1}{4[f(0)]^2}$. Hence

$$\begin{pmatrix} \hat{\alpha} \\ \\ \hat{\beta} \end{pmatrix} \approx N_2\left[\begin{pmatrix} \alpha \\ \\ \beta \end{pmatrix}, \frac{1}{n}\frac{1}{4[f(0)]^2}\begin{pmatrix} \frac{\frac{1}{n}\sum X_i^2}{S_x^2} & \frac{-\overline{X}}{S_x^2} \\ \\ \frac{-\overline{X}}{S_x^2} & \frac{1}{S_x^2} \end{pmatrix}\right].$$

You should get an answer like $0.0648/n$.

Problems from old quizzes and exams. Problems from old qualifying exams are marked with a Q.

8.10. Let $X_1,\ldots,X_n$ be iid Bernoulli(p) random variables.

a) Find $I_1(p)$.

b) Find the FCRLB for estimating p.

c) Find the limiting distribution of $\sqrt{n}(\,\overline{X}_n - p\,)$.

d) Find the limiting distribution of $\sqrt{n}\,[\,(\overline{X}_n)^2 - c\,]$ for an appropriate constant c.

8.11. Let $X_1, \ldots, X_n$ be iid exponential(β) random variables.

a) Find the FCRLB for estimating β.

b) Find the limiting distribution of $\sqrt{n}(\,\overline{X}_n - \beta\,)$.

c) Find the limiting distribution of $\sqrt{n}\,[\,(\overline{X}_n)^2 - c\,]$ for an appropriate constant c.

8.12. Let $Y_1, \ldots, Y_n$ be iid Poisson (λ) random variables.
a) Find the limiting distribution of $\sqrt{n}(\,\overline{Y}_n - \lambda\,)$.

b) Find the limiting distribution of $\sqrt{n}\,[\,(\overline{Y}_n)^2 - c\,]$ for an appropriate constant c.

8.13. Let $Y_n \sim \chi^2_n$.

a) Find the limiting distribution of $\sqrt{n}\left(\frac{Y_n}{n} - 1\right)$.

b) Find the limiting distribution of $\sqrt{n}\left[\left(\frac{Y_n}{n}\right)^3 - 1\right]$.

8.14. Let $X_1, \ldots, X_n$ be iid with cdf $F(x) = P(X \le x)$. Let $Y_i = I(X_i \le x)$ where the indicator equals 1 if $X_i \le x$ and 0, otherwise.
a) Find $E(Y_i)$.

b) Find VAR(Y_i).

c) Let $\hat{F}_n(x) = \frac{1}{n}\sum_{i=1}^{n} I(X_i \le x)$ for some fixed real number x. Find the limiting distribution of $\sqrt{n}\,\left(\hat{F}_n(x) - c_x\right)$ for an appropriate constant c_x.

8.15. Suppose X_n has cdf

$$F_n(x) = 1 - \left(1 - \frac{x}{\theta n}\right)^n$$

for $x \ge 0$ and $F_n(x) = 0$ for $x < 0$. Show that $X_n \xrightarrow{D} X$ by finding the cdf of X.

8.16. Let X_n be a sequence of random variables such that $P(X_n = 1/n) = 1$. Does X_n converge in distribution? If yes, prove it by finding X and the cdf of X. If no, prove it.

8.17. Suppose that $Y_1, \ldots, Y_n$ are iid with $E(Y) = (1-\rho)/\rho$ and VAR(Y) $= (1-\rho)/\rho^2$ where $0 < \rho < 1$.

a) Find the limiting distribution of

$$\sqrt{n}\left(\overline{Y}_n - \frac{1-\rho}{\rho}\right).$$

b) Find the limiting distribution of $\sqrt{n}\,\left[\,g(\overline{Y}_n) - \rho\right]$ for appropriate function g.

8.18. Let $X_n \sim$ Binomial(n, p) where the positive integer n is large and $0 < p < 1$.

a) Find the limiting distribution of $\sqrt{n}\left(\frac{X_n}{n} - p\right)$.

b) Find the limiting distribution of $\sqrt{n}\left[\left(\frac{X_n}{n}\right)^2 - p^2\right]$.

8.19. Let $Y_1, \ldots, Y_n$ be iid exponential (λ) so that $E(Y) = \lambda$ and MED$(Y) = \log(2)\lambda$.

a) Let $T_{1,n} = \log(2)\overline{Y}$ and find the limiting distribution of $\sqrt{n}(T_{1,n} - \log(2)\lambda)$.

b) Let $T_{2,n} = \text{MED}(n)$ be the sample median and find the limiting distribution of $\sqrt{n}(T_{2,n} - \log(2)\lambda)$.

c) Find ARE$(T_{1,n}, T_{2,n})$.

8.20. Suppose that $\eta = g(\theta)$, $\theta = g^{-1}(\eta)$ and $g'(\theta) > 0$ exists. If X has pdf or pmf $f(x|\theta)$, then in terms of η, the pdf or pmf is $f^*(x|\eta) = f(x|g^{-1}(\eta))$. Now

$$A = \frac{\partial}{\partial \eta} \log[f(x|g^{-1}(\eta))] = \frac{1}{f(x|g^{-1}(\eta))} \frac{\partial}{\partial \eta} f(x|g^{-1}(\eta)) =$$

$$\left[\frac{1}{f(x|g^{-1}(\eta))}\right] \left[\frac{\partial}{\partial \theta} f(x|\theta)\bigg|_{\theta = g^{-1}(\eta)}\right] \left[\frac{\partial}{\partial \eta} g^{-1}(\eta)\right]$$

using the chain rule twice. Since $\theta = g^{-1}(\eta)$,

$$A = \left[\frac{1}{f(x|\theta)}\right] \left[\frac{\partial}{\partial \theta} f(x|\theta)\right] \left[\frac{\partial}{\partial \eta} g^{-1}(\eta)\right].$$

Hence

$$A = \frac{\partial}{\partial \eta} \log[f(x|g^{-1}(\eta))] = \left[\frac{\partial}{\partial \theta} \log[f(x|\theta)]\right] \left[\frac{\partial}{\partial \eta} g^{-1}(\eta)\right].$$

Now show that

$$I_1^*(\eta) = \frac{I_1(\theta)}{[g'(\theta)]^2}.$$

8.21. Let $Y_1, \ldots, Y_n$ be iid exponential (1) so that $P(Y \le y) = F(y) = 1 - e^{-y}$ for $y \ge 0$. Let $Y_{(n)} = \max(Y_1, \ldots, Y_n)$.

a) Show that $F_{Y_{(n)}}(t) = P(Y_{(n)} \le t) = [1 - e^{-t}]^n$ for $t \ge 0$.

b) Show that $P(Y_{(n)} - \log(n) \le t) \to \exp(-e^{-t})$ (for all $t \in (-\infty, \infty)$ since $t + \log(n) > 0$ implies $t \in \mathbb{R}$ as $n \to \infty$).

8.22. Let $Y_1, \ldots, Y_n$ be iid uniform $(0, 2\theta)$.

a) Let $T_{1,n} = \overline{Y}$ and find the limiting distribution of $\sqrt{n}(T_{1,n} - \theta)$.

b) Let $T_{2,n} = \text{MED}(n)$ be the sample median and find the limiting distribution of $\sqrt{n}(T_{2,n} - \theta)$.

c) Find $\text{ARE}(T_{1,n}, T_{2,n})$. Which estimator is better, asymptotically?

8.23. Suppose that $Y_1, \ldots, Y_n$ are iid from a distribution with pdf $f(y|\theta)$ and that the integral and differentiation operators of all orders can be interchanged (e.g., the data is from a one-parameter exponential family).

a) Show that $0 = E\left[\frac{\partial}{\partial\theta}\log(f(Y|\theta))\right]$ by showing that

$$\frac{\partial}{\partial\theta}1 = 0 = \frac{\partial}{\partial\theta}\int f(y|\theta)dy = \int\left[\frac{\partial}{\partial\theta}\log(f(y|\theta))\right]f(y|\theta)dy. \quad (*)$$

b) Take second derivatives of (*) to show that

$$I_1(\theta) = E_\theta\left[\left(\frac{\partial}{\partial\theta}\log f(Y|\theta)\right)^2\right] = -E_\theta\left[\frac{\partial^2}{\partial\theta^2}\log(f(Y|\theta))\right].$$

8.24. Suppose that $X_1, \ldots, X_n$ are iid $N(\mu, \sigma^2)$.

a) Find the limiting distribution of $\sqrt{n}\ \left(\overline{X}_n - \ \mu\ \right)$.

b) Let $g(\theta) = [\log(1+\theta)]^2$. Find the limiting distribution of $\sqrt{n}\ \left(g(\overline{X}_n) - \ g(\mu)\ \right)$ for $\mu > 0$.

c) Let $g(\theta) = [\log(1+\theta)]^2$. Find the limiting distribution of $n\ \left(g(\overline{X}_n) - \ g(\mu)\ \right)$ for $\mu = 0$. Hint: Use Theorem 8.30.

8.25. Let $W_n = X_n - X$ and let $r > 0$. Notice that for any $\varepsilon > 0$,

$$E|X_n - X|^r \geq E[|X_n - X|^r\, I(|X_n - X| \geq \varepsilon)] \geq \varepsilon^r P(|X_n - X| \geq \varepsilon).$$

Show that $W_n \xrightarrow{P} 0$ if $E|X_n - X|^r \to 0$ as $n \to \infty$.

8.26. Let $X_1, \ldots, X_n$ be iid with $E(X) = \mu$ and $V(X) = \sigma^2$. What is the limiting distribution of $n[(\overline{X})^2 - \mu^2]$ for the value or values of μ where the delta method does not apply? Hint: use Theorem 8.30.

8.27^Q. Let $X \sim$ Binomial(n, p) where the positive integer n is large and $0 < p < 1$.

a) Find the limiting distribution of $\sqrt{n}\left(\frac{X}{n} - p\right)$.

b) Find the limiting distribution of $\sqrt{n}\left[\left(\frac{X}{n}\right)^2 - p^2\right]$.

c) Show how to find the limiting distribution of $\left[\left(\frac{X}{n}\right)^3 - \frac{X}{n}\right]$ when $p = \frac{1}{\sqrt{3}}$.

(Actually want the limiting distribution of

$$n\left(\left[\left(\frac{X}{n}\right)^3 - \frac{X}{n}\right] - g(p)\right)$$

where $g(\theta) = \theta^3 - \theta$.)

8.28^{Q}**.** Let $X_1, \ldots, X_n$ be independent and identically distributed (iid) from a Poisson(λ) distribution.

a) Find the limiting distribution of $\sqrt{n}\ (\ \overline{X} - \lambda\ )$.

b) Find the limiting distribution of $\sqrt{n}\ [\ (\overline{X})^3 - (\lambda)^3\ ]$.

8.29^{Q}**.** Let $X_1, \ldots, X_n$ be iid from a normal distribution with unknown mean μ and known variance σ^2. Let $\overline{X} = \frac{\sum_{i=1}^n X_i}{n}$ and $S^2 = \frac{1}{n-1}\sum_{i=1}^n (X_i - \overline{X})^2$.

a) Show that $\overline{X}$ and S^2 are independent.

b) Find the limiting distribution of $\sqrt{n}((\overline{X})^3 - c)$ for an appropriate constant c.

8.30. Suppose that $Y_1, \ldots, Y_n$ are iid logistic$(\theta, 1)$ with pdf

$$f(y) = \frac{\exp(-(y-\theta))}{[1 + \exp(-(y-\theta))]^2}$$

where and y and θ are real.

a) $I_1(\theta) = 1/3$ and the family is regular so the "standard limit theorem" for the MLE $\hat{\theta}_n$ holds. Using this standard theorem, what is the limiting distribution of $\sqrt{n}(\hat{\theta}_n - \theta)$?

b) Find the limiting distribution of $\sqrt{n}(\overline{Y}_n - \theta)$.

c) Find the limiting distribution of $\sqrt{n}(\text{MED}(n) - \theta)$.

d) Consider the estimators $\hat{\theta}_n$, $\overline{Y}_n$ and MED(n). Which is the best estimator and which is the worst?

8.31. Let $Y_n \sim$ binomial(n,p). Find the limiting distribution of

$$\sqrt{n}\left(\arcsin\left(\sqrt{\frac{Y_n}{n}}\right) - \arcsin(\sqrt{p})\right).$$

(Hint:

$$\frac{d}{dx}\arcsin(x) = \frac{1}{\sqrt{1-x^2}}.)$$

8.32. Suppose $Y_n \sim$ uniform$(-n,n)$. Let $F_n(y)$ be the cdf of Y_n.
a) Find $F(y)$ such that $F_n(y) \to F(y)$ for all y as $n \to \infty$.

b) Does $Y_n \xrightarrow{L} Y$? Explain briefly.

8.33. Suppose $Y_n \sim$ uniform$(0,n)$. Let $F_n(y)$ be the cdf of Y_n.
a) Find $F(y)$ such that $F_n(y) \to F(y)$ for all y as $n \to \infty$.

b) Does $Y_n \xrightarrow{L} Y$? Explain briefly.

8.34^Q. Let $Y_1,\ldots,Y_n$ be independent and identically distributed (iid) from a distribution with probability mass function $f(y) = \rho(1-\rho)^y$ for $y = 0,1,2,\ldots$ and $0 < \rho < 1$. Then $E(Y) = (1-\rho)/\rho$ and VAR$(Y) = (1-\rho)/\rho^2$.

a) Find the limiting distribution of $\sqrt{n}\left(\overline{Y} - \dfrac{1-\rho}{\rho}\right)$.

b) Show how to find the limiting distribution of $g(\overline{Y}) = \frac{1}{1+\overline{Y}}$. Deduce it completely. (This bad notation means find the limiting distribution of $\sqrt{n}(g(\overline{Y}) - c)$ for some constant c.)

c) Find the method of moments estimator of ρ.

d) Find the limiting distribution of $\sqrt{n}\ (\ (1+\overline{Y}) - d\ )$ for appropriate constant d.

e) Note that $1 + E(Y) = 1/\rho$. Find the method of moments estimator of $1/\rho$.

8.35^Q. Let $X_1,\ldots,X_n$ be independent identically distributed random variables from a normal distribution with mean μ and variance σ^2.

a) Find the approximate distribution of $1/\bar{X}$. Is this valid for all values of μ?

b) Show that $1/\bar{X}$ is asymptotically efficient for $1/\mu$, provided $\mu \neq \mu^*$. Identify μ^*.

8.36^Q. Let $Y_1,\ldots,Y_n$ be independent and identically distributed (iid) from a distribution with probability density function

$$f(y) = \frac{2y}{\theta^2}$$

for $0 < y \le \theta$ and $f(y) = 0$, otherwise.

a) Find the limiting distribution of $\sqrt{n}\ \left(\overline{Y} - c\right)$ for appropriate constant c.

b) Find the limiting distribution of $\sqrt{n}\ \left(\log(\ \overline{Y}\ ) - d\right)$ for appropriate constant d.

c) Find the method of moments estimator of θ^k.

8.37^Q. Let $Y_1,\ldots,Y_n$ be independent identically distributed discrete random variables with probability mass function

$$f(y) = P(Y = y) = \binom{r+y-1}{y}\rho^r(1-\rho)^y$$

for $y = 0, 1, \ldots$ where positive integer r **is known** and $0 < \rho < 1$. Then $E(Y) = r(1-\rho)/\rho$, and $V(Y) = r(1-\rho)/\rho^2$.

a) Find the limiting distribution of $\sqrt{n} \left(\overline{Y} - \frac{r(1-\rho)}{\rho} \right)$.

b) Let $g(\overline{Y}) = \frac{r}{r+\overline{Y}}$. Find the limiting distribution of $\sqrt{n} \left(g(\overline{Y}) - c \right)$ for appropriate constant c.

c) Find the method of moments estimator of ρ.

8.38^Q. Let $X_1, \ldots, X_n$ be independent identically distributed uniform $(0, \theta)$ random variables where $\theta > 0$.

a) Find the limiting distribution of $\sqrt{n}(\overline{X} - c_\theta)$ for an appropriate constant c_θ that may depend on θ.

b) Find the limiting distribution of $\sqrt{n}[(\overline{X})^2 - k_\theta]$ for an appropriate constant k_θ that may depend on θ.

Chapter 9
Confidence Intervals

Point estimators give a single reasonable (value) estimate of θ and were covered in Chaps. 5 and 6. Interval estimators, such as confidence intervals, give an interval of "reasonable" estimated values of the parameter. Large sample confidence intervals and tests are also discussed in this chapter. Section 9.3 suggests that bootstrap and randomization confidence intervals and tests should use $m = \max(B, \lceil n\log(n)\rceil)$ samples instead of a fixed number of samples such as $B = 1{,}000$.

9.1 Introduction

Definition 9.1. Let the data $Y_1,\ldots,Y_n$ have joint pdf or pmf $f(\boldsymbol{y}|\theta)$ with parameter space Θ and support $\mathscr{Y}$. Let $L_n(\boldsymbol{Y})$ and $U_n(\boldsymbol{Y})$ be statistics such that $L_n(\boldsymbol{y}) \le U_n(\boldsymbol{y})$, $\forall \boldsymbol{y} \in \mathscr{Y}$. Then $(L_n(\boldsymbol{y}), U_n(\boldsymbol{y}))$ is a $100\,(1-\alpha)\,\%$ **confidence interval** (CI) for θ if

$$P_\theta(L_n(\boldsymbol{Y}) < \theta < U_n(\boldsymbol{Y})) = 1 - \alpha$$

for all $\boldsymbol{\theta} \in \Theta$. The interval $(L_n(\boldsymbol{y}), U_n(\boldsymbol{y}))$ is a large sample $100\,(1-\alpha)\,\%$ CI for $\boldsymbol{\theta}$ if

$$P_\theta(L_n(\boldsymbol{Y}) < \theta < U_n(\boldsymbol{Y})) \to 1 - \alpha$$

for all $\theta \in \Theta$ as $n \to \infty$.

Definition 9.2. Let the data $Y_1,\ldots,Y_n$ have joint pdf or pmf $f(\boldsymbol{y}|\boldsymbol{\theta})$ with parameter space Θ and support $\mathscr{Y}$. The random variable $R(\boldsymbol{Y}|\boldsymbol{\theta})$ is a **pivot** or pivotal quantity if the distribution of $R(\boldsymbol{Y}|\boldsymbol{\theta})$ is independent $\boldsymbol{\theta}$. The quantity $R(\boldsymbol{Y},\boldsymbol{\theta})$ is an **asymptotic pivot** or asymptotic pivotal quantity if the limiting distribution of $R(\boldsymbol{Y},\boldsymbol{\theta})$ is independent of $\boldsymbol{\theta}$.

The first CI in Definition 9.1 is sometimes called an exact CI. The words "exact" and "large sample" are often omitted. In the following definition, the scaled asymptotic length is closely related to asymptotic relative efficiency of an estimator and high power of a test of hypotheses.

D.J. Olive, *Statistical Theory and Inference*, DOI 10.1007/978-3-319-04972-4_9,

Definition 9.3. Let (L_n, U_n) be a $100\,(1-\alpha)\,\%$ CI or large sample CI for θ. If

$$n^{\delta}(U_n - L_n) \xrightarrow{P} A_{\alpha},$$

then A_α is the *scaled asymptotic length* of the CI. Typically $\delta = 0.5$ but superefficient CIs have $\delta = 1$. For fixed δ and fixed coverage $1-\alpha$, a CI with smaller A_α is "better" than a CI with larger A_α. If $A_{1,\alpha}$ and $A_{2,\alpha}$ are for two competing CIs with the same δ, then $(A_{2,\alpha}/A_{1,\alpha})^{1/\delta}$ is a measure of "asymptotic relative efficiency."

Definition 9.4. Suppose a nominal $100(1-\alpha)\,\%$ CI for θ has actual coverage $1-\delta$, so that $P_\theta(L_n(\boldsymbol{Y}) < \theta < U_n(\boldsymbol{Y})) = 1-\delta$ for all $\theta \in \Theta$. If $1-\delta > 1-\alpha$, then the CI is *conservative*. If $1-\delta < 1-\alpha$, then the CI is *liberal*. Conservative CIs are generally considered better than liberal CIs. Suppose a nominal $100(1-\alpha)\,\%$ large sample CI for θ has actual coverage $1-\delta_n$ where $\delta_n \to \delta$ as $n \to \infty$ for all $\theta \in \Theta$. If $1-\delta > 1-\alpha$, then the CI is *asymptotically conservative*. If $1-\delta < 1-\alpha$, then the CI is *asymptotically liberal*. It is possible that $\delta \equiv \delta(\theta)$ depends on θ, and that the CI is (asymptotically) conservative or liberal for different values of θ, in that the (asymptotic) coverage is higher or lower than the nominal coverage, depending on θ.

Example 9.1. a) Let $Y_1, \ldots, Y_n$ be iid $N(\mu, \sigma^2)$ where $\sigma^2 > 0$. Then

$$R(\boldsymbol{Y}|\mu,\sigma^2) = \frac{\overline{Y} - \mu}{S/\sqrt{n}} \sim t_{n-1}$$

is a pivot or pivotal quantity.

To use this pivot to find a CI for μ, let $t_{p,\alpha}$ be the α percentile of the t_p distribution. Hence $P(T \le t_{p,\alpha}) = \alpha$ if $T \sim t_p$. Using $t_{p,\alpha} = -t_{p,1-\alpha}$ for $0 < \alpha < 0.5$, note that

$$
\begin{aligned}
lll1 - \alpha &= P\left(-t_{n-1,1-\alpha/2} \le \frac{\overline{Y} - \mu}{S/\sqrt{n}} \le t_{n-1,1-\alpha/2}\right) \\
&= P(-t_{n-1,1-\alpha/2}\ S/\sqrt{n} \le \overline{Y} - \mu \le t_{n-1,1-\alpha/2}\ S/\sqrt{n}) \\
&= P(-\overline{Y} - t_{n-1,1-\alpha/2}\ S/\sqrt{n} \le -\mu \le -\overline{Y} + t_{n-1,1-\alpha/2}\ S/\sqrt{n}) \\
&= P(\overline{Y} - t_{n-1,1-\alpha/2}\ S/\sqrt{n} \le \mu \le \overline{Y} + t_{n-1,1-\alpha/2}\ S/\sqrt{n}).
\end{aligned}
$$

Thus

$$\overline{Y} \pm t_{n-1,1-\alpha/2}\ S/\sqrt{n}$$

is a $100(1-\alpha)\%$ CI for μ.

b) If $Y_1, \ldots, Y_n$ are iid with $E(Y) = \mu$ and $\text{VAR}(Y) = \sigma^2 > 0$, then, by the CLT and Slutsky's Theorem,

$$R(\boldsymbol{Y}|\mu,\sigma^2) = \frac{\overline{Y} - \mu}{S/\sqrt{n}} = \frac{\sigma}{S}\,\frac{\overline{Y} - \mu}{\sigma/\sqrt{n}} \xrightarrow{D} N(0,1)$$

is an asymptotic pivot or asymptotic pivotal quantity.

To use this asymptotic pivot to find a large sample CI for μ, let z_α be the α percentile of the $N(0,1)$ distribution. Hence $P(Z \le z_\alpha) = \alpha$ if $Z \sim N(0,1)$. Using $z_\alpha = -z_{1-\alpha}$ for $0 < \alpha < 0.5$, note that for large n,

$$\begin{aligned}1-\alpha &\approx P\left(-z_{1-\alpha/2} \le \frac{\overline{Y}-\mu}{S/\sqrt{n}} \le z_{1-\alpha/2}\right)\\ &= P\left(-z_{1-\alpha/2}\ S/\sqrt{n} \le \overline{Y}-\mu \le z_{1-\alpha/2}\ S/\sqrt{n}\right)\\ &= P\left(-\overline{Y}-z_{1-\alpha/2} S/\sqrt{n} \le -\mu \le -\overline{Y}+z_{1-\alpha/2}\ S/\sqrt{n}\right)\\ &= P\left(\overline{Y}-z_{1-\alpha/2}\ S/\sqrt{n} \le \mu \le \overline{Y}+z_{1-\alpha/2}\ S/\sqrt{n}\right).\end{aligned}$$

Thus

$$\overline{Y} \pm z_{1-\alpha/2}\ S/\sqrt{n}$$

is a large sample $100(1-\alpha)\%$ CI for μ.

Since $t_{n-1,1-\alpha/2} > z_{1-\alpha/2}$ but $t_{n-1,1-\alpha/2} \to z_{1-\alpha/2}$ as $n \to \infty$,

$$\overline{Y} \pm t_{n-1,1-\alpha/2}\ S/\sqrt{n}$$

is also a large sample $100(1-\alpha)\%$ CI for μ. This t interval is the same as that in a) and is the most widely used confidence interval in statistics. Replacing $z_{1-\alpha/2}$ by $t_{n-1,1-\alpha/2}$ makes the CI longer and hence less likely to be liberal.

Large sample theory can be used to find a CI from the asymptotic pivot. Suppose that $\boldsymbol{Y} = (Y_1,\ldots,Y_n)$ and that $W_n \equiv W_n(\boldsymbol{Y})$ is an estimator of some parameter μ_W such that

$$\sqrt{n}(W_n - \mu_W) \xrightarrow{D} N(0, \sigma_W^2)$$

where σ_W^2/n is the asymptotic variance of the estimator W_n. The above notation means that if n is large, then for probability calculations

$$W_n - \mu_W \approx N(0, \sigma_W^2/n).$$

Suppose that S_W^2 is a consistent estimator of σ_W^2 so that the (asymptotic) *standard error* of W_n is $\mathrm{SE}(W_n) = S_W/\sqrt{n}$. As in Example 9.1, let $P(Z \le z_\alpha) = \alpha$ if $Z \sim N(0,1)$. Then for large n

$$1-\alpha \approx P\left(-z_{1-\alpha/2} \le \frac{W_n - \mu_W}{\mathrm{SE}(W_n)} \le z_{1-\alpha/2}\right),$$

and an approximate or large sample $100(1-\alpha)\%$ CI for μ_W is given by

$$(W_n - z_{1-\alpha/2}\mathrm{SE}(W_n), W_n + z_{1-\alpha/2}\mathrm{SE}(W_n)). \tag{9.1}$$

Since

$$\frac{t_{p,1-\alpha/2}}{z_{1-\alpha/2}} \to 1$$

if $p \equiv p_n \to \infty$ as $n \to \infty$, another large sample $100(1-\alpha)\%$ CI for μ_W is

$$(W_n - t_{p,1-\alpha/2}\text{SE}(W_n), W_n + t_{p,1-\alpha/2}\text{SE}(W_n)). \tag{9.2}$$

The CI (9.2) often performs better than the CI (9.1) in small samples. The quantity $t_{p,1-\alpha/2}/z_{1-\alpha/2}$ can be regarded as a small sample correction factor. The CI (9.2) is longer than the CI (9.1). Hence the CI (9.2) is more *conservative* than the CI (9.1).

Suppose that there are two independent samples $Y_1,\ldots,Y_n$ and $X_1,\ldots,X_m$ and that

$$\begin{pmatrix} \sqrt{n}(W_n(\boldsymbol{Y}) - \mu_W(Y)) \\ \sqrt{m}(W_m(\boldsymbol{X}) - \mu_W(X)) \end{pmatrix} \xrightarrow{D} N_2\left(\begin{pmatrix} 0 \\ 0 \end{pmatrix}, \begin{pmatrix} \sigma_W^2(Y) & 0 \\ 0 & \sigma_W^2(X) \end{pmatrix}\right).$$

Then

$$\begin{pmatrix} (W_n(\boldsymbol{Y}) - \mu_W(Y)) \\ (W_m(\boldsymbol{X}) - \mu_W(X)) \end{pmatrix} \approx N_2\left(\begin{pmatrix} 0 \\ 0 \end{pmatrix}, \begin{pmatrix} \sigma_W^2(Y)/n & 0 \\ 0 & \sigma_W^2(X)/m \end{pmatrix}\right),$$

and

$$W_n(\boldsymbol{Y}) - W_m(\boldsymbol{X}) - (\mu_W(Y) - \mu_W(X)) \approx N\left(0, \frac{\sigma_W^2(Y)}{n} + \frac{\sigma_W^2(X)}{m}\right).$$

Hence $\text{SE}(W_n(\boldsymbol{Y}) - W_m(\boldsymbol{X})) =$

$$\sqrt{\frac{S_W^2(\boldsymbol{Y})}{n} + \frac{S_W^2(\boldsymbol{X})}{m}} = \sqrt{[\text{SE}(W_n(\boldsymbol{Y}))]^2 + [\text{SE}(W_m(\boldsymbol{X}))]^2},$$

and the large sample $100(1-\alpha)\%$ CI for $\mu_W(Y) - \mu_W(X)$ is given by

$$(W_n(\boldsymbol{Y}) - W_m(\boldsymbol{X})) \pm z_{1-\alpha/2}\text{SE}(W_n(\boldsymbol{Y}) - W_m(\boldsymbol{X})). \tag{9.3}$$

If p_n is the degrees of freedom used for a single sample procedure when the sample size is n, let $p = \min(p_n, p_m)$. Then another large sample $100(1-\alpha)\%$ CI for $\mu_W(Y) - \mu_W(X)$ is given by

$$(W_n(\boldsymbol{Y}) - W_m(\boldsymbol{X})) \pm t_{p,1-\alpha/2}\text{SE}(W_n(\boldsymbol{Y}) - W_m(\boldsymbol{X})). \tag{9.4}$$

These CIs are known as *Welch intervals*. See Welch (1937) and Yuen (1974).

Example 9.2. Consider the single sample procedures where $W_n = \overline{Y}_n$. Then $\mu_W = E(Y)$, $\sigma_W^2 = \text{VAR}(Y)$, $S_W = S_n$, and $p = n-1$. Let t_p denote a random variable with a t distribution with p degrees of freedom and let the α percentile $t_{p,\alpha}$ satisfy $P(t_p \le t_{p,\alpha}) = \alpha$. Then the classical *t-interval* for $\mu \equiv E(Y)$ is

$$\overline{Y}_n \pm t_{n-1,1-\alpha/2}\frac{S_n}{\sqrt{n}}$$

and the *t-test statistic* for $Ho: \mu = \mu_o$ is

$$t_o = \frac{\overline{Y} - \mu_o}{S_n/\sqrt{n}}.$$

The right tailed p-value is given by $P(t_{n-1} > t_o)$.

Now suppose that there are two samples where $W_n(\boldsymbol{Y}) = \overline{Y}_n$ and $W_m(\boldsymbol{X}) = \overline{X}_m$. Then $\mu_W(Y) = E(Y) \equiv \mu_Y$, $\mu_W(X) = E(X) \equiv \mu_X$, $\sigma^2_W(Y) = \text{VAR}(Y) \equiv \sigma^2_Y$, $\sigma^2_W(X) = \text{VAR}(X) \equiv \sigma^2_X$, and $p_n = n-1$. Let $p = \min(n-1, m-1)$. Since

$$\text{SE}(W_n(\boldsymbol{Y}) - W_m(\boldsymbol{X})) = \sqrt{\frac{S^2_n(\boldsymbol{Y})}{n} + \frac{S^2_m(\boldsymbol{X})}{m}},$$

the *two sample t-interval* for $\mu_Y - \mu_X$

$$(\overline{Y}_n - \overline{X}_m) \pm t_{p,1-\alpha/2}\sqrt{\frac{S^2_n(\boldsymbol{Y})}{n} + \frac{S^2_m(\boldsymbol{X})}{m}}$$

and *two sample t-test statistic*

$$t_o = \frac{\overline{Y}_n - \overline{X}_m}{\sqrt{\frac{S^2_n(\boldsymbol{Y})}{n} + \frac{S^2_m(\boldsymbol{X})}{m}}}.$$

The right tailed p-value is given by $P(t_p > t_o)$. For sample means, values of the degrees of freedom that are more accurate than $p = \min(n-1, m-1)$ can be computed. See Moore (2007, p. 474).

The remainder of this section follows Olive (2008b, Section 2.4) closely. Let $\lfloor x \rfloor$ denote the "greatest integer function" (so $\lfloor 7.7 \rfloor = 7$). Let $\lceil x \rceil$ denote the smallest integer greater than or equal to x (so $\lceil 7.7 \rceil = 8$).

Example 9.3: Inference with the sample median. Let $U_n = n - L_n$ where $L_n = \lfloor n/2 \rfloor - \lceil \sqrt{n/4}\, \rceil$ and use

$$\text{SE}(\text{MED}(n)) = 0.5(Y_{(U_n)} - Y_{(L_n+1)}).$$

Let $p = U_n - L_n - 1$. Then a large sample $100(1-\alpha)\%$ confidence interval for the population median MED(Y) is

$$\text{MED}(n) \pm t_{p,1-\alpha/2}\text{SE}(\text{MED}(n)). \tag{9.5}$$

Example 9.4: Inference with the trimmed mean. The symmetrically trimmed mean or the δ *trimmed mean*

$$T_n = T_n(L_n, U_n) = \frac{1}{U_n - L_n}\sum_{i=L_n+1}^{U_n} Y_{(i)} \tag{9.6}$$

where $L_n = \lfloor n\delta \rfloor$ and $U_n = n - L_n$. If $\delta = 0.25$, say, then the δ trimmed mean is called the 25 % trimmed mean.

The trimmed mean is estimating a truncated mean μ_T. Assume that Y has a probability density function $f_Y(y)$ that is continuous and positive on its support. Let y_δ be the number satisfying $P(Y \le y_\delta) = \delta$. Then

$$\mu_T = \frac{1}{1-2\delta} \int_{y_\delta}^{y_{1-\delta}} y f_Y(y) dy. \tag{9.7}$$

Notice that the 25 % trimmed mean is estimating

$$\mu_T = \int_{y_{0.25}}^{y_{0.75}} 2y f_Y(y) dy.$$

To perform inference, find $d_1, \ldots, d_n$ where

$$d_i = \begin{cases} Y_{(L_n+1)}, & i \le L_n \\ Y_{(i)}, & L_n + 1 \le i \le U_n \\ Y_{(U_n)}, & i \ge U_n + 1. \end{cases}$$

Then the Winsorized variance is the sample variance $S_n^2(d_1, \ldots, d_n)$ of $d_1, \ldots, d_n$, and the scaled Winsorized variance

$$V_{SW}(L_n, U_n) = \frac{S_n^2(d_1, \ldots, d_n)}{([U_n - L_n]/n)^2}. \tag{9.8}$$

The standard error of T_n is $\mathrm{SE}(T_n) = \sqrt{V_{SW}(L_n, U_n)/n}$.

A large sample 100 $(1-\alpha)$% confidence interval (CI) for μ_T is

$$T_n \pm t_{p,1-\frac{\alpha}{2}} \mathrm{SE}(T_n) \tag{9.9}$$

where $P(t_p \le t_{p,1-\frac{\alpha}{2}}) = 1 - \alpha/2$ if t_p is from a t distribution with $p = U_n - L_n - 1$ degrees of freedom. This interval is the classical t-interval when $\delta = 0$, but $\delta = 0.25$ gives a robust CI.

Example 9.5. Suppose the data below is from a symmetric distribution with mean μ. Find a 95 % CI for μ.

6, 9, 9, 7, 8, 9, 9, 7

Solution. When computing small examples by hand, the steps are to sort the data from smallest to largest value, find n, L_n, U_n, $Y_{(L_n)}$, $Y_{(U_n)}$, p, MED(n) and SE(MED(n)). After finding $t_{p,1-\alpha/2}$, plug the relevant quantities into the formula for the CI. The sorted data are 6, 7, 7, 8, 9, 9, 9, 9. Thus $\mathrm{MED}(n) = (8+9)/2 = 8.5$. Since $n = 8$, $L_n = \lfloor 4 \rfloor - \lceil \sqrt{2} \rceil = 4 - \lceil 1.414 \rceil = 4 - 2 = 2$ and $U_n = n - L_n = 8 - 2 = 6$. Hence $\mathrm{SE}(\mathrm{MED}(n)) = 0.5(Y_{(6)} - Y_{(3)}) = 0.5 * (9 - 7) = 1$. The degrees of freedom $p = U_n - L_n - 1 = 6 - 2 - 1 = 3$. The cutoff $t_{3,0.975} = 3.182$. Thus the 95 % CI for MED(Y) is

$$\text{MED}(n) \pm t_{3,0.975}\text{SE}(\text{MED}(n))$$

$= 8.5 \pm 3.182(1) = (5.318, 11.682)$. The classical t-interval uses $\overline{Y} = (6+7+7+8+9+9+9+9)/8$ and $S_n^2 = (1/7)[(\sum_{i=1}^n Y_i^2) - 8(8^2)] = (1/7)[(522 - 8(64)] = 10/7 \approx 1.4286$, and $t_{7,0.975} \approx 2.365$. Hence the 95 % CI for μ is $8 \pm 2.365(\sqrt{1.4286/8}) = (7.001, 8.999)$. Notice that the t-cutoff $= 2.365$ for the classical interval is less than the t-cutoff $= 3.182$ for the median interval and that $\text{SE}(\overline{Y}) < \text{SE}(\text{MED}(n))$.

Example 9.6. In the last example, what happens if the 6 becomes 66 and a 9 becomes 99?

Solution. Then the ordered data are 7, 7, 8, 9, 9, 9, 66, 99. Hence $\text{MED}(n) = 9$. Since L_n and U_n only depend on the sample size, they take the same values as in the previous example and $\text{SE}(\text{MED}(n)) = 0.5(Y_{(6)} - Y_{(3)}) = 0.5 * (9-8) = 0.5$. Hence the 95 % CI for MED(Y) is $\text{MED}(n) \pm t_{3,0.975}\text{SE}(\text{MED}(n)) = 9 \pm 3.182(0.5) = (7.409, 10.591)$. Notice that with discrete data, it is possible to drive SE(MED(n)) to 0 with a few outliers if n is small. The classical confidence interval $\overline{Y} \pm t_{7,0.975}S/\sqrt{n}$ blows up and is equal to $(-2.955, 56.455)$.

Example 9.7. The Buxton (1920) data contains 87 heights of men, but five of the men were recorded to be about 0.75 in. tall! The mean height is $\overline{Y} = 1598.862$ and the classical 95 % CI is (1514.206, 1683.518). $\text{MED}(n) = 1693.0$ and the resistant 95 % CI based on the median is (1678.517, 1707.483). The 25 % trimmed mean $T_n = 1689.689$ with 95 % CI (1672.096, 1707.282).

The heights for the five men were recorded under their head lengths, so the outliers can be corrected. Then $\overline{Y} = 1692.356$ and the classical 95 % CI is (1678.595, 1706.118). Now $\text{MED}(n) = 1694.0$ and the 95 % CI based on the median is (1678.403, 1709.597). The 25 % trimmed mean $T_n = 1693.200$ with 95 % CI (1676.259, 1710.141). Notice that when the outliers are corrected, the three intervals are very similar although the classical interval length is slightly shorter. Also notice that the outliers roughly shifted the median confidence interval by about 1 mm while the outliers greatly increased the length of the classical t-interval.

9.2 Some Examples

Example 9.8. Suppose that $Y_1, \ldots, Y_n$ are iid from a one-parameter exponential family with parameter τ. Assume that $T_n = \sum_{i=1}^n t(Y_i)$ is a complete sufficient statistic. Then from Theorems 3.6 and 3.7, often $T_n \sim G(na, 2b\,\tau)$ where a and b are known positive constants. Then

$$\hat{\tau} = \frac{T_n}{2nab}$$

is the UMVUE and often the MLE of τ. Since $T_n/(b\ \tau) \sim G(na, 2)$, a $100(1-\alpha)\%$ confidence interval for τ is

$$\left(\frac{T_n/b}{G(na,2,1-\alpha/2)}, \frac{T_n/b}{G(na,2,\alpha/2)}\right) \approx \left(\frac{T_n/b}{\chi_d^2(1-\alpha/2)}, \frac{T_n/b}{\chi_d^2(\alpha/2)}\right) \tag{9.10}$$

where $d = \lfloor 2na \rfloor$, $\lfloor x \rfloor$ is the greatest integer function (e.g., $\lfloor 7.7 \rfloor = \lfloor 7 \rfloor = 7$), $P[G \leq G(\nu,\lambda,\alpha)] = \alpha$ if $G \sim G(\nu,\lambda)$, and $P[X \leq \chi_d^2(\alpha)] = \alpha$ if X has a chi-square χ_d^2 distribution with d degrees of freedom.

This confidence interval can be inverted to perform two tail tests of hypotheses. By Theorem 7.3, if $w(\theta)$ is increasing, then the uniformly most powerful (UMP) test of $H_o: \tau \leq \tau_o$ versus $H_A: \tau > \tau_o$ rejects H_o if and only if $T_n > k$ where $P[G > k] = \alpha$ when $G \sim G(na, 2b\ \tau_o)$. Hence

$$k = G(na, 2b\ \tau_o, 1-\alpha). \tag{9.11}$$

A good approximation to this test rejects H_o if and only if

$$T_n > b\ \tau_o \chi_d^2(1-\alpha)$$

where $d = \lfloor 2na \rfloor$.

Example 9.9. If Y is half normal HN(μ,σ), then the pdf of Y is

$$f(y) = \frac{2}{\sqrt{2\pi}\ \sigma} \exp\left(\frac{-(y-\mu)^2}{2\sigma^2}\right)$$

where $\sigma > 0$ and $y > \mu$ and μ is real. Since

$$f(y) = \frac{2}{\sqrt{2\pi}\ \sigma} I[y > \mu] \exp\left[\left(\frac{-1}{2\sigma^2}\right)(y-\mu)^2\right],$$

Y is a 1P-REF if μ is known.

Since $T_n = \sum (Y_i - \mu)^2 \sim G(n/2, 2\sigma^2)$, in Example 9.8 take $a = 1/2, b = 1, d = n$ and $\tau = \sigma^2$. Then a $100(1-\alpha)\%$ confidence interval for σ^2 is

$$\left(\frac{T_n}{\chi_n^2(1-\alpha/2)}, \frac{T_n}{\chi_n^2(\alpha/2)}\right). \tag{9.12}$$

The UMP test of $H_o: \sigma^2 \leq \sigma_o^2$ versus $H_A: \sigma^2 > \sigma_o^2$ rejects H_o if and only if

$$T_n/\sigma_o^2 > \chi_n^2(1-\alpha).$$

Now consider inference when both μ and σ are unknown. Then the family is no longer an exponential family since the support depends on μ. Let

$$D_n = \sum_{i=1}^n (Y_i - Y_{1:n})^2. \tag{9.13}$$

Pewsey (2002) showed that $(\hat{\mu}, \hat{\sigma}^2) = (Y_{1:n}, \frac{1}{n} D_n)$ is the MLE of (μ, σ^2), and that

$$\frac{Y_{1:n} - \mu}{\sigma \Phi^{-1}(\frac{1}{2} + \frac{1}{2n})} \xrightarrow{D} \text{EXP}(1).$$

Since $(\sqrt{\pi/2})/n$ is an approximation to $\Phi^{-1}(\frac{1}{2} + \frac{1}{2n})$ based on a first order Taylor series expansion such that

$$\frac{\Phi^{-1}(\frac{1}{2} + \frac{1}{2n})}{(\sqrt{\pi/2})/n} \to 1,$$

it follows that

$$\frac{n(Y_{1:n} - \mu)}{\sigma \sqrt{\frac{\pi}{2}}} \xrightarrow{D} \text{EXP}(1). \tag{9.14}$$

Using this fact, it can be shown that a large sample $100(1-\alpha)\%$ CI for μ is

$$\left(\hat{\mu} + \hat{\sigma} \log(\alpha)\, \Phi^{-1}\left(\frac{1}{2} + \frac{1}{2n}\right) (1 + 13/n^2),\ \hat{\mu} \right) \tag{9.15}$$

where the term $(1 + 13/n^2)$ is a small sample correction factor. See Abuhassan and Olive (2008).

Note that

$$D_n = \sum_{i=1}^n (Y_i - Y_{1:n})^2 = \sum_{i=1}^n (Y_i - \mu + \mu - Y_{1:n})^2 =$$

$$\sum_{i=1}^n (Y_i - \mu)^2 + n(\mu - Y_{1:n})^2 + 2(\mu - Y_{1:n}) \sum_{i=1}^n (Y_i - \mu).$$

Hence

$$D_n = T_n + \frac{1}{n}[n(Y_{1:n} - \mu)]^2 - 2[n(Y_{1:n} - \mu)] \frac{\sum_{i=1}^n (Y_i - \mu)}{n},$$

or

$$\frac{D_n}{\sigma^2} = \frac{T_n}{\sigma^2} + \frac{1}{n}\frac{1}{\sigma^2}[n(Y_{1:n} - \mu)]^2 - 2\left[\frac{n(Y_{1:n} - \mu)}{\sigma}\right] \frac{\sum_{i=1}^n (Y_i - \mu)}{n\sigma}. \tag{9.16}$$

Consider the three terms on the right-hand side of (9.16). The middle term converges to 0 in distribution while the third term converges in distribution to a $-2\text{EXP}(1)$ or $-\chi^2_2$ distribution since $\sum_{i=1}^n (Y_i - \mu)/(\sigma n)$ is the sample mean of HN(0,1) random variables and $E(X) = \sqrt{2/\pi}$ when $X \sim \text{HN}(0,1)$.

Let $T_{n-p} = \sum_{i=1}^{n-p} (Y_i - \mu)^2$. Then

$$D_n = T_{n-p} + \sum_{i=n-p+1}^{n} (Y_i - \mu)^2 - V_n \tag{9.17}$$

where

$$\frac{V_n}{\sigma^2} \xrightarrow{D} \chi^2_2.$$

Hence

$$\frac{D_n}{T_{n-p}} \xrightarrow{D} 1$$

and D_n/σ^2 is asymptotically equivalent to a χ^2_{n-p} random variable where p is an arbitrary nonnegative integer. Pewsey (2002) used $p = 1$.

Thus when both μ and σ^2 are unknown, a large sample $100(1-\alpha)\%$ confidence interval for σ^2 is

$$\left(\frac{D_n}{\chi^2_{n-1}(1-\alpha/2)}, \frac{D_n}{\chi^2_{n-1}(\alpha/2)}\right). \tag{9.18}$$

It can be shown that $\sqrt{n}$ CI length converges in probability to $\sigma^2\sqrt{2}(z_{1-\alpha/2} - z_{\alpha/2})$ for CIs (9.12) and (9.18) while n length CI (9.15) converges in probability to $-\sigma\log(\alpha)\sqrt{\pi/2}$.

When μ and σ^2 are unknown, an approximate α level test of $H_o : \sigma^2 \leq \sigma_o^2$ versus $H_A : \sigma^2 > \sigma_o^2$ that rejects H_o if and only if

$$D_n/\sigma_o^2 > \chi^2_{n-1}(1-\alpha) \tag{9.19}$$

has nearly as much power as the α level UMP test when μ is known if n is large.

Example 9.10. Following Mann et al. (1974, p. 176), let $W_1, \ldots, W_n$ be iid $EXP(\theta, \lambda)$ random variables. Let

$$W_{1:n} = \min(W_1, \ldots, W_n).$$

Then the MLE

$$(\hat{\theta}, \hat{\lambda}) = \left(W_{1:n}, \frac{1}{n}\sum_{i=1}^{n}(W_i - W_{1:n})\right) = (W_{1:n}, \overline{W} - W_{1:n}).$$

Let $D_n = n\hat{\lambda}$. For $n > 1$, a $100(1-\alpha)\%$ confidence interval (CI) for θ is

$$(W_{1:n} - \hat{\lambda}[(\alpha)^{-1/(n-1)} - 1], W_{1:n}) \tag{9.20}$$

while a $100(1-\alpha)\%$ CI for λ is

$$\left(\frac{2D_n}{\chi^2_{2(n-1),1-\alpha/2}}, \frac{2D_n}{\chi^2_{2(n-1),\alpha/2}}\right). \tag{9.21}$$

Let $T_n = \sum_{i=1}^{n}(W_i - \theta) = n(\overline{W} - \theta)$. If θ is known, then

$$\hat{\lambda}_\theta = \frac{\sum_{i=1}^{n}(W_i - \theta)}{n} = \overline{W} - \theta$$

is the UMVUE and MLE of λ, and a $100(1-\alpha)\%$ CI for λ is

$$\left(\frac{2T_n}{\chi^2_{2n,1-\alpha/2}}, \frac{2T_n}{\chi^2_{2n,\alpha/2}} \right). \tag{9.22}$$

Using $\chi^2_{n,\alpha}/\sqrt{n} \approx \sqrt{2}z_\alpha + \sqrt{n}$, it can be shown that $\sqrt{n}$ CI length converges to $\lambda(z_{1-\alpha/2} - z_{\alpha/2})$ for CIs (9.21) and (9.22) (in probability). It can be shown that n length CI (9.20) converges to $-\lambda \log(\alpha)$.

When a random variable is a simple transformation of a distribution that has an easily computed CI, the transformed random variable will often have an easily computed CI. Similarly the MLEs of the two distributions are often closely related. See the discussion above Example 5.11. The first three of the following four examples are from Abuhassan and Olive (2008).

Example 9.11. If Y has a Pareto distribution, $Y \sim \text{PAR}(\sigma, \lambda)$, then $W = \log(Y) \sim \text{EXP}(\theta = \log(\sigma), \lambda)$. If $\theta = \log(\sigma)$ so $\sigma = e^\theta$, then a 100 $(1-\alpha)\%$ CI for θ is (9.20). A 100 $(1-\alpha)\%$ CI for σ is obtained by exponentiating the endpoints of (9.20), and a 100 $(1-\alpha)\%$ CI for λ is (9.21). The fact that the Pareto distribution is a log-location–scale family ($W = \log(Y)$ is from a location–scale family) and hence has simple inference does not seem to be well known.

Example 9.12. If Y has a power distribution, $Y \sim \text{POW}(\lambda)$, then $W = -\log(Y)$ is $\text{EXP}(0, \lambda)$. A 100 $(1-\alpha)\%$ CI for λ is (9.22).

If Y has a two-parameter power distribution, $Y \sim \text{power}(\tau, \lambda)$, then

$$F(y) = \left(\frac{y}{\tau}\right)^{1/\lambda}$$

for $0 < y \leq \tau$. The pdf

$$f(y) = \frac{1}{\tau\lambda}\left(\frac{y}{\tau}\right)^{\frac{1}{\lambda}-1} I(0 < y \leq \tau).$$

Then $W = -\log(Y) \sim \text{EXP}(-\log(\tau), \lambda)$. Thus (9.21) is an exact $100(1-\alpha)\%$ CI for λ, and (9.20) $= (L_n, U_n)$ is an exact $100(1-\alpha)\%$ CI for $-\log(\tau)$. Hence (e^{L_n}, e^{U_n}) is a $100(1-\alpha)\%$ CI for $1/\tau$, and (e^{-U_n}, e^{-L_n}) is a $100(1-\alpha)\%$ CI for τ.

Example 9.13. If Y has a truncated extreme value distribution, $Y \sim \text{TEV}(\lambda)$, then $W = e^Y - 1$ is $\text{EXP}(0, \lambda)$. A 100 $(1-\alpha)\%$ CI for λ is (9.22).

Example 9.14. If Y has a lognormal distribution, $Y \sim \text{LN}(\mu, \sigma^2)$, then $W_i = \log(Y_i) \sim N(\mu, \sigma^2)$. Thus a $(1-\alpha)100\%$ CI for μ when σ is unknown is

$$\left(\overline{W}_n - t_{n-1,1-\frac{\alpha}{2}} \frac{S_W}{\sqrt{n}}, \overline{W}_n + t_{n-1,1-\frac{\alpha}{2}} \frac{S_W}{\sqrt{n}} \right)$$

where

$$S_W = \frac{n}{n-1}\hat{\sigma} = \sqrt{\frac{1}{n-1}\sum_{i=1}^{n}(W_i - \overline{W})^2},$$

and $P(t \leq t_{n-1,1-\frac{\alpha}{2}}) = 1 - \alpha/2$ when $t \sim t_{n-1}$.

Example 9.15. Let $X_1, \ldots, X_n$ be iid Poisson(θ) random variables. The classical large sample 100 $(1-\alpha)\%$ CI for θ is

$$\overline{X} \pm z_{1-\alpha/2}\sqrt{\overline{X}/n}$$

where $P(Z \leq z_{1-\alpha/2}) = 1 - \alpha/2$ if $Z \sim N(0,1)$.

Following Byrne and Kabaila (2005), a modified large sample 100 $(1-\alpha)\%$ CI for θ is (L_n, U_n) where

$$L_n = \frac{1}{n}\left(\sum_{i=1}^{n} X_i - 0.5 + 0.5z^2_{1-\alpha/2} - z_{1-\alpha/2}\sqrt{\sum_{i=1}^{n} X_i - 0.5 + 0.25z^2_{1-\alpha/2}}\right)$$

and

$$U_n = \frac{1}{n}\left(\sum_{i=1}^{n} X_i + 0.5 + 0.5z^2_{1-\alpha/2} + z_{1-\alpha/2}\sqrt{\sum_{i=1}^{n} X_i + 0.5 + 0.25z^2_{1-\alpha/2}}\right).$$

Following Grosh (1989, pp. 59, 197–200), let $W = \sum_{i=1}^{n} X_i$ and suppose that $W = w$ is observed. Let $P(T < \chi^2_d(\alpha)) = \alpha$ if $T \sim \chi^2_d$. Then an "exact" 100 $(1-\alpha)\%$ CI for θ is

$$\left(\frac{\chi^2_{2w}(\frac{\alpha}{2})}{2n}, \frac{\chi^2_{2w+2}(1-\frac{\alpha}{2})}{2n}\right)$$

for $w \neq 0$ and

$$\left(0, \frac{\chi^2_2(1-\alpha)}{2n}\right)$$

for $w = 0$.

The "exact" CI is conservative: the actual coverage $(1-\delta_n) \geq 1 - \alpha =$ the nominal coverage. This interval performs well if θ is very close to 0. See Problem 9.3.

Example 9.16. Let $Y_1, \ldots, Y_n$ be iid bin$(1, \rho)$. Let $\hat{\rho} = \sum_{i=1}^{n} Y_i/n =$ number of "successes"$/n$. The classical large sample 100 $(1-\alpha)\%$ CI for ρ is

$$\hat{\rho} \pm z_{1-\alpha/2}\sqrt{\frac{\hat{\rho}(1-\hat{\rho})}{n}}$$

where $P(Z \leq z_{1-\alpha/2}) = 1 - \alpha/2$ if $Z \sim N(0,1)$.

The Agresti Coull CI takes $\tilde{n} = n + z^2_{1-\alpha/2}$ and

$$\tilde{\rho} = \frac{n\hat{\rho} + 0.5z^2_{1-\alpha/2}}{n + z^2_{1-\alpha/2}}.$$

(The method "adds" $0.5z^2_{1-\alpha/2}$ "0's" and $0.5z^2_{1-\alpha/2}$ "1's" to the sample, so the "sample size" increases by $z^2_{1-\alpha/2}$.) Then the large sample 100 $(1-\alpha)$% Agresti Coull CI for ρ is

$$\tilde{\rho} \pm z_{1-\alpha/2}\sqrt{\frac{\tilde{\rho}(1-\tilde{\rho})}{\tilde{n}}}.$$

Now let $Y_1, \ldots, Y_n$ be independent bin(m_i, ρ) random variables, let $W = \sum_{i=1}^n Y_i \sim$ bin$(\sum_{i=1}^n m_i, \rho)$ and let $n_w = \sum_{i=1}^n m_i$. Often $m_i \equiv 1$ and then $n_w = n$. Let $P(F_{d_1,d_2} \leq F_{d_1,d_2}(\alpha)) = \alpha$ where F_{d_1,d_2} has an F distribution with d_1 and d_2 degrees of freedom. Assume $W = w$ is observed. Then the Clopper Pearson "exact" 100 $(1-\alpha)$% CI for ρ is

$$\left(0, \frac{1}{1 + n_w\, F_{2n_w,2}(\alpha)}\right) \quad \text{for } w = 0,$$

$$\left(\frac{n_w}{n_w + F_{2,2n_w}(1-\alpha)}, 1\right) \quad \text{for } w = n_w,$$

and (ρ_L, ρ_U) for $0 < w < n_w$ with

$$\rho_L = \frac{w}{w + (n_w - w + 1)F_{2(n_w-w+1),2w}(1-\alpha/2)}$$

and

$$\rho_U = \frac{w+1}{w + 1 + (n_w - w)F_{2(n_w-w),2(w+1)}(\alpha/2)}.$$

The "exact" CI is conservative: the actual coverage $(1-\delta_n) \geq 1-\alpha =$ the nominal coverage. This interval performs well if ρ is very close to 0 or 1. The classical interval should only be used if it agrees with the Agresti Coull interval. See Problem 9.2.

Example 9.17. Let $\hat{\rho} =$ number of "successes"$/n$. Consider taking a simple random sample of size n from a finite population of known size N. Then the classical finite population large sample 100 $(1-\alpha)$% CI for ρ is

$$\hat{\rho} \pm z_{1-\alpha/2}\sqrt{\frac{\hat{\rho}(1-\hat{\rho})}{n-1}\left(\frac{N-n}{N}\right)} = \hat{\rho} \pm z_{1-\alpha/2}\text{SE}(\hat{\rho}) \tag{9.23}$$

where $P(Z \leq z_{1-\alpha/2}) = 1-\alpha/2$ if $Z \sim N(0,1)$.

Following DasGupta (2008, p. 121), suppose the number of successes Y has a hypergeometric $(C, N-C, n)$ where $p = C/N$. If $n/N \approx \lambda \in (0,1)$ where n and N are both large, then

$$\hat{\rho} \approx N\left(\rho, \frac{\rho(1-\rho)(1-\lambda)}{n}\right).$$

Hence CI (9.23) should be good if the above normal approximation is good.

Let $\tilde{n} = n + z^2_{1-\alpha/2}$ and

$$\tilde{\rho} = \frac{n\hat{\rho} + 0.5z^2_{1-\alpha/2}}{n + z^2_{1-\alpha/2}}.$$

(Heuristically, the method adds $0.5z^2_{1-\alpha/2}$ "0's" and $0.5z^2_{1-\alpha/2}$ "1's" to the sample, so the "sample size" increases by $z^2_{1-\alpha/2}$.) Then a large sample $100\,(1-\alpha)\%$ Agresti Coull type (ACT) finite population CI for ρ is

$$\tilde{\rho} \pm z_{1-\alpha/2}\sqrt{\frac{\tilde{\rho}(1-\tilde{\rho})}{\tilde{n}}\left(\frac{N-n}{N}\right)} = \tilde{\rho} \pm z_{1-\alpha/2}\mathrm{SE}(\tilde{\rho}). \tag{9.24}$$

Notice that a 95 % CI uses $z_{1-\alpha/2} = 1.96 \approx 2$.

For data from a finite population, large sample theory gives useful approximations as N and $n \to \infty$ and $n/N \to 0$. Hence theory suggests that the ACT CI should have better coverage than the classical CI if the p is near 0 or 1, if the sample size n is moderate, and if n is small compared to the population size N. The coverage of the classical and ACT CIs should be very similar if n is large enough but small compared to N (which may only be possible if N is enormous). As n increases to N, $\hat{\rho}$ goes to p, SE($\hat{\rho}$) goes to 0, and the classical CI may perform well. SE($\tilde{\rho}$) also goes to 0, but $\tilde{\rho}$ is a biased estimator of ρ and the ACT CI will not perform well if n/N is too large.

Want an interval that gives good coverage even if ρ is near 0 or 1 or if n/N is large. A simple method is to combine the two intervals. Let (L_C, U_C) and (L_A, U_A) be the classical and ACT $100(1-\alpha)\,\%$ intervals. Let the modified $100(1-\alpha)\%$ interval be

$$(\max[0, \min(L_C, L_U)], \min[1, \max(U_C, U_A)]). \tag{9.25}$$

The modified interval seems to perform well. See Problem 9.4.

Example 9.18. If $Y_1, \ldots, Y_n$ are iid Weibull (ϕ, λ), then the MLE $(\hat{\phi}, \hat{\lambda})$ must be found before obtaining CIs. The likelihood

$$L(\phi, \lambda) = \frac{\phi^n}{\lambda^n} \prod_{i=1}^{n} y_i^{\phi-1} \exp\left[\frac{-1}{\lambda} \sum y_i^{\phi}\right],$$

and the log likelihood

$$\log(L(\phi,\lambda)) = n\log(\phi) - n\log(\lambda) + (\phi-1)\sum_{i=1}^n \log(y_i) - \frac{1}{\lambda}\sum y_i^\phi.$$

Hence

$$\frac{\partial}{\partial\lambda}\log(L(\phi,\lambda)) = \frac{-n}{\lambda} + \frac{\sum y_i^\phi}{\lambda^2} \stackrel{set}{=} 0,$$

or $\sum y_i^\phi = n\lambda$, or

$$\hat{\lambda} = \frac{\sum y_i^{\hat{\phi}}}{n}.$$

Now

$$\frac{\partial}{\partial\phi}\log(L(\phi,\lambda)) = \frac{n}{\phi} + \sum_{i=1}^n \log(y_i) - \frac{1}{\lambda}\sum y_i^\phi \log(y_i) \stackrel{set}{=} 0,$$

so

$$n + \phi\left[\sum_{i=1}^n \log(y_i) - \frac{1}{\lambda}\sum y_i^\phi \log(y_i)\right] = 0,$$

or

$$\hat{\phi} = \frac{n}{\frac{1}{\hat{\lambda}}\sum y_i^{\hat{\phi}}\log(y_i) - \sum_{i=1}^n \log(y_i)}.$$

One way to find the MLE is to use iteration

$$\hat{\lambda}_k = \frac{\sum y_i^{\hat{\phi}_{k-1}}}{n}$$

and

$$\hat{\phi}_k = \frac{n}{\frac{1}{\hat{\lambda}_k}\sum y_i^{\hat{\phi}_{k-1}}\log(y_i) - \sum_{i=1}^n \log(y_i)}.$$

Since $W = \log(Y) \sim \text{SEV}(\theta = \log(\lambda^{1/\phi}), \sigma = 1/\phi)$, let

$$\hat{\sigma}_R = \text{MAD}(W_1,\ldots,W_n)/0.767049$$

and

$$\hat{\theta}_R = \text{MED}(W_1,\ldots,W_n) - \log(\log(2))\hat{\sigma}_R.$$

Then $\hat{\phi}_0 = 1/\hat{\sigma}_R$ and $\hat{\lambda}_0 = \exp(\hat{\theta}_R/\hat{\sigma}_R)$. The iteration might be run until both $|\hat{\phi}_k - \hat{\phi}_{k-1}| < 10^{-6}$ and $|\hat{\lambda}_k - \hat{\lambda}_{k-1}| < 10^{-6}$. Then take $(\hat{\phi},\hat{\lambda}) = (\hat{\phi}_k,\hat{\lambda}_k)$.

By Example 8.13,

$$\sqrt{n}\left(\begin{pmatrix}\hat{\lambda}\\ \hat{\phi}\end{pmatrix} - \begin{pmatrix}\lambda\\ \phi\end{pmatrix}\right) \xrightarrow{D} N_2(\mathbf{0},\boldsymbol{\Sigma})$$

where $\boldsymbol{\Sigma} =$

$$\begin{bmatrix} 1.109\lambda^2(1+0.4635\log(\lambda)+0.5482(\log(\lambda))^2) & 0.257\phi\lambda+0.608\lambda\phi\log(\lambda) \\ 0.257\phi\lambda+0.608\lambda\phi\log(\lambda) & 0.608\phi^2 \end{bmatrix}.$$

Thus $1-\alpha \approx P(-z_{1-\alpha/2}\sqrt{0.608}\,\hat{\phi} < \sqrt{n}(\hat{\phi}-\phi) < z_{1-\alpha/2}\sqrt{0.608}\,\hat{\phi})$ and a large sample $100(1-\alpha)\%$ CI for ϕ is

$$\hat{\phi} \pm z_{1-\alpha/2}\,\hat{\phi}\,\sqrt{0.608/n}. \tag{9.26}$$

Similarly, a large sample $100(1-\alpha)\%$ CI for λ is

$$\hat{\lambda} \pm \frac{z_{1-\alpha/2}}{\sqrt{n}}\sqrt{1.109\hat{\lambda}^2[1+0.4635\log(\hat{\lambda})+0.5824(\log(\hat{\lambda}))^2]}. \tag{9.27}$$

In simulations, for small n the number of iterations for the MLE to converge could be in the thousands, and the coverage of the large sample CIs is poor for $n < 50$. See Problem 9.7.

Iterating the likelihood equations until "convergence" to a point $\hat{\boldsymbol{\theta}}$ is called a fixed point algorithm. Such algorithms may not converge, so check that $\hat{\boldsymbol{\theta}}$ satisfies the likelihood equations. Other methods such as Newton's method may perform better.

Newton's method is used to solve $\boldsymbol{g}(\boldsymbol{\theta}) = \boldsymbol{0}$ for $\boldsymbol{\theta}$, where the solution is called $\hat{\boldsymbol{\theta}}$, and uses

$$\boldsymbol{\theta}_{k+1} = \boldsymbol{\theta}_k - \left[\boldsymbol{D}_{\boldsymbol{g}(\boldsymbol{\theta}_k)}\right]^{-1}\boldsymbol{g}(\boldsymbol{\theta}_k) \tag{9.28}$$

where

$$\boldsymbol{D}_{\boldsymbol{g}(\boldsymbol{\theta})} = \begin{bmatrix} \frac{\partial}{\partial\theta_1}g_1(\boldsymbol{\theta}) & \dots & \frac{\partial}{\partial\theta_p}g_1(\boldsymbol{\theta}) \\ \vdots & & \vdots \\ \frac{\partial}{\partial\theta_1}g_p(\boldsymbol{\theta}) & \dots & \frac{\partial}{\partial\theta_p}g_p(\boldsymbol{\theta}) \end{bmatrix}.$$

If the MLE is the solution of the likelihood equations, then use $\boldsymbol{g}(\boldsymbol{\theta}) = (g_1(\boldsymbol{\theta}),\dots,g_p(\boldsymbol{\theta}))^T$ where

$$g_i(\boldsymbol{\theta}) = \frac{\partial}{\partial\theta_i}\log(L(\boldsymbol{\theta})).$$

Let $\boldsymbol{\theta}_0$ be an initial estimator, such as the method of moments estimator of $\boldsymbol{\theta}$. Let $\boldsymbol{D} = \boldsymbol{D}_{\boldsymbol{g}(\boldsymbol{\theta})}$. Then

$$D_{ij} = \frac{\partial}{\partial\theta_j}g_i(\boldsymbol{\theta}) = \frac{\partial^2}{\partial\theta_i\partial\theta_j}\log(L(\boldsymbol{\theta})) = \sum_{k=1}^n \frac{\partial^2}{\partial\theta_i\partial\theta_j}\log(f(x_k|\boldsymbol{\theta})),$$

and

$$\frac{1}{n}D_{ij} = \frac{1}{n}\sum_{k=1}^{n} \frac{\partial^2}{\partial\theta_i\partial\theta_j}\log(f(X_k|\boldsymbol{\theta})) \xrightarrow{D} E\left[\frac{\partial^2}{\partial\theta_i\partial\theta_j}\log(f(X|\boldsymbol{\theta}))\right].$$

Newton's method converges if the initial estimator is sufficiently close to $\boldsymbol{\theta}$, but may diverge otherwise. Hence $\sqrt{n}$ consistent initial estimators are recommended. Newton's method is also popular because if the partial derivative and integration operations can be interchanged, then

$$\frac{1}{n}\boldsymbol{D}_{\boldsymbol{g}(\boldsymbol{\theta})} \xrightarrow{D} -\boldsymbol{I}(\boldsymbol{\theta}). \tag{9.29}$$

For example, the regularity conditions hold for a kP-REF by Proposition 8.20. Then a 100 $(1-\alpha)$% large sample CI for θ_i is

$$\hat{\theta}_i \pm z_{1-\alpha/2}\sqrt{-\boldsymbol{D}_{ii}^{-1}} \tag{9.30}$$

where

$$\boldsymbol{D}^{-1} = \left[\boldsymbol{D}_{\boldsymbol{g}(\hat{\boldsymbol{\theta}})}\right]^{-1}.$$

This result follows because

$$\sqrt{-\boldsymbol{D}_{ii}^{-1}} \approx \sqrt{[\boldsymbol{I}^{-1}(\hat{\boldsymbol{\theta}})]_{ii}/n}.$$

Example 9.19. Problem 9.8 simulates CIs for the Rayleigh (μ,σ) distribution of the form (9.30) although no check has been made on whether (9.29) holds for the Rayleigh distribution (which is not a 2P-REF).

$$L(\mu,\sigma) = \left(\prod \frac{y_i-\mu}{\sigma^2}\right)\exp\left[-\frac{1}{2\sigma^2}\sum(y_i-\mu)^2\right].$$

Notice that for fixed σ, $L(Y_{(1)},\sigma) = 0$. Hence the MLE $\hat{\mu} < Y_{(1)}$. Now the log likelihood

$$\log(L(\mu,\sigma)) = \sum_{i=1}^{n}\log(y_i-\mu) - 2n\log(\sigma) - \frac{1}{2}\sum\frac{(y_i-\mu)^2}{\sigma^2}.$$

Hence $g_1(\mu,\sigma) =$

$$\frac{\partial}{\partial\mu}\log(L(\mu,\sigma)) = -\sum_{i=1}^{n}\frac{1}{y_i-\mu} + \frac{1}{\sigma^2}\sum_{i=1}^{n}(y_i-\mu) \overset{\text{set}}{=} 0,$$

and $g_2(\mu,\sigma) =$

$$\frac{\partial}{\partial\sigma}\log(L(\mu,\sigma)) = \frac{-2n}{\sigma} + \frac{1}{\sigma^3}\sum_{i=1}^{n}(y_i-\mu)^2 \overset{\text{set}}{=} 0,$$

which has solution

$$\hat{\sigma}^2 = \frac{1}{2n} \sum_{i=1}^{n} (Y_i - \hat{\mu})^2. \tag{9.31}$$

To obtain initial estimators, let $\hat{\sigma}_M = \sqrt{S^2/0.429204}$ and $\hat{\mu}_M = \overline{Y} - 1.253314\hat{\sigma}_M$. These would be the method of moments estimators if S_M^2 was used instead of the sample variance S^2. Then use $\mu_0 = \min(\hat{\mu}_M, 2Y_{(1)} - \hat{\mu}_M)$ and $\sigma_0 = \sqrt{\sum (Y_i - \mu_0)^2/(2n)}$. Now $\boldsymbol{\theta} = (\mu, \sigma)^T$ and

$$\boldsymbol{D} \equiv \boldsymbol{D}_{\boldsymbol{g}(\boldsymbol{\theta})} = \begin{bmatrix} \frac{\partial}{\partial \mu} g_1(\boldsymbol{\theta}) & \frac{\partial}{\partial \sigma} g_1(\boldsymbol{\theta}) \\ \\ \frac{\partial}{\partial \mu} g_2(\boldsymbol{\theta}) & \frac{\partial}{\partial \sigma} g_2(\boldsymbol{\theta}) \end{bmatrix} =$$

$$\begin{bmatrix} -\sum_{i=1}^n \frac{1}{(y_i-\mu)^2} - \frac{n}{\sigma^2} & -\frac{2}{\sigma^3} \sum_{i=1}^n (y_i - \mu) \\ \\ -\frac{2}{\sigma^3} \sum_{i=1}^n (y_i - \mu) & \frac{2n}{\sigma^2} - \frac{3}{\sigma^4} \sum_{i=1}^n (y_i - \mu)^2 \end{bmatrix}.$$

So

$$\boldsymbol{\theta}_{k+1} = \boldsymbol{\theta}_k - \begin{bmatrix} -\sum_{i=1}^n \frac{1}{(y_i-\mu_k)^2} - \frac{n}{\sigma_k^2} & -\frac{2}{\sigma_k^3} \sum_{i=1}^n (y_i - \mu_k) \\ \\ -\frac{2}{\sigma_k^3} \sum_{i=1}^n (y_i - \mu_k) & \frac{2n}{\sigma_k^2} - \frac{3}{\sigma_k^4} \sum_{i=1}^n (y_i - \mu_k)^2 \end{bmatrix}^{-1} \boldsymbol{g}(\boldsymbol{\theta}_k)$$

where

$$\boldsymbol{g}(\boldsymbol{\theta}_k) = \begin{pmatrix} -\sum_{i=1}^n \frac{1}{(y_i-\mu_k)} + \frac{1}{\sigma_k^2} \sum_{i=1}^n (y_i - \mu_k) \\ \\ \frac{-2n}{\sigma_k} + \frac{1}{\sigma_k^3} \sum_{i=1}^n (y_i - \mu_k)^2 \end{pmatrix}.$$

This formula could be iterated for 100 steps resulting in $\boldsymbol{\theta}_{101} = (\mu_{101}, \sigma_{101})^T$. Then take $\hat{\mu} = \min(\mu_{101}, 2Y_{(1)} - \mu_{101})$ and

$$\hat{\sigma} = \sqrt{\frac{1}{2n} \sum_{i=1}^{n} (Y_i - \hat{\mu})^2}.$$

Then $\hat{\boldsymbol{\theta}} = (\hat{\mu}, \hat{\sigma})^T$ and compute $\boldsymbol{D} \equiv \boldsymbol{D}_{\boldsymbol{g}(\hat{\boldsymbol{\theta}})}$. Then (assuming (9.29) holds) a 100 $(1-\alpha)\%$ large sample CI for μ is

$$\hat{\mu} \pm z_{1-\alpha/2} \sqrt{-\boldsymbol{D}_{11}^{-1}}$$

and a 100 $(1-\alpha)\%$ large sample CI for σ is

$$\hat{\sigma} \pm z_{1-\alpha/2} \sqrt{-\boldsymbol{D}_{22}^{-1}}.$$

Example 9.20. Assume that $Y_1, \ldots, Y_n$ are iid discrete uniform $(1, \eta)$ where η is an integer. For example, each Y_i could be drawn with replacement from a population of η tanks with serial numbers 1, 2, ..., η. The Y_i would be the serial number observed, and the goal would be to estimate the population size η = number of tanks. Then $P(Y_i = i) = 1/\eta$ for $i = 1, \ldots, \eta$. Then the cdf of Y is

$$F(y) = \sum_{i=1}^{\lfloor y \rfloor} \frac{1}{\eta} = \frac{\lfloor y \rfloor}{\eta}$$

for $1 \le y \le \eta$. Here $\lfloor y \rfloor$ is the greatest integer function, e.g., $\lfloor 7.7 \rfloor = 7$.

Now let $Z_i = Y_i/\eta$ which has cdf

$$F_Z(t) = P(Z \le t) = P(Y \le t\eta) = \frac{\lfloor t\eta \rfloor}{\eta} \approx t$$

for $0 < t < 1$. Let $Z_{(n)} = Y_{(n)}/\eta = \max(Z_1, \ldots, Z_n)$. Then

$$F_{Z_{(n)}}(t) = P\left(\frac{Y_{(n)}}{\eta} \le t\right) = \left(\frac{\lfloor t\eta \rfloor}{\eta}\right)^n$$

for $1/\eta < t < 1$.

Want c_n so that

$$P\left(c_n \le \frac{Y_{(n)}}{\eta} \le 1\right) = 1 - \alpha$$

for $0 < \alpha < 1$. So

$$1 - F_{Z_{(n)}}(c_n) = 1 - \alpha \text{ or } 1 - \left(\frac{\lfloor c_n\eta \rfloor}{\eta}\right)^n = 1 - \alpha$$

or

$$\frac{\lfloor c_n\eta \rfloor}{\eta} = \alpha^{1/n}.$$

The solution may not exist, but $c_n - 1/\eta \le \alpha^{1/n} \le c_n$. Take $c_n = \alpha^{1/n}$ then

$$\left[Y_{(n)}, \frac{Y_{(n)}}{\alpha^{1/n}}\right)$$

is a CI for η that has coverage slightly less than $100(1-\alpha)\%$ for small n, but the coverage converges in probability to 1 as $n \to \infty$.

For small n the midpoint of the 95 % CI might be a better estimator of η than $Y_{(n)}$. The left endpoint is closed since $Y_{(n)}$ is a consistent estimator of η. If the endpoint was open, coverage would go to 0 instead of 1. It can be shown that n (length CI) converges to $-\eta \log(\alpha)$ in probability. Hence n (length 95 % CI) $\approx 3\eta$. Problem 9.9 provides simulations that suggest that the 95 % CI coverage and length is close to the asymptotic values for $n \ge 10$.

Example 9.21. Assume that $Y_1,\ldots,Y_n$ are iid uniform $(0,\theta)$. Let $Z_i = Y_i/\theta \sim U(0,1)$ which has cdf $F_Z(t) = t$ for $0 < t < 1$. Let $Z_{(n)} = Y_{(n)}/\theta = \max(Z_1,\ldots,Z_n)$. Then

$$F_{Z_{(n)}}(t) = P\left(\frac{Y_{(n)}}{\theta} \le t\right) = t^n$$

for $0 < t < 1$.

Want c_n so that

$$P\left(c_n \le \frac{Y_{(n)}}{\theta} \le 1\right) = 1 - \alpha$$

for $0 < \alpha < 1$. So

$$1 - F_{Z_{(n)}}(c_n) = 1 - \alpha \text{ or } 1 - c_n^n = 1 - \alpha$$

or

$$c_n = \alpha^{1/n}.$$

Then

$$\left(Y_{(n)}, \frac{Y_{(n)}}{\alpha^{1/n}}\right)$$

is an exact $100(1-\alpha)\%$ CI for θ. It can be shown that n (length CI) converges to $-\theta\log(\alpha)$ in probability.

If $Y_1,\ldots,Y_n$ are iid $U(\theta_1,\theta_2)$ where θ_1 is known, then $Y_i - \theta_1$ are iid $U(0,\theta_2-\theta_1)$ and

$$\left(Y_{(n)} - \theta_1, \frac{Y_{(n)} - \theta_1}{\alpha^{1/n}}\right)$$

is a $100(1-\alpha)\%$ CI for $\theta_2 - \theta_1$. Thus if θ_1 is known, then

$$\left(Y_{(n)}, \theta_1\left(1 - \frac{1}{\alpha^{1/n}}\right) + \frac{Y_{(n)}}{\alpha^{1/n}}\right)$$

is a $100(1-\alpha)\%$ CI for θ_2.

Example 9.22. Assume $Y_1,\ldots,Y_n$ are iid with mean μ and variance σ^2. Bickel and Doksum (2007, p. 279) suggest that

$$W_n = n^{-1/2}\left[\frac{(n-1)S^2}{\sigma^2} - n\right]$$

can be used as an asymptotic pivot for σ^2 if $E(Y^4) < \infty$. Notice that $W_n =$

$$n^{-1/2}\left[\frac{\sum(Y_i-\mu)^2}{\sigma^2} - \frac{n(\overline{Y}-\mu)^2}{\sigma^2} - n\right] =$$

$$\sqrt{n}\left[\frac{\sum\left(\frac{Y_i-\mu}{\sigma}\right)^2}{n} - 1\right] - \frac{1}{\sqrt{n}}\, n\left(\frac{\overline{Y}-\mu}{\sigma}\right)^2 = X_n - Z_n.$$

Since $\sqrt{n}Z_n \xrightarrow{D} \chi_1^2$, the term $Z_n \xrightarrow{D} 0$. Now $X_n = \sqrt{n}(\overline{U} - 1) \xrightarrow{D} N(0, \tau)$ by the CLT since $U_i = [(Y_i - \mu)/\sigma]^2$ has mean $E(U_i) = 1$ and variance

$$V(U_i) = \tau = E(U_i^2) - (E(U_i))^2 = \frac{E[(Y_i - \mu)^4]}{\sigma^4} - 1 = \kappa + 2$$

where κ is the kurtosis of Y_i. Thus $W_n \xrightarrow{D} N(0, \tau)$.

Hence

$$1 - \alpha \approx P\left(-z_{1-\alpha/2} < \frac{W_n}{\sqrt{\tau}} < z_{1-\alpha/2}\right) = P(-z_{1-\alpha/2}\sqrt{\tau} < W_n < z_{1-\alpha/2}\sqrt{\tau})$$

$$= P\left(-z_{1-\alpha/2}\sqrt{n\tau} < \frac{(n-1)S^2}{\sigma^2} - n < z_{1-\alpha/2}\sqrt{n\tau}\right)$$

$$= P\left(n - z_{1-\alpha/2}\sqrt{n\tau} < \frac{(n-1)S^2}{\sigma^2} < n + z_{1-\alpha/2}\sqrt{n\tau}\right).$$

Hence a large sample $100(1 - \alpha)\%$ CI for σ^2 is

$$\left(\frac{(n-1)S^2}{n + z_{1-\alpha/2}\sqrt{n\hat{\tau}}}, \frac{(n-1)S^2}{n - z_{1-\alpha/2}\sqrt{n\hat{\tau}}}\right)$$

where

$$\hat{\tau} = \frac{\frac{1}{n}\sum_{i=1}^n (Y_i - \overline{Y})^4}{S^4} - 1.$$

Notice that this CI needs $n > z_{1-\alpha/2}\sqrt{n\hat{\tau}}$ for the right endpoint to be positive. It can be shown that $\sqrt{n}$ (length CI) converges to $2\sigma^2 z_{1-\alpha/2}\sqrt{\tau}$ in probability.

Problem 9.10 uses an asymptotically equivalent $100(1 - \alpha)\%$ CI of the form

$$\left(\frac{(n-a)S^2}{n + t_{n-1,1-\alpha/2}\sqrt{n\hat{\tau}}}, \frac{(n+b)S^2}{n - t_{n-1,1-\alpha/2}\sqrt{n\hat{\tau}}}\right)$$

where a and b depend on $\hat{\tau}$. The goal was to make a 95 % CI with good coverage for a wide variety of distributions (with 4th moments) for $n \geq 100$. The price is that the CI is too long for some of the distributions with small kurtosis. The $N(\mu, \sigma^2)$ distribution has $\tau = 2$, while the EXP(λ) distribution has $\sigma^2 = \lambda^2$ and $\tau = 8$. The quantity τ is small for the uniform distribution but large for the lognormal LN(0,1) distribution.

By the binomial theorem, if $E(Y^4)$ exists and $E(Y) = \mu$ then

$$E(Y - \mu)^4 = \sum_{j=0}^{4} \binom{4}{j} E[Y^j](-\mu)^{4-j} =$$

$$\mu^4 - 4\mu^3 E(Y) + 6\mu^2(V(Y) + [E(Y)]^2) - 4\mu E(Y^3) + E(Y^4).$$

This fact can be useful for computing

$$\tau = \frac{E[(Y_i - \mu)^4]}{\sigma^4} - 1 = \kappa + 2.$$

Example 9.23. Following DasGupta (2008, pp. 402–404), consider the pooled t CI for $\mu_1 - \mu_2$. Let $X_1, \ldots, X_{n_1}$ be iid with mean μ_1 and variance σ_1^2. Let $Y_1, \ldots, Y_{n_2}$ be iid with mean μ_2 and variance σ_2^2. Assume that the two samples are independent and that $n_i \to \infty$ for $i = 1, 2$ in such a way that $\hat{\rho} = \frac{n_1}{n_1+n_2} \to \rho \in (0,1)$. Let $\theta = \sigma_2^2/\sigma_1^2$, and let the pooled sample variance

$$S_p^2 = \frac{(n_1 - 1)S_1^2 + (n_2 - 1)S_2^2}{n_1 + n_2 - 2}.$$

Then

$$\begin{pmatrix} \sqrt{n_1}(\overline{X} - \mu_1) \\ \sqrt{n_2}(\overline{Y} - \mu_2) \end{pmatrix} \xrightarrow{D} N_2(\mathbf{0}, \boldsymbol{\Sigma})$$

where $\boldsymbol{\Sigma} = \text{diag}(\sigma_1^2, \sigma_2^2)$. Hence

$$\begin{pmatrix} \frac{1}{\sqrt{n_1}} & \frac{-1}{\sqrt{n_2}} \end{pmatrix} \begin{pmatrix} \sqrt{n_1}(\overline{X} - \mu_1) \\ \sqrt{n_2}(\overline{Y} - \mu_2) \end{pmatrix} = \overline{X} - \overline{Y} - (\mu_1 - \mu_2) \xrightarrow{D} N\left(0, \frac{\sigma_1^2}{n_1} + \frac{\sigma_2^2}{n_2}\right).$$

So

$$\frac{\overline{X} - \overline{Y} - (\mu_1 - \mu_2)}{\sqrt{\frac{S_1^2}{n_1} + \frac{S_2^2}{n_2}}} \xrightarrow{D} N(0,1).$$

Thus

$$\frac{\sqrt{\frac{S_1^2}{n_1} + \frac{S_2^2}{n_2}}}{S_p\sqrt{\frac{1}{n_1} + \frac{1}{n_2}}} \; \frac{\overline{X} - \overline{Y} - (\mu_1 - \mu_2)}{\sqrt{\frac{S_1^2}{n_1} + \frac{S_2^2}{n_2}}} = \frac{\overline{X} - \overline{Y} - (\mu_1 - \mu_2)}{S_p\sqrt{\frac{1}{n_1} + \frac{1}{n_2}}} \xrightarrow{D} N(0, \tau^2)$$

where

$$\frac{\frac{\sigma_1^2}{n_1} + \frac{\sigma_2^2}{n_2}}{\left(\frac{1}{n_1} + \frac{1}{n_2}\right)\frac{n_1\sigma_1^2 + n_2\sigma_2^2}{n_1+n_2}} = \frac{\frac{\sigma_1^2}{n_1} + \frac{\sigma_2^2}{n_2}}{\hat{\rho}\sigma_1^2 + (1-\hat{\rho})\sigma_2^2} \; \frac{1/\sigma_1^2}{1/\sigma_1^2} \; \frac{n_1 n_2}{n_1 + n_2}$$

$$= \frac{\frac{1}{n_1} + \frac{\theta}{n_2}}{\hat{\rho} + (1-\hat{\rho})\theta} \; \frac{n_1 n_2}{n_1 + n_2} \xrightarrow{D} \frac{1 - \rho + \rho\theta}{\rho + (1-\rho)\theta} = \tau^2.$$

Now let $\hat{\theta} = S_2^2/S_1^2$ and

$$\hat{\tau}^2 = \frac{1 - \hat{\rho} + \hat{\rho}\,\hat{\theta}}{\hat{\rho} + (1-\hat{\rho})\,\hat{\theta}}.$$

Notice that $\hat{\tau} = 1$ if $\hat{\rho} = 1/2$, and $\hat{\tau} = 1$ if $\hat{\theta} = 1$.

The usual large sample $(1-\alpha)100\%$ pooled t CI for $(\mu_1 - \mu_2)$ is

$$\overline{X} - \overline{Y} \pm t_{n_1+n_2-2,1-\alpha/2} \; S_p \sqrt{\frac{1}{n_1} + \frac{1}{n_2}}. \tag{9.32}$$

The large sample theory says that this CI is valid if $\tau = 1$, and that

$$\frac{\overline{X} - \overline{Y} - (\mu_1 - \mu_2)}{\hat{\tau} \, S_p \sqrt{\frac{1}{n_1} + \frac{1}{n_2}}} \xrightarrow{D} N(0,1).$$

Hence a large sample $(1-\alpha)100\%$ CI for $(\mu_1 - \mu_2)$ is

$$\overline{X} - \overline{Y} \pm z_{1-\alpha/2} \, \hat{\tau} \, S_p \sqrt{\frac{1}{n_1} + \frac{1}{n_2}}.$$

Then the large sample $(1-\alpha)100\%$ modified pooled t CI for $(\mu_1 - \mu_2)$ is

$$\overline{X} - \overline{Y} \pm t_{n_1+n_2-4,1-\alpha/2} \, \hat{\tau} \, S_p \sqrt{\frac{1}{n_1} + \frac{1}{n_2}}. \tag{9.33}$$

The large sample $(1-\alpha)100\%$ Welch CI for $(\mu_1 - \mu_2)$ is

$$\overline{X} - \overline{Y} \pm t_{d,1-\alpha/2} \sqrt{\frac{S_1^2}{n_1} + \frac{S_2^2}{n_2}} \tag{9.34}$$

where $d = \max(1, [d_0])$, and

$$d_0 = \frac{\left(\frac{S_1^2}{n_1} + \frac{S_2^2}{n_2}\right)^2}{\frac{1}{n_1 - 1}\left(\frac{S_1^2}{n_1}\right)^2 + \frac{1}{n_2 - 1}\left(\frac{S_2^2}{n_2}\right)^2}.$$

Suppose $n_1/(n_1+n_2) \to \rho$. It can be shown that if the CI length is multiplied by $\sqrt{n_1}$, then the scaled length of the pooled t CI converges in probability to $2z_{1-\alpha/2}\sqrt{\frac{\rho}{1-\rho}\sigma_1^2 + \sigma_2^2}$ while the scaled lengths of the modified pooled t CI and Welch CI both converge in probability to $2z_{1-\alpha/2}\sqrt{\sigma_1^2 + \frac{\rho}{1-\rho}\sigma_2^2}$.

9.3 Bootstrap and Randomization Tests

Randomization tests and bootstrap tests and confidence intervals are resampling algorithms used to provide information about the sampling distribution of a statistic $T_n \equiv T_n(F) \equiv T_n(\boldsymbol{Y}_n)$ where $\boldsymbol{Y}_n = (Y_1, \ldots, Y_n)$ and the Y_i are iid from a distribution with cdf $F(y) = P(Y \le y)$. Then T_n has a cdf $H_n(y) = P(T_n \le y)$. If $F(y)$ is known, then m independent samples $\boldsymbol{Y}^*_{j,n} = (Y^*_{j,1}, \ldots, Y^*_{j,n})$ of size n could be generated, where the $Y^*_{j,k}$ are iid from a distribution with cdf F and $j = 1, \ldots, m$. Then the statistic T_n is computed for each sample, resulting in m statistics $T_{1,n}(F), \ldots, T_{m,n}(F)$ which are iid from a distribution with cdf $H_n(y)$. Equivalent notation $T_{i,n}(F) \equiv T^*_{i,n}(\boldsymbol{Y}^*_{i,n})$ is often used, where $i = 1, \ldots, m$.

If $W_1, \ldots, W_m$ are iid from a distribution with cdf F_W, then the empirical cdf F_m corresponding to F_W is given by

$$F_m(y) = \frac{1}{m} \sum_{i=1}^{m} I(W_i \leq y)$$

where the indicator $I(W_i \leq y) = 1$ if $W_i \leq y$ and $I(W_i \leq y) = 0$ if $W_i > y$. Fix m and y. Then $mF_m(y) \sim$ binomial $(m, F_W(y))$. Thus $E[F_m(y)] = F_W(y)$ and $V[F_m(y)] = F_W(y)[1 - F_W(y)]/m$. By the central limit theorem,

$$\sqrt{m}(F_m(y) - F_W(y)) \xrightarrow{D} N(0, F_W(y)[1 - F_W(y)]).$$

Thus $F_m(y) - F_W(y) = O_P(m^{-1/2})$, and F_m is a reasonable estimator of F_W if the number of samples m is large.

Let $W_i = T_{i,n}(F)$. Then $F_m \equiv \tilde{H}_{m,n}$ is an empirical cdf corresponding to H_n. Let $W_i = Y_i$ and $m = n$. Then F_n is the empirical cdf corresponding to F. Let $\boldsymbol{y}_n = (y_1, \ldots, y_n)$ be the observed data. Now F_n is the cdf of the population that consists of $y_1, \ldots, y_n$ where the probability of selecting y_i is $1/n$. Hence an iid sample of size d from F_n is obtained by drawing a sample of size d with replacement from $y_1, \ldots, y_n$. If $d = n$, let $\boldsymbol{Y}^*_{j,n} = (Y^*_{j,1}, \ldots, Y^*_{j,n})$ be an iid sample of size n from the empirical cdf F_n. Hence each $Y^*_{j,k}$ is one of the $y_1, \ldots, y_n$ where repetition is allowed. Take m independent samples from F_n and compute the statistic T_n for each sample, resulting in m statistics $T_{1,n}(F_n), \ldots, T_{m,n}(F_n)$ where $T_{i,n}(F_n) \equiv T^*_{i,n}(\boldsymbol{Y}^*_{i,n})$ for $i = 1, \ldots, m$. This type of sampling can be done even if F is unknown, and if $T_n(F_n) \approx T_n(F)$, then the empirical cdf based on the $T_{i,n}(F_n)$ may be a useful approximation for H_n.

For general resampling algorithms let $T^*_{i,n}(\boldsymbol{Y}^*_{i,n})$ be the statistic based on a randomly chosen sample $\boldsymbol{Y}^*_{i,n}$ used by the resampling algorithm. Let $H_{A,n}$ be the cdf of the $T^*_{i,n}$ based on all J_n possible samples, and let $H_{m,n}$ be the cdf of the $T^*_{i,n}$ based on m randomly chosen samples. Often theoretical results are given for $H_{A,n}$ but are not known for $H_{m,n}$. Let $G_{N,n}$ be a cdf based on a normal approximation for H_n. Central limit type theorems are used and $G_{N,n}$ is often first order accurate: $H_n(y) - G_{N,n}(y) = O_P(n^{-1/2})$. Approximations $G_{E,n}$ based on the Edgeworth expansion (which is not a cdf) and $H_{A,n}$ are sometimes second order accurate: $H_n(y) - H_{A,n}(y) = O_P(n^{-1})$. The following two examples follow DasGupta (2008, pp. 462, 469, 513).

Example 9.24. Let $Y_1, \ldots, Y_n$ be iid with cdf F. Then the *ordinary bootstrap distribution* of T_n is $H_{A,n}(y) = P_{F_n}(T_n(\boldsymbol{Y}^*_{i,n}) \leq y)$ where $\boldsymbol{Y}^*_{i,n} = (Y^*_{i,1}, \ldots, Y^*_{i,n})$ is an iid sample of size n from the empirical cdf F_n obtained by selecting with replacement from $Y_1, \ldots, Y_n$. Here $T^*_{i,n}(\boldsymbol{Y}^*_{i,n}) = T_n(\boldsymbol{Y}^*_{i,n})$. Note that there are $J_n = n^n$ ordered samples and $n^n/n!$ unordered samples from F_n. The bootstrap distribution

$H_{m,n}$ typically used in practice is based on m samples randomly selected with replacement. Both $H_{A,n}$ and $H_{m,n}$ are estimators of H_n, the cdf of T_n.

For example, suppose the data is 1, 2, 3, 4, 5, 6, 7. Then $n = 7$ and the sample median T_n is 4. Using R, we drew $m = 2$ bootstrap samples (samples of size n drawn with replacement from the original data) and computed the sample median $T^*_{1,n} = 3$ and $T^*_{2,n} = 4$.

```
b1 <- sample(1:7,replace=T)
b1
[1] 3 2 3 2 5 2 6
median(b1)
[1] 3
b2 <- sample(1:7,replace=T)
b2
[1] 3 5 3 4 3 5 7
median(b2)
[1] 4
```

Heuristically, suppose $T_n(F_n)$ is an unbiased estimator of θ. Let $T^*_{i,n} = T^*_{i,n}(\boldsymbol{Y}^*_{i,n})$. Then $T^*_{1,n}, \ldots, T^*_{m,n}$ each gives an unbiased estimator of θ. If m is large, then typical values of $T^*_{i,n}$ should provide information about θ. For example, the middle 95 % of the T^*_i should be an approximate 95 % *percentile method* CI for θ. Then reject $H_0 : \theta = \theta_0$ if θ_0 is not in the CI. This bootstrap inference has two sources of error. First, n needs to be large enough so that $T_n(F_n) \approx T_n(F)$. Second, the $T^*_{i,n}$ are used to form an empirical cdf $H_{m,n}$ corresponding to $H_{A,n}$, so m needs to be large enough so that the empirical cdf $H_{m,n}$ is a good estimator of $H_{A,n}$.

Example 9.25. Let $X_1, \ldots, X_{k_1}$ be iid with pdf $f(y)$ while $Y_1, \ldots, Y_{k_2}$ are iid with pdf $f(y - \mu)$. Let $n = k_1 + k_2$ and consider testing $H_0 : \mu = 0$. Let $T_n \equiv T_{k_1,k_2}$ be the two sample t-statistic. Under H_0, the random variables in the combined sample $X_1, \ldots, X_{k_1}, Y_1, \ldots, Y_{k_2}$ are iid with pdf $f(y)$. Let $\boldsymbol{Z}_n$ be any permutation of $(X_1, \ldots, X_{k_1}, Y_1, \ldots, Y_{k_2})$ and compute $T_n(\boldsymbol{Z}_n)$ for each permutation. Then $H_{A,n}$ is the cdf based on all of the $T_n(\boldsymbol{Z}_n)$. H_0 is rejected if T_n is in the extreme tails of $H_{A,n}$. The number of ordered samples is $J_n = n!$ while the number of unordered samples is $\binom{n}{k_1}$. Such numbers get enormous quickly. Usually m randomly drawn permutations are selected with replacement, resulting in a cdf $H_{m,n}$ used to choose the appropriate cutoffs c_L and c_U.

For randomization tests that used a fixed number $m = B$ of permutations, calculations using binomial approximations suggest that $B = 999$ to 5,000 will give a test similar to those based on using all permutations. See Efron and Tibshirani (1993, pp. 208–210). Jöckel (1986) shows, under regularity conditions, that the power of a randomization test is increasing and converges as $m \to \infty$. It is suggested that the tests have good power if $m = 999$, but the p-value of such a test is bounded below by 0.001 since the p-value = (1 + the number of the m test statistics at least as extreme as the observed statistic)$/(m+1)$. Buckland (1984) shows that the expected coverage of the nominal $100(1-\alpha)\%$ percentile method confidence interval is approximately correct, but the standard deviation of the coverage is proportional to $1/\sqrt{m}$. Hence the percentile method is a large sample confidence interval, in that the true coverage converges in probability to the nominal coverage, only if $m \to \infty$ as $n \to \infty$. These results are good reasons for using $m = \max(B, [n\log(n)])$ samples.

The key observation for theory is that $H_{m,n}$ is an empirical cdf. To see this claim, recall that $H_{A,n}(y) \equiv H_{A,n}(y|\boldsymbol{Y}_n)$ is a random cdf: it depends on the data $\boldsymbol{Y}_n$. Hence $H_{A,n}(y) \equiv H_{A,n}(y|\boldsymbol{y}_n)$ is the observed cdf based on the observed data. $H_{A,n}(y|\boldsymbol{y}_n)$ can be computed by finding $T_{i,n}^*(\boldsymbol{Y}_{i,n}^*)$ for all J_n possible samples $\boldsymbol{Y}_{i,n}^*$. If m samples are selected with replacement from all possible samples, then the samples are iid and $T_{1,n}^*, \ldots, T_{m,n}^*$ are iid with cdf $H_{A,n}(y|\boldsymbol{y}_n)$. Hence $F_m \equiv H_{m,n}$ is an empirical cdf corresponding to $F \equiv H_{A,n}(y|\boldsymbol{y}_n)$.

Thus empirical cdf theory can be applied to $H_{m,n}$. Fix n and y. Then $mH_{m,n}(y) \sim$ binomial $(m, H_{A,n}(y|\boldsymbol{y}_n))$. Thus $E[H_{m,n}(y)] = H_{A,n}(y|\boldsymbol{y}_n)$ and $V[H_{m,n}(y)] = H_{A,n}(y|\boldsymbol{y}_n)[1-H_{A,n}(y|\boldsymbol{y}_n)]/m$. Also

$$\sqrt{m}(H_{m,n}(y) - H_{A,n}(y|\boldsymbol{y}_n)) \xrightarrow{D} N(0, H_{A,n}(y|\boldsymbol{y}_n)[1-H_{A,n}(y|\boldsymbol{y}_n)]).$$

Thus $H_{m,n}(y) - H_{A,n}(y|\boldsymbol{y}_n) = O_P(m^{-1/2})$. Note that the probabilities and expectations depend on m and on the observed data $\boldsymbol{y}_n$.

This result suggests that if $H_{A,n}$ is a first order accurate estimator of H_n, then $H_{m,n}$ cannot be a first order accurate estimator of H_n unless m is proportional to n. If $m = \max(1{,}000, [n\log(n)])$, then $H_{m,n}$ is asymptotically equivalent to $H_{A,n}$ up to terms of order $n^{-1/2}$. If $m = \max(1{,}000, [0.1n^2\log(n)])$, then $H_{m,n}$ asymptotically equivalent to $H_{A,n}$ up to terms of order n^{-1}.

As an application, Efron and Tibshirani (1993, pp. 187, 275) state that percentile method for bootstrap confidence intervals is first order accurate and that the coefficient of variation of a bootstrap percentile is proportional to $\sqrt{\frac{1}{n}+\frac{1}{m}}$. If $m = 1{,}000$, then the percentile bootstrap is not first order accurate. If $m = \max(1{,}000, [n\log(n)])$, then the percentile bootstrap is first order accurate. Similarly, claims that a bootstrap method is second order accurate are false unless m is proportional to n^2. See a similar result in Robinson (1988).

Practical resampling algorithms often use $m = B = 1{,}000$, 5,000, or 10,000. The choice of $m = 10{,}000$ works well for small n and for simulation studies since the cutoffs based on $H_{m,n}$ will be close to those based on $H_{A,n}$ with high probability since $V[H_{10{,}000,n}(y)] \le 1/40{,}000$. For the following theorem, also see Serfling (1980, pp. 59–61).

Theorem 9.1. Let $\boldsymbol{Y}_1, \ldots, \boldsymbol{Y}_n$ be iid $k \times 1$ random vectors from a distribution with cdf $F(\boldsymbol{y}) = P(Y_1 \le y_1, \ldots, Y_k \le y_k)$. Let

$$D_n = \sup_{\boldsymbol{y} \in \mathbb{R}^k} |F_n(\boldsymbol{y}) - F(\boldsymbol{y})|.$$

a) Massart (1990) $k=1$: $P(D_n > d) \le 2\exp(-2nd^2)$ if $nd^2 \ge 0.5\log(2)$.

b) Kiefer (1961) $k \geq 2 : P(D_n > d) \leq C\exp(-(2-\varepsilon)nd^2)$ where $\varepsilon > 0$ is fixed and the positive constant C depends on ε and k but not on F.

To use Theorem 9.1a, fix n (and suppressing the dependence on $\boldsymbol{y}_n$), take $F = H_{A,n}$ computed from the observed data and take $F_m = H_{m,n}$. Then

$$D_m = \sup_{y \in \mathbb{R}} |H_{m,n}(y) - H_{A,n}(y)|.$$

Recalling that the probability is with respect to the observed data, consider the following choices of m.

i) If $m = 10{,}000$, then $P(D_m > 0.01) \leq 2e^{-2} \approx 0.271$.

ii) If $m = \max(10{,}000, [0.25n\log(n)])$, then for $n > 5{,}000$

$P\left(D_m > \frac{1}{\sqrt{n}}\right) \leq 2\exp(-2[0.25n\log(n)]/n) \approx 2/\sqrt{n}.$

iii) If $m = \max(10{,}000, [0.5n^2\log(n)])$, then for $n > 70$

$P\left(D_m > \frac{1}{n}\right) \leq 2\exp(-2[0.5n^2\log(n)]/n^2) \approx 2/n.$

Two tail tests with nominal level α and confidence intervals with nominal coverage $1-\alpha$ tend to use the lower and upper $\alpha/2$ percentiles from $H_{m,n}$. This procedure corresponds to an interval covering $100(1-\alpha)\%$ of the mass. The interval is short if the distribution corresponding to $H_{m,n}$ is approximately symmetric. Skewness or approximate symmetry can be checked by plotting the $T^*_{i,n}$. Shorter intervals can be found if the distribution is skewed by using the shorth(c) estimator where $c = \lceil m(1-\alpha) \rceil$ and $\lceil x \rceil$ is the smallest integer $\geq x$, e.g., $\lceil 7.7 \rceil = 8$. See Grübel (1988). That is, let $T^*_{(1)}, \ldots, T^*_{(m)}$ be the order statistics of the $T^*_{1,n}, \ldots, T^*_{m,n}$ computed by the resampling algorithm. Compute $T^*_{(c)} - T^*_{(1)}, T^*_{(c+1)} - T^*_{(2)}, \ldots, T^*_{(m)} - T^*_{(m-c+1)}$. Let $[T^*_{(s)}, T^*_{(s+c-1)}]$ correspond to the closed interval with the smallest distance. Then reject $H_0 : \theta = \theta_0$ if θ_0 is not in the interval.

Resampling methods can be used in courses on resampling methods, nonparametric statistics, and experimental design. In such courses it can be stated that it is well known that $H_{m,n}$ has good statistical properties (under regularity conditions) if $m \to \infty$ as $n \to \infty$, but algorithms tend to use $m = B$ between 999 and 10,000. Such algorithms may perform well in simulations, but lead to tests with p-value bounded away from 0, confidence intervals with coverage that fails to converge to the nominal coverage, and fail to take advantage of the theory derived for the impractical all subset algorithms. Since $H_{m,n}$ is the empirical cdf corresponding to the all subset algorithm cdf $H_{A,n}$, taking $m = \max(B, [n\log(n)])$ leads to a practical algorithm with good theoretical properties (under regularity conditions) that performs well in simulations.

Although theory for resampling algorithms given in Lehmann (1999, p. 425) and Sen and Singer (1993, p. 365) has the number of samples $m \to \infty$ as the sample size $n \to \infty$, much of the literature suggests using $m = B$ between 999 and 10,000. This choice is often justified using simulations and binomial approximations. An exception is Shao (1989) where $n/m \to 0$ as $n \to \infty$. Let $[x]$ be the integer part of x, so $[7.7] = 7$. Then $m = [n^{1.01}]$ may give poor results for $n < 900$. To combine theory with empirical results, we suggest using $m = \max(B, [n\log(n)])$.

Theory for resampling algorithms such as first order accuracy of the bootstrap and the power of randomization tests is usually for the impractical algorithm that uses all J_n samples. Practical algorithms use B randomly drawn samples where B is chosen to give good performance when n is small. We suggest using $m = \max(B, [n\log(n)])$ randomly drawn samples results in a practical algorithm that is asymptotically equivalent to the impractical algorithm up to terms of order $n^{-1/2}$ while also having good small sample performance.

Example 9.26. Suppose F is the cdf of the $N(\mu, \sigma^2)$ distribution and $T_n(F) = \overline{Y}_n \sim N(\mu, \sigma^2/n)$. Suppose m independent samples $(Y^*_{j,1}, \ldots, Y^*_{j,n}) = \boldsymbol{Y}^*_{j,n}$ of size n are generated, where the $Y^*_{j,k}$ are iid $N(\mu, \sigma^2)$ and $j = 1, \ldots, m$. Then let the sample mean $T^*_{j,n} = \overline{Y}^*_{j,n} \sim N(\mu, \sigma^2/n)$ for $j = 1, \ldots, m$.

We want to examine, for a given m and n, how well do the sample quantiles $T^*_{(\lceil m\,\rho \rceil)} = \overline{Y}^*_{(\lceil m\,\rho \rceil),n}$ of the $\overline{Y}^*_{j,n}$ estimate the quantiles $\xi_{\rho,n}$ of the $N(\mu, \sigma^2/n)$ distribution and how well does $(T^*_{(\lceil m\,0.025 \rceil)}, T^*_{(\lceil m\,0.975 \rceil)})$ perform as a 95 % CI for μ. Here $P(X \leq \xi_{\rho,n}) = \rho$ if $X \sim N(\mu, \sigma^2/n)$. Note that $\xi_{\rho,n} = \mu + z_\rho \sigma/\sqrt{n}$ where $P(Z \leq z_\rho) = \rho$ if $Z \sim N(0,1)$.

Fix n and let f_n be the pdf of the $N(\mu, \sigma^2/n)$ distribution. By Theorem 8.27, as $m \to \infty$

$$\sqrt{m}(\overline{Y}^*_{(\lceil m\,\rho \rceil),n} - \xi_{\rho,n}) \xrightarrow{D} N(0, \tau_n^2)$$

where

$$\tau_n^2 \equiv \tau_n^2(\rho) = \frac{\rho(1-\rho)}{[f_n(\xi_\rho)]^2} = \frac{\rho(1-\rho)2\pi\sigma^2}{n\exp(-z_\rho^2)}.$$

Since the quantile $\xi_{\rho,n} = \mu + z_\rho \sigma/\sqrt{n}$, need m fairly large for the estimated quantile to be good. To see this claim, suppose we want m so that

$$P(\xi_{0.975,n} - 0.04\sigma/\sqrt{n} < \overline{Y}^*_{(\lceil m\,0.975 \rceil),n} < \xi_{0.975,n} - 0.04\sigma/\sqrt{n}) > 0.9.$$

(For $N(0,1)$ data, this would be similar to wanting the estimated 0.975 quantile to be between 1.92 and 2.00 with high probability.) Then $0.9 \approx$

$$P\left(\frac{-0.04\sigma\sqrt{m}}{\tau_n\sqrt{n}} < Z < \frac{0.04\sigma\sqrt{m}}{\tau_n\sqrt{n}}\right) \approx P(-0.01497\sqrt{m} < Z < 0.01497\sqrt{m}\,)$$

or

$$m \approx \left(\frac{z_{0.05}}{-0.01497}\right)^2 \approx 12076.$$

With $m = 1{,}000$, the above probability is only about 0.36. To have the probability go to one, need $m \to \infty$ as $n \to \infty$.

Note that if $m = B = 1{,}000$, say, then the sample quantile is not a consistent estimator of the population quantile $\xi_{\rho,n}$. Also, $(\overline{Y}^*_{(\lceil m\,\rho \rceil),n} - \xi_{\rho,n}) = O_P(n^{-\delta})$ needs

$m \propto n^{2\delta}$ where $\delta = 1/2$ or 1 are the most interesting cases. For good simulation results, typically need m larger than a few hundred, e.g., $B = 1{,}000$, for small n. Hence $m = \max(B, [n\log(n)])$ combines theory with good simulation results.

The CI length behaves fairly well for large n. For example, the 95 % CI length will be close to $3.92/\sqrt{n}$ since roughly 95 % of the $\overline{Y}^*_{j,n}$ are between $\mu - 1.96\sigma/\sqrt{n}$ and $\mu + 1.96\sigma/\sqrt{n}$. The coverage is conservative (higher than 95 %) for moderate m. To see this, note that the 95 % CI contains μ if $T^*_{(\lceil m\ 0.025 \rceil)} < \mu$ and $T^*_{(\lceil m\ 0.975 \rceil)} > \mu$. Let $W \sim$ binomial $(m, 0.5)$. Then

$$P(T^*_{(\lceil m\ 0.975 \rceil)}) > \mu) \approx P(W > 0.025m) \approx P\left(Z > \frac{0.025m - 0.5m}{0.5\sqrt{m}}\right) =$$

$P(Z > -0.95\sqrt{m}) \to 1$ as $m \to \infty$. (Note that if $m = 1{,}000$, then $T^*_{(\lceil m\ 0.975 \rceil)} > \mu$ if 225 or more $\overline{Y}^*_{j,n} > \mu$ or if fewer than 975 $\overline{Y}^*_{j,n} < \mu$.)

Since F is not known, we cannot sample from $T_n(F)$, but sampling from $T_n(F_n)$ can at least be roughly approximated using computer-generated random numbers. The bootstrap replaces m samples from $T_n(F)$ by m samples from $T_n(F_n)$, that is, there is a single sample $Y_1, \ldots, Y_n$ of data. Take a sample of size n with replacement from $Y_1, \ldots, Y_n$ and compute the sample mean $\overline{Y}^*_{1,n}$. Repeat to obtain the bootstrap sample $\overline{Y}^*_{1,n}, \ldots, \overline{Y}^*_{m,n}$. Expect the bootstrap estimator of the quantile to perform less well than that based on samples from $T_n(F)$. So still need m large so that the estimated quantiles are near the population quantiles.

Simulated coverage for the bootstrap percentile 95 % CI tends to be near 0.95 for moderate m, and we expect the length of the 95 % CI to again be near $3.92/\sqrt{n}$. The bootstrap sample tends to be centered about the observed value of $\overline{Y}$. If there is a "bad sample" so that $\overline{Y}$ is in the left or right tail of the sampling distribution, say $\overline{Y} > \mu + 1.96\sigma/\sqrt{n}$ or $\overline{Y} < \mu - 1.96\sigma/\sqrt{n}$, then the coverage may be much less that 95 %. But the probability of a "bad sample" is 0.05 for this example.

9.4 Complements

Guenther (1969) is a useful reference for confidence intervals. Agresti and Coull (1998), Brown et al. (2001, 2002), and Pires and Amado (2008) discuss CIs for a binomial proportion. Agresti and Caffo (2000) discuss CIs for the difference of two binomial proportions $\rho_1 - \rho_2$ obtained from two independent samples. Barker (2002), Byrne and Kabaila (2005), Garwood (1936), and Swift (2009) discuss CIs for Poisson (θ) data. Brown et al. (2003) discuss CIs for several discrete exponential families. Abuhassan and Olive (2008) consider CIs for some transformed random variables. Also see Brownstein and Pensky (2008).

A comparison of CIs with other intervals (such as prediction intervals) is given in Vardeman (1992). Also see Frey (2013), Olive (2013), Hahn and Meeker (1991), and Krishnamoorthy and Mathew (2009). Frey (2013) and Olive (2013) note that the shorth intervals are too short (liberal) if the number of bootstrap samples B is not large, and suggest small sample correction factors.

Newton's method is described, for example, in Peressini et al. (1988, p. 85).

9.5 Problems

PROBLEMS WITH AN ASTERISK * ARE ESPECIALLY USEFUL.

Refer to Chap. 10 for the pdf or pmf of the distributions in the problems below.

9.1^Q. Suppose that $X_1,\ldots,X_n$ are iid with the Weibull distribution, that is the common pdf is

$$f(x) = \begin{cases} \frac{b}{a} x^{b-1} e^{-\frac{x^b}{a}} & 0 < x \\ 0 & \text{elsewhere} \end{cases}$$

where a is the unknown parameter, but $b(> 0)$ is assumed known.

a) Find a minimal sufficient statistic for a.

b) Assume $n = 10$. Use the chi-square table and the minimal sufficient statistic to find a 95 % two-sided confidence interval for a.

R Problems

Use a command like *source("G:/sipack.txt")* **to download the functions**. See Sect. 12.1. Typing the name of the `sipack` function, e.g., *accisimf*, will display the code for the function. Use the `args` command, e.g., *args(accisimf)*, to display the needed arguments for the function.

9.2. Let $Y_1,\ldots,Y_n$ be iid binomial$(1,\rho)$ random variables.

From the website (http://lagrange.math.siu.edu/Olive/sipack.txt), enter the *R* function `bcisim` into *R*. This function simulates the three CIs (classical, Agresti Coull, and exact) from Example 9.16, but changes the CI (L,U) to (max(0,L),min(1,U)) to get shorter lengths.

To run the function for $n = 10$ and $\rho \equiv p = 0.001$, enter the *R* command `bcisim(n=10,p=0.001)`. Make a table with header "n p ccov clen accov aclen ecov elen." Fill the table for $n = 10$ and $p = 0.001, 0.01, 0.5, 0.99, 0.999$ and then repeat for $n = 100$. The "cov" is the proportion of 500 runs where the CI contained p and the nominal coverage is 0.95. A coverage between 0.92 and 0.98 gives little evidence that the true coverage differs from the nominal coverage of 0.95. A coverage greater that 0.98 suggests that the CI is conservative while a coverage less than 0.92 suggests that the CI is liberal. Typically want the true coverage $\geq$ the nominal coverage, so conservative intervals are better than liberal CIs. The "len" is the average scaled length of the CI and for large n should be near $2(1.96)\sqrt{p(1-p)}$.

From your table, is the classical estimator or the Agresti Coull CI better? When is the "exact" interval good? Explain briefly.

9.3. Let $X_1, \ldots, X_n$ be iid Poisson(θ) random variables.

From the website (http://lagrange.math.siu.edu/Olive/sipack.txt), enter the *R* function `poiscisim` into *R*. This function simulates the three CIs (classical, modified, and exact) from Example 9.15. To run the function for $n = 100$ and $\theta = 5$, enter the *R* command `poiscisim(theta=5)`. Make a table with header "theta ccov clen mcov mlen ecov elen." Fill the table for $theta = 0.001, 0.1, 1.0$, and 5.

The "cov" is the proportion of 500 runs where the CI contained θ and the nominal coverage is 0.95. A coverage between 0.92 and 0.98 gives little evidence that the true coverage differs from the nominal coverage of 0.95. A coverage greater that 0.98 suggests that the CI is conservative while a coverage less than 0.92 suggests that the CI is liberal (too short). Typically want the true coverage $\geq$ the nominal coverage, so conservative intervals are better than liberal CIs. The "len" is the average scaled length of the CI and for large $n\theta$ should be near $2(1.96)\sqrt{\theta}$ for the classical and modified CIs.

From your table, is the classical CI or the modified CI or the "exact" CI better? Explain briefly. (Warning: in a 1999 version of *R*, there was a bug for the Poisson random number generator for $\theta \geq 10$. The 2011 version of *R* seems to work.)

9.4. This problem simulates the CIs from Example 9.17.

a) Download the function `accisimf` into *R*.

b) The function will be used to compare the classical, ACT, and modified 95 % CIs when the population size $N = 500$ and p is close to 0.01. The function generates such a population, then selects 5,000 independent simple random samples from the population. The 5,000 CIs are made for both types of intervals, and the number of times the true population p is in the ith CI is counted. The simulated coverage is this count divided by 5,000 (the number of CIs). The nominal coverage is 0.95. To run the function for $n = 50$ and $p \approx 0.01$, enter the command `accisimf(n=50,p=0.01)`. Make a table with header "n p ccov clen accov aclen mcov mlen." Fill the table for $n = 50$ and then repeat for $n = 100, 150, 200, 250, 300, 350, 400$, and 450. The "len" is $\sqrt{n}$ times the mean length from the 5,000 runs. The "cov" is the proportion of 5,000 runs where the CI contained p and the nominal coverage is 0.95. For 5,000 runs, an observed coverage between 0.94 and 0.96 gives little evidence that the true coverage differs from the nominal coverage of 0.95. A coverage greater that 0.96 suggests that the CI is conservative while a coverage less than 0.94 suggests that the CI is liberal. Typically want the true coverage $\geq$ the nominal coverage, so conservative intervals are better than liberal CIs. The "ccov" is for the classical CI, "accov" is for the Agresti Coull type (ACT) CI, and "mcov" is for the modified interval. Given good coverage > 0.94, want short length.

c) First compare the classical and ACT intervals. From your table, for what values of n is the ACT CI better, for what values of n are the three intervals about the same, and for what values of n is the classical CI better?

d) Was the modified CI ever good?

9.5. This problem simulates the CIs from Example 9.10.

a) Download the function `expsim` into *R*.

The output from this function are the coverages scov, lcov, and ccov of the CI for λ, θ and of λ if θ is known. The scaled average lengths of the CIs are also given. The lengths of the CIs for λ are multiplied by $\sqrt{n}$ while the length of the CI for θ is multiplied by n.

b) The 5,000 CIs are made for three intervals, and the number of times the true population parameter λ or θ is in the ith CI is counted. The simulated coverage is this count divided by 5,000 (the number of CIs). The nominal coverage is 0.95. To run the function for $n = 5$, $\theta = 0$ and $\lambda = 1$ enter the command `expsim(n=5)`. Make a table with header

"CI for λ CI for θ CI for λ, θ known."

Then make a second header "n cov slen cov slen cov slen" where "cov slen" is below each of the three CI headers. Fill the table for $n = 5$ and then repeat for $n = 10, 20, 50, 100$, and 1,000. The "cov" is the proportion of 5,000 runs where the CI contained λ or θ and the nominal coverage is 0.95. For 5,000 runs, an observed coverage between 0.94 and 0.96 gives little evidence that the true coverage differs from the nominal coverage of 0.95. A coverage greater that 0.96 suggests that the CI is conservative, while a coverage less than 0.94 suggests that the CI is liberal. As n gets large, the values of slen should get closer to 3.92, 2.9957, and 3.92.

9.6. This problem simulates the CIs from Example 9.9.

a) Download the function `hnsim` into *R*.

The output from this function are the coverages scov, lcov, and ccov of the CI for σ^2, μ and of σ^2 if μ is known. The scaled average lengths of the CIs are also given. The lengths of the CIs for σ^2 are multiplied by $\sqrt{n}$, while the length of the CI for μ is multiplied by n.

b) The 5,000 CIs are made for three intervals, and the number of times the true population parameter $\theta = \mu$ or σ^2 is in the ith CI is counted. The simulated coverage is this count divided by 5,000 (the number of CIs). The nominal coverage is 0.95. To run the function for $n = 5$, $\mu = 0$ and $\sigma^2 = 1$ enter the command `hnsim(n=5)`. Make a table with header

"CI for σ^2 CI for μ CI for σ^2, μ known."

Then make a second header "n cov slen cov slen cov slen" where "cov slen" is below each of the three CI headers. Fill the table for $n = 5$ and then repeat for $n = 10, 20, 50, 100$, and 1,000. The "cov" is the proportion of 5,000 runs where the CI contained θ and the nominal coverage is 0.95. For 5,000 runs, an observed coverage between 0.94 and 0.96 gives little evidence that the true coverage differs from the nominal coverage of 0.95. A coverage greater that 0.96 suggests that the

CI is conservative while a coverage less than 0.94 suggests that the CI is liberal. As n gets large, the values of slen should get closer to 5.5437, 3.7546, and 5.5437.

9.7. a) Download the function `wcisim` into *R*.

The output from this function includes the coverages `pcov` and `lcov` of the CIs for ϕ and λ if the simulated data $Y_1, \dots, Y_n$ are iid Weibull (ϕ, λ). The scaled average lengths of the CIs are also given. The values `pconv` and `lconv` should be less than 10^{-5}. If this is not the case, increase `iter`. 100 samples of size $n = 100$ are used to create the 95 % large sample CIs for ϕ and λ given in Example 9.18. If the sample size is large, then `sdphihat`, the sample standard deviation of the 100 values of the MLE $\hat{\phi}$, should be close to `phiasd` $= \phi\sqrt{.608}$. Similarly, `sdlamhat` should be close to the asymptotic standard deviation `lamasd` $= \sqrt{1.109\lambda^2(1 + 0.4635\log(\lambda) + 0.5282(\log(\lambda))^2)}$.

b) Type the command
`wcisim(n = 100, phi = 1, lam = 1, iter = 100)`
and record the coverages for the CIs for ϕ and λ.

c) Type the command
`wcisim(n = 100, phi = 20, lam = 20, iter = 100)`
and record the coverages for the CIs for ϕ and λ.

9.8. a) Download the function `raysim` into *R* to simulate the CI of Example 9.19.

b) Type the command
`raysim(n = 100, mu = 20, sigma = 20, iter = 100)`
and record the coverages for the CIs for μ and σ.

9.9. a) Download the function `ducisim` into *R* to simulate the CI of Example 9.20.

b) Type the command
`ducisim(n=10,nruns=1000,eta=1000).`
Repeat for $n = 50, 100, 500$ and make a table with header
"n coverage n 95 % CI length."
Fill in the table for $n = 10, 50, 100$ and 500.

c) Are the coverages close to or higher than 0.95 and is the scaled length close to $3\eta = 3{,}000$?

9.10. a) Download the function `varcisim` into *R* to simulate a modified version of the CI of Example 9.22.

b) Type the command `varcisim(n = 100, nruns = 1000, type = 1)` to simulate the 95 % CI for the variance for iid N(0,1) data. Is the coverage *vcov* close to or higher than 0.95? Is the scaled length $vlen = \sqrt{n}$ (CI length) $= 2(1.96)\sigma^2\sqrt{\tau} = 5.554\sigma^2$ close to 5.554?

c) Type the command `varcisim(n = 100, nruns = 1000, type = 2)` to simulate the 95 % CI for the variance for iid EXP(1) data. Is the coverage *vcov* close to or higher than 0.95? Is the scaled length $vlen = \sqrt{n}$ (CI length) $= 2(1.96)\sigma^2\sqrt{\tau} = 2(1.96)\lambda^2\sqrt{8} = 11.087\lambda^2$ close to 11.087?

d) Type the command `varcisim(n = 100, nruns = 1000, type = 3)` to simulate the 95 % CI for the variance for iid LN(0,1) data. Is the coverage *vcov* close to or higher than 0.95? Is the scaled length *vlen* long?

9.11. a) Download the function `pcisim` into *R* to simulate the three CIs of Example 9.23. The modified pooled *t* CI is almost the same as the Welch CI, but uses degrees of freedom $= n_1 + n_2 - 4$ instead of the more complicated formula for the Welch CI. The pooled *t* CI should have coverage that is too low if

$$\frac{\rho}{1-\rho}\sigma_1^2 + \sigma_2^2 < \sigma_1^2 + \frac{\rho}{1-\rho}\sigma_2^2.$$

b) Type the command `pcisim(n1=100,n2=200,var1=10,var2=1)` to simulate the CIs for $N(\mu_i, \sigma_i^2)$ data for $i = 1,2$. The terms *pcov*, *mpcov*, and *wcov* are the simulated coverages for the pooled, modified pooled, and Welch 95 % CIs. Record these quantities. Are they near 0.95?

Problems from old qualifying exams are marked with a Q.

9.12^Q. Let $X_1,\ldots,X_n$ be a random sample from a uniform$(0,\theta)$ distribution. Let $Y = \max(X_1, X_2, \ldots, X_n)$.

a) Find the pdf of Y/θ.

b) To find a confidence interval for θ, can Y/θ be used as a pivot?

c) Find the shortest $(1-\alpha)\%$ confidence interval for θ.

9.13. Let $Y_1,\ldots,Y_n$ be iid from a distribution with fourth moments and let S_n^2 be the sample variance. Then

$$\sqrt{n}(S_n^2 - \sigma^2) \xrightarrow{D} N(0, M_4 - \sigma^4)$$

where M_4 is the fourth central moment $E[(Y-\mu)^4]$. Let $\hat{M}_{4,n} = \frac{1}{n}\sum_{i=1}^n (Y_i - \overline{Y})^4$.

a) Use the asymptotic pivot

$$\frac{\sqrt{n}(S_n^2 - \sigma^2)}{\sqrt{\hat{M}_{4,n} - S_n^4}} \xrightarrow{D} N(0,1)$$

to find a large sample $100(1-\alpha)\%$ CI for σ^2.

b) Use equation (9.4) to find a large sample $100(1-\alpha)\%$ CI for $\sigma_1^2 - \sigma_2^2$.

Chapter 10
Some Useful Distributions

This chapter contains many useful examples of parametric distributions, one- and two-parameter exponential families, location–scale families, maximum likelihood estimators, method of moment estimators, transformations $t(Y)$, $E(Y)$, $V(Y)$, moment generating functions, and confidence intervals. Many of the distributions can be used to create exam questions on the above topics as well as the kernel method, MSE, and hypothesis testing. Using the population median and median absolute deviation, robust estimators of parameters can often be found using the sample median and median absolute deviation.

Definition 10.1. The *population median* is any value $\text{MED}(Y)$ such that

$$P(Y \leq \text{MED}(Y)) \geq 0.5 \text{ and } P(Y \geq \text{MED}(Y)) \geq 0.5. \tag{10.1}$$

Definition 10.2. The *population median absolute deviation* is

$$\text{MAD}(Y) = \text{MED}(|Y - \text{MED}(Y)|). \tag{10.2}$$

Finding $\text{MED}(Y)$ and $\text{MAD}(Y)$ for symmetric distributions and location–scale families is made easier by the following lemma. Let $F(y_\alpha) = P(Y \leq y_\alpha) = \alpha$ for $0 < \alpha < 1$ where the cdf $F(y) = P(Y \leq y)$. Let $D = \text{MAD}(Y)$, $M = \text{MED}(Y) = y_{0.5}$ and $U = y_{0.75}$.

Lemma 10.1. a) If $W = a + bY$, then $\text{MED}(W) = a + b\text{MED}(Y)$ and $\text{MAD}(W) = |b|\text{MAD}(Y)$.

b) If Y has a pdf that is continuous and positive on its support and symmetric about μ, then $\text{MED}(Y) = \mu$ and $\text{MAD}(Y) = y_{0.75} - \text{MED}(Y)$. Find $M = \text{MED}(Y)$ by solving the equation $F(M) = 0.5$ for M, and find U by solving $F(U) = 0.75$ for U. Then $D = \text{MAD}(Y) = U - M$.

c) Suppose that W is from a location–scale family with standard pdf $f_Y(y)$ that is continuous and positive on its support. Then $W = \mu + \sigma Y$ where $\sigma > 0$. First find M by solving $F_Y(M) = 0.5$. After finding M, find D by solving $F_Y(M + D) - F_Y(M - D) = 0.5$. Then $\text{MED}(W) = \mu + \sigma M$ and $\text{MAD}(W) = \sigma D$.

D.J. Olive, *Statistical Theory and Inference*, DOI 10.1007/978-3-319-04972-4_10,

Definition 10.3. The *gamma function* $\Gamma(x) = \int_0^\infty t^{x-1}e^{-t}dt$ for $x > 0$.

Some properties of the gamma function follow.
i) $\Gamma(k) = (k-1)!$ for integer $k \geq 1$.
ii) $\Gamma(x+1) = x\,\Gamma(x)$ for $x > 0$.
iii) $\Gamma(x) = (x-1)\,\Gamma(x-1)$ for $x > 1$.
iv) $\Gamma(0.5) = \sqrt{\pi}$.

Some lower case Greek letters are alpha: α, beta: β, gamma: γ, delta: δ, epsilon: ε, zeta: ζ, eta: η, theta: θ, iota: ι, kappa: κ, lambda: λ, mu: μ, nu: ν, xi: ξ, omicron: o, pi: π, rho: ρ, sigma: σ, upsilon: υ, phi: ϕ, chi: χ, psi: ψ, and omega: ω.

Some capital Greek letters are gamma: Γ, theta: Θ, sigma: Σ, and phi: Φ.

For the discrete uniform and geometric distributions, the following facts on series are useful.

Lemma 10.2. Let n, n_1, and n_2 be integers with $n_1 \leq n_2$, and let a be a constant. Notice that $\sum_{i=n_1}^{n_2} a^i = n_2 - n_1 + 1$ if $a = 1$.

$$a)\ \sum_{i=n_1}^{n_2} a^i = \frac{a^{n_1} - a^{n_2+1}}{1-a}, \quad a \neq 1.$$

$$b)\ \sum_{i=0}^{\infty} a^i = \frac{1}{1-a}, \quad |a| < 1.$$

$$c)\ \sum_{i=1}^{\infty} a^i = \frac{a}{1-a}, \quad |a| < 1.$$

$$d)\ \sum_{i=n_1}^{\infty} a^i = \frac{a^{n_1}}{1-a}, \quad |a| < 1.$$

$$e)\ \sum_{i=1}^{n} i = \frac{n(n+1)}{2}.$$

$$f)\ \sum_{i=1}^{n} i^2 = \frac{n(n+1)(2n+1)}{6}.$$

See Gabel and Roberts (1980, pp. 473–476) for the proof of a)–d).

For the special case of $0 \leq n_1 \leq n_2$, notice that

$$\sum_{i=0}^{n_2} a^i = \frac{1-a^{n_2+1}}{1-a}, \quad a \neq 1.$$

To see this, multiply both sides by $(1-a)$. Then

$$(1-a)\sum_{i=0}^{n_2} a^i = (1-a)(1+a+a^2+\cdots+a^{n_2-1}+a^{n_2}) =$$

$$1 + a + a^2 + \cdots + a^{n_2 - 1} + a^{n_2}$$
$$- a - a^2 - \cdots - a^{n_2} - a^{n_2+1}$$

$= 1 - a^{n_2+1}$ and the result follows. Hence for $a \neq 1$,

$$\sum_{i=n_1}^{n_2} a^i = \sum_{i=0}^{n_2} a^i - \sum_{i=0}^{n_1-1} a^i = \frac{1 - a^{n_2+1}}{1-a} - \frac{1 - a^{n_1}}{1-a} = \frac{a^{n_1} - a^{n_2+1}}{1-a}.$$

The binomial theorem below is sometimes useful.

Theorem 10.3, The Binomial Theorem. For any real numbers x and y and for any integer $n \geq 0$,

$$(x+y)^n = \sum_{i=0}^{n} \binom{n}{i} x^i y^{n-i} = (y+x)^n = \sum_{i=0}^{n} \binom{n}{i} y^i x^{n-i}.$$

For the following theorem, see Marshall and Olkin (2007, pp. 15–16). Note that part b) follows from part a).

Theorem 10.4. If $E(X)$ exists, then
a) $E(X) = \int_0^\infty (1 - F(x))\, dx - \int_{-\infty}^0 F(x)\, dx$.
b) If $F(x) = 0$ for $x < 0$, then $E(X) = \int_0^\infty (1 - F(x))\, dx$.

10.1 The Beta Distribution

If Y has a beta distribution, $Y \sim \text{beta}(\delta, \nu)$, then the probability density function (pdf) of Y is

$$f(y) = \frac{\Gamma(\delta + \nu)}{\Gamma(\delta)\Gamma(\nu)} y^{\delta-1} (1-y)^{\nu-1}$$

where $\delta > 0$, $\nu > 0$ and $0 \leq y \leq 1$.

$$E(Y) = \frac{\delta}{\delta + \nu}.$$

$$\text{VAR}(Y) = \frac{\delta\nu}{(\delta+\nu)^2(\delta+\nu+1)}.$$

Notice that

$$f(y) = \frac{\Gamma(\delta + \nu)}{\Gamma(\delta)\Gamma(\nu)} I_{[0,1]}(y) \exp[(\delta - 1)\log(y) + (\nu - 1)\log(1-y)]$$

$$= I_{[0,1]}(y) \frac{\Gamma(\delta + \nu)}{\Gamma(\delta)\Gamma(\nu)} \exp[(1-\delta)(-\log(y)) + (1-\nu)(-\log(1-y))]$$

is a **2P–REF** (two-parameter regular exponential family). Hence $\Theta = (0,\infty) \times (0,\infty)$, $\eta_1 = 1-\delta$, $\eta_2 = 1-\nu$ and $\Omega = (-\infty,1) \times (-\infty,1)$.

If $\delta = 1$, then $W = -\log(1-Y) \sim \text{EXP}(1/\nu)$. Hence $T_n = -\sum \log(1-Y_i) \sim G(n,1/\nu)$, and if $r > -n$ then T_n^r is the UMVUE of

$$E(T_n^r) = \frac{1}{\nu^r}\frac{\Gamma(r+n)}{\Gamma(n)}.$$

If $\nu = 1$, then $W = -\log(Y) \sim \text{EXP}(1/\delta)$. Hence $T_n = -\sum \log(Y_i) \sim G(n,1/\delta)$, and if $r > -n$ then $[T_n]^r$ is the UMVUE of

$$E([T_n]^r) = \frac{1}{\delta^r}\frac{\Gamma(r+n)}{\Gamma(n)}.$$

10.2 The Beta-Binomial Distribution

If Y has a beta-binomial distribution, $Y \sim \text{BB}(m,\rho,\theta)$, then the probability mass function (pmf) of Y is

$$f(y) = P(Y=y) = \binom{m}{y}\frac{B(\delta+y, \nu+m-y)}{B(\delta,\nu)}$$

for $y = 0,1,2,\ldots,m$ where $0 < \rho < 1$ and $\theta > 0$. Here $\delta = \rho/\theta$ and $\nu = (1-\rho)/\theta$, so $\rho = \delta/(\delta+\nu)$ and $\theta = 1/(\delta+\nu)$. Also

$$B(\delta,\nu) = \frac{\Gamma(\delta)\Gamma(\nu)}{\Gamma(\delta+\nu)}.$$

Hence $\delta > 0$ and $\nu > 0$. Then $E(Y) = m\delta/(\delta+\nu) = m\rho$ and $\text{V}(Y) = m\rho(1-\rho)[1+(m-1)\theta/(1+\theta)]$. If $Y|\pi \sim \text{binomial}(m,\pi)$ and $\pi \sim \text{beta}(\delta,\nu)$, then $Y \sim \text{BB}(m,\rho,\theta)$.

As $\theta \to 0$, the beta-binomial (m,ρ,θ) distribution converges to a binomial(m,ρ) distribution.

10.3 The Binomial and Bernoulli Distributions

If Y has a binomial distribution, $Y \sim \text{BIN}(k,\rho)$, then the pmf of Y is

$$f(y) = P(Y=y) = \binom{k}{y}\rho^y(1-\rho)^{k-y}$$

for $y = 0,1,\ldots,k$ where $0 < \rho < 1$.

If $\rho = 0$, $P(Y=0) = 1 = (1-\rho)^k$ while if $\rho = 1$, $P(Y=k) = 1 = \rho^k$.

The moment generating function

$$m(t) = [(1-\rho) + \rho e^t]^k,$$

and the characteristic function $c(t) = [(1-\rho) + \rho e^{it}]^k$.

$$E(Y) = k\rho.$$

$$\text{VAR}(Y) = k\rho(1-\rho).$$

The Bernoulli (ρ) distribution is the binomial ($k = 1, \rho$) distribution.

Pourahmadi (1995) showed that the moments of a binomial (k, ρ) random variable can be found recursively. If $r \geq 1$ is an integer, $\binom{0}{0} = 1$ and the last term below is 0 for $r = 1$, then

$$E(Y^r) = k\rho \sum_{i=0}^{r-1} \binom{r-1}{i} E(Y^i) - \rho \sum_{i=0}^{r-2} \binom{r-1}{i} E(Y^{i+1}).$$

The following normal approximation is often used.

$$Y \approx N(k\rho, k\rho(1-\rho))$$

when $k\rho(1-\rho) > 9$. Hence

$$P(Y \leq y) \approx \Phi\left(\frac{y + 0.5 - k\rho}{\sqrt{k\rho(1-\rho)}}\right).$$

Also

$$P(Y = y) \approx \frac{1}{\sqrt{k\rho(1-\rho)}} \frac{1}{\sqrt{2\pi}} \exp\left(-\frac{1}{2}\frac{(y-k\rho)^2}{k\rho(1-\rho)}\right).$$

See Johnson et al. (1992, p. 115). This approximation suggests that $\text{MED}(Y) \approx k\rho$, and $\text{MAD}(Y) \approx 0.674\sqrt{k\rho(1-\rho)}$. Hamza (1995) states that $|E(Y) - \text{MED}(Y)| \leq \max(\rho, 1-\rho)$ and shows that

$$|E(Y) - \text{MED}(Y)| \leq \log(2).$$

If k is large and $k\rho$ small, then $Y \approx \text{Poisson}(k\rho)$.

If $Y_1, \ldots, Y_n$ are independent $\text{BIN}(k_i, \rho)$, then $\sum_{i=1}^n Y_i \sim \text{BIN}(\sum_{i=1}^n k_i, \rho)$.

Notice that

$$f(y) = \binom{k}{y} I[y \in \{0, \ldots, k\}](1-\rho)^k \exp\left[\log\left(\frac{\rho}{1-\rho}\right) y\right]$$

is a **1P-REF** (one-parameter regular exponential family) in ρ if k is known. Thus $\Theta = (0, 1)$,

$$\eta = \log\left(\frac{\rho}{1-\rho}\right)$$

and $\Omega = (-\infty, \infty)$.

Assume that $Y_1, \ldots, Y_n$ are iid BIN(k, ρ), then

$$T_n = \sum_{i=1}^{n} Y_i \sim \text{BIN}(nk, \rho).$$

If k is known, then the likelihood

$$L(\rho) = c\ \rho^{\sum_{i=1}^n y_i}\ (1-\rho)^{nk - \sum_{i=1}^n y_i},$$

and the log likelihood

$$\log(L(\rho)) = d + \log(\rho)\sum_{i=1}^{n} y_i + \left(nk - \sum_{i=1}^{n} y_i\right)\log(1-\rho).$$

Hence

$$\frac{d}{d\rho}\log(L(\rho)) = \frac{\sum_{i=1}^n y_i}{\rho} + \frac{nk - \sum_{i=1}^n y_i}{1-\rho}(-1) \stackrel{\text{set}}{=} 0,$$

or $(1-\rho)\sum_{i=1}^n y_i = \rho(nk - \sum_{i=1}^n y_i)$, or $\sum_{i=1}^n y_i = \rho nk$ or

$$\hat{\rho} = \sum_{i=1}^{n} y_i/(nk).$$

This solution is unique and

$$\frac{d^2}{d\rho^2}\log(L(\rho)) = \frac{-\sum_{i=1}^n y_i}{\rho^2} - \frac{nk - \sum_{i=1}^n y_i}{(1-\rho)^2} < 0$$

if $0 < \sum_{i=1}^n y_i < nk$. Hence $k\hat{\rho} = \overline{Y}$ is the UMVUE, MLE, and MME (method of moments estimator) of $k\rho$ if k is known.

Let $\hat{\rho}$ = number of "successes"$/n$ and let $P(Z \le z_{1-\alpha/2}) = 1 - \alpha/2$ if $Z \sim N(0,1)$. Let $\tilde{n} = n + z_{1-\alpha/2}^2$ and

$$\tilde{\rho} = \frac{n\hat{\rho} + 0.5 z_{1-\alpha/2}^2}{n + z_{1-\alpha/2}^2}.$$

Then the large sample 100 $(1-\alpha)$% Agresti Coull confidence interval (CI) for ρ is

$$\tilde{p} \pm z_{1-\alpha/2}\sqrt{\frac{\tilde{\rho}(1-\tilde{\rho})}{\tilde{n}}}.$$

Let $W = \sum_{i=1}^n Y_i \sim \text{bin}(\sum_{i=1}^n k_i, \rho)$ and let $n_w = \sum_{i=1}^n k_i$. Often $k_i \equiv 1$ and then $n_w = n$. Let $P(F_{d_1,d_2} \le F_{d_1,d_2}(\alpha)) = \alpha$ where F_{d_1,d_2} has an F distribution with d_1 and d_2 degrees of freedom. Then the Clopper Pearson "exact" 100 $(1-\alpha)$% CI for ρ is

$$\left(0, \frac{1}{1 + n_w \, F_{2n_w,2}(\alpha)}\right) \text{ for } W = 0,$$

$$\left(\frac{n_w}{n_w + F_{2,2n_w}(1-\alpha)}, 1\right) \text{ for } W = n_w,$$

and (ρ_L, ρ_U) for $0 < W < n_w$ with

$$\rho_L = \frac{W}{W + (n_w - W + 1)F_{2(n_w-W+1),2W}(1-\alpha/2)}$$

and

$$\rho_U = \frac{W+1}{W+1+(n_w - W)F_{2(n_w-W),2(W+1)}(\alpha/2)}.$$

10.4 The Birnbaum Saunders Distribution

If Y has a Birnbaum Saunders distribution, $Y \sim BS(\nu, \theta)$, then the pdf of Y is

$$f(y) = \frac{1}{2\sqrt{2\pi}} \left(\frac{1}{\theta}\sqrt{\frac{\theta}{y}} + \frac{\theta}{y^2}\sqrt{\frac{y}{\theta}} \right) \frac{1}{\nu} \exp\left[\frac{-1}{2\nu^2} \left(\frac{y}{\theta} + \frac{\theta}{y} - 2 \right) \right] =$$

$$\frac{1}{2\sqrt{2\pi} \ \theta y^2} \frac{y^2 - \theta^2}{\sqrt{\frac{y}{\theta}} - \sqrt{\frac{\theta}{y}}} \frac{1}{\nu} \exp\left[\frac{-1}{2\nu^2} \left(\frac{y}{\theta} + \frac{\theta}{y} - 2 \right) \right]$$

$y > 0, \theta > 0$ and $\nu > 0$.

The cdf of Y is

$$F(y) = \Phi\left[\frac{1}{\nu} \left(\sqrt{\frac{y}{\theta}} - \sqrt{\frac{\theta}{y}} \right) \right]$$

where $\Phi(x)$ is the N(0,1) cdf and $y > 0$. Hence MED$(Y) = \theta$. Let $W = t(Y) = \frac{Y}{\theta} + \frac{\theta}{Y} - 2$, then $W \sim \nu^2 \chi^2_1$.
$E(Y) = \theta \ (1 + \nu^2/2)$ and $V(Y) = (\nu\theta)^2 \ (1 + 5\nu^2/6)$.

Suppose θ is known and $Y_1, \ldots, Y_i$ are iid. Let $W_i = t(Y_i)$. Then the likelihood

$$L(\nu) = d \frac{1}{\nu^n} \exp\left(\frac{-1}{2\nu^2} \sum_{i=1}^n w_i \right),$$

and the log likelihood

$$\log(L(\nu)) = c - n\log(\nu) - \frac{1}{2\nu^2} \sum_{i=1}^n w_i.$$

Hence

$$\frac{d}{d\nu}\log(L(\nu)) = \frac{-n}{\nu} + \frac{1}{\nu^3}\sum_{i=1}^n w_i \stackrel{\text{set}}{=} 0,$$

or

$$\hat{\nu} = \sqrt{\frac{\sum_{i=1}^n w_i}{n}}.$$

This solution is unique and

$$\frac{d^2}{d\nu^2}\log(L(\nu)) = \frac{n}{\nu^2} - \left.\frac{3\sum_{i=1}^n w_i}{\nu^4}\right|_{\nu=\hat{\nu}} = \frac{n}{\hat{\nu}^2} - \frac{3n\hat{\nu}^2}{\hat{\nu}^4} = \frac{-2n}{\hat{\nu}^2} < 0.$$

Thus

$$\hat{\nu} = \sqrt{\frac{\sum_{i=1}^n W_i}{n}}$$

is the MLE of ν if $\hat{\nu} > 0$.

If ν is fixed, this distribution is a scale family with scale parameter θ. If θ is known this distribution is a **1P-REF**. This family of distributions is similar to the lognormal family.

10.5 The Burr Type III Distribution

If Y has a Burr Type III distribution, $Y \sim \text{BTIII}(\phi,\lambda)$, then the pdf of Y is

$$f(y) = I(y>0)\,\frac{\phi}{\lambda}\; y^{-(\phi+1)}\,(1+y^{-\phi})^{-(\frac{1}{\lambda}+1)}$$

$$= I(y>0)\;\phi\; y^{-(\phi+1)}\frac{1}{1+y^{-\phi}}\,\frac{1}{\lambda}\,\exp\left[\frac{-1}{\lambda}\log(1+y^{-\phi})\right]$$

where $\phi > 0$ and $\lambda > 0$. This family is a **1P-REF** if ϕ is known.

The cdf

$$F(y) = (1+y^{-\phi})^{-1/\lambda}$$

for $y > 0$.

$X = 1/Y \sim \text{BTXII}(\lambda,\phi)$.

$W = \log(1+Y^{-\phi}) \sim \text{EXP}(\lambda)$.

If $Y_1,\ldots,Y_n$ are iid $\text{BTIII}(\lambda,\phi)$, then

$$T_n = \sum_{i=1}^n \log(1+Y_i^{-\phi}) \sim G(n,\lambda).$$

If ϕ is known, then the likelihood

$$L(\lambda) = c\frac{1}{\lambda^n}\exp\left[-\frac{1}{\lambda}\sum_{i=1}^n \log(1+y_i^{-\phi})\right],$$

and the log likelihood $\log(L(\lambda)) = d - n\log(\lambda) - \frac{1}{\lambda}\sum_{i=1}^n \log(1+y_i^{-\phi})$. Hence

$$\frac{d}{d\lambda}\log(L(\lambda)) = \frac{-n}{\lambda} + \frac{\sum_{i=1}^n \log(1+y_i^{-\phi})}{\lambda^2} \stackrel{\text{set}}{=} 0,$$

or $\sum_{i=1}^n \log(1+y_i^{-\phi}) = n\lambda$ or

$$\hat{\lambda} = \frac{\sum_{i=1}^n \log(1+y_i^{-\phi})}{n}.$$

This solution is unique and

$$\frac{d^2}{d\lambda^2}\log(L(\lambda)) = \frac{n}{\lambda^2} - \left.\frac{2\sum_{i=1}^n \log(1+y_i^{-\phi})}{\lambda^3}\right|_{\lambda=\hat{\lambda}} = \frac{n}{\hat{\lambda}^2} - \frac{2n\hat{\lambda}}{\hat{\lambda}^3} = \frac{-n}{\hat{\lambda}^2} < 0.$$

Thus

$$\hat{\lambda} = \frac{\sum_{i=1}^n \log(1+Y_i^{-\phi})}{n}$$

is the UMVUE and MLE of λ if ϕ is known.

If ϕ is known and $r > -n$, then T_n^r is the UMVUE of

$$E(T_n^r) = \lambda^r \frac{\Gamma(r+n)}{\Gamma(n)}.$$

10.6 The Burr Type X Distribution

If Y has a Burr Type X distribution, $Y \sim \text{BTX}(\tau)$, then the pdf of Y is

$$f(y) = I(y>0)\; 2\;\tau\; y\; e^{-y^2}\; (1-e^{-y^2})^{\tau-1} =$$

$$I(y>0)\; \frac{2y\; e^{-y^2}}{1-e^{-y^2}}\; \tau\; \exp[\tau\log(1-e^{-y^2})]$$

where $\tau > 0$. This family is a **1P-REF**. $W = -\log(1-e^{-Y^2}) \sim \text{EXP}(1/\tau)$ and $\text{MED}(W) = \log(2)/\tau$.

Given data $Y_1,\ldots,Y_n$, a robust estimator of τ is $\hat{\tau} = \log(2)/\text{MED}(n)$ where $\text{MED}(n)$ is the sample median of $W_1,\ldots,W_n$ and $W_i = -\log(1-e^{-Y_i^2})$.

If $Y_1,\ldots,Y_n$ are iid BTX(τ), then

$$T_n = -\sum_{i=1}^n \log(1-e^{-Y_i^2}) \sim G(n, 1/\tau).$$

The likelihood $L(\tau) = c\ \tau^n\ \exp[\tau \sum_{i=1}^n \log(1 - e^{-y_i^2})]$, and the log likelihood $\log(L(\tau)) = d + n\log(\tau) + \tau \sum_{i=1}^n \log(1 - e^{-y_i^2})$.

Hence

$$\frac{d}{d\tau}\log(L(\tau)) = \frac{n}{\tau} + \sum_{i=1}^n \log(1 - e^{-y_i^2}) \stackrel{\text{set}}{=} 0,$$

or

$$\hat{\tau} = \frac{n}{-\sum_{i=1}^n \log(1 - e^{-y_i^2})}.$$

This solution is unique and

$$\frac{d^2}{d\tau^2}\log(L(\tau)) = \frac{-n}{\tau^2} < 0.$$

Thus

$$\hat{\tau} = \frac{n}{-\sum_{i=1}^n \log(1 - e^{-Y_i^2})}$$

is the MLE of τ.

Now $(n-1)\hat{\tau}/n$ is the UMVUE of τ, and if $r > -n$, then $[T_n]^r$ is the UMVUE of

$$E([T_n]^r) = \frac{\Gamma(r+n)}{\tau^r\ \Gamma(n)}.$$

10.7 The Burr Type XII Distribution

If Y has a Burr Type XII distribution, $Y \sim \text{BTXII}(\phi, \lambda)$, then the pdf of Y is

$$f(y) = \frac{1}{\lambda} \frac{\phi y^{\phi-1}}{(1+y^\phi)^{\frac{1}{\lambda}+1}}$$

where y, ϕ, and λ are all positive. The cdf of Y is

$$F(y) = 1 - \exp\left[\frac{-\log(1+y^\phi)}{\lambda}\right] = 1 - (1+y^\phi)^{-1/\lambda} \ \text{ for } \ y > 0.$$

$\text{MED}(Y) = [e^{\lambda \log(2)} - 1]^{1/\phi}$. See Patel et al. (1976, p. 195).
$W = \log(1+Y^\phi)$ is $\text{EXP}(\lambda)$.

Notice that

$$f(y) = \phi y^{\phi-1} \frac{1}{1+y^\phi} \frac{1}{\lambda} \exp\left[-\frac{1}{\lambda}\log(1+y^\phi)\right] I(y>0)$$

is a **1P-REF** if ϕ is known.

If $Y_1,\ldots,Y_n$ are iid BTXII(λ,ϕ), then

$$T_n = \sum_{i=1}^n \log(1+Y_i^\phi) \sim G(n,\lambda).$$

If ϕ is known, then the likelihood

$$L(\lambda) = c\frac{1}{\lambda^n}\exp\left[-\frac{1}{\lambda}\sum_{i=1}^n \log(1+y_i^\phi)\right],$$

and the log likelihood $\log(L(\lambda)) = d - n\log(\lambda) - \frac{1}{\lambda}\sum_{i=1}^n \log(1+y_i^\phi)$. Hence

$$\frac{d}{d\lambda}\log(L(\lambda)) = \frac{-n}{\lambda} + \frac{\sum_{i=1}^n \log(1+y_i^\phi)}{\lambda^2} \stackrel{\text{set}}{=} 0,$$

or $\sum_{i=1}^n \log(1+y_i^\phi) = n\lambda$ or

$$\hat{\lambda} = \frac{\sum_{i=1}^n \log(1+y_i^\phi)}{n}.$$

This solution is unique and

$$\frac{d^2}{d\lambda^2}\log(L(\lambda)) = \left.\frac{n}{\lambda^2} - \frac{2\sum_{i=1}^n \log(1+y_i^\phi)}{\lambda^3}\right|_{\lambda=\hat{\lambda}} = \frac{n}{\hat{\lambda}^2} - \frac{2n\hat{\lambda}}{\hat{\lambda}^3} = \frac{-n}{\hat{\lambda}^2} < 0.$$

Thus

$$\hat{\lambda} = \frac{\sum_{i=1}^n \log(1+Y_i^\phi)}{n}$$

is the UMVUE and MLE of λ if ϕ is known.

If ϕ is known and $r > -n$, then T_n^r is the UMVUE of

$$E(T_n^r) = \lambda^r \frac{\Gamma(r+n)}{\Gamma(n)}.$$

10.8 The Cauchy Distribution

If Y has a Cauchy distribution, $Y \sim C(\mu,\sigma)$, then the pdf of Y is

$$f(y) = \frac{\sigma}{\pi}\frac{1}{\sigma^2+(y-\mu)^2} = \frac{1}{\pi\sigma\left[1+\left(\frac{y-\mu}{\sigma}\right)^2\right]}$$

where y and μ are real numbers and $\sigma > 0$. The cumulative distribution function (cdf) of Y is

$$F(y) = \frac{1}{\pi}\left[\arctan\left(\frac{y-\mu}{\sigma}\right) + \pi/2\right].$$

See Ferguson (1967, p. 102). This family is a location–scale family that is symmetric about μ.

The moments of Y do not exist, but the characteristic function of Y is

$$c(t) = \exp(it\mu - |t|\sigma).$$

$\text{MED}(Y) = \mu$, the upper quartile $= \mu + \sigma$, and the lower quartile $= \mu - \sigma$.
$\text{MAD}(Y) = F^{-1}(3/4) - \text{MED}(Y) = \sigma$.
If $Y_1, \ldots, Y_n$ are independent $C(\mu_i, \sigma_i)$, then

$$\sum_{i=1}^n a_i Y_i \sim C\left(\sum_{i=1}^n a_i \mu_i, \sum_{i=1}^n |a_i| \sigma_i\right).$$

In particular, if $Y_1, \ldots, Y_n$ are iid $C(\mu, \sigma)$, then $\overline{Y} \sim C(\mu, \sigma)$.

If $W \sim U(-\pi/2, \pi/2)$, then $Y = \tan(W) \sim C(0, 1)$.

10.9 The Chi Distribution

If Y has a chi distribution (also called a p-dimensional Rayleigh distribution), $Y \sim \text{chi}(p, \sigma)$, then the pdf of Y is

$$f(y) = \frac{y^{p-1} e^{\frac{-1}{2\sigma^2} y^2}}{\sigma^p 2^{\frac{p}{2}-1} \Gamma(p/2)}$$

where $y > 0$ and $\sigma, p > 0$. This is a scale family if p is known.

$$E(Y) = \sigma\sqrt{2}\frac{\Gamma(\frac{1+p}{2})}{\Gamma(p/2)}.$$

$$\text{VAR}(Y) = 2\sigma^2\left[\frac{\Gamma(\frac{2+p}{2})}{\Gamma(p/2)} - \left(\frac{\Gamma(\frac{1+p}{2})}{\Gamma(p/2)}\right)^2\right],$$

and

$$E(Y^r) = 2^{r/2}\sigma^r \frac{\Gamma(\frac{r+p}{2})}{\Gamma(p/2)}$$

for $r > -p$.

The mode is at $\sigma\sqrt{p-1}$ for $p \geq 1$. See Cohen and Whitten (1988, ch. 10). Note that $W = Y^2 \sim G(p/2, 2\sigma^2)$.

$Y \sim$ generalized gamma $(\nu = p/2, \lambda = \sigma\sqrt{2}, \phi = 2)$.

If $p = 1$, then Y has a half normal distribution, $Y \sim \text{HN}(0, \sigma^2)$.

If $p = 2$, then Y has a Rayleigh distribution, $Y \sim \text{R}(0, \sigma)$.

If $p = 3$, then Y has a Maxwell–Boltzmann distribution, $Y \sim \text{MB}\ (0, \sigma)$.

If p is an integer and $Y \sim \text{chi}(p, 1)$, then $Y^2 \sim \chi^2_p$.

Since

$$f(y) = I(y > 0)\ \frac{1}{2^{\frac{p}{2}-1}\Gamma(p/2)\sigma^p}\ \exp\left[(p-1)\log(y) - \frac{1}{2\sigma^2}y^2\right],$$

this family is a **2P–REF**. Notice that $\Theta = (0, \infty) \times (0, \infty)$, $\eta_1 = p - 1$, $\eta_2 = -1/(2\sigma^2)$, and $\Omega = (-1, \infty) \times (-\infty, 0)$.

If p is known, then

$$f(y) = \frac{y^{p-1}}{2^{\frac{p}{2}-1}\Gamma(p/2)} I(y > 0)\ \frac{1}{\sigma^p} \exp\left[\frac{-1}{2\sigma^2} y^2\right]$$

is a 1P-REF.

If $Y_1, \ldots, Y_n$ are iid chi(p, σ), then

$$T_n = \sum_{i=1}^n Y_i^2 \sim G(np/2, 2\sigma^2).$$

If p is known, then the likelihood

$$L(\sigma^2) = c\ \frac{1}{\sigma^{np}} \exp\left[\frac{-1}{2\sigma^2} \sum_{i=1}^n y_i^2\right],$$

and the log likelihood

$$\log(L(\sigma^2)) = d - \frac{np}{2} \log(\sigma^2) - \frac{1}{2\sigma^2} \sum_{i=1}^n y_i^2.$$

Hence

$$\frac{d}{d(\sigma^2)} \log(\sigma^2) = \frac{-np}{2\sigma^2} + \frac{1}{2(\sigma^2)^2} \sum_{i=1}^n y_i^2 \stackrel{\text{set}}{=} 0,$$

or $\sum_{i=1}^n y_i^2 = np\sigma^2$ or

$$\hat{\sigma}^2 = \frac{\sum_{i=1}^n y_i^2}{np}.$$

This solution is unique and

$$\frac{d^2}{d(\sigma^2)^2}\log(L(\sigma^2)) = \frac{np}{2(\sigma^2)^2} - \frac{\sum_{i=1}^n y_i^2}{(\sigma^2)^3}\bigg|_{\sigma^2=\hat{\sigma}^2} = \frac{np}{2(\hat{\sigma}^2)^2} - \frac{np\hat{\sigma}}{(\hat{\sigma}^2)^3}\frac{2}{2} =$$

$$\frac{-np}{2(\hat{\sigma}^2)^2} < 0.$$

Thus

$$\hat{\sigma}^2 = \frac{\sum_{i=1}^n Y_i^2}{np}$$

is the UMVUE and MLE of σ^2 when p is known.

If p is known and $r > -np/2$, then T_n^r is the UMVUE of

$$E(T_n^r) = \frac{2^r \sigma^{2r}\Gamma(r+np/2)}{\Gamma(np/2)}.$$

10.10 The Chi-Square Distribution

If Y has a chi-square distribution, $Y \sim \chi_p^2$, then the pdf of Y is

$$f(y) = \frac{y^{\frac{p}{2}-1}e^{-\frac{y}{2}}}{2^{\frac{p}{2}}\Gamma(\frac{p}{2})}$$

where $y \geq 0$ and p is a positive integer. The mgf of Y is

$$m(t) = \left(\frac{1}{1-2t}\right)^{p/2} = (1-2t)^{-p/2}$$

for $t < 1/2$. The characteristic function

$$c(t) = \left(\frac{1}{1-i2t}\right)^{p/2}.$$

$E(Y) = p$.
$\text{VAR}(Y) = 2p$.

Since Y is gamma $G(\nu = p/2, \lambda = 2)$,

$$E(Y^r) = \frac{2^r\Gamma(r+p/2)}{\Gamma(p/2)},\ r > -p/2.$$

$\text{MED}(Y) \approx p - 2/3$. See Pratt (1968, p. 1470) for more terms in the expansion of $\text{MED}(Y)$. Empirically,

$$\text{MAD}(Y) \approx \frac{\sqrt{2p}}{1.483}\left(1 - \frac{2}{9p}\right)^2 \approx 0.9536\sqrt{p}.$$

There are several normal approximations for this distribution. The Wilson–Hilferty approximation is

$$\left(\frac{Y}{p}\right)^{\frac{1}{3}} \approx N\left(1 - \frac{2}{9p}, \frac{2}{9p}\right).$$

See Bowman and Shenton (1988, p. 6). This approximation gives

$$P(Y \le x) \approx \Phi\left[\left(\left(\frac{x}{p}\right)^{1/3} - 1 + 2/9p\right)\sqrt{9p/2}\right],$$

and

$$\chi^2_{p,\alpha} \approx p\left(z_\alpha\sqrt{\frac{2}{9p}} + 1 - \frac{2}{9p}\right)^3$$

where z_α is the standard normal percentile, $\alpha = \Phi(z_\alpha)$. The last approximation is good if $p > -1.24\log(\alpha)$. See Kennedy and Gentle (1980, p. 118).

This family is a one-parameter exponential family, but is not a REF since the set of integers does not contain an open interval.

10.11 The Discrete Uniform Distribution

If Y has a discrete uniform distribution, $Y \sim \text{DU}(\theta_1, \theta_2)$, then the pmf of Y is

$$f(y) = P(Y = y) = \frac{1}{\theta_2 - \theta_1 + 1}$$

for $\theta_1 \le y \le \theta_2$ where y and the θ_i are integers. Let $\theta_2 = \theta_1 + \tau - 1$ where $\tau = \theta_2 - \theta_1 + 1$. The cdf for Y is

$$F(y) = \frac{\lfloor y \rfloor - \theta_1 + 1}{\theta_2 - \theta_1 + 1}$$

for $\theta_1 \le y \le \theta_2$. Here $\lfloor y \rfloor$ is the greatest integer function, e.g., $\lfloor 7.7 \rfloor = 7$. This result holds since for $\theta_1 \le y \le \theta_2$,

$$F(y) = \sum_{i=\theta_1}^{\lfloor y \rfloor} \frac{1}{\theta_2 - \theta_1 + 1}.$$

$E(Y) = (\theta_1 + \theta_2)/2 = \theta_1 + (\tau - 1)/2$ while $V(Y) = (\tau^2 - 1)/12$. The result for $E(Y)$ follows by symmetry, or because

$$E(Y) = \sum_{y=\theta_1}^{\theta_2} \frac{y}{\theta_2 - \theta_1 + 1} = \frac{\theta_1(\theta_2 - \theta_1 + 1) + [0 + 1 + 2 + \cdots + (\theta_2 - \theta_1)]}{\theta_2 - \theta_1 + 1}$$

where last equality follows by adding and subtracting θ_1 to y for each of the $\theta_2 - \theta_1 + 1$ terms in the middle sum. Thus

$$E(Y) = \theta_1 + \frac{(\theta_2 - \theta_1)(\theta_2 - \theta_1 + 1)}{2(\theta_2 - \theta_1 + 1)} = \frac{2\theta_1}{2} + \frac{\theta_2 - \theta_1}{2} = \frac{\theta_1 + \theta_2}{2}$$

since $\sum_{i=1}^n i = n(n+1)/2$ by Lemma 10.2 e) with $n = \theta_2 - \theta_1$.

To see the result for $V(Y)$, let $W = Y - \theta_1 + 1$. Then $V(Y) = V(W)$ and $f(w) = 1/\tau$ for $w = 1, \ldots, \tau$. Hence $W \sim \text{DU}(1, \tau)$,

$$E(W) = \frac{1}{\tau} \sum_{w=1}^{\tau} w = \frac{\tau(\tau+1)}{2\tau} = \frac{1+\tau}{2},$$

and

$$E(W^2) = \frac{1}{\tau} \sum_{w=1}^{\tau} w^2 = \frac{\tau(\tau+1)(2\tau+1)}{6\tau} = \frac{(\tau+1)(2\tau+1)}{6}$$

by Lemma 10.2 e) and f). So

$$V(Y) = V(W) = E(W^2) - (E(W))^2 = \frac{(\tau+1)(2\tau+1)}{6} - \left(\frac{1+\tau}{2}\right)^2 =$$

$$\frac{2(\tau+1)(2\tau+1) - 3(\tau+1)^2}{12} = \frac{2(\tau+1)[2(\tau+1) - 1] - 3(\tau+1)^2}{12} =$$

$$\frac{4(\tau+1)^2 - 2(\tau+1) - 3(\tau+1)^2}{12} = \frac{(\tau+1)^2 - 2\tau - 2}{12} =$$

$$\frac{\tau^2 + 2\tau + 1 - 2\tau - 2}{12} = \frac{\tau^2 - 1}{12}.$$

Let $\mathbb{Z}$ be the set of integers and let $Y_1, \ldots, Y_n$ be iid $\text{DU}(\theta_1, \theta_2)$. Then the likelihood function $L(\theta_1, \theta_2) =$

$$\frac{1}{(\theta_2 - \theta_1 + 1)^n} I(\theta_1 \le Y_{(1)}) I(\theta_2 \ge Y_{(n)}) I(\theta_1 \le \theta_2) I(\theta_1 \in \mathbb{Z}) I(\theta_2 \in \mathbb{Z})$$

is maximized by making $\theta_2 - \theta_1 - 1$ as small as possible where integers $\theta_2 \ge \theta_1$. So need θ_2 as small as possible and θ_1 as large as possible, and the MLE of (θ_1, θ_2) is $(Y_{(1)}, Y_{(n)})$.

10.12 The Double Exponential Distribution

If Y has a double exponential distribution (or Laplace distribution), $Y \sim \text{DE}(\theta, \lambda)$, then the pdf of Y is

$$f(y) = \frac{1}{2\lambda} \exp\left(\frac{-|y-\theta|}{\lambda}\right)$$

where y is real and $\lambda > 0$. The cdf of Y is

$$F(y) = 0.5 \exp\left(\frac{y-\theta}{\lambda}\right) \quad \text{if } y \le \theta,$$

and

$$F(y) = 1 - 0.5 \exp\left(\frac{-(y-\theta)}{\lambda}\right) \quad \text{if } y \ge \theta.$$

This family is a location–scale family which is symmetric about θ.
The mgf

$$m(t) = \exp(\theta t)/(1 - \lambda^2 t^2)$$

for $|t| < 1/\lambda$, and the characteristic function $c(t) = \exp(\theta i t)/(1 + \lambda^2 t^2)$.
$E(Y) = \theta$, and
$\text{MED}(Y) = \theta$.
$\text{VAR}(Y) = 2\lambda^2$, and
$\text{MAD}(Y) = \log(2)\lambda \approx 0.693\lambda$.
Hence $\lambda = \text{MAD}(Y)/\log(2) \approx 1.443\text{MAD}(Y)$.
To see that $\text{MAD}(Y) = \lambda \log(2)$, note that $F(\theta + \lambda \log(2)) = 1 - 0.25 = 0.75$.

The maximum likelihood estimators are $\hat{\theta}_{\text{MLE}} = \text{MED}(n)$ and

$$\hat{\lambda}_{\text{MLE}} = \frac{1}{n} \sum_{i=1}^{n} |Y_i - \text{MED}(n)|.$$

A $100(1-\alpha)\%$ CI for λ is

$$\left(\frac{2\sum_{i=1}^{n} |Y_i - \text{MED}(n)|}{\chi^2_{2n-1, 1-\frac{\alpha}{2}}}, \frac{2\sum_{i=1}^{n} |Y_i - \text{MED}(n)|}{\chi^2_{2n-1, \frac{\alpha}{2}}} \right),$$

and a $100(1-\alpha)\%$ CI for θ is

$$\left(\text{MED}(n) \pm \frac{z_{1-\alpha/2} \sum_{i=1}^{n} |Y_i - \text{MED}(n)|}{n\sqrt{n - z^2_{1-\alpha/2}}} \right)$$

where $\chi^2_{p,\alpha}$ and z_α are the α percentiles of the χ^2_p and standard normal distributions, respectively. See Patel et al. (1976, p. 194).
$W = |Y - \theta| \sim \text{EXP}(\lambda)$.

Notice that

$$f(y) = \frac{1}{2\lambda} \exp\left[\frac{-1}{\lambda}|y - \theta|\right]$$

is a one-parameter exponential family in λ if θ is known.

If $Y_1, \ldots, Y_n$ are iid $DE(\theta, \lambda)$ then

$$T_n = \sum_{i=1}^n |Y_i - \theta| \sim G(n, \lambda).$$

If θ is known, then the likelihood

$$L(\lambda) = c \frac{1}{\lambda^n} \exp\left[\frac{-1}{\lambda} \sum_{i=1}^n |y_i - \theta|\right],$$

and the log likelihood

$$\log(L(\lambda)) = d - n\log(\lambda) - \frac{1}{\lambda} \sum_{i=1}^n |y_i - \theta|.$$

Hence

$$\frac{d}{d\lambda} \log(L(\lambda)) = \frac{-n}{\lambda} + \frac{1}{\lambda^2} \sum_{i=1}^n |y_i - \theta| \stackrel{\text{set}}{=} 0$$

or $\sum_{i=1}^n |y_i - \theta| = n\lambda$ or

$$\hat{\lambda} = \frac{\sum_{i=1}^n |y_i - \theta|}{n}.$$

This solution is unique and

$$\frac{d^2}{d\lambda^2} \log(L(\lambda)) = \left. \frac{n}{\lambda^2} - \frac{2\sum_{i=1}^n |y_i - \theta|}{\lambda^3} \right|_{\lambda=\hat{\lambda}} = \frac{n}{\hat{\lambda}^2} - \frac{2n\hat{\lambda}}{\hat{\lambda}^3} = \frac{-n}{\hat{\lambda}^2} < 0.$$

Thus

$$\hat{\lambda} = \frac{\sum_{i=1}^n |Y_i - \theta|}{n}$$

is the UMVUE and MLE of λ if θ is known.

10.13 The Exponential Distribution

If Y has an exponential distribution, $Y \sim \text{EXP}(\lambda)$, then the pdf of Y is

$$f(y) = \frac{1}{\lambda} \exp\left(\frac{-y}{\lambda}\right) I(y \geq 0)$$

where $\lambda > 0$. The cdf of Y is

$$F(y) = 1 - \exp(-y/\lambda),\ y \geq 0.$$

This distribution is a scale family with scale parameter λ.
The mgf

$$m(t) = 1/(1 - \lambda t)$$

for $t < 1/\lambda$, and the characteristic function $c(t) = 1/(1 - i\lambda t)$.
$E(Y) = \lambda$,
and $\text{VAR}(Y) = \lambda^2$.
$W = 2Y/\lambda \sim \chi^2_2$.

Since Y is gamma $G(\nu = 1, \lambda)$, $E(Y^r) = \lambda^r \Gamma(r+1)$ for $r > -1$.
$\text{MED}(Y) = \log(2)\lambda$ and
$\text{MAD}(Y) \approx \lambda/2.0781$ since it can be shown that

$$\exp(\text{MAD}(Y)/\lambda) = 1 + \exp(-\text{MAD}(Y)/\lambda).$$

Hence $2.0781\ \text{MAD}(Y) \approx \lambda$.

The classical estimator is $\hat{\lambda} = \overline{Y}_n$ and the $100(1-\alpha)\%$ CI for $E(Y) = \lambda$ is

$$\left(\frac{2\sum_{i=1}^n Y_i}{\chi^2_{2n,1-\frac{\alpha}{2}}}, \frac{2\sum_{i=1}^n Y_i}{\chi^2_{2n,\frac{\alpha}{2}}} \right)$$

where $P(Y \leq \chi^2_{2n,\frac{\alpha}{2}}) = \alpha/2$ if Y is χ^2_{2n}. See Patel et al. (1976, p. 188).

Notice that

$$f(y) = I(y \geq 0)\ \frac{1}{\lambda}\ \exp\left[\frac{-1}{\lambda} y\right]$$

is a **1P-REF**. Hence $\Theta = (0, \infty)$, $\eta = -1/\lambda$, and $\Omega = (-\infty, 0)$.

Suppose that $Y_1, \ldots, Y_n$ are iid $\text{EXP}(\lambda)$, then

$$T_n = \sum_{i=1}^n Y_i \sim G(n, \lambda).$$

The likelihood

$$L(\lambda) = \frac{1}{\lambda^n} \exp\left[\frac{-1}{\lambda} \sum_{i=1}^n y_i\right],$$

and the log likelihood

$$\log(L(\lambda)) = -n\log(\lambda) - \frac{1}{\lambda} \sum_{i=1}^n y_i.$$

Hence

$$\frac{d}{d\lambda}\log(L(\lambda)) = \frac{-n}{\lambda} + \frac{1}{\lambda^2}\sum_{i=1}^{n} y_i \stackrel{\text{set}}{=} 0,$$

or $\sum_{i=1}^{n} y_i = n\lambda$ or

$$\hat{\lambda} = \overline{y}.$$

This solution is unique and

$$\frac{d^2}{d\lambda^2}\log(L(\lambda)) = \frac{n}{\lambda^2} - \left.\frac{2\sum_{i=1}^{n} y_i}{\lambda^3}\right|_{\lambda=\hat{\lambda}} = \frac{n}{\hat{\lambda}^2} - \frac{2n\hat{\lambda}}{\hat{\lambda}^3} = \frac{-n}{\hat{\lambda}^2} < 0.$$

Thus $\hat{\lambda} = \overline{Y}$ is the UMVUE, MLE, and MME of λ.

If $r > -n$, then T_n^r is the UMVUE of

$$E(T_n^r) = \frac{\lambda^r \Gamma(r+n)}{\Gamma(n)}.$$

10.14 The Two-Parameter Exponential Distribution

If Y has a two-parameter exponential distribution, $Y \sim \text{EXP}(\theta, \lambda)$, then the pdf of Y is

$$f(y) = \frac{1}{\lambda}\exp\left(\frac{-(y-\theta)}{\lambda}\right) I(y \geq \theta)$$

where $\lambda > 0$ and θ is real. The cdf of Y is

$$F(y) = 1 - \exp[-(y-\theta)/\lambda)],\ y \geq \theta.$$

This family is an asymmetric location–scale family.

The mgf

$$m(t) = \exp(t\theta)/(1-\lambda t)$$

for $t < 1/\lambda$, and the characteristic function $c(t) = \exp(it\theta)/(1-i\lambda t)$.

$E(Y) = \theta + \lambda$,

and $\text{VAR}(Y) = \lambda^2$.

$$\text{MED}(Y) = \theta + \lambda\log(2)$$

and

$$\text{MAD}(Y) \approx \lambda/2.0781.$$

Hence $\theta \approx \text{MED}(Y) - 2.0781\log(2)\text{MAD}(Y)$. See Rousseeuw and Croux (1993) for similar results. Note that $2.0781\log(2) \approx 1.44$.

To see that $2.0781\text{MAD}(Y) \approx \lambda$, note that

$$0.5 = \int_{\theta+\lambda\log(2)-\text{MAD}}^{\theta+\lambda\log(2)+\text{MAD}} \frac{1}{\lambda}\exp(-(y-\theta)/\lambda)dy$$

$$= 0.5[-e^{-\text{MAD}/\lambda} + e^{\text{MAD}/\lambda}]$$

assuming $\lambda\log(2) > \text{MAD}$. Plug in $\text{MAD} = \lambda/2.0781$ to get the result.

If θ is known, then

$$f(y) = I(y \geq \theta)\frac{1}{\lambda}\exp\left[\frac{-1}{\lambda}(y-\theta)\right]$$

is a 1P-REF in λ. Notice that $Y - \theta \sim \text{EXP}(\lambda)$. Let

$$\hat{\lambda} = \frac{\sum_{i=1}^{n}(Y_i - \theta)}{n}.$$

Then $\hat{\lambda}$ is the UMVUE and MLE of λ if θ is known.

If $Y_1, \ldots, Y_n$ are iid $\text{EXP}(\theta, \lambda)$, then the likelihood

$$L(\theta,\lambda) = \frac{1}{\lambda^n}\exp\left[\frac{-1}{\lambda}\sum_{i=1}^{n}(y_i - \theta)\right] I(y_{(1)} \geq \theta).$$

Need $y_{(1)} \geq \hat{\theta}$, and for $y_{(1)} \geq \theta$, the log likelihood

$$\log(L(\theta,\lambda)) = \left[-n\log(\lambda) - \frac{1}{\lambda}\sum_{i=1}^{n}(y_i - \theta)\right] I(y_{(1)} \geq \theta).$$

For any fixed $\lambda > 0$, the log likelihood is maximized by maximizing θ. Hence $\hat{\theta} = Y_{(1)}$, and the profile log likelihood is

$$\log(L(\lambda|y_{(1)})) = -n\log(\lambda) - \frac{1}{\lambda}\sum_{i=1}^{n}(y_i - y_{(1)})$$

is maximized by $\hat{\lambda} = \frac{1}{n}\sum_{i=1}^{n}(y_i - y_{(1)})$. Hence the MLE

$$(\hat{\theta},\hat{\lambda}) = \left(Y_{(1)}, \frac{1}{n}\sum_{i=1}^{n}(Y_i - Y_{(1)})\right) = (Y_{(1)}, \overline{Y} - Y_{(1)}).$$

Let $D_n = \sum_{i=1}^{n}(Y_i - Y_{(1)}) = n\hat{\lambda}$. Then for $n \geq 2$,

$$\left(\frac{2D_n}{\chi^2_{2(n-1),1-\alpha/2}}, \frac{2D_n}{\chi^2_{2(n-1),\alpha/2}}\right) \tag{10.3}$$

is a $100(1-\alpha)\%$ CI for λ, while

$$\left(Y_{(1)} - \hat{\lambda}[(\alpha)^{-1/(n-1)} - 1], Y_{(1)}\right) \tag{10.4}$$

is a $100\,(1-\alpha)\%$ CI for θ. See Mann et al. (1974, p. 176).

If θ is known and $T_n = \sum_{i=1}^n (Y_i - \theta)$, then a $100(1-\alpha)\%$ CI for λ is

$$\left(\frac{2T_n}{\chi^2_{2n,1-\alpha/2}}, \frac{2T_n}{\chi^2_{2n,\alpha/2}}\right). \tag{10.5}$$

10.15 The *F* Distribution

If Y has an F distribution, $Y \sim F(\nu_1, \nu_2)$, then the pdf of Y is

$$f(y) = \frac{\Gamma(\frac{\nu_1+\nu_2}{2})}{\Gamma(\nu_1/2)\Gamma(\nu_2/2)} \left(\frac{\nu_1}{\nu_2}\right)^{\nu_1/2} \frac{y^{(\nu_1-2)/2}}{\left(1 + \left(\frac{\nu_1}{\nu_2}\right) y\right)^{(\nu_1+\nu_2)/2}}$$

where $y > 0$ and ν_1 and ν_2 are positive integers.

$$E(Y) = \frac{\nu_2}{\nu_2 - 2}, \quad \nu_2 > 2$$

and

$$\text{VAR}(Y) = 2\left(\frac{\nu_2}{\nu_2-2}\right)^2 \frac{(\nu_1+\nu_2-2)}{\nu_1(\nu_2-4)}, \quad \nu_2 > 4.$$

$$E(Y^r) = \frac{\Gamma(\frac{\nu_1+2r}{2})\Gamma(\frac{\nu_2-2r}{2})}{\Gamma(\nu_1/2)\Gamma(\nu_2/2)} \left(\frac{\nu_2}{\nu_1}\right)^r, \quad r < \nu_2/2.$$

Suppose that X_1 and X_2 are independent where $X_1 \sim \chi^2_{\nu_1}$ and $X_2 \sim \chi^2_{\nu_2}$. Then

$$W = \frac{(X_1/\nu_1)}{(X_2/\nu_2)} \sim F(\nu_1, \nu_2).$$

Notice that $E(Y^r) = E(W^r) = \left(\frac{\nu_2}{\nu_1}\right)^r E(X_1^r)E(X_2^{-r})$.

If $W \sim t_\nu$, then $Y = W^2 \sim F(1, \nu)$.

10.16 The Gamma Distribution

If Y has a gamma distribution, $Y \sim G(\nu, \lambda)$, then the pdf of Y is

$$f(y) = \frac{y^{\nu-1} e^{-y/\lambda}}{\lambda^\nu \Gamma(\nu)}$$

where ν, λ, and y are positive. The mgf of Y is

$$m(t) = \left(\frac{1/\lambda}{\frac{1}{\lambda} - t}\right)^{\nu} = \left(\frac{1}{1-\lambda t}\right)^{\nu}$$

for $t < 1/\lambda$. The characteristic function

$$c(t) = \left(\frac{1}{1-i\lambda t}\right)^{\nu}.$$

$E(Y) = \nu\lambda$.
$\text{VAR}(Y) = \nu\lambda^2$.

$$E(Y^r) = \frac{\lambda^r \Gamma(r+\nu)}{\Gamma(\nu)} \text{ if } r > -\nu. \tag{10.6}$$

Chen and Rubin (1986) show that $\lambda(\nu - 1/3) < \text{MED}(Y) < \lambda\nu = E(Y)$. Empirically, for $\nu > 3/2$,

$$\text{MED}(Y) \approx \lambda(\nu - 1/3),$$

and

$$\text{MAD}(Y) \approx \frac{\lambda\sqrt{\nu}}{1.483}.$$

This family is a scale family for fixed ν, so if Y is $G(\nu,\lambda)$ then cY is $G(\nu, c\lambda)$ for $c > 0$. If W is EXP(λ), then W is $G(1,\lambda)$. If W is χ^2_p, then W is $G(p/2, 2)$.

Some classical estimators are given next. Let

$$w = \log\left[\frac{\overline{y}_n}{\text{geometric mean}(n)}\right]$$

where geometric mean$(n) = (y_1 y_2 \ldots y_n)^{1/n} = \exp[\frac{1}{n}\sum_{i=1}^n \log(y_i)]$. Then Thom's estimator (Johnson and Kotz 1970a, p. 188) is

$$\hat{\nu} \approx \frac{0.25(1 + \sqrt{1 + 4w/3}\,)}{w}.$$

Also

$$\hat{\nu}_{\text{MLE}} \approx \frac{0.5000876 + 0.1648852w - 0.0544274w^2}{w}$$

for $0 < w \le 0.5772$, and

$$\hat{\nu}_{\text{MLE}} \approx \frac{8.898919 + 9.059950w + 0.9775374w^2}{w(17.79728 + 11.968477w + w^2)}$$

for $0.5772 < w \le 17$. If $W > 17$, then estimation is much more difficult, but a rough approximation is $\hat{\nu} \approx 1/w$ for $w > 17$. See Bowman and Shenton (1988, p. 46) and Greenwood and Durand (1960). Finally, $\hat{\lambda} = \overline{Y}_n/\hat{\nu}$. Notice that $\hat{\lambda}$ may not be very good if $\hat{\nu} < 1/17$.

Several normal approximations are available. The Wilson–Hilferty approximation says that for $\nu > 0.5$,

$$Y^{1/3} \approx N\left((\nu\lambda)^{1/3}\left(1 - \frac{1}{9\nu}\right), (\nu\lambda)^{2/3}\frac{1}{9\nu}\right).$$

Hence if Y is $G(\nu,\lambda)$ and

$$\alpha = P[Y \le G_\alpha],$$

then

$$G_\alpha \approx \nu\lambda\left[z_\alpha\sqrt{\frac{1}{9\nu}} + 1 - \frac{1}{9\nu}\right]^3$$

where z_α is the standard normal percentile, $\alpha = \Phi(z_\alpha)$. Bowman and Shenton (1988, p. 101) include higher order terms.

Notice that

$$f(y) = I(y > 0)\,\frac{1}{\lambda^\nu \Gamma(\nu)}\,\exp\left[\frac{-1}{\lambda}y + (\nu - 1)\log(y)\right]$$

is a **2P–REF**. Hence $\Theta = (0,\infty) \times (0,\infty)$, $\eta_1 = -1/\lambda$, $\eta_2 = \nu - 1$ and $\Omega = (-\infty,0) \times (-1,\infty)$.

If $Y_1,\ldots,Y_n$ are independent $G(\nu_i,\lambda)$, then $\sum_{i=1}^n Y_i \sim G(\sum_{i=1}^n \nu_i,\lambda)$.

If $Y_1,\ldots,Y_n$ are iid $G(\nu,\lambda)$, then

$$T_n = \sum_{i=1}^n Y_i \sim G(n\nu,\lambda).$$

Since

$$f(y) = \frac{1}{\Gamma(\nu)}\exp[(\nu - 1)\log(y)]I(y > 0)\frac{1}{\lambda^\nu}\exp\left[\frac{-1}{\lambda}y\right],$$

Y is a 1P-REF when ν is known.

If ν is known, then the likelihood

$$L(\beta) = c\,\frac{1}{\lambda^{n\nu}}\exp\left[\frac{-1}{\lambda}\sum_{i=1}^n y_i\right].$$

The log likelihood

$$\log(L(\lambda)) = d - n\nu\log(\lambda) - \frac{1}{\lambda}\sum_{i=1}^n y_i.$$

Hence

$$\frac{d}{d\lambda}\log(L(\lambda)) = \frac{-n\nu}{\lambda} + \frac{\sum_{i=1}^n y_i}{\lambda^2} \stackrel{\text{set}}{=} 0,$$

or $\sum_{i=1}^n y_i = n\nu\lambda$ or

$$\hat{\lambda} = \overline{y}/\nu.$$

This solution is unique and

$$\frac{d^2}{d\lambda^2}\log(L(\lambda)) = \frac{n\nu}{\lambda^2} - \frac{2\sum_{i=1}^n y_i}{\lambda^3}\Bigg|_{\lambda=\hat{\lambda}} = \frac{n\nu}{\hat{\lambda}^2} - \frac{2n\nu\hat{\lambda}}{\hat{\lambda}^3} = \frac{-n\nu}{\hat{\lambda}^2} < 0.$$

Thus $\overline{Y}$ is the UMVUE, MLE, and MME of $\nu\lambda$ if ν is known.

The (lower) incomplete gamma function is

$$\gamma_\nu(x) = \frac{1}{\Gamma(\nu)}\int_0^x t^{\nu-1}e^{-t}dt.$$

If ν is a positive integer, then

$$\gamma_\nu(x) = 1 - e^{-x}\sum_{k=0}^{\nu-1}\frac{x^k}{k!}.$$

If $Y \sim G(\nu,\lambda)$, then $F(y) = \gamma_\nu(y/\lambda)$. Hence if $\nu > 0$ is an integer,

$$F(y) = \gamma_\nu(y/\lambda) = 1 - \sum_{k=0}^{\nu-1} e^{-y/\lambda}\frac{(y/\lambda)^k}{k!}.$$

For example, if $Y \sim G(\nu = 4, \lambda = 2)$, then $P(Y \le 14) = F(14) = \gamma_4(7)$ and $P(Y \le 10) = F(10) = \gamma_4(5)$.

10.17 The Generalized Gamma Distribution

If Y has a generalized gamma distribution, $Y \sim \text{GG}(\nu,\lambda,\phi)$, then the pdf of Y is

$$f(y) = \frac{\phi y^{\phi\nu-1}}{\lambda^{\phi\nu}\Gamma(\nu)}\exp(-y^\phi/\lambda^\phi)$$

where ν,λ,ϕ, and y are positive.
This family is a scale family with scale parameter λ if ϕ and ν are known.

$$E(Y^k) = \frac{\lambda^k\Gamma(\nu+\frac{k}{\phi})}{\Gamma(\nu)} \quad \text{if } k > -\phi\nu. \tag{10.7}$$

If ϕ and ν are known, then

$$f(y) = \frac{\phi y^{\phi\nu-1}}{\Gamma(\nu)}I(y>0)\frac{1}{\lambda^{\phi\nu}}\exp\left[\frac{-1}{\lambda^\phi}y^\phi\right],$$

which is a one-parameter exponential family.

Notice that $W = Y^\phi \sim G(\nu, \lambda^\phi)$. If $Y_1, \ldots, Y_n$ are iid GG(ν, λ, ϕ) where ϕ and ν are known, then $T_n = \sum_{i=1}^n Y_i^\phi \sim G(n\nu, \lambda^\phi)$, and T_n^r is the UMVUE of

$$E(T_n^r) = \lambda^{\phi r} \frac{\Gamma(r + n\nu)}{\Gamma(n\nu)}$$

for $r > -n\nu$.

10.18 The Generalized Negative Binomial Distribution

If Y has a generalized negative binomial distribution, $Y \sim \text{GNB}(\mu, \kappa)$, then the pmf of Y is

$$f(y) = P(Y = y) = \frac{\Gamma(y+\kappa)}{\Gamma(\kappa)\Gamma(y+1)} \left(\frac{\kappa}{\mu+\kappa}\right)^\kappa \left(1 - \frac{\kappa}{\mu+\kappa}\right)^y$$

for $y = 0, 1, 2, \ldots$ where $\mu > 0$ and $\kappa > 0$. This distribution is a generalization of the negative binomial (κ, ρ) distribution with $\rho = \kappa/(\mu + \kappa)$ and $\kappa > 0$ is an unknown real parameter rather than a known integer.

The mgf is

$$m(t) = \left[\frac{\kappa}{\kappa + \mu(1 - e^t)}\right]^\kappa$$

for $t < -\log(\mu/(\mu + \kappa))$.
$E(Y) = \mu$ and
$\text{VAR}(Y) = \mu + \mu^2/\kappa$.

If $Y_1, \ldots, Y_n$ are iid GNB(μ, κ), then $\sum_{i=1}^n Y_i \sim \text{GNB}(n\mu, n\kappa)$.

When κ is known, this distribution is a **1P-REF**. If $Y_1, \ldots, Y_n$ are iid GNB(μ, κ) where κ is known, then $\hat{\mu} = \overline{Y}$ is the MLE, UMVUE, and MME of μ.

Let $\tau = 1/\kappa$. As $\tau \to 0$, the GNB(μ, κ), distribution converges to the Poisson$(\lambda = \mu)$ distribution.

10.19 The Geometric Distribution

If Y has a geometric distribution, $Y \sim \text{geom}(\rho)$ then the pmf of Y is

$$f(y) = P(Y = y) = \rho(1 - \rho)^y$$

for $y = 0, 1, 2, \ldots$ and $0 < \rho < 1$.

The cdf for Y is $F(y) = 1 - (1 - \rho)^{\lfloor y \rfloor + 1}$ for $y \geq 0$ and $F(y) = 0$ for $y < 0$. Here $\lfloor y \rfloor$ is the greatest integer function, e.g., $\lfloor 7.7 \rfloor = 7$. To see this, note that for $y \geq 0$,

$$F(y) = \rho \sum_{i=0}^{\lfloor y \rfloor} (1 - \rho)^y = \rho \frac{1 - (1 - \rho)^{\lfloor y \rfloor + 1}}{1 - (1 - \rho)}$$

by Lemma 10.2a with $n_1 = 0$, $n_2 = \lfloor y \rfloor$ and $a = 1 - \rho$.

$E(Y) = (1-\rho)/\rho$.

$\text{VAR}(Y) = (1-\rho)/\rho^2$.

$Y \sim \text{NB}(1,\rho)$.

Hence the mgf of Y is

$$m(t) = \frac{\rho}{1-(1-\rho)e^t}$$

for $t < -\log(1-\rho)$.

Notice that

$$f(y) = I[y \in \{0,1,\ldots\}] \ \rho \ \exp[\log(1-\rho)y]$$

is a **1P-REF**. Hence $\Theta = (0,1)$, $\eta = \log(1-\rho)$, and $\Omega = (-\infty, 0)$.

If $Y_1, \ldots, Y_n$ are iid geom(ρ), then

$$T_n = \sum_{i=1}^n Y_i \sim \text{NB}(n,\rho).$$

The likelihood

$$L(\rho) = \rho^n \exp\left[\log(1-\rho)\sum_{i=1}^n y_i\right],$$

and the log likelihood

$$\log(L(\rho)) = n\log(\rho) + \log(1-\rho)\sum_{i=1}^n y_i.$$

Hence

$$\frac{d}{d\rho}\log(L(\rho)) = \frac{n}{\rho} - \frac{1}{1-\rho}\sum_{i=1}^n y_i \stackrel{\text{set}}{=} 0$$

or $n(1-\rho)/\rho = \sum_{i=1}^n y_i$ or $n - n\rho - \rho\sum_{i=1}^n y_i = 0$ or

$$\hat{\rho} = \frac{n}{n + \sum_{i=1}^n y_i}.$$

This solution is unique and

$$\frac{d^2}{d\rho^2}\log(L(\rho)) = \frac{-n}{\rho^2} - \frac{\sum_{i=1}^n y_i}{(1-\rho)^2} < 0.$$

Thus

$$\hat{\rho} = \frac{n}{n + \sum_{i=1}^n Y_i}$$

is the MLE of ρ.

The UMVUE, MLE, and MME of $(1-\rho)/\rho$ is $\overline{Y}$.

10.20 The Gompertz Distribution

If Y has a Gompertz distribution, $Y \sim \text{Gomp}(\theta, \nu)$, then the pdf of Y is

$$f(y) = \nu e^{\theta y} \exp\left[\frac{-\nu}{\theta}(e^{\theta y} - 1)\right]$$

for $\theta > 0$ where $\nu > 0$ and $y > 0$. If $\theta = 0$, the Gomp($\theta = 0, \nu$) distribution is the exponential $(1/\nu)$ distribution. The cdf is

$$F(y) = 1 - \exp\left[\frac{-\nu}{\theta}(e^{\theta y} - 1)\right]$$

for $\theta > 0$ and $y > 0$. For fixed θ this distribution is a scale family with scale parameter $1/\nu$.

This family is a 1P-REF if $\theta \in (0, \infty)$ is known, and $W = e^{\theta Y} - 1 \sim \text{EXP}(\theta/\nu)$. Thus $e^{tY} \sim \text{EXP}(1, t/\nu)$ for $t > 0$ and $E(e^{tY}) = 1 + t/\nu$ for $t \geq 0$, but the mgf for Y does not exist. Note that the kth derivative of $(1 + t/\nu)$ is 0 for integer $k \geq 2$.

10.21 The Half Cauchy Distribution

If Y has a half Cauchy distribution, $Y \sim \text{HC}(\mu, \sigma)$, then the pdf of Y is

$$f(y) = \frac{2}{\pi\sigma\left[1 + \left(\frac{y-\mu}{\sigma}\right)^2\right]}$$

where $y \geq \mu$, μ is a real number and $\sigma > 0$. The cdf of Y is

$$F(y) = \frac{2}{\pi} \arctan\left(\frac{y-\mu}{\sigma}\right)$$

for $y \geq \mu$ and is 0, otherwise. This distribution is a right skewed location–scale family.

MED$(Y) = \mu + \sigma$.
MAD$(Y) = 0.73205\sigma$.

10.22 The Half Logistic Distribution

If Y has a half logistic distribution, $Y \sim \text{HL}(\mu, \sigma)$, then the pdf of Y is

$$f(y) = \frac{2\exp(-(y-\mu)/\sigma)}{\sigma[1 + \exp(-(y-\mu)/\sigma)]^2}$$

where $\sigma > 0$, $y \geq \mu$, and μ are real. The cdf of Y is

$$F(y) = \frac{\exp[(y-\mu)/\sigma] - 1}{1 + \exp[(y-\mu)/\sigma)]}$$

for $y \geq \mu$ and 0 otherwise. This family is a right skewed location–scale family.
MED(Y) $= \mu + \log(3)\sigma$.
MAD(Y) $= 0.67346\sigma$.

10.23 The Half Normal Distribution

If Y has a half normal distribution, $Y \sim \text{HN}(\mu, \sigma^2)$, then the pdf of Y is

$$f(y) = \frac{2}{\sqrt{2\pi}\,\sigma} \exp\left(\frac{-(y-\mu)^2}{2\sigma^2}\right)$$

where $\sigma > 0$ and $y \geq \mu$ and μ is real. Let $\Phi(y)$ denote the standard normal cdf. Then the cdf of Y is

$$F(y) = 2\Phi\left(\frac{y-\mu}{\sigma}\right) - 1$$

for $y > \mu$ and $F(y) = 0$, otherwise.
$E(Y) = \mu + \sigma\sqrt{2/\pi} \approx \mu + 0.797885\sigma$.

$$\text{VAR}(Y) = \frac{\sigma^2(\pi - 2)}{\pi} \approx 0.363380\sigma^2.$$

This is an asymmetric location–scale family that has the same distribution as $\mu + \sigma|Z|$ where $Z \sim N(0,1)$. Note that $Z^2 \sim \chi^2_1$. Hence the formula for the rth moment of the χ^2_1 random variable can be used to find the moments of Y.
MED(Y) $= \mu + 0.6745\sigma$.
MAD(Y) $= 0.3990916\sigma$.
Notice that

$$f(y) = I(y \geq \mu)\,\frac{2}{\sqrt{2\pi}\,\sigma}\,\exp\left[\left(\frac{-1}{2\sigma^2}\right)(y-\mu)^2\right]$$

is a **1P-REF** if μ is known. Hence $\Theta = (0,\infty)$, $\eta = -1/(2\sigma^2)$, and $\Omega = (-\infty, 0)$.
$W = (Y-\mu)^2 \sim G(1/2, 2\sigma^2)$.
If $Y_1, \ldots, Y_n$ are iid HN(μ, σ^2), then

$$T_n = \sum_{i=1}^{n}(Y_i - \mu)^2 \sim G(n/2, 2\sigma^2).$$

If μ is known, then the likelihood

$$L(\sigma^2) = c\,\frac{1}{\sigma^n}\exp\left[\left(\frac{-1}{2\sigma^2}\right)\sum_{i=1}^n (y_i-\mu)^2\right],$$

and the log likelihood

$$\log(L(\sigma^2)) = d - \frac{n}{2}\log(\sigma^2) - \frac{1}{2\sigma^2}\sum_{i=1}^n (y_i-\mu)^2.$$

Hence

$$\frac{d}{d(\sigma^2)}\log(L(\sigma^2)) = \frac{-n}{2(\sigma^2)} + \frac{1}{2(\sigma^2)^2}\sum_{i=1}^n (y_i-\mu)^2 \stackrel{\text{set}}{=} 0,$$

or $\sum_{i=1}^n (y_i-\mu)^2 = n\sigma^2$ or

$$\hat{\sigma}^2 = \frac{1}{n}\sum_{i=1}^n (y_i-\mu)^2.$$

This solution is unique and

$$\frac{d^2}{d(\sigma^2)^2}\log(L(\sigma^2)) =$$

$$\frac{n}{2(\sigma^2)^2} - \left.\frac{\sum_{i=1}^n (y_i-\mu)^2}{(\sigma^2)^3}\right|_{\sigma^2=\hat{\sigma}^2} = \frac{n}{2(\hat{\sigma}^2)^2} - \frac{n\hat{\sigma}^2}{(\hat{\sigma}^2)^3}\frac{2}{2} = \frac{-n}{2(\hat{\sigma}^2)^2} < 0.$$

Thus

$$\hat{\sigma}^2 = \frac{1}{n}\sum_{i=1}^n (Y_i-\mu)^2$$

is the UMVUE and MLE of σ^2 if μ is known.

If $r > -n/2$ and if μ is known, then T_n^r is the UMVUE of

$$E(T_n^r) = 2^r\sigma^{2r}\Gamma(r+n/2)/\Gamma(n/2).$$

Example 5.3 shows that $(\hat{\mu},\hat{\sigma}^2) = (Y_{(1)}, \frac{1}{n}\sum_{i=1}^n (Y_i - Y_{(1)})^2)$ is the MLE of (μ,σ^2). Following Abuhassan and Olive (2008), a large sample $100(1-\alpha)\%$ confidence interval for σ^2 is

$$\left(\frac{n\hat{\sigma}^2}{\chi^2_{n-1}(1-\alpha/2)}, \frac{n\hat{\sigma}^2}{\chi^2_{n-1}(\alpha/2)}\right) \tag{10.8}$$

while a large sample $100(1-\alpha)\%$ CI for μ is

$$\left(\hat{\mu} + \hat{\sigma}\log(\alpha)\,\Phi^{-1}\left(\frac{1}{2}+\frac{1}{2n}\right)(1+13/n^2),\ \hat{\mu}\right). \tag{10.9}$$

If μ is known, then a $100(1-\alpha)\%$ CI for σ^2 is

$$\left(\frac{T_n}{\chi^2_n(1-\alpha/2)}, \frac{T_n}{\chi^2_n(\alpha/2)}\right). \tag{10.10}$$

10.24 The Hypergeometric Distribution

If Y has a hypergeometric distribution, $Y \sim \text{HG}(C, N-C, n)$, then the data set contains N objects of two types. There are C objects of the first type (that you wish to count) and $N-C$ objects of the second type. Suppose that n objects are selected at random without replacement from the N objects. Then Y counts the number of the n selected objects that were of the first type. The pmf of Y is

$$f(y) = P(Y=y) = \frac{\binom{C}{y}\binom{N-C}{n-y}}{\binom{N}{n}}$$

where the integer y satisfies $\max(0, n-N+C) \le y \le \min(n, C)$. The right inequality is true since if n objects are selected, then the number of objects y of the first type must be less than or equal to both n and C. The first inequality holds since $n-y$ counts the number of objects of second type. Hence $n-y \le N-C$.

Let $p = C/N$. Then

$$E(Y) = \frac{nC}{N} = np$$

and

$$\text{VAR}(Y) = \frac{nC(N-C)}{N^2}\frac{N-n}{N-1} = np(1-p)\frac{N-n}{N-1}.$$

If n is small compared to both C and $N-C$, then $Y \approx \text{BIN}(n, p)$. If n is large but n is small compared to both C and $N-C$, then $Y \approx N(np, np(1-p))$.

10.25 The Inverse Exponential Distribution

If Y has an inverse exponential distribution, $Y \sim \text{IEXP}(\theta)$, then the pdf of Y is

$$f(y) = \frac{\theta}{y^2}\exp\left(\frac{-\theta}{y}\right)$$

where $y > 0$ and $\theta > 0$. The cdf $F(y) = \exp(-\theta/y)$ for $y > 0$. $E(Y)$ and $V(Y)$ do not exist. $\text{MED}(Y) = \theta/\log(2)$. This distribution is a **1P-REF** and a scale family with scale parameter θ. This distribution is the inverted gamma($\nu = 1, \lambda = 1/\theta$) distribution.

$W = 1/Y \sim \text{EXP}(1/\theta)$.

Suppose that $Y_1, \ldots, Y_n$ are iid IEXP(θ), then

$$T_n = \sum_{i=1}^{n} \frac{1}{Y_i} \sim G(n, 1/\theta).$$

The likelihood

$$L(\theta) = c\ \theta^n \exp\left[-\theta \sum_{i=1}^{n} \frac{1}{y_i}\right],$$

and the log likelihood

$$\log(L(\theta)) = d + n\log(\theta) - \theta \sum_{i=1}^{n} \frac{1}{y_i}.$$

Hence

$$\frac{d}{d\theta}\log(L(\theta)) = \frac{n}{\theta} - \sum_{i=1}^{n} \frac{1}{y_i} \stackrel{\text{set}}{=} 0,$$

or

$$\hat{\theta} = \frac{n}{\sum_{i=1}^{n} \frac{1}{y_i}}.$$

Since this solution is unique and

$$\frac{d^2}{d\theta^2}\log(L(\theta)) = \frac{-n}{\theta^2} < 0,$$

$$\hat{\theta} = \frac{n}{\sum_{i=1}^{n} \frac{1}{Y_i}}$$

is the MLE of θ.

If $r > -n$, then $[T_n]^r$ is the UMVUE of

$$E([T_n]^r) = \frac{\Gamma(r+n)}{\theta^r \Gamma(n)}.$$

10.26 The Inverse Gaussian Distribution

If Y has an inverse Gaussian distribution (also called a Wald distribution), $Y \sim$ IG(θ, λ), then the pdf of Y is

$$f(y) = \sqrt{\frac{\lambda}{2\pi y^3}} \exp\left[\frac{-\lambda(y-\theta)^2}{2\theta^2 y}\right]$$

where $y, \theta, \lambda > 0$. The mgf is

$$m(t) = \exp\left[\frac{\lambda}{\theta}\left(1 - \sqrt{1 - \frac{2\theta^2 t}{\lambda}}\right)\right]$$

for $t < \lambda/(2\theta^2)$. The characteristic function is

$$\phi(t) = \exp\left[\frac{\lambda}{\theta}\left(1 - \sqrt{1 - \frac{2\theta^2 it}{\lambda}}\right)\right].$$

$E(Y) = \theta$ and

$$\text{VAR}(Y) = \frac{\theta^3}{\lambda}.$$

See Datta (2005) and Schwarz and Samanta (1991) for additional properties.

Notice that

$$f(y) = \sqrt{\frac{1}{y^3} I(y > 0)}\ \sqrt{\frac{\lambda}{2\pi}}\ e^{\lambda/\theta} \exp\left[\frac{-\lambda}{2\theta^2} y - \frac{\lambda}{2}\frac{1}{y}\right]$$

is a two-parameter exponential family.

If $Y_1, \ldots, Y_n$ are iid $\text{IG}(\theta, \lambda)$, then

$$\sum_{i=1}^n Y_i \sim \text{IG}(n\theta, n^2\lambda) \ \text{ and } \ \overline{Y} \sim \text{IG}(\theta, n\lambda).$$

If λ is known, then the likelihood

$$L(\theta) = c\, e^{n\lambda/\theta} \exp\left[\frac{-\lambda}{2\theta^2} \sum_{i=1}^n y_i\right],$$

and the log likelihood

$$\log(L(\theta)) = d + \frac{n\lambda}{\theta} - \frac{\lambda}{2\theta^2} \sum_{i=1}^n y_i.$$

Hence

$$\frac{d}{d\theta} \log(L(\theta)) = \frac{-n\lambda}{\theta^2} + \frac{\lambda}{\theta^3} \sum_{i=1}^n y_i \stackrel{\text{set}}{=} 0,$$

or $\sum_{i=1}^n y_i = n\theta$ or

$$\hat{\theta} = \overline{y}.$$

This solution is unique and

$$\frac{d^2}{d\theta^2} \log(L(\theta)) = \frac{2n\lambda}{\theta^3} - \frac{3\lambda \sum_{i=1}^n y_i}{\theta^4}\bigg|_{\theta=\hat{\theta}} = \frac{2n\lambda}{\hat{\theta}^3} - \frac{3n\lambda\hat{\theta}}{\hat{\theta}^4} = \frac{-n\lambda}{\hat{\theta}^3} < 0.$$

Thus $\overline{Y}$ is the UMVUE, MLE, and MME of θ if λ is known.

If θ is known, then the likelihood

$$L(\lambda) = c\,\lambda^{n/2} \exp\left[\frac{-\lambda}{2\theta^2}\sum_{i=1}^{n}\frac{(y_i-\theta)^2}{y_i}\right],$$

and the log likelihood

$$\log(L(\lambda)) = d + \frac{n}{2}\log(\lambda) - \frac{\lambda}{2\theta^2}\sum_{i=1}^{n}\frac{(y_i-\theta)^2}{y_i}.$$

Hence

$$\frac{d}{d\lambda}\log(L(\lambda)) = \frac{n}{2\lambda} - \frac{1}{2\theta^2}\sum_{i=1}^{n}\frac{(y_i-\theta)^2}{y_i} \stackrel{\text{set}}{=} 0$$

or

$$\hat{\lambda} = \frac{n\theta^2}{\sum_{i=1}^{n}\frac{(y_i-\theta)^2}{y_i}}.$$

This solution is unique and

$$\frac{d^2}{d\lambda^2}\log(L(\lambda)) = \frac{-n}{2\lambda^2} < 0.$$

Thus

$$\hat{\lambda} = \frac{n\theta^2}{\sum_{i=1}^{n}\frac{(Y_i-\theta)^2}{Y_i}}$$

is the MLE of λ if θ is known.

Another parameterization of the inverse Gaussian distribution takes $\theta = \sqrt{\lambda/\psi}$ so that

$$f(y) = \sqrt{\frac{1}{y^3}}I[y>0]\,\sqrt{\frac{\lambda}{2\pi}}e^{\sqrt{\lambda\psi}}\exp\left[\frac{-\psi}{2}y - \frac{\lambda}{2}\frac{1}{y}\right],$$

where $\lambda > 0$ and $\psi \geq 0$. Here $\Theta = (0,\infty)\times[0,\infty)$, $\eta_1 = -\psi/2$, $\eta_2 = -\lambda/2$ and $\Omega = (-\infty,0]\times(-\infty,0)$. Since Ω is not an open set, this is a **two-parameter full exponential family that is not regular**. If ψ is known then Y is a 1P-REF, but if λ is known then Y is a one-parameter full exponential family. When $\psi = 0$, Y has a one-sided stable distribution with index $1/2$. See Barndorff–Nielsen (1978, p. 117).

10.27 The Inverse Weibull Distribution

Y has an inverse Weibull distribution, $Y \sim \text{IW}(\phi,\lambda)$, if the pdf of Y is

$$f(y) = \frac{\phi}{\lambda}\frac{1}{y^{\phi+1}}\exp\left(\frac{-1}{\lambda}\frac{1}{y^{\phi}}\right)$$

where $y > 0$, $\lambda > 0$ and $\phi > 0$ is known. The cdf is

$$F(y) = \exp\left(\frac{-1}{\lambda}\frac{1}{y^{\phi}}\right)$$

for $y > 0$. This family is a **1P-REF** if ϕ is known.

$1/Y \sim W(\phi,\lambda)$ and $1/Y^{\phi} \sim \text{EXP}(\lambda)$. See Mahmoud et al. (2003).

If $Y_1,\ldots,Y_n$ are iid IW(ϕ,λ), then

$$T_n = \sum_{i=1}^{n}\frac{1}{Y_i^{\phi}} \sim G(n,\lambda).$$

If ϕ is known, then the likelihood

$$L(\lambda) = c\,\frac{1}{\lambda^n}\exp\left[\frac{-1}{\lambda}\sum_{i=1}^{n}\frac{1}{y_i^{\phi}}\right],$$

and the log likelihood

$$\log(L(\lambda)) = d - n\log(\lambda) - \frac{1}{\lambda}\sum_{i=1}^{n}\frac{1}{y_i^{\phi}}.$$

Hence

$$\frac{d}{d\lambda}\log(L(\lambda)) = \frac{-n}{\lambda} + \frac{1}{\lambda^2}\sum_{i=1}^{n}\frac{1}{y_i^{\phi}} \stackrel{\text{set}}{=} 0,$$

or

$$\hat{\lambda} = \frac{1}{n}\sum_{i=1}^{n}\frac{1}{y_i^{\phi}}.$$

This solution is unique and

$$\frac{d^2}{d\lambda^2}\log(L(\lambda)) = \left.\frac{n}{\lambda^2} - \frac{2}{\lambda^3}\sum_{i=1}^{n}\frac{1}{y_i^{\phi}}\right|_{\lambda=\hat{\lambda}} = \frac{n}{\hat{\lambda}^2} - \frac{2n\hat{\lambda}}{\hat{\lambda}^3} = \frac{-n}{\hat{\lambda}^2} < 0.$$

Thus

$$\hat{\lambda} = \frac{1}{n}\sum_{i=1}^{n}\frac{1}{Y_i^{\phi}}$$

is the MLE of λ if ϕ is known.

If $r > -n$ and ϕ is known, then $[T_n]^r$ is the UMVUE of

$$E([T_n]^r) = \frac{\lambda^r\Gamma(r+n)}{\Gamma(n)}.$$

10.28 The Inverted Gamma Distribution

If Y has an inverted gamma distribution, $Y \sim \text{INVG}(\nu, \lambda)$, then the pdf of Y is

$$f(y) = \frac{1}{y^{\nu+1}\Gamma(\nu)} I(y > 0) \frac{1}{\lambda^\nu} \exp\left(\frac{-1}{\lambda}\frac{1}{y}\right)$$

where λ, ν, and y are all positive. It can be shown that $W = 1/Y \sim G(\nu, \lambda)$. This family is a **2P-REF**. This family is a scale family with scale parameter $\tau = 1/\lambda$ if ν is known.

$$E(Y^r) = \frac{\Gamma(\nu - r)}{\lambda^r \Gamma(\nu)}$$

for $\nu > r$.

If ν is known, this family is a one-parameter exponential family. If $Y_1, \ldots, Y_n$ are iid INVG(ν, λ) and ν is known, then $T_n = \sum_{i=1}^n \frac{1}{Y_i} \sim G(n\nu, \lambda)$ and T_n^r is the UMVUE of

$$\lambda^r \frac{\Gamma(r + n\nu)}{\Gamma(n\nu)}$$

for $r > -n\nu$.

If W has an inverse exponential distribution, $W \sim \text{IEXP}(\theta)$, then $W \sim \text{INVG}(\nu = 1, \lambda = 1/\theta)$.

10.29 The Largest Extreme Value Distribution

If Y has a largest extreme value distribution (or Gumbel distribution), $Y \sim \text{LEV}(\theta, \sigma)$, then the pdf of Y is

$$f(y) = \frac{1}{\sigma} \exp\left(-\left(\frac{y-\theta}{\sigma}\right)\right) \exp\left[-\exp\left(-\left(\frac{y-\theta}{\sigma}\right)\right)\right]$$

where y and θ are real and $\sigma > 0$. The cdf of Y is

$$F(y) = \exp\left[-\exp\left(-\left(\frac{y-\theta}{\sigma}\right)\right)\right].$$

This family is an asymmetric location–scale family with a mode at θ.
The mgf

$$m(t) = \exp(t\theta)\Gamma(1 - \sigma t)$$

for $|t| < 1/\sigma$.
$E(Y) \approx \theta + 0.57721\sigma$, and
$\text{VAR}(Y) = \sigma^2\pi^2/6 \approx 1.64493\sigma^2$.

$$\text{MED}(Y) = \theta - \sigma\log(\log(2)) \approx \theta + 0.36651\sigma$$

and

$$\text{MAD}(Y) \approx 0.767049\sigma.$$

$W = \exp(-(Y-\theta)/\sigma) \sim \text{EXP}(1)$.

Notice that

$$f(y) = e^{-y/\sigma} \frac{1}{\sigma} e^{\theta/\sigma} \exp\left[-e^{\theta/\sigma} e^{-y/\sigma}\right]$$

is a one-parameter exponential family in θ if σ is known.

If $Y_1, \ldots, Y_n$ are iid LEV(θ, σ) where σ is known, then the likelihood

$$L(\sigma) = c\, e^{n\theta/\sigma} \exp\left[-e^{\theta/\sigma} \sum_{i=1}^{n} e^{-y_i/\sigma}\right],$$

and the log likelihood

$$\log(L(\theta)) = d + \frac{n\theta}{\sigma} - e^{\theta/\sigma} \sum_{i=1}^{n} e^{-y_i/\sigma}.$$

Hence

$$\frac{d}{d\theta} \log(L(\theta)) = \frac{n}{\sigma} - e^{\theta/\sigma} \frac{1}{\sigma} \sum_{i=1}^{n} e^{-y_i/\sigma} \stackrel{\text{set}}{=} 0,$$

or

$$e^{\theta/\sigma} \sum_{i=1}^{n} e^{-y_i/\sigma} = n,$$

or

$$e^{\theta/\sigma} = \frac{n}{\sum_{i=1}^{n} e^{-y_i/\sigma}},$$

or

$$\hat{\theta} = \sigma \log\left(\frac{n}{\sum_{i=1}^{n} e^{-y_i/\sigma}}\right).$$

Since this solution is unique and

$$\frac{d^2}{d\theta^2} \log(L(\theta)) = \frac{-1}{\sigma^2} e^{\theta/\sigma} \sum_{i=1}^{n} e^{-y_i/\sigma} < 0,$$

$$\hat{\theta} = \sigma \log\left(\frac{n}{\sum_{i=1}^{n} e^{-Y_i/\sigma}}\right)$$

is the MLE of θ.

10.30 The Logarithmic Distribution

If Y has a logarithmic distribution, then the pmf of Y is

$$f(y) = P(Y = y) = \frac{-1}{\log(1-\theta)} \frac{\theta^y}{y}$$

for $y = 1, 2, \ldots$ and $0 < \theta < 1$. This distribution is sometimes called the logarithmic series distribution or the log-series distribution. The mgf

$$m(t) = \frac{\log(1 - \theta e^t)}{\log(1 - \theta)}$$

for $t < -\log(\theta)$.

$$E(Y) = \frac{-1}{\log(1-\theta)} \frac{\theta}{1-\theta}.$$

$$V(Y) = \frac{-\theta}{(1-\theta)^2 \log(\theta)} \left(1 + \frac{\theta}{\log(1-\theta)}\right).$$

Notice that

$$f(y) = \frac{-1}{\log(1-\theta)} \frac{1}{y} \exp(\log(\theta) y)$$

is a **1P-REF**. Hence $\Theta = (0,1)$, $\eta = \log(\theta)$ and $\Omega = (-\infty, 0)$.

If $Y_1, \ldots, Y_n$ are iid logarithmic (θ), then $\overline{Y}$ is the UMVUE of $E(Y)$.

10.31 The Logistic Distribution

If Y has a logistic distribution, $Y \sim L(\mu, \sigma)$, then the pdf of Y is

$$f(y) = \frac{\exp(-(y-\mu)/\sigma)}{\sigma[1 + \exp(-(y-\mu)/\sigma)]^2}$$

where $\sigma > 0$ and y and μ are real. The cdf of Y is

$$F(y) = \frac{1}{1 + \exp(-(y-\mu)/\sigma)} = \frac{\exp((y-\mu)/\sigma)}{1 + \exp((y-\mu)/\sigma)}.$$

This family is a symmetric location–scale family.
The mgf of Y is $m(t) = \pi\sigma t e^{\mu t} \csc(\pi\sigma t)$ for $|t| < 1/\sigma$, and
the chf is $c(t) = \pi i \sigma t e^{i\mu t} \csc(\pi i \sigma t)$ where $csc(t)$ is the cosecant of t.
$E(Y) = \mu$, and
$\text{MED}(Y) = \mu$.
$\text{VAR}(Y) = \sigma^2\pi^2/3$, and
$\text{MAD}(Y) = \log(3)\sigma \approx 1.0986\,\sigma$.
Hence $\sigma = \text{MAD}(Y)/\log(3)$.

The estimators $\hat{\mu} = \overline{Y}_n$ and $S^2 = \frac{1}{n-1}\sum_{i=1}^n (Y_i - \overline{Y}_n)^2$ are sometimes used.

Note that if

$$q = F_{L(0,1)}(c) = \frac{e^c}{1 + e^c} \quad \text{then} \quad c = \log\left(\frac{q}{1-q}\right).$$

Taking $q = .9995$ gives $c = \log(1999) \approx 7.6$.
To see that $\text{MAD}(Y) = \log(3)\sigma$, note that $F(\mu + \log(3)\sigma) = 0.75$, $F(\mu - \log(3)\sigma) = 0.25$, and $0.75 = \exp(\log(3))/(1 + \exp(\log(3)))$.

10.32 The Log-Cauchy Distribution

If Y has a log-Cauchy distribution, $Y \sim \text{LC}(\mu, \sigma)$, then the pdf of Y is

$$f(y) = \frac{1}{\pi\sigma y[1 + (\frac{\log(y)-\mu}{\sigma})^2]}$$

where $y > 0$, $\sigma > 0$, and μ is a real number. This family is a scale family with scale parameter $\tau = e^{\mu}$ if σ is known. It can be shown that $W = \log(Y)$ has a Cauchy(μ, σ) distribution.

10.33 The Log-Gamma Distribution

Y has a log-gamma distribution, $Y \sim LG(\nu, \lambda)$, if the pdf of Y is

$$f(y) = \frac{1}{\lambda^{\nu}\Gamma(\nu)} \exp\left(\nu y + \left(\frac{-1}{\lambda}\right) e^{y}\right)$$

where y is real, $\nu > 0$, and $\lambda > 0$. The mgf

$$m(t) = \frac{\Gamma(t + \nu)\lambda^t}{\Gamma(\nu)}$$

for $t > -\nu$. This family is a **2P-REF**.
$W = e^Y \sim$ gamma (ν, λ).

10.34 The Log-Logistic Distribution

If Y has a log-logistic distribution, $Y \sim \text{LL}(\phi, \tau)$, then the pdf of Y is

$$f(y) = \frac{\phi\tau(\phi y)^{\tau-1}}{[1 + (\phi y)^{\tau}]^2}$$

where $y > 0$, $\phi > 0$, and $\tau > 0$. The cdf of Y is

$$F(y) = 1 - \frac{1}{1 + (\phi y)^{\tau}}$$

for $y > 0$. This family is a scale family with scale parameter ϕ^{-1} if τ is known.

MED$(Y) = 1/\phi$.

It can be shown that $W = \log(Y)$ has a logistic$(\mu = -\log(\phi), \sigma = 1/\tau)$ distribution. Hence $\phi = e^{-\mu}$ and $\tau = 1/\sigma$. Kalbfleisch and Prentice (1980, pp. 27–28) suggest that the log-logistic distribution is a competitor of the lognormal distribution.

10.35 The Lognormal Distribution

If Y has a lognormal distribution, $Y \sim \text{LN}(\mu, \sigma^2)$, then the pdf of Y is

$$f(y) = \frac{1}{y\sqrt{2\pi\sigma^2}} \exp\left(\frac{-(\log(y) - \mu)^2}{2\sigma^2}\right)$$

where $y > 0$ and $\sigma > 0$ and μ is real. The cdf of Y is

$$F(y) = \Phi\left(\frac{\log(y) - \mu}{\sigma}\right) \quad \text{for } y > 0$$

where $\Phi(y)$ is the standard normal $N(0,1)$ cdf. This family is a scale family with scale parameter $\tau = e^{\mu}$ if σ^2 is known.

$$E(Y) = \exp(\mu + \sigma^2/2)$$

and

$$\text{VAR}(Y) = \exp(\sigma^2)(\exp(\sigma^2) - 1)\exp(2\mu).$$

For any r,

$$E(Y^r) = \exp(r\mu + r^2\sigma^2/2).$$

MED$(Y) = \exp(\mu)$ and
$\exp(\mu)[1 - \exp(-0.6744\sigma)] \le \text{MAD}(Y) \le \exp(\mu)[1 + \exp(0.6744\sigma)]$.

Notice that

$$f(y) = \frac{1}{y} I(y \ge 0) \frac{1}{\sqrt{2\pi}} \frac{1}{\sigma} \exp\left(\frac{-\mu^2}{2\sigma^2}\right) \exp\left[\frac{-1}{2\sigma^2}(\log(y))^2 + \frac{\mu}{\sigma^2}\log(y)\right]$$

is a **2P–REF**. Hence $\Theta = (-\infty, \infty) \times (0, \infty)$, $\eta_1 = -1/(2\sigma^2)$, $\eta_2 = \mu/\sigma^2$ and $\Omega = (-\infty, 0) \times (-\infty, \infty)$.

Note that $W = \log(Y) \sim N(\mu, \sigma^2)$.
Notice that

$$f(y) = \frac{1}{y} I(y \ge 0) \frac{1}{\sqrt{2\pi}} \frac{1}{\sigma} \exp\left[\frac{-1}{2\sigma^2}(\log(y) - \mu)^2\right]$$

is a 1P-REF if μ is known,.

If $Y_1, \ldots, Y_n$ are iid LN(μ, σ^2) where μ is known, then the likelihood

$$L(\sigma^2) = c \frac{1}{\sigma^n} \exp\left[\frac{-1}{2\sigma^2} \sum_{i=1}^n (\log(y_i) - \mu)^2\right],$$

and the log likelihood

$$\log(L(\sigma^2)) = d - \frac{n}{2} \log(\sigma^2) - \frac{1}{2\sigma^2} \sum_{i=1}^n (\log(y_i) - \mu)^2.$$

Hence

$$\frac{d}{d(\sigma^2)} \log(L(\sigma^2)) = \frac{-n}{2\sigma^2} + \frac{1}{2(\sigma^2)^2} \sum_{i=1}^n (\log(y_i) - \mu)^2 \overset{\text{set}}{=} 0,$$

or $\sum_{i=1}^n (\log(y_i) - \mu)^2 = n\sigma^2$ or

$$\hat{\sigma}^2 = \frac{\sum_{i=1}^n (\log(y_i) - \mu)^2}{n}.$$

Since this solution is unique and

$$\frac{d^2}{d(\sigma^2)^2} \log(L(\sigma^2)) =$$

$$\left.\frac{n}{2(\sigma^2)^2} - \frac{\sum_{i=1}^n (\log(y_i) - \mu)^2}{(\sigma^2)^3}\right|_{\sigma^2 = \hat{\sigma}^2} = \frac{n}{2(\hat{\sigma}^2)^2} - \frac{n\hat{\sigma}^2}{(\hat{\sigma}^2)^3} \frac{2}{2} = \frac{-n}{2(\hat{\sigma}^2)^2} < 0,$$

$$\hat{\sigma}^2 = \frac{\sum_{i=1}^n (\log(Y_i) - \mu)^2}{n}$$

is the UMVUE and MLE of σ^2 if μ is known.

Since $T_n = \sum_{i=1}^n [\log(Y_i) - \mu]^2 \sim G(n/2, 2\sigma^2)$, if μ is known and $r > -n/2$ then T_n^r is UMVUE of

$$E(T_n^r) = 2^r \sigma^{2r} \frac{\Gamma(r + n/2)}{\Gamma(n/2)}.$$

If σ^2 is known,

$$f(y) = \frac{1}{\sqrt{2\pi}} \frac{1}{\sigma} \frac{1}{y} I(y \ge 0) \exp\left(\frac{-1}{2\sigma^2}(\log(y))^2\right) \exp\left(\frac{-\mu^2}{2\sigma^2}\right) \exp\left[\frac{\mu}{\sigma^2} \log(y)\right]$$

is a 1P-REF.

If $Y_1,\ldots,Y_n$ are iid LN(μ,σ^2), where σ^2 is known, then the likelihood

$$L(\mu) = c\ \exp\left(\frac{-n\mu^2}{2\sigma^2}\right) \exp\left[\frac{\mu}{\sigma^2}\sum_{i=1}^n \log(y_i)\right],$$

and the log likelihood

$$\log(L(\mu)) = d - \frac{n\mu^2}{2\sigma^2} + \frac{\mu}{\sigma^2}\sum_{i=1}^n \log(y_i).$$

Hence

$$\frac{d}{d\mu}\log(L(\mu)) = \frac{-2n\mu}{2\sigma^2} + \frac{\sum_{i=1}^n \log(y_i)}{\sigma^2} \stackrel{\text{set}}{=} 0,$$

or $\sum_{i=1}^n \log(y_i) = n\mu$ or

$$\hat{\mu} = \frac{\sum_{i=1}^n \log(y_i)}{n}.$$

This solution is unique and

$$\frac{d^2}{d\mu^2}\log(L(\mu)) = \frac{-n}{\sigma^2} < 0.$$

Since $T_n = \sum_{i=1}^n \log(Y_i) \sim N(n\mu, n\sigma^2)$,

$$\hat{\mu} = \frac{\sum_{i=1}^n \log(Y_i)}{n}$$

is the UMVUE and MLE of μ if σ^2 is known.

When neither μ nor σ are known, the log likelihood

$$\log(L(\mu,\sigma^2)) = d - \frac{n}{2}\log(\sigma^2) - \frac{1}{2\sigma^2}\sum_{i=1}^n (\log(y_i)-\mu)^2.$$

Let $w_i = \log(y_i)$. Then the log likelihood is

$$\log(L(\mu,\sigma^2)) = d - \frac{n}{2}\log(\sigma^2) - \frac{1}{2\sigma^2}\sum_{i=1}^n (w_i-\mu)^2,$$

which has the same form as the normal $N(\mu,\sigma^2)$ log likelihood. Hence the MLE

$$(\hat{\mu},\hat{\sigma}) = \left(\frac{1}{n}\sum_{i=1}^n W_i, \sqrt{\frac{1}{n}\sum_{i=1}^n (W_i-\overline{W})^2}\right).$$

Hence inference for μ and σ is simple. Use the fact that $W_i = \log(Y_i) \sim N(\mu, \sigma^2)$ and then perform the corresponding normal based inference on the W_i. For example, the classical $(1-\alpha)100\,\%$ CI for μ when σ is unknown is

$$\left(\overline{W}_n - t_{n-1,1-\frac{\alpha}{2}} \frac{S_W}{\sqrt{n}}, \overline{W}_n + t_{n-1,1-\frac{\alpha}{2}} \frac{S_W}{\sqrt{n}}\right)$$

where

$$S_W = \frac{n}{n-1}\hat{\sigma} = \sqrt{\frac{1}{n-1}\sum_{i=1}^{n}(W_i - \overline{W})^2},$$

and $P(t \le t_{n-1,1-\frac{\alpha}{2}}) = 1 - \alpha/2$ when t is from a t distribution with $n-1$ degrees of freedom. Compare Meeker and Escobar (1998, p. 175).

10.36 The Maxwell–Boltzmann Distribution

If Y has a Maxwell–Boltzmann distribution (also known as a Boltzmann distribution or a Maxwell distribution), $Y \sim \text{MB}(\mu, \sigma)$, then the pdf of Y is

$$f(y) = \frac{\sqrt{2}(y-\mu)^2 e^{\frac{-1}{2\sigma^2}(y-\mu)^2}}{\sigma^3\sqrt{\pi}}$$

where μ is real, $y \ge \mu$ and $\sigma > 0$. This is a location–scale family.

$$E(Y) = \mu + \sigma\sqrt{2}\frac{1}{\Gamma(3/2)} = \mu + \sigma\,\frac{2\sqrt{2}}{\sqrt{\pi}}.$$

$$\text{VAR}(Y) = 2\sigma^2\left[\frac{\Gamma(\frac{5}{2})}{\Gamma(3/2)} - \left(\frac{1}{\Gamma(3/2)}\right)^2\right] = \sigma^2\left(3 - \frac{8}{\pi}\right).$$

$\text{MED}(Y) = \mu + 1.5381722\sigma$ and $\text{MAD}(Y) = 0.460244\sigma$.
This distribution a one-parameter exponential family when μ is known.
Note that $W = (Y-\mu)^2 \sim G(3/2, 2\sigma^2)$.
If $Z \sim \text{MB}(0, \sigma)$, then $Z \sim \text{chi}(p = 3, \sigma)$, and

$$E(Z^r) = 2^{r/2}\sigma^r\frac{\Gamma(\frac{r+3}{2})}{\Gamma(3/2)}$$

for $r > -3$.
The mode of Z is at $\sigma\sqrt{2}$.

10.37 The Modified DeMoivre's Law Distribution

If Y has a modified DeMoivre's law distribution, $Y \sim \text{MDL}(\theta, \phi)$, then the pdf of Y is

$$f(y) = \frac{\phi}{\theta}\left(\frac{\theta - y}{\theta}\right)^{\phi-1}$$

for $0 < y < \theta$ where $\phi > 0$ and $\theta > 0$. The cdf of Y is

$$F(y) = 1 - \left(1 - \frac{y}{\theta}\right)^{\phi}$$

for $0 < y < \theta$.

$$E(Y) = \frac{\theta}{\phi + 1}.$$

Notice that

$$f(y) = \frac{\phi}{\theta} I(0 < y < \theta) \frac{\theta}{\theta - y} \exp\left[-\phi \log\left(\frac{\theta}{\theta - y}\right)\right]$$

is a **1P-REF** if θ is known. Thus $\Theta = (0, \infty)$, $\eta = -\phi$ and $\Omega = (-\infty, 0)$.

$$W = t(Y) = -\log\left(\frac{\theta - Y}{\theta}\right) = \log\left(\frac{\theta}{\theta - Y}\right) \sim \text{EXP}(1/\phi).$$

If $Y_1, \ldots, Y_n$ are iid $\text{MDL}(\theta, \phi)$, then

$$T_n = -\sum_{i=1}^n \log\left(\frac{\theta - Y}{\theta}\right) = \sum_{i=1}^n \log\left(\frac{\theta}{\theta - Y}\right) \sim G(n, 1/\phi).$$

If θ is known, then the likelihood

$$L(\phi) = c\, \phi^n \exp\left[-\phi \sum_{i=1}^n \log\left(\frac{\theta}{\theta - y_i}\right)\right],$$

and the log likelihood

$$\log(L(\phi)) = d + n\log(\phi) - \phi \sum_{i=1}^n \log\left(\frac{\theta}{\theta - y_i}\right).$$

Hence

$$\frac{d}{d\phi}\log(L(\phi)) = \frac{n}{\phi} - \sum_{i=1}^n \log\left(\frac{\theta}{\theta - y_i}\right) \overset{\text{set}}{=} 0,$$

or

$$\hat{\phi} = \frac{n}{\sum_{i=1}^n \log\left(\frac{\theta}{\theta - y_i}\right)}.$$

This solution is unique and

$$\frac{d^2}{d\phi^2}\log(L(\phi)) = \frac{-n}{\phi^2} < 0.$$

Hence if θ is known, then

$$\hat{\phi} = \frac{n}{\sum_{i=1}^{n}\log\left(\frac{\theta}{\theta - Y_i}\right)}$$

is the MLE of ϕ. If $r > -n$ and θ is known, then T_n^r is the UMVUE of

$$E(T_n^r) = \frac{1}{\phi^r}\frac{\Gamma(r+n)}{\Gamma(n)}.$$

10.38 The Negative Binomial Distribution

If Y has a negative binomial distribution (also called the Pascal distribution), $Y \sim \text{NB}(r,\rho)$, then the pmf of Y is

$$f(y) = P(Y = y) = \binom{r+y-1}{y}\rho^r(1-\rho)^y$$

for $y = 0, 1, \ldots$ where $0 < \rho < 1$. The moment generating function

$$m(t) = \left[\frac{\rho}{1-(1-\rho)e^t}\right]^r$$

for $t < -\log(1-\rho)$.
$E(Y) = r(1-\rho)/\rho$, and

$$\text{VAR}(Y) = \frac{r(1-\rho)}{\rho^2}.$$

Notice that

$$f(y) = I[y \in \{0,1,\ldots\}]\binom{r+y-1}{y}\rho^r\exp[\log(1-\rho)y]$$

is a **1P-REF** in ρ for known r. Thus $\Theta = (0,1)$, $\eta = \log(1-\rho)$ and $\Omega = (-\infty, 0)$.

If $Y_1, \ldots, Y_n$ are independent $\text{NB}(r_i, \rho)$, then

$$\sum_{i=1}^{n} Y_i \sim \text{NB}\left(\sum_{i=1}^{n} r_i, \rho\right).$$

If $Y_1, \ldots, Y_n$ are iid NB(r, ρ), then

$$T_n = \sum_{i=1}^n Y_i \sim \text{NB}(nr, \rho).$$

If r is known, then the likelihood

$$L(p) = c\, \rho^{nr} \exp\left[\log(1-\rho) \sum_{i=1}^n y_i\right],$$

and the log likelihood

$$\log(L(\rho)) = d + nr\log(\rho) + \log(1-\rho)\sum_{i=1}^n y_i.$$

Hence

$$\frac{d}{d\rho}\log(L(\rho)) = \frac{nr}{\rho} - \frac{1}{1-\rho}\sum_{i=1}^n y_i \stackrel{\text{set}}{=} 0,$$

or

$$\frac{1-\rho}{\rho} nr = \sum_{i=1}^n y_i,$$

or $nr - \rho nr - \rho \sum_{i=1}^n y_i = 0$ or

$$\hat{\rho} = \frac{nr}{nr + \sum_{i=1}^n y_i}.$$

This solution is unique and

$$\frac{d^2}{d\rho^2}\log(L(\rho)) = \frac{-nr}{\rho^2} - \frac{1}{(1-\rho)^2}\sum_{i=1}^n y_i < 0.$$

Thus

$$\hat{\rho} = \frac{nr}{nr + \sum_{i=1}^n Y_i}$$

is the MLE of ρ if r is known.

Notice that $\overline{Y}$ is the UMVUE, MLE, and MME of $r(1-\rho)/\rho$ if r is known.

10.39 The Normal Distribution

If Y has a normal distribution (or Gaussian distribution), $Y \sim N(\mu, \sigma^2)$, then the pdf of Y is

$$f(y) = \frac{1}{\sqrt{2\pi\sigma^2}} \exp\left(\frac{-(y-\mu)^2}{2\sigma^2}\right)$$

where $\sigma > 0$ and μ and y are real. Let $\Phi(y)$ denote the standard normal cdf. Recall that $\Phi(y) = 1 - \Phi(-y)$. The cdf $F(y)$ of Y does not have a closed form, but

$$F(y) = \Phi\left(\frac{y-\mu}{\sigma}\right),$$

and

$$\Phi(y) \approx 0.5(1 + \sqrt{1 - \exp(-2y^2/\pi)}\ )$$

for $y \geq 0$. See Johnson and Kotz (1970a, p. 57).
The moment generating function is

$$m(t) = \exp(t\mu + t^2\sigma^2/2).$$

The characteristic function is $c(t) = \exp(it\mu - t^2\sigma^2/2)$.
$E(Y) = \mu$ and
$\text{VAR}(Y) = \sigma^2$.

$$E[|Y - \mu|^r] = \sigma^r \, \frac{2^{r/2}\Gamma((r+1)/2)}{\sqrt{\pi}} \quad \text{for } r > -1.$$

If $k \geq 2$ is an integer, then $E(Y^k) = (k-1)\sigma^2 E(Y^{k-2}) + \mu E(Y^{k-1})$. See Stein (1981) and Casella and Berger (2002, p. 125) .
$\text{MED}(Y) = \mu$ and

$$\text{MAD}(Y) = \Phi^{-1}(0.75)\sigma \approx 0.6745\sigma.$$

Hence $\sigma = [\Phi^{-1}(0.75)]^{-1}\text{MAD}(Y) \approx 1.483\text{MAD}(Y)$.
This family is a location–scale family which is symmetric about μ.

Suggested estimators are

$$\overline{Y}_n = \hat{\mu} = \frac{1}{n}\sum_{i=1}^{n} Y_i \text{ and } S^2 = S_Y^2 = \frac{1}{n-1}\sum_{i=1}^{n}(Y_i - \overline{Y}_n)^2.$$

The classical $(1-\alpha)100\%$ CI for μ when σ is unknown is

$$\left(\overline{Y}_n - t_{n-1,1-\frac{\alpha}{2}}\frac{S_Y}{\sqrt{n}}, \overline{Y}_n + t_{n-1,1-\frac{\alpha}{2}}\frac{S_Y}{\sqrt{n}}\right)$$

where $P(t \leq t_{n-1,1-\frac{\alpha}{2}}) = 1 - \alpha/2$ when t is from a t distribution with $n-1$ degrees of freedom.

If $\alpha = \Phi(z_\alpha)$, then

$$z_\alpha \approx m - \frac{c_o + c_1 m + c_2 m^2}{1 + d_1 m + d_2 m^2 + d_3 m^3}$$

where

$$m = [-2\log(1-\alpha)]^{1/2},$$

$c_0 = 2.515517, c_1 = 0.802853, c_2 = 0.010328, d_1 = 1.432788, d_2 = 0.189269, d_3 = 0.001308$, and $0.5 \le \alpha$. For $0 < \alpha < 0.5$,

$$z_\alpha = -z_{1-\alpha}.$$

See Kennedy and Gentle (1980, p. 95) .

To see that $\text{MAD}(Y) = \Phi^{-1}(0.75)\sigma$, note that $3/4 = F(\mu + \text{MAD})$ since Y is symmetric about μ. However,

$$F(y) = \Phi\left(\frac{y-\mu}{\sigma}\right)$$

and

$$\frac{3}{4} = \Phi\left(\frac{\mu + \Phi^{-1}(3/4)\sigma - \mu}{\sigma}\right).$$

So $\mu + \text{MAD} = \mu + \Phi^{-1}(3/4)\sigma$. Cancel μ from both sides to get the result.

Notice that

$$f(y) = \frac{1}{\sqrt{2\pi\sigma^2}} \exp\left(\frac{-\mu^2}{2\sigma^2}\right) \exp\left[\frac{-1}{2\sigma^2}y^2 + \frac{\mu}{\sigma^2}y\right]$$

is a **2P–REF**. Hence $\Theta = (0,\infty) \times (-\infty,\infty)$, $\eta_1 = -1/(2\sigma^2)$, $\eta_2 = \mu/\sigma^2$ and $\Omega = (-\infty,0) \times (-\infty,\infty)$.

If σ^2 is known,

$$f(y) = \frac{1}{\sqrt{2\pi\sigma^2}} \exp\left[\frac{-1}{2\sigma^2}y^2\right] \exp\left(\frac{-\mu^2}{2\sigma^2}\right) \exp\left[\frac{\mu}{\sigma^2}y\right]$$

is a 1P-REF. Also the likelihood

$$L(\mu) = c \exp\left(\frac{-n\mu^2}{2\sigma^2}\right) \exp\left[\frac{\mu}{\sigma^2}\sum_{i=1}^n y_i\right]$$

and the log likelihood

$$\log(L(\mu)) = d - \frac{n\mu^2}{2\sigma^2} + \frac{\mu}{\sigma^2}\sum_{i=1}^n y_i.$$

Hence

$$\frac{d}{d\mu}\log(L(\mu)) = \frac{-2n\mu}{2\sigma^2} + \frac{\sum_{i=1}^n y_i}{\sigma^2} \overset{\text{set}}{=} 0,$$

or $n\mu = \sum_{i=1}^n y_i$, or

$$\hat{\mu} = \overline{y}.$$

This solution is unique and

$$\frac{d^2}{d\mu^2}\log(L(\mu)) = \frac{-n}{\sigma^2} < 0.$$

Since $T_n = \sum_{i=1}^n Y_i \sim N(n\mu, n\sigma^2)$, $\overline{Y}$ is the UMVUE, MLE, and MME of μ if σ^2 is known.

If μ is known,

$$f(y) = \frac{1}{\sqrt{2\pi\sigma^2}} \exp\left[\frac{-1}{2\sigma^2}(y-\mu)^2\right]$$

is a 1P-REF. Also the likelihood

$$L(\sigma^2) = c\,\frac{1}{\sigma^n}\exp\left[\frac{-1}{2\sigma^2}\sum_{i=1}^n (y_i-\mu)^2\right]$$

and the log likelihood

$$\log(L(\sigma^2)) = d - \frac{n}{2}\log(\sigma^2) - \frac{1}{2\sigma^2}\sum_{i=1}^n (y_i-\mu)^2.$$

Hence

$$\frac{d}{d\sigma^2}\log(L(\sigma^2)) = \frac{-n}{2\sigma^2} + \frac{1}{2(\sigma^2)^2}\sum_{i=1}^n (y_i-\mu)^2 \overset{\text{set}}{=} 0,$$

or $n\sigma^2 = \sum_{i=1}^n (y_i-\mu)^2$, or

$$\hat{\sigma}^2 = \frac{\sum_{i=1}^n (y_i-\mu)^2}{n}.$$

This solution is unique and

$$\frac{d^2}{d(\sigma^2)^2}\log(L(\sigma^2)) = \left.\frac{n}{2(\sigma^2)^2} - \frac{\sum_{i=1}^n (y_i-\mu)^2}{(\sigma^2)^3}\right|_{\sigma^2=\hat{\sigma}^2} = \frac{n}{2(\hat{\sigma}^2)^2} - \frac{n\hat{\sigma}^2}{(\hat{\sigma}^2)^3}\frac{2}{2}$$

$$= \frac{-n}{2(\hat{\sigma}^2)^2} < 0.$$

Since $T_n = \sum_{i=1}^n (Y_i-\mu)^2 \sim G(n/2, 2\sigma^2)$,

$$\hat{\sigma}^2 = \frac{\sum_{i=1}^n (Y_i-\mu)^2}{n}$$

is the UMVUE and MLE of σ^2 if μ is known.

Note that if μ is known and $r > -n/2$, then T_n^r is the UMVUE of

$$E(T_n^r) = 2^r\sigma^{2r}\frac{\Gamma(r+n/2)}{\Gamma(n/2)}.$$

10.40 The One-Sided Stable Distribution

If Y has a one-sided stable distribution (with index $1/2$, also called a Lévy distribution), $Y \sim OSS(\sigma)$, then the pdf of Y is

$$f(y) = \frac{1}{\sqrt{2\pi y^3}} \sqrt{\sigma} \exp\left(\frac{-\sigma}{2} \frac{1}{y}\right)$$

for $y > 0$ and $\sigma > 0$. The cdf

$$F(y) = 2\left[1 - \Phi\left(\sqrt{\frac{\sigma}{y}}\right)\right]$$

for $y > 0$ where $\Phi(x)$ is the cdf of a $N(0,1)$ random variable.

$$\text{MED}(Y) = \frac{\sigma}{[\Phi^{-1}(3/4)]^2}.$$

This distribution is a scale family with scale parameter σ and a **1P-REF**. When $\sigma = 1$, $Y \sim \text{INVG}(\nu = 1/2, \lambda = 2)$ where INVG stands for inverted gamma. This family is a special case of the inverse Gaussian IG distribution. It can be shown that $W = 1/Y \sim G(1/2, 2/\sigma)$. This distribution is even more outlier prone than the Cauchy distribution. See Feller (1971, p. 52) and Lehmann (1999, p. 76). For applications see Besbeas and Morgan (2004) .

If $Y_1, \ldots, Y_n$ are iid OSS(σ) then $T_n = \sum_{i=1}^n \frac{1}{Y_i} \sim G(n/2, 2/\sigma)$. The likelihood

$$L(\sigma) = \prod_{i=1}^n f(y_i) = \left(\prod_{i=1}^n \frac{1}{\sqrt{2\pi y_i^3}}\right) \sigma^{n/2} \exp\left(\frac{-\sigma}{2} \sum_{i=1}^n \frac{1}{y_i}\right),$$

and the log likelihood

$$\log(L(\sigma)) = \log\left(\prod_{i=1}^n \frac{1}{\sqrt{2\pi y_i^3}}\right) + \frac{n}{2}\log(\sigma) - \frac{\sigma}{2} \sum_{i=1}^n \frac{1}{y_i}.$$

Hence

$$\frac{d}{d\sigma} \log(L(\sigma)) = \frac{n}{2}\frac{1}{\sigma} - \frac{1}{2}\sum_{i=1}^n \frac{1}{y_i} \stackrel{\text{set}}{=} 0,$$

or

$$\frac{n}{2} = \sigma \frac{1}{2} \sum_{i=1}^n \frac{1}{y_i},$$

or

$$\hat{\sigma} = \frac{n}{\sum_{i=1}^n \frac{1}{y_i}}.$$

This solution is unique and

$$\frac{d^2}{d\sigma^2}\log(L(\sigma)) = -\frac{n}{2}\frac{1}{\sigma^2} < 0.$$

Hence the MLE

$$\hat{\sigma} = \frac{n}{\sum_{i=1}^n \frac{1}{Y_i}}.$$

Notice that T_n/n is the UMVUE and MLE of $1/\sigma$ and T_n^r is the UMVUE of

$$\frac{1}{\sigma^r}\frac{2^r\Gamma(r+n/2)}{\Gamma(n/2)}$$

for $r > -n/2$.

10.41 The Pareto Distribution

If Y has a Pareto distribution, $Y \sim \text{PAR}(\sigma, \lambda)$, then the pdf of Y is

$$f(y) = \frac{\frac{1}{\lambda}\sigma^{1/\lambda}}{y^{1+1/\lambda}}$$

where $y \geq \sigma$, $\sigma > 0$, and $\lambda > 0$. The mode is at $Y = \sigma$. The cdf of Y is $F(y) = 1 - (\sigma/y)^{1/\lambda}$ for $y > \sigma$.

This family is a scale family with scale parameter σ when λ is fixed.

$$E(Y) = \frac{\sigma}{1-\lambda}$$

for $\lambda < 1$.

$$E(Y^r) = \frac{\sigma^r}{1-r\lambda} \quad \text{for } r < 1/\lambda.$$

$\text{MED}(Y) = \sigma 2^{\lambda}$.

$X = \log(Y/\sigma)$ is $\text{EXP}(\lambda)$ and $W = \log(Y)$ is $\text{EXP}(\theta = \log(\sigma), \lambda)$.

Notice that

$$f(y) = \frac{1}{y} I[y \geq \sigma] \frac{1}{\lambda} \exp\left[\frac{-1}{\lambda}\log(y/\sigma)\right]$$

is a one-parameter exponential family if σ is known.

If $Y_1, \ldots, Y_n$ are iid $\text{PAR}(\sigma, \lambda)$ then

$$T_n = \sum_{i=1}^n \log(Y_i/\sigma) \sim G(n, \lambda).$$

If σ is known, then the likelihood

$$L(\lambda) = c\frac{1}{\lambda^n}\exp\left[-\left(1+\frac{1}{\lambda}\right)\sum_{i=1}^n \log(y_i/\sigma)\right],$$

and the log likelihood

$$\log(L(\lambda)) = d - n\log(\lambda) - \left(1+\frac{1}{\lambda}\right)\sum_{i=1}^n \log(y_i/\sigma).$$

Hence

$$\frac{d}{d\lambda}\log(L(\lambda)) = \frac{-n}{\lambda} + \frac{1}{\lambda^2}\sum_{i=1}^n \log(y_i/\sigma) \stackrel{\text{set}}{=} 0,$$

or $\sum_{i=1}^n \log(y_i/\sigma) = n\lambda$ or

$$\hat{\lambda} = \frac{\sum_{i=1}^n \log(y_i/\sigma)}{n}.$$

This solution is unique and

$$\frac{d^2}{d\lambda^2}\log(L(\lambda)) = \frac{n}{\lambda^2} - \left.\frac{2\sum_{i=1}^n \log(y_i/\sigma)}{\lambda^3}\right|_{\lambda=\hat{\lambda}} =$$

$$\frac{n}{\hat{\lambda}^2} - \frac{2n\hat{\lambda}}{\hat{\lambda}^3} = \frac{-n}{\hat{\lambda}^2} < 0.$$

Hence

$$\hat{\lambda} = \frac{\sum_{i=1}^n \log(Y_i/\sigma)}{n}$$

is the UMVUE and MLE of λ if σ is known.

If σ is known and $r > -n$, then T_n^r is the UMVUE of

$$E(T_n^r) = \lambda^r \frac{\Gamma(r+n)}{\Gamma(n)}.$$

If neither σ nor λ are known, notice that

$$f(y) = \frac{1}{y}\frac{1}{\lambda}\exp\left[-\left(\frac{\log(y)-\log(\sigma)}{\lambda}\right)\right] I(y \ge \sigma).$$

Hence the likelihood

$$L(\lambda,\sigma) = c\frac{1}{\lambda^n}\exp\left[-\sum_{i=1}^n\left(\frac{\log(y_i)-\log(\sigma)}{\lambda}\right)\right] I(y_{(1)} \ge \sigma),$$

and the log likelihood is

$$\log L(\lambda,\sigma) = \left[d - n\log(\lambda) - \sum_{i=1}^{n}\left(\frac{\log(y_i)-\log(\sigma)}{\lambda}\right)\right] I(y_{(1)} \geq \sigma).$$

Let $w_i = \log(y_i)$ and $\theta = \log(\sigma)$, so $\sigma = e^{\theta}$. Then the log likelihood is

$$\log L(\lambda,\theta) = \left[d - n\log(\lambda) - \sum_{i=1}^{n}\left(\frac{w_i-\theta}{\lambda}\right)\right] I(w_{(1)} \geq \theta),$$

which has the same form as the log likelihood of the EXP(θ,λ) distribution. Hence $(\hat{\lambda},\hat{\theta}) = (\overline{W} - W_{(1)}, W_{(1)})$, and by invariance, the MLE

$$(\hat{\lambda},\hat{\sigma}) = (\overline{W} - W_{(1)}, Y_{(1)}).$$

Following Abuhassan and Olive (2008) , let $D_n = \sum_{i=1}^n (W_i - W_{1:n}) = n\hat{\lambda}$ where $W_{(1)} = W_{1:n}$. For $n > 1$, a 100 $(1-\alpha)\%$ CI for θ is

$$(W_{1:n} - \hat{\lambda}[(\alpha)^{-1/(n-1)} - 1], W_{1:n}). \tag{10.11}$$

Exponentiate the endpoints for a 100 $(1-\alpha)\%$ CI for σ. A 100 $(1-\alpha)\%$ CI for λ is

$$\left(\frac{2D_n}{\chi^2_{2(n-1),1-\alpha/2}}, \frac{2D_n}{\chi^2_{2(n-1),\alpha/2}}\right). \tag{10.12}$$

This distribution is used to model economic data such as national yearly income data and size of loans made by a bank.

10.42 The Poisson Distribution

If Y has a Poisson distribution, $Y \sim$ POIS(θ), then the pmf of Y is

$$f(y) = P(Y = y) = \frac{e^{-\theta}\theta^y}{y!}$$

for $y = 0, 1, \ldots,$ where $\theta > 0$. The mgf of Y is

$$m(t) = \exp(\theta(e^t - 1)),$$

and the characteristic function of Y is $c(t) = \exp(\theta(e^{it} - 1))$.

$E(Y) = \theta$, and
$\text{VAR}(Y) = \theta$.

Chen and Rubin (1986) and Adell and Jodrá (2005) show that
$-1 < \text{MED}(Y) - E(Y) < 1/3$.

Pourahmadi (1995) showed that the moments of a Poisson (θ) random variable can be found recursively. If $k \geq 1$ is an integer and $\binom{0}{0} = 1$, then

$$E(Y^k) = \theta \sum_{i=0}^{k-1} \binom{k-1}{i} E(Y^i).$$

The classical estimator of θ is $\hat{\theta} = \overline{Y}_n$.
The approximations $Y \approx N(\theta, \theta)$ and $2\sqrt{Y} \approx N(2\sqrt{\theta}, 1)$ are sometimes used.

Notice that

$$f(y) = \frac{1}{y!} I[y \in \{0, 1, \ldots\}] \, e^{-\theta} \, \exp[\log(\theta) y]$$

is a **1P-REF**. Thus $\Theta = (0, \infty)$, $\eta = \log(\theta)$ and $\Omega = (-\infty, \infty)$.

If $Y_1, \ldots, Y_n$ are independent POIS(θ_i), then $\sum_{i=1}^n Y_i \sim \text{POIS}(\sum_{i=1}^n \theta_i)$.

If $Y_1, \ldots, Y_n$ are iid $POIS(\theta)$, then

$$T_n = \sum_{i=1}^n Y_i \sim \text{POIS}(n\theta).$$

The likelihood

$$L(\theta) = c \, e^{-n\theta} \exp[\log(\theta) \sum_{i=1}^n y_i],$$

and the log likelihood

$$\log(L(\theta)) = d - n\theta + \log(\theta) \sum_{i=1}^n y_i.$$

Hence

$$\frac{d}{d\theta} \log(L(\theta)) = -n + \frac{1}{\theta} \sum_{i=1}^n y_i \stackrel{\text{set}}{=} 0,$$

or $\sum_{i=1}^n y_i = n\theta$, or

$$\hat{\theta} = \overline{y}.$$

This solution is unique and

$$\frac{d^2}{d\theta^2} \log(L(\theta)) = \frac{-\sum_{i=1}^n y_i}{\theta^2} < 0$$

unless $\sum_{i=1}^n y_i = 0$.

Hence $\overline{Y}$ is the UMVUE and MLE of θ.

Let $W = \sum_{i=1}^n Y_i$ and suppose that $W = w$ is observed. Let $P(T < \chi^2_d(\alpha)) = \alpha$ if $T \sim \chi^2_d$. Then an "exact" $100\,(1-\alpha)\%$ CI for θ is

$$\left(\frac{\chi^2_{2w}(\frac{\alpha}{2})}{2n}, \frac{\chi^2_{2w+2}(1-\frac{\alpha}{2})}{2n} \right)$$

for $w \neq 0$ and

$$\left(0, \frac{\chi^2_2(1-\alpha)}{2n} \right)$$

for $w = 0$.

10.43 The Power Distribution

If Y has a power distribution, $Y \sim \text{POW}(\lambda)$, then the pdf of Y is

$$f(y) = \frac{1}{\lambda} y^{\frac{1}{\lambda}-1},$$

where $\lambda > 0$ and $0 < y \leq 1$. The cdf of Y is $F(y) = y^{1/\lambda}$ for $0 < y \leq 1$. $\text{MED}(Y) = (1/2)^\lambda$.

$W = -\log(Y)$ is $\text{EXP}(\lambda)$. Notice that $Y \sim \text{beta}(\delta = 1/\lambda, \nu = 1)$.

Notice that

$$f(y) = \frac{1}{\lambda} I_{(0,1]}(y) \exp\left[\left(\frac{1}{\lambda} - 1 \right) \log(y) \right]$$

$$= \frac{1}{y} I_{(0,1]}(y)\ \frac{1}{\lambda}\ \exp\left[\frac{-1}{\lambda} (-\log(y)) \right]$$

is a **1P-REF**. Thus $\Theta = (0, \infty)$, $\eta = -1/\lambda$ and $\Omega = (-\infty, 0)$.

If $Y_1, \ldots, Y_n$ are iid $\text{POW}(\lambda)$, then

$$T_n = -\sum_{i=1}^n \log(Y_i) \sim G(n, \lambda).$$

The likelihood

$$L(\lambda) = \frac{1}{\lambda^n} \exp\left[\left(\frac{1}{\lambda} - 1 \right) \sum_{i=1}^n \log(y_i) \right],$$

and the log likelihood

$$\log(L(\lambda)) = -n \log(\lambda) + \left(\frac{1}{\lambda} - 1 \right) \sum_{i=1}^n \log(y_i).$$

Hence

$$\frac{d}{d\lambda}\log(L(\lambda)) = \frac{-n}{\lambda} - \frac{\sum_{i=1}^n \log(y_i)}{\lambda^2} \stackrel{\text{set}}{=} 0,$$

or $-\sum_{i=1}^n \log(y_i) = n\lambda$, or

$$\hat{\lambda} = \frac{-\sum_{i=1}^n \log(y_i)}{n}.$$

This solution is unique and

$$\frac{d^2}{d\lambda^2}\log(L(\lambda)) = \frac{n}{\lambda^2} + \left.\frac{2\sum_{i=1}^n \log(y_i)}{\lambda^3}\right|_{\lambda=\hat{\lambda}}$$

$$= \frac{n}{\hat{\lambda}^2} - \frac{2n\hat{\lambda}}{\hat{\lambda}^3} = \frac{-n}{\hat{\lambda}^2} < 0.$$

Hence

$$\hat{\lambda} = \frac{-\sum_{i=1}^n \log(Y_i)}{n}$$

is the UMVUE and MLE of λ.

If $r > -n$, then T_n^r is the UMVUE of

$$E(T_n^r) = \lambda^r \frac{\Gamma(r+n)}{\Gamma(n)}.$$

A $100\,(1-\alpha)\%$ CI for λ is

$$\left(\frac{2T_n}{\chi^2_{2n,1-\alpha/2}}, \frac{2T_n}{\chi^2_{2n,\alpha/2}}\right). \tag{10.13}$$

10.44 The Two-Parameter Power Distribution

Y has a two-parameter power distribution, $Y \sim \text{power}(\tau,\lambda)$, if the pdf of Y is

$$f(y) = \frac{1}{\tau\lambda}\left(\frac{y}{\tau}\right)^{\frac{1}{\lambda}-1} I(0 < y \le \tau)$$

where $\tau > 0$ and $\lambda > 0$. The cdf

$$F(y) = \left(\frac{y}{\tau}\right)^{1/\lambda}$$

for $0 < y \le \tau$. This is a scale family for fixed λ.

$W = -\log(Y) \sim \text{EXP}(-\log(\tau), \lambda)$.

If $Y_1, \ldots, Y_n$ are iid $power(\tau, \lambda)$, then the likelihood is

$$L(\tau,\lambda) = \frac{1}{\tau^{n/\lambda}} I(0 < y_{(n)} \le \tau) \frac{1}{\lambda^n} \exp\left[\left(\frac{1}{\lambda} - 1\right) \sum_{i=1}^n \log(y_i)\right].$$

For fixed λ, the likelihood L is maximized by minimizing τ. So $\hat{\tau} = Y_{(n)}$.

Then the profile likelihood

$$L_p(\lambda) = L(y_{(n)}, \lambda) = \frac{1}{y_{(n)}^{n/\lambda}} \frac{1}{\lambda^n} \exp\left[\left(\frac{1}{\lambda} - 1\right) \sum_{i=1}^n \log(y_i)\right],$$

and the log profile likelihood

$$\log(L_p(\lambda)) = \log(L(y_{(n)}, \lambda) = -\frac{n}{\lambda} \log(y_{(n)}) - n\log(\lambda) + \left(\frac{1}{\lambda} - 1\right) \sum_{i=1}^n \log(y_i).$$

Hence

$$\frac{d}{d\lambda} \log(L_p(\lambda)) = \frac{n}{\lambda^2} \log(y_{(n)}) + \frac{-n}{\lambda} - \frac{\sum_{i=1}^n \log(y_i)}{\lambda^2} \stackrel{\text{set}}{=} 0,$$

or $-n\lambda + n\log(y_{(n)}) - \sum_{i=1}^n \log(y_i) = 0$ or $-\sum_{i=1}^n \log\left(\frac{y_i}{y_{(n)}}\right) = n\lambda$, or

$$\hat{\lambda} = \frac{-1}{n} \sum_{i=1}^n \log\left(\frac{y_i}{y_{(n)}}\right).$$

This solution is unique and

$$\frac{d^2}{d\lambda^2} \log(L_p(\lambda)) = \frac{n}{\lambda^2} + \left.\frac{2\sum_{i=1}^n \log\left(\frac{y_i}{y_{(n)}}\right)}{\lambda^3}\right|_{\lambda=\hat{\lambda}}$$

$$= \frac{n}{\hat{\lambda}^2} - \frac{2n\hat{\lambda}}{\hat{\lambda}^3} = \frac{-n}{\hat{\lambda}^2} < 0.$$

Hence

$$(\hat{\tau}, \hat{\lambda}) = \left(Y_{(n)}, \frac{-1}{n} \sum_{i=1}^n \log\left(\frac{Y_i}{Y_{(n)}}\right)\right)$$

is the MLE of (τ, λ).

Confidence intervals for τ and λ are given in Example 9.12.

10.45 The Rayleigh Distribution

If Y has a Rayleigh distribution, $Y \sim R(\mu, \sigma)$, then the pdf of Y is

$$f(y) = \frac{y-\mu}{\sigma^2} \exp\left[-\frac{1}{2}\left(\frac{y-\mu}{\sigma}\right)^2\right]$$

where $\sigma > 0$, μ is real, and $y \geq \mu$. See Cohen and Whitten (1988, ch. 10) . This is an asymmetric location–scale family. The cdf of Y is

$$F(y) = 1 - \exp\left[-\frac{1}{2}\left(\frac{y-\mu}{\sigma}\right)^2\right]$$

for $y \geq \mu$, and $F(y) = 0$, otherwise.

$$E(Y) = \mu + \sigma\sqrt{\pi/2} \approx \mu + 1.253314\sigma.$$

$$\text{VAR}(Y) = \sigma^2(4-\pi)/2 \approx 0.429204\sigma^2.$$

$\text{MED}(Y) = \mu + \sigma\sqrt{\log(4)} \approx \mu + 1.17741\sigma.$
Hence $\mu \approx \text{MED}(Y) - 2.6255\text{MAD}(Y)$ and $\sigma \approx 2.230\text{MAD}(Y)$.
Let $\sigma D = \text{MAD}(Y)$. If $\mu = 0$, and $\sigma = 1$, then

$$0.5 = \exp[-0.5(\sqrt{\log(4)} - D)^2] - \exp[-0.5(\sqrt{\log(4)} + D)^2].$$

Hence $D \approx 0.448453$ and $\text{MAD}(Y) \approx 0.448453\sigma$.
It can be shown that $W = (Y-\mu)^2 \sim \text{EXP}(2\sigma^2)$.

Other parameterizations for the Rayleigh distribution are possible.

Note that

$$f(y) = (y-\mu)I(y \geq \mu)\,\frac{1}{\sigma^2}\,\exp\left[-\frac{1}{2\sigma^2}(y-\mu)^2\right]$$

is a **1P-REF** if μ is known.

If $Y_1, \ldots, Y_n$ are iid $R(\mu, \sigma)$, then

$$T_n = \sum_{i=1}^{n}(Y_i - \mu)^2 \sim G(n, 2\sigma^2).$$

If μ is known, then the likelihood

$$L(\sigma^2) = c\,\frac{1}{\sigma^{2n}}\exp\left[-\frac{1}{2\sigma^2}\sum_{i=1}^{n}(y_i-\mu)^2\right],$$

and the log likelihood

$$\log(L(\sigma^2)) = d - n\log(\sigma^2) - \frac{1}{2\sigma^2}\sum_{i=1}^{n}(y_i - \mu)^2.$$

Hence

$$\frac{d}{d(\sigma^2)}\log(L(\sigma^2)) = \frac{-n}{\sigma^2} + \frac{1}{2(\sigma^2)^2}\sum_{i=1}^{n}(y_i - \mu)^2 \stackrel{\text{set}}{=} 0,$$

or $\sum_{i=1}^{n}(y_i - \mu)^2 = 2n\sigma^2$, or

$$\hat{\sigma}^2 = \frac{\sum_{i=1}^{n}(y_i - \mu)^2}{2n}.$$

This solution is unique and

$$\frac{d^2}{d(\sigma^2)^2}\log(L(\sigma^2)) = \frac{n}{(\sigma^2)^2} - \left.\frac{\sum_{i=1}^{n}(y_i - \mu)^2}{(\sigma^2)^3}\right|_{\sigma^2=\hat{\sigma}^2} =$$

$$\frac{n}{(\hat{\sigma}^2)^2} - \frac{2n\hat{\sigma}^2}{(\hat{\sigma}^2)^3} = \frac{-n}{(\hat{\sigma}^2)^2} < 0.$$

Hence

$$\hat{\sigma}^2 = \frac{\sum_{i=1}^{n}(Y_i - \mu)^2}{2n}$$

is the UMVUE and MLE of σ^2 if μ is known.

If μ is known and $r > -n$, then T_n^r is the UMVUE of

$$E(T_n^r) = 2^r\sigma^{2r}\frac{\Gamma(r+n)}{\Gamma(n)}.$$

10.46 The Smallest Extreme Value Distribution

If Y has a smallest extreme value distribution (or log-Weibull distribution), $Y \sim \text{SEV}(\theta, \sigma)$, then the pdf of Y is

$$f(y) = \frac{1}{\sigma}\exp\left(\frac{y-\theta}{\sigma}\right)\exp\left[-\exp\left(\frac{y-\theta}{\sigma}\right)\right]$$

where y and θ are real and $\sigma > 0$. The cdf of Y is

$$F(y) = 1 - \exp\left[-\exp\left(\frac{y-\theta}{\sigma}\right)\right].$$

This family is an asymmetric location–scale family with a longer left tail than right.
$E(Y) \approx \theta - 0.57721\sigma$, and
$\mathrm{VAR}(Y) = \sigma^2\pi^2/6 \approx 1.64493\sigma^2$.
$\mathrm{MED}(Y) = \theta - \sigma\log(\log(2))$.
$\mathrm{MAD}(Y) \approx 0.767049\sigma$.
Y is a one-parameter exponential family in θ if σ is known.
If Y has a SEV(θ, σ) distribution, then $W = -Y$ has an LEV$(-\theta, \sigma)$ distribution.

10.47 The Student's *t* Distribution

If Y has a Student's t distribution, $Y \sim t_p$, then the pdf of Y is

$$f(y) = \frac{\Gamma(\frac{p+1}{2})}{(p\pi)^{1/2}\Gamma(p/2)} \left(1 + \frac{y^2}{p}\right)^{-(\frac{p+1}{2})}$$

where p is a positive integer and y is real. This family is symmetric about 0. The t_1 distribution is the Cauchy(0, 1) distribution. If Z is $N(0,1)$ and is independent of $W \sim \chi^2_p$, then

$$\frac{Z}{(\frac{W}{p})^{1/2}}$$

is t_p.
$E(Y) = 0$ for $p \geq 2$.
$\mathrm{MED}(Y) = 0$.
$\mathrm{VAR}(Y) = p/(p-2)$ for $p \geq 3$, and
$\mathrm{MAD}(Y) = t_{p,0.75}$ where $P(t_p \leq t_{p,0.75}) = 0.75$.
If $\alpha = P(t_p \leq t_{p,\alpha})$, then Cooke et al. (1982, p. 84) suggest the approximation

$$t_{p,\alpha} \approx \sqrt{p\left[\exp\left(\frac{w_\alpha^2}{p}\right) - 1\right]}$$

where

$$w_\alpha = \frac{z_\alpha(8p+3)}{8p+1},$$

z_α is the standard normal cutoff: $\alpha = \Phi(z_\alpha)$, and $0.5 \leq \alpha$. If $0 < \alpha < 0.5$, then

$$t_{p,\alpha} = -t_{p,1-\alpha}.$$

This approximation seems to get better as the degrees of freedom increase.

10.48 The Topp–Leone Distribution

If Y has a Topp–Leone distribution, $Y \sim \text{TL}(\nu)$, then pdf of Y is

$$f(y) = \nu(2-2y)(2y-y^2)^{\nu-1}$$

for $\nu > 0$ and $0 < y < 1$. The cdf of Y is $F(y) = (2y-y^2)^\nu$ for $0 < y < 1$. This distribution is a 1P-REF since

$$f(y) = (2-2y)I_{(0,1)}(y)\ \nu\ \exp[(1-\nu)(-\log(2y-y^2))].$$

$\text{MED}(Y) = 1 - \sqrt{1-(1/2)^{1/\nu}}$, and Example 2.15 showed that $W = -\log(2Y-Y^2) \sim \text{EXP}(1/\nu)$.

The likelihood

$$L(\nu) = c\ \nu^n \prod_{i=1}^n (2y_i - y_i^2)^{\nu-1},$$

and the log likelihood

$$\log(L(\nu)) = d + n\log(\nu) + (\nu-1)\sum_{i=1}^n \log(2y_i - y_i^2).$$

Hence

$$\frac{d}{d\nu}\log(L(\nu)) = \frac{n}{\nu} + \sum_{i=1}^n \log(2y_i - y_i^2) \overset{\text{set}}{=} 0,$$

or $n + \nu\sum_{i=1}^n \log(2y_i - y_i^2) = 0$, or

$$\hat{\nu} = \frac{-n}{\sum_{i=1}^n \log(2y_i - y_i^2)}.$$

This solution is unique and

$$\frac{d^2}{d\nu^2}\log(L(\nu)) = \frac{-n}{\nu^2} < 0.$$

Hence

$$\hat{\nu} = \frac{-n}{\sum_{i=1}^n \log(2Y_i - Y_i^2)} = \frac{n}{-\sum_{i=1}^n \log(2Y_i - Y_i^2)}$$

is the MLE of ν.

If $T_n = -\sum_{i=1}^n \log(2Y_i - Y_i^2) \sim G(n, 1/\nu)$, then T_n^r is the UMVUE of

$$E(T_n^r) = \frac{1}{\nu^r}\frac{\Gamma(r+n)}{\Gamma(n)}$$

for $r > -n$. In particular, $\hat{\nu} = \frac{n}{T_n}$ is the MLE and $\frac{n-1}{T_n}$ is the UMVUE of ν for $n > 1$.

10.49 The Truncated Extreme Value Distribution

If Y has a truncated extreme value distribution, $Y \sim \text{TEV}(\lambda)$, then the pdf of Y is

$$f(y) = \frac{1}{\lambda} \exp\left(y - \frac{e^y - 1}{\lambda}\right)$$

where $y > 0$ and $\lambda > 0$. This distribution is also called the modified extreme value distribution, and $Y \sim \text{Gomp}(\theta = 1, \nu = 1/\lambda)$. The cdf of Y is

$$F(y) = 1 - \exp\left[\frac{-(e^y - 1)}{\lambda}\right]$$

for $y > 0$.
$\text{MED}(Y) = \log(1 + \lambda \log(2))$.
$W = e^Y - 1$ is $\text{EXP}(\lambda)$.

Notice that

$$f(y) = e^y I(y \geq 0) \; \frac{1}{\lambda} \; \exp\left[\frac{-1}{\lambda}(e^y - 1)\right]$$

is a **1P-REF**. Hence $\Theta = (0, \infty)$, $\eta = -1/\lambda$ and $\Omega = (-\infty, 0)$.

If $Y_1, \ldots, Y_n$ are iid $\text{TEV}(\lambda)$, then

$$T_n = \sum_{i=1}^{n} (e^{Y_i} - 1) \sim G(n, \lambda).$$

The likelihood

$$L(\lambda) = c \; \frac{1}{\lambda^n} \exp\left[\frac{-1}{\lambda} \sum_{i=1}^{n} (e^{y_i} - 1)\right],$$

and the log likelihood

$$\log(L(\lambda)) = d - n\log(\lambda) - \frac{1}{\lambda} \sum_{i=1}^{n} (e^{y_i} - 1).$$

Hence

$$\frac{d}{d\lambda} \log(L(\lambda)) = \frac{-n}{\lambda} + \frac{\sum_{i=1}^{n} (e^{y_i} - 1)}{\lambda^2} \stackrel{\text{set}}{=} 0,$$

or $\sum_{i=1}^{n} (e^{y_i} - 1) = n\lambda$, or

$$\hat{\lambda} = \frac{-\sum_{i=1}^{n} (e^{y_i} - 1)}{n}.$$

This solution is unique and

$$\frac{d^2}{d\lambda^2}\log(L(\lambda)) = \frac{n}{\lambda^2} - \left.\frac{2\sum_{i=1}^n (e^{y_i}-1)}{\lambda^3}\right|_{\lambda=\hat{\lambda}}$$

$$= \frac{n}{\hat{\lambda}^2} - \frac{2n\hat{\lambda}}{\hat{\lambda}^3} = \frac{-n}{\hat{\lambda}^2} < 0.$$

Hence

$$\hat{\lambda} = \frac{-\sum_{i=1}^n (e^{Y_i}-1)}{n}$$

is the UMVUE and MLE of λ.

If $r > -n$, then T_n^r is the UMVUE of

$$E(T_n^r) = \lambda^r \frac{\Gamma(r+n)}{\Gamma(n)}.$$

A $100\,(1-\alpha)\%$ CI for λ is

$$\left(\frac{2T_n}{\chi^2_{2n,1-\alpha/2}}, \frac{2T_n}{\chi^2_{2n,\alpha/2}} \right). \tag{10.14}$$

10.50 The Uniform Distribution

If Y has a uniform distribution, $Y \sim U(\theta_1, \theta_2)$, then the pdf of Y is

$$f(y) = \frac{1}{\theta_2 - \theta_1} I(\theta_1 \leq y \leq \theta_2).$$

The cdf of Y is $F(y) = (y - \theta_1)/(\theta_2 - \theta_1)$ for $\theta_1 \leq y \leq \theta_2$.
This family is a location–scale family which is symmetric about $(\theta_1 + \theta_2)/2$. By definition, $m(0) = c(0) = 1$. For $t \neq 0$, the mgf of Y is

$$m(t) = \frac{e^{t\theta_2} - e^{t\theta_1}}{(\theta_2 - \theta_1)t},$$

and the characteristic function of Y is

$$c(t) = \frac{e^{it\theta_2} - e^{it\theta_1}}{(\theta_2 - \theta_1)it}.$$

$E(Y) = (\theta_1 + \theta_2)/2$, and
$\text{MED}(Y) = (\theta_1 + \theta_2)/2$.
$\text{VAR}(Y) = (\theta_2 - \theta_1)^2/12$, and

$\text{MAD}(Y) = (\theta_2 - \theta_1)/4$.
Note that $\theta_1 = \text{MED}(Y) - 2\text{MAD}(Y)$ and $\theta_2 = \text{MED}(Y) + 2\text{MAD}(Y)$.
Some classical estimators are $\hat{\theta}_1 = Y_{(1)}$ and $\hat{\theta}_2 = Y_{(n)}$.

10.51 The Weibull Distribution

If Y has a Weibull distribution, $Y \sim W(\phi, \lambda)$, then the pdf of Y is

$$f(y) = \frac{\phi}{\lambda} y^{\phi-1} e^{-\frac{y^\phi}{\lambda}}$$

where λ, y, and ϕ are all positive. For fixed ϕ, this is a scale family in $\sigma = \lambda^{1/\phi}$.
The cdf of Y is $F(y) = 1 - \exp(-y^\phi/\lambda)$ for $y > 0$.
$E(Y) = \lambda^{1/\phi}\, \Gamma(1 + 1/\phi)$.
$\text{VAR}(Y) = \lambda^{2/\phi}\Gamma(1 + 2/\phi) - (E(Y))^2$.

$$E(Y^r) = \lambda^{r/\phi}\, \Gamma\left(1 + \frac{r}{\phi}\right) \quad \text{for } r > -\phi.$$

$\text{MED}(Y) = (\lambda \log(2))^{1/\phi}$. Note that

$$\lambda = \frac{(\text{MED}(Y))^\phi}{\log(2)}.$$

$W = Y^\phi$ is $\text{EXP}(\lambda)$.

$W = \log(Y)$ has a smallest extreme value $\text{SEV}(\theta = \log(\lambda^{1/\phi}), \sigma = 1/\phi)$ distribution.

Notice that

$$f(y) = y^{\phi-1} I(y \geq 0)\, \frac{\phi}{\lambda}\, \exp\left[\frac{-1}{\lambda} y^\phi\right]$$

is a one-parameter exponential family in λ if ϕ is known.

If $Y_1, \ldots, Y_n$ are iid $W(\phi, \lambda)$, then

$$T_n = \sum_{i=1}^n Y_i^\phi \sim G(n, \lambda).$$

If ϕ is known, then the likelihood

$$L(\lambda) = c\, \frac{1}{\lambda^n} \exp\left[\frac{-1}{\lambda} \sum_{i=1}^n y_i^\phi\right],$$

and the log likelihood

$$\log(L(\lambda)) = d - n\log(\lambda) - \frac{1}{\lambda}\sum_{i=1}^{n} y_i^{\phi}.$$

Hence

$$\frac{d}{d\lambda}\log(L(\lambda)) = \frac{-n}{\lambda} + \frac{\sum_{i=1}^{n} y_i^{\phi}}{\lambda^2} \stackrel{\text{set}}{=} 0,$$

or $\sum_{i=1}^{n} y_i^{\phi} = n\lambda$, or

$$\hat{\lambda} = \frac{\sum_{i=1}^{n} y_i^{\phi}}{n}.$$

This solution was unique and

$$\frac{d^2}{d\lambda^2}\log(L(\lambda)) = \left.\frac{n}{\lambda^2} - \frac{2\sum_{i=1}^{n} y_i^{\phi}}{\lambda^3}\right|_{\lambda=\hat{\lambda}}$$

$$= \frac{n}{\hat{\lambda}^2} - \frac{2n\hat{\lambda}}{\hat{\lambda}^3} = \frac{-n}{\hat{\lambda}^2} < 0.$$

Hence

$$\hat{\lambda} = \frac{\sum_{i=1}^{n} Y_i^{\phi}}{n}$$

is the UMVUE and MLE of λ.

If $r > -n$ and ϕ is known, then T_n^r is the UMVUE of

$$E(T_n^r) = \lambda^r \frac{\Gamma(r+n)}{\Gamma(n)}.$$

MLEs and CIs for ϕ and λ are discussed in Example 9.18.

10.52 The Zero Truncated Poisson Distribution

If Y has a zero truncated Poisson distribution, $Y \sim \text{ZTP}(\theta)$, then the pmf of Y is

$$f(y) = \frac{e^{-\theta}\,\theta^y}{(1-e^{-\theta})\,y!}$$

for $y = 1, 2, 3, \ldots$ where $\theta > 0$. This distribution is a **1P-REF**. The mgf of Y is

$$m(t) = \exp(\theta(e^t - 1))\,\frac{1 - e^{-\theta e^t}}{1 - e^{-\theta}}$$

for all t.

$$E(Y) = \frac{\theta}{(1-e^{-\theta})},$$

and

$$V(Y) = \frac{\theta^2+\theta}{1-e^{-\theta}} - \left(\frac{\theta}{1-e^{-\theta}}\right)^2.$$

The ZTP pmf is obtained from a Poisson distribution where $y = 0$ values are truncated, so not allowed. If $W \sim \text{Poisson}(\theta)$ with pmf $f_W(y)$, then $P(W = 0) = e^{-\theta}$, so $\sum_{y=1}^{\infty} f_W(y) = 1 - e^{-\theta} = \sum_{y=0}^{\infty} f_W(y) - \sum_{y=1}^{\infty} f_W(y)$. So the ZTP pmf $f(y) = f_W(y)/(1-e^{-\theta})$ for $y \neq 0$.

Now $E(Y) = \sum_{y=1}^{\infty} y f(y) = \sum_{y=0}^{\infty} y f(y) = \sum_{y=0}^{\infty} y f_W(y)/(1-e^{-\theta}) = E(W)/(1-e^{-\theta}) = \theta/(1-e^{-\theta})$.

Similarly, $E(Y^2) = \sum_{y=1}^{\infty} y^2 f(y) = \sum_{y=0}^{\infty} y^2 f(y) = \sum_{y=0}^{\infty} y^2 f_W(y)/(1-e^{-\theta}) = E(W^2)/(1-e^{-\theta}) = [\theta^2+\theta]/(1-e^{-\theta})$. So

$$V(Y) = E(Y^2) - (E(Y))^2 = \frac{\theta^2+\theta}{1-e^{-\theta}} - \left(\frac{\theta}{1-e^{-\theta}}\right)^2.$$

10.53 The Zeta Distribution

If Y has a Zeta distribution, $Y \sim \text{Zeta}(\nu)$, then the pmf of Y is

$$f(y) = P(Y = y) = \frac{1}{y^\nu \zeta(\nu)}$$

where $\nu > 1$ and $y = 1, 2, 3, \ldots$. Here the zeta function

$$\zeta(\nu) = \sum_{y=1}^{\infty} \frac{1}{y^\nu}$$

for $\nu > 1$. This distribution is a one-parameter exponential family.

$$E(Y) = \frac{\zeta(\nu-1)}{\zeta(\nu)}$$

for $\nu > 2$, and

$$\text{VAR}(Y) = \frac{\zeta(\nu-2)}{\zeta(\nu)} - \left[\frac{\zeta(\nu-1)}{\zeta(\nu)}\right]^2$$

for $\nu > 3$.

$$E(Y^r) = \frac{\zeta(\nu-r)}{\zeta(\nu)}$$

for $\nu > r+1$.

This distribution is sometimes used for count data, especially by linguistics for word frequency. See Lindsey (2004, p. 154).

10.54 The Zipf Distribution

Y has a Zipf distribution, $Y \sim \text{Zipf}(\nu)$, if the pmf of Y is

$$f(y) = \frac{1}{y^{\nu} z(\nu)}$$

where $y \in \{1, \ldots, m\}$ and m is known, ν is real and

$$z(\nu) = \sum_{y=1}^{m} \frac{1}{y^{\nu}}.$$

$$E(Y^r) = \frac{z(\nu - r)}{z(\nu)}$$

for real r. This family is a **1P-REF** if m is known.

10.55 Complements

Many of the distribution results used in this chapter came from Johnson and Kotz (1970ab) and Patel et al. (1976) . Bickel and Doksum (2007), Castillo (1988), Cohen and Whitten (1988), Cramér (1946), DeGroot and Schervish (2012), Ferguson (1967), Forbes et al. (2011) , Kennedy and Gentle (1980), Kotz and van Dorp (2004), Leemis and McQueston (2008), Lehmann (1983), and Meeker and Escobar (1998) also have useful results on distributions. Also see articles in Kotz and Johnson (1982ab, 1983ab, 1985ab, 1986, 1988ab). Often an entire book is devoted to a single distribution, see for example, Bowman and Shenton (1988).

Abuhassan and Olive (2008) discuss confidence intervals for the two-parameter exponential, two-parameter power, half normal, and Pareto distributions.

A recent discussion of Burr Type III and XII distributions is given in Headrick et al. (2010).

Brownstein and Pensky (2008) show that if $Y = t(X)$ where X has simple inference and t is a monotone transformation that does not depend on any unknown parameters, then Y also has simple inference.

Other distributions in this text (including the hburr, hlev, hpar, hpow, hray, hsev, htev, hweib, inverse half normal, Lindley, and truncated exponential distributions) are the distributions in Problems 1.29, 1.30, 2.52–2.59, 3.8, 3.9, 3.17, 4.9a, 5.47, 5.51, 5.52, 6.25, and 6.41. The Lindley distribution is discussed in Al-Mutairi et al. (2013). See Problems 1.35, 1.36, 3.15, and 5.49.

Chapter 11
Bayesian Methods

Two large classes of parametric inference are frequentist and Bayesian methods. Frequentist methods assume that $\boldsymbol{\theta}$ are constant parameters "generated by nature," while Bayesian methods assume that the parameters $\boldsymbol{\theta}$ are random variables. Chapters 1–10 consider frequentist methods with an emphasis on exponential families, but Bayesian methods also tie in nicely with exponential family theory.

This chapter discusses Bayes' theorem, prior and posterior distributions, Bayesian point estimation, credible intervals, estimating the highest density region with the shorth, Bayesian hypothesis testing, and Bayesian computation.

11.1 Bayes' Theorem

There will be a change of notation in this chapter. Context will be used to determine whether y or θ are random variables, observed random variables, or arguments in a pdf or pmf. For frequentist methods, the notation $f(y)$ and $f(y|x)$ was used for pdfs, pmfs, and conditional pdfs and pmfs. For now, assume that f is replaced by p, $\boldsymbol{y} = (y_1, \ldots, y_n)$ and $\boldsymbol{\theta} = (\theta_1, .., \theta_k)$ where the k parameters are random variables. Then the conditional pdf or pmf $p(\boldsymbol{\theta}|\boldsymbol{y})$ is of great interest, and

$$p(\boldsymbol{\theta}|\boldsymbol{y}) = \frac{p(\boldsymbol{y},\boldsymbol{\theta})}{p(\boldsymbol{y})} = \frac{p(\boldsymbol{y}|\boldsymbol{\theta})p(\boldsymbol{\theta})}{p(\boldsymbol{y})} = \frac{p(\boldsymbol{y}|\boldsymbol{\theta})p(\boldsymbol{\theta})}{\int p(\boldsymbol{y},\boldsymbol{\theta})d\boldsymbol{\theta}} = \frac{p(\boldsymbol{y}|\boldsymbol{\theta})p(\boldsymbol{\theta})}{\int p(\boldsymbol{y},\boldsymbol{t})d\boldsymbol{t}}$$

$$= \frac{p(\boldsymbol{y}|\boldsymbol{\theta})p(\boldsymbol{\theta})}{\int p(\boldsymbol{y}|\boldsymbol{t})p(\boldsymbol{t})d\boldsymbol{t}}$$

where $d\boldsymbol{\theta} = d\boldsymbol{t}$ are dummy variables and the integral is used for a pdf. Bayes' theorem follows by plugging in the following quantities into the above argument. Replace the integral by a sum for a pmf.

Definition 11.1. The *prior* pdf or pmf $\pi(\boldsymbol{\theta}) \equiv \pi_{\boldsymbol{\eta}}(\boldsymbol{\theta})$ of $\boldsymbol{\theta}$ corresponds to the *prior distribution* of $\boldsymbol{\theta}$ where $\boldsymbol{\eta} = (\eta_1, \ldots, \eta_d)$ is a vector of unknown constant parameters called *hyperparameters*. The pdf or pmf $p(\boldsymbol{y}|\boldsymbol{\theta})$ is the *likelihood function* corresponding to the *sampling distribution* and will be denoted by $f(\boldsymbol{y}|\boldsymbol{\theta})$. The *posterior* pdf or pmf $p(\boldsymbol{\theta}\,|\,\boldsymbol{y})$ corresponds to the *posterior distribution*. The marginal pdf or pmf of $\boldsymbol{y}$ is denoted by $m_{\boldsymbol{\eta}}(\boldsymbol{y})$.

Theorem 11.1. Bayes' Theorem: The posterior distribution is given by

$$p(\boldsymbol{\theta}\,|\,\boldsymbol{y}) = \frac{f(\boldsymbol{y}|\boldsymbol{\theta})\pi(\boldsymbol{\theta})}{\int f(\boldsymbol{y}|\boldsymbol{t})\pi(\boldsymbol{t})d\boldsymbol{t}} = \frac{f(\boldsymbol{y}|\boldsymbol{\theta})\pi(\boldsymbol{\theta})}{m_{\boldsymbol{\eta}}(\boldsymbol{y})}.$$

The prior distribution can be interpreted as the prior information about $\boldsymbol{\theta}$ before gathering data. After gathering data, the likelihood function is used to update the prior information, resulting in the posterior distribution for $\boldsymbol{\theta}$. Either the prior distribution or the likelihood function can be multiplied by a constant $c > 0$ and the posterior distribution will be the same since the constant c cancels in the denominator and numerator of Bayes' theorem.

As in Chaps. 1–10, conditional on $\boldsymbol{\theta}$, the likelihood $f(\boldsymbol{y}|\boldsymbol{\theta})$ comes from a parametric distribution with k parameters $\theta_1, \ldots, \theta_k$. However, for Bayesian methods, the parameters are random variables with prior distribution $\pi(\boldsymbol{\theta})$. Note that conditional on the observed data $\boldsymbol{y}$, the marginal $m_{\boldsymbol{\eta}}(\boldsymbol{y})$ is a constant. Hence $p(\boldsymbol{\theta}\,|\,\boldsymbol{y}) \propto f(\boldsymbol{y}|\boldsymbol{\theta})\pi(\boldsymbol{\theta})$. Often $m_{\boldsymbol{\eta}}(\boldsymbol{y})$ is difficult to compute, although there is a large literature on computational methods. See Sect. 11.4. For exponential families, there is often a *conjugate prior*: the prior and posterior come from the same brand name distribution. In this case sometimes the posterior distribution can be computed with the kernel method of Sect. 1.5.

Note that if the likelihood function corresponds to a brand name pdf or pmf, then θ almost always takes on values in an interval. Hence the prior and posterior distributions for θ usually have pdfs instead of pmfs.

Example 11.1, Conjugate Prior for an Exponential Family. Suppose that conditional on $\boldsymbol{\theta}$, y comes from a k-parameter exponential family with pdf or pmf

$$f(y|\boldsymbol{\theta}) = f(y|\boldsymbol{\theta}) = h(y)c(\boldsymbol{\theta})\exp\left[\sum_{j=1}^{k} w_j(\boldsymbol{\theta})t_j(y)\right].$$

If $y_1, \ldots, y_n$ are a random sample, then the likelihood function

$$f(\boldsymbol{y}|\boldsymbol{\theta}) \propto [c(\boldsymbol{\theta})]^n \exp\left[\sum_{j=1}^{k} w_j(\boldsymbol{\theta})\left(\sum_{i=1}^{n} t_j(y_i)\right)\right] = [c(\boldsymbol{\theta})]^n \exp\left[\sum_{j=1}^{k} w_j(\boldsymbol{\theta})\nu_j\right]$$

where $\nu_j = \sum_{i=1}^{n} t_j(y_i)$. Then the conjugate prior pdf

$$\pi(\boldsymbol{\theta}) \propto [c(\boldsymbol{\theta})]^{\eta} \exp\left[\sum_{j=1}^{k} w_j(\boldsymbol{\theta})\gamma_j\right]$$

where the hyperparameters are $\gamma_1,\ldots,\gamma_k$, and η. Hence the posterior pdf

$$p(\boldsymbol{\theta}|\boldsymbol{y}) \propto [c(\boldsymbol{\theta})]^{\eta+n} \exp\left[\sum_{j=1}^{k} w_j(\boldsymbol{\theta})[\gamma_j + v_j]\right].$$

Let $\boldsymbol{\eta} = (\gamma_1 + v_1,\ldots,\gamma_k + v_k, \eta + n)$. Let $\tilde{t}_j(\boldsymbol{\theta}) = w_j(\boldsymbol{\theta})$ and $\eta_j = \gamma_j + v_j$ for $j = 1,\ldots,k$. Let $\tilde{t}_{k+1}(\boldsymbol{\theta}) = \log[c(\boldsymbol{\theta})]$, and $\eta_{k+1} = \eta + n$. Hence the prior and posterior distributions are from a $(k+1)$-parameter exponential family. So $p(\boldsymbol{\theta}|\boldsymbol{y}) = \tilde{h}(\boldsymbol{\theta})b(\boldsymbol{\eta})\exp\left[\sum_{j=1}^{k+1}\tilde{t}_j(\boldsymbol{\theta})\eta_j\right]$.

Example 11.2. Suppose y is a random variable and the likelihood function of y is the binomial(n,ρ) pmf where n is known. Hence $y|\rho \sim$ binomial(n,ρ). Equivalently, conditional on ρ, there are n iid binomial $(1,\rho)$ random variables $w_1,\ldots,w_n$ with $y = \sum_{i=1}^n w_i$. Then the conjugate prior pdf is the beta(δ_π, ν_π) pdf. Hence the posterior pdf $p(\rho|y) \propto f(y|\rho)\pi(\rho) \propto \rho^y(1-\rho)^{n-y}\rho^{\delta_\pi - 1}(1-\rho)^{\nu_\pi - 1} = \rho^{\delta_\pi + y - 1}(1-\rho)^{\nu_\pi + n - y - 1}$. Thus the posterior pdf is the beta$(\delta_\pi + y, \nu_\pi + n - y)$ pdf.

Example 11.3. Suppose that conditional on θ, $y_1,\ldots,y_n \sim$ Poisson(θ) where the conjugate prior is the gamma$(\nu_\pi, 1/\lambda_\pi)$ pdf. Hence the posterior pdf $p(\theta|\boldsymbol{y}) \propto f(\boldsymbol{y}|\theta)\pi(\theta) \propto e^{-n\theta}\theta^{\sum_{i=1}^n y_i}\theta^{\nu_\pi - 1}e^{-\lambda_\pi\theta} = \theta^{(\nu_\pi + \sum_{i=1}^n y_i) - 1}e^{-(\lambda_\pi + n)\theta}$. Thus the posterior pdf is the gamma $(\nu_\pi + \sum_{i=1}^n y_i, 1/(\lambda_\pi + n))$ pdf.

Example 11.4. Suppose that conditional on μ, $y_1,\ldots,y_n \sim$ EXP$(1/\mu)$ where the conjugate prior is the gamma$(\nu_\pi, 1/\lambda_\pi)$ pdf. Hence the posterior pdf $p(\mu|\boldsymbol{y}) \propto f(\boldsymbol{y}|\mu)\pi(\mu) \propto \mu^n e^{-\mu\sum_{i=1}^n y_i}\mu^{\nu_\pi - 1}e^{-\lambda_\pi\mu} = \mu^{(n+\nu_\pi - 1)}e^{-(\lambda_\pi + \sum_{i=1}^n y_i)\mu}$. Thus the posterior pdf is the gamma $(n + \nu_\pi, 1/(\lambda_\pi + \sum_{i=1}^n y_i))$ pdf.

11.2 Bayesian Point Estimation

In frequentist statistics, a point estimator of θ is a one number summary $\hat{\theta}$. Given a posterior distribution $p(\theta|\boldsymbol{y})$, a Bayesian point estimator or Bayes estimator is a one number summary of the posterior distribution. Let $T \equiv T_n(\boldsymbol{y})$ be the Bayes estimator. Let $L(\theta,a)$ be the loss function if a is used as the point estimator of θ. The squared error loss is $L(\theta,a) = (\theta - a)^2$ while the absolute error loss is $L(\theta,a) = |\theta - a|$.

Definition 11.2. Given a loss function $L(\theta,a)$ and a posterior distribution $p(\theta|\boldsymbol{y})$, the *Bayes estimator* T is the value of a that minimizes $E_{\theta|\boldsymbol{y}}[L(\theta,a)]$ where the expectation is with respect to the posterior distribution.

Theorem 11.2. The Bayes estimator T is
a) the mean of the posterior distribution if $L(\theta,a) = (\theta - a)^2$,
b) the median of the posterior distribution if $L(\theta,a) = |\theta - a|$.

Note that $E[(\theta - a)^2]$ is the second moment of θ about a and is minimized by taking a equal to the population mean μ of θ. This result follows since $E(\theta-\mu)^2 = E(\theta - a + a - \mu)^2 = E(\theta-a)^2 + 2E(\theta-a)(a-\mu) + (a-\mu)^2 = E(\theta-a)^2 - (a-\mu)^2$. Hence $E(\theta-\mu)^2 < E(\theta-a)^2$ unless $a = \mu$. It can be shown that taking a equal to the population median minimizes $E[|\theta - a|]$.

Example 11.5. Assume squared error loss is used. For Example 11.2, the Bayes estimator is $T(\boldsymbol{y}) = \frac{\delta_\pi + y}{\delta_\pi + \nu_\pi + n}$. For Example 11.3, the Bayes estimator is $T(\boldsymbol{y}) = \frac{\nu_\pi + \sum_{i=1}^n y_i}{\lambda_\pi + n}$. For Example 11.4, the Bayes estimator is $T(\boldsymbol{y}) = \frac{\nu_\pi + n}{\lambda_\pi + \sum_{i=1}^n y_i}$.

To compare Bayesian and frequentist estimators, use large sample theory and the mean square error $\text{MSE}_T(\theta) = V_{\boldsymbol{y}|\theta}(T) + E_{\boldsymbol{y}|\theta}(T-\theta)^2$ where the variance and expectation are taken with respect to the distribution of $\boldsymbol{y}|\theta$ and θ is treated as a constant: the unknown value of θ that resulted in the sample $y_1, \ldots, y_n$.

Example 11.6. In Example 11.2, condition on ρ. Then $y \sim$ binomial(n, ρ) and the MLE is $M = y/n$. Then $\text{MSE}_M(\rho) = V(y/n) = \rho(1-\rho)/n$. Under squared error loss, the Bayes estimator is $T(y) = \dfrac{\delta_\pi + y}{\delta_\pi + \nu_\pi + n} = \left(\dfrac{n}{\delta_\pi + \nu_\pi + n}\right)\dfrac{y}{n} +$ $(1 - \dfrac{n}{\delta_\pi + \nu_\pi + n})\dfrac{\delta_\pi}{\delta_\pi + \nu_\pi}$, a linear combination of the MLE y/n and the mean of the prior distribution $\dfrac{\delta_\pi}{\delta_\pi + \nu_\pi}$. Then

$$\text{MSE}_T(\rho) = \frac{n\rho(1-\rho)}{(\delta_\pi + \nu_\pi + n)^2} + \left(\frac{\delta_\pi + n\rho}{\delta_\pi + \nu_\pi + n} - \frac{(\delta_\pi + \nu_\pi + n)\rho}{\delta_\pi + \nu_\pi + n}\right)^2 =$$

$$\frac{n\rho(1-\rho) + (\delta_\pi - \delta_\pi\rho - \nu_\pi\rho)^2}{(\delta_\pi + \nu_\pi + n)^2}.$$

Since

$$\sqrt{n}(T - y/n) = \sqrt{n}\,\frac{\delta_\pi - \frac{y}{n}(\delta_\pi + \nu_\pi)}{\delta_\pi + \nu_\pi + n} \xrightarrow{P} 0$$

as $n \to \infty$, the Bayes estimator T and the MLE y/n are asymptotically equivalent. Hence, conditional on ρ, $\sqrt{n}(T-\rho) \xrightarrow{D} X$ and $\sqrt{n}(\frac{y}{n} - \rho) \xrightarrow{D} X$ as $n \to \infty$ where $X \sim N(0, \rho(1-\rho))$ by the CLT applied to y/n where $y = \sum_{i=1}^n w_i$ as in Example 11.2. Note that conditional on ρ, if the likelihood $y|\rho \sim$ binomial(n, ρ) is correct, the above results hold even if the prior is misspecified.

11.3 Bayesian Inference

Let the pdf of the random variable Z be equal to the posterior pdf $p(\theta|\boldsymbol{y})$ of θ. Let ξ_α be the α percentile of Z so that $P(Z \le \xi_\alpha) = P(\theta \le \xi_\alpha|\boldsymbol{y}) = \alpha$ where $0 < \alpha < 1$.

Definition 11.3. Let θ have posterior pdf $p(\theta|\boldsymbol{y})$. A $100(1-\alpha)\%$ *one-sided Bayesian credible interval* (CIU) for θ with upper credible bound U_n is $(-\infty, U_n] = (-\infty, \xi_{1-\alpha}]$. A $100(1-\alpha)\%$ *one-sided Bayesian credible interval* (CIL) for θ with lower credible bound L_n is $(L_n, \infty) = (\xi_\alpha, \infty)$. A $100(1-\alpha)\%$ *Bayesian credible interval* (BCI) for θ is $(L_n, U_n] = (\xi_{\alpha_1}, \xi_{1-\alpha_2}]$ where $\alpha_1 + \alpha_2 = \alpha$. A $100(1-\alpha)\%$ *highest density Bayesian credible interval* (HCI) for θ is a $100(1-\alpha)\%$ BCI with shortest length. The $100(1-\alpha)\%$ *highest density region* (HDR) for θ is the set R_θ which is a region with the smallest volume such that $P(\theta \in R_\theta|\boldsymbol{y}) = 1-\alpha$.

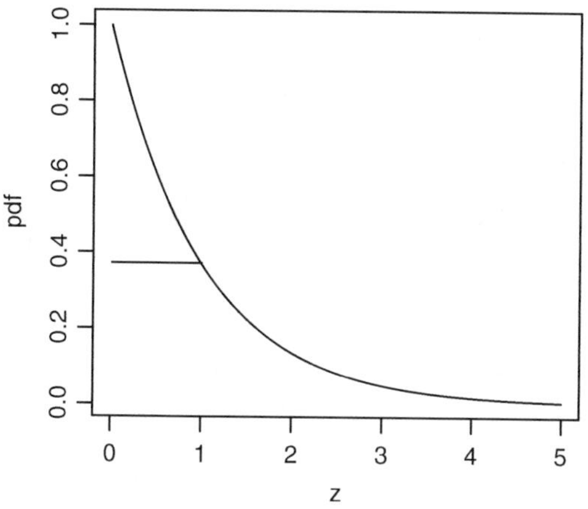

Fig. 11.1 The 36.8 % highest density region is (0,1)

For a random variable θ, the highest density region is a union of $k \geq 1$ disjoint intervals such that sum of the k interval lengths is as small as possible. Suppose that $p(\theta|\boldsymbol{y})$ is a unimodal pdf that has interval support, and that the pdf $p(\theta|\boldsymbol{y})$ decreases rapidly as θ moves away from the mode. Let (a,b) be the shortest interval such that $F_{\theta|\boldsymbol{y}}(b) - F_{\theta|\boldsymbol{y}}(a) = 1-\alpha$ where the cumulative distribution function $F_{\theta|\boldsymbol{y}}(x) = P(\theta \leq x|\boldsymbol{y})$. Then the interval (a,b) is the $100(1-\alpha)$ highest density region and is also the $100(1-\alpha)\%$ HCI. To find the $(1-\alpha)100\%$ highest density region of a posterior pdf, move a horizontal line down from the top of the posterior pdf. The line will intersect the posterior pdf or the boundaries of the support of the posterior pdf at $(a_1,b_1),\ldots,(a_k,b_k)$ for some $k \geq 1$. Stop moving the line when the areas under the posterior pdf corresponding to the intervals is equal to $1-\alpha$. As an example, let $p(z|\boldsymbol{y}) = e^{-z}$ for $z > 0$. See Fig. 11.1 where the area under the posterior pdf from 0 to 1 is 0.368. Hence (0,1) is the 36.8 %, highest density region. Often the highest density region is an interval (a,b) where $p(a|\boldsymbol{y}) = p(b|\boldsymbol{y})$, especially if the support where $p(\theta|\boldsymbol{y}) > 0$ is $(-\infty,\infty)$.

The credible intervals are Bayesian analogs of confidence intervals. Similar definitions apply for a posterior pmf, but make the coverage $\geq (1-\alpha)$ since equality holds only for certain values of α. For example, suppose $p(\theta|\boldsymbol{y}) = 3/6, 2/6$ and $1/6$

for $\theta = 1, 2$, and 3, respectively. Then the 50 % highest density region is $\{1\} = [1, 1]$. The $(500/6)\%$ highest credible interval is [1,2] while the $(500/6)\%$ highest density region is $\{1, 2\}$.

Definition 11.4. Assume θ has posterior pdf $p(\theta|\boldsymbol{y})$.
a) For the *right tailed test* $H_0 : \theta \leq \theta_0$ versus $H_1 : \theta > \theta_0$, reject H_0 if $P(\theta \leq \theta_0|\boldsymbol{y}) = \int_{-\infty}^{\theta_0} p(\theta|\boldsymbol{y})d\theta < \alpha$. Equivalently, reject H_0 if θ is not in the $100(1-\alpha)\%$ CIU $(-\infty, \xi_{1-\alpha}]$. Otherwise, fail to reject H_0.
b) For the *left tailed test* $H_0 : \theta \geq \theta_0$ versus $H_1 : \theta < \theta_0$, reject H_0 if $P(\theta \geq \theta_0|\boldsymbol{y}) = \int_{\theta_0}^{\infty} p(\theta|\boldsymbol{y})d\theta < \alpha$. Equivalently, reject H_0 if θ is not in the $100(1-\alpha)\%$ CIL (ξ_α, ∞). Otherwise, fail to reject H_0.
c) For the *two tailed test* $H_0 : \theta = \theta_0$ versus $H_1 : \theta \neq \theta_0$, reject H_0 if θ_0 is not in a $100(1-\alpha)\%$ credible interval or highest posterior density region for θ. Otherwise, fail to reject H_0.

For the two tailed test, often the $100(1-\alpha)\%$ credible interval $(\xi_{\alpha/2}, \xi_{1-\alpha/2})$ is used. This test is good if the posterior pdf is unimodal and approximately symmetric since then the BCI approximates the highest posterior density region. The $100(1-\alpha)\%$ highest posterior density region should be used whenever possible.

The shorth(c) estimator or interval can be used to approximate the $100(1-\alpha)\%$ highest density credible interval which is the $100(1-\alpha)\%$ highest density region if the posterior pdf is unimodal and the pdf $p(\theta|\boldsymbol{y})$ decreases rapidly as θ moves away from the mode. Let $m = \max(100{,}000, n)$ and use random numbers to generate $z_1, \ldots, z_m$ from a distribution with pdf $p(\theta|\boldsymbol{y})$. Let $c = \lceil m(1-\alpha) \rceil$ where $\lceil x \rceil$ is the smallest integer $\geq x$, e.g., $\lceil 7.7 \rceil = 8$. See Grübel (1988). Let $z_{(1)}, \ldots, z_{(m)}$ be the order statistics of $z_1, \ldots, z_m$. Compute $z_{(c)} - z_{(1)}, z_{(c+1)} - z_{(2)}, \ldots, z_{(m)} - z_{(m-c+1)}$. Then the shorth($c$) interval is the closed interval $[z_{(s)}, z_{(s+c-1)}] = [\tilde{\xi}_{\alpha_1}, \tilde{\xi}_{1-\alpha_2}]$ corresponding to the closed interval with the smallest distance. Then the two tailed test rejects H_0 if θ_0 is not in the shorth interval. The discussion below Theorem 9.1 suggested using the shorth(c) estimator for bootstrap tests.

Remark 11.1. It is often useful to plot the prior pdf $p(\theta)$ to see if the prior pdf represents the researcher's prior information, and to plot the posterior pdf $p(\theta|\boldsymbol{y})$ to see if the posterior pdf is unimodal so that the highest density region will be an interval.

Example 11.7. Following Bolstad (2004, pp. 121, 159), if $Z \sim \text{beta}(a, b)$ with $a > 10$ and $b > 10$, then

$$Z \approx N\left(\frac{a}{a+b}, \frac{ab}{(a+b)^2(a+b+1)}\right),$$

the normal distribution with the same mean and variance as the beta distribution. Then an approximate $100(1-\alpha)\%$ BCI is

$$\frac{a}{a+b} \pm z_{1-\alpha/2}\sqrt{\frac{ab}{(a+b)^2(a+b+1)}}$$

where $P(W \le z_\alpha) = \alpha$ if $W \sim N(0,1)$. In Example 11.2, suppose a beta(1,1) prior is used for ρ. This prior is also the $U(0,1)$ noninformative prior. Suppose $n = 10$ and $y = 8$ so the posterior distribution is beta(9,3). Consider testing $H_0 : \rho \le 0.6$ versus $H_1 : \rho > 0.6$ using $\alpha = 0.05$. Then

$$P(\rho \le 0.6|y) = \int_0^{0.6} \frac{\Gamma(12)}{\Gamma(3)\Gamma(9)} \rho^2 (1-\rho)^8 d\rho = 0.1189 > 0.05 = \alpha.$$

Hence we fail to reject H_0. Use the following *R* command to evaluate the integral.

```
pbeta(q=0.6,shape1=9,shape2=3)
[1] 0.1189168
```

The following *R* commands shows that the 95 % CIU is [0,0.921) while the 95 % CIL is (0.530,1]. The qbeta function is used to find the percentile ξ_δ for $\delta = 0.95$ and $\delta = 0.05$.

```
qbeta(0.95,shape1=9,shape2=3)
[1] 0.92118
qbeta(0.05,shape1=9,shape2=3)
[1] 0.5299132
```

Use the text's *sipack* function shorth and the following commands to estimate the 95 % highest density region $\approx [0.5176, 0.9613]$. The pbeta command computes the cdf $F(z) = P(Z \le z)$ when $Z \sim$ beta($\nu =$ shape1, $\lambda =$ shape 2).

```
z <- rbeta(100000,shape1=9,shape2=3)
shorth(z)
$Ln
[1] 0.5175672
$Un
[1] 0.9613357
pbeta(0.9613,shape1=9,shape2=3) -
pbeta(0.5176,shape1=9,shape2=3)
[1] 0.9502639
```

Now suppose the posterior distribution is beta(10,10). Using the normal approximation, the 95 % highest posterior density region $\approx$

$$\frac{a}{a+b} \pm z_{1-\alpha/2} \sqrt{\frac{ab}{(a+b)^2(a+b+1)}} = 0.5 \pm 1.96 \sqrt{\frac{100}{400(21)}} = (0.2861, 0.7139).$$

The 95 % highest density region using the shorth is $\approx [0.2878, 0.7104]$. See the *R* commands below.

```
pbeta(0.7139,shape1=10,shape2=10) -
pbeta(0.2861,shape1=10,shape2=10)
```

```
[1] 0.952977
z <- rbeta(100000,shape1=10,shape2=10)
shorth(z)
$Ln
[1] 0.287823
$Un
[1] 0.7104014
pbeta(0.7104,shape1=10,shape2=10) -
pbeta(0.2878,shape1=10,shape2=10)
[1] 0.9499
```

Now consider testing $H_0 : f(\boldsymbol{y}|\theta) = f(\boldsymbol{y}|0)$ versus $H_1 : f(\boldsymbol{y}|\theta) = f(\boldsymbol{y}|1)$ which is equivalent to testing $H_0 : \theta = 0$ versus $H_1 : \theta = 1$ where θ is a random variable taking on values 0 and 1 such that H_0 occurs if $\theta = 0$ and H_1 occurs if $\theta = 1$. This test is the Bayesian analog of the frequentist test corresponding to the Neyman–Pearson lemma. Now θ is discrete so the prior pmf $\pi(\theta)$ is the Bernoulli(π_1) pmf with $\pi(i) \equiv \pi_i$ for $i = 0, 1$. The posterior pmf is the Bernoulli($p(1|\boldsymbol{y})$) pmf. The previous discussion usually used prior and posterior distributions that had pdfs. Note that we can think of H_0 and H_1 being disjoint events such that H_1 is the complement of H_0. Bayes' theorem for a pmf replaces the integral by a sum in the denominator. Hence the posterior pmf

$$p(\theta|\boldsymbol{y}) = \frac{\pi(\theta) f(\boldsymbol{y}|\theta)}{\sum_{t=0}^{1} \pi(t) f(\boldsymbol{y}|t)}. \text{ So } p(0|\boldsymbol{y}) = P(H_0|\boldsymbol{y}) = \frac{\pi_0 f(\boldsymbol{y}|0)}{\pi_0 f(\boldsymbol{y}|0) + \pi_1 f(\boldsymbol{y}|1)},$$

and

$$p(1|\boldsymbol{y}) = P(H_1|\boldsymbol{y}) = \frac{\pi_1 f(\boldsymbol{y}|1)}{\pi_0 f(\boldsymbol{y}|0) + \pi_1 f(\boldsymbol{y}|1)}.$$

Let the complement of an event A be denoted by $\overline{A}$. Let D denote the "data." Since the posterior $\propto$ (likelihood)(prior), write $P(A|D) \propto P(D|A)P(A)$, where $P(A|D) = P(D|A)P(A)/P(D)$. Using A and $\overline{A}$,

$$\frac{P(A|D)}{P(\overline{A}|D)} = \frac{P(D|A)}{P(D|\overline{A})} \frac{P(A)}{P(\overline{A})},$$

in the definition below, shows that the posterior odds equals the likelihood ratio times the prior odds. Let the *odds ratio* = the posterior odds divided by the prior odds.

Definition 11.5. The *odds* for an event A equals the probability of the event occurring divided by the probability of the event not occurring:

$$odds(A) = \frac{P(A)}{1 - P(A)} = \frac{P(A)}{P(\overline{A})}.$$

The likelihood ratio is the *Bayes factor* (in favor of A)

$$B(D) = \frac{P(D|A)}{P(D|\overline{A})} = \frac{posterior\ odds}{prior\ odds}.$$

For testing $H_0 : f(\boldsymbol{y}|\theta) = f(\boldsymbol{y}|0)$ versus $H_1 : f(\boldsymbol{y}|\theta) = f(\boldsymbol{y}|1)$, use $A = H_0$ and $\overline{A} = H_1$ with Bayes' theorem. Then the *prior odds*

$$odds_\pi(\theta = 0) = \frac{\pi_0}{\pi_1} = \frac{\pi(0)}{\pi(1)} = \frac{P(H_0)}{P(H_1)},$$

while the *posterior odds*

$$odds_{\theta|\boldsymbol{y}}(\theta = 0) = \frac{p(0|\boldsymbol{y})}{p(1|\boldsymbol{y})} = \frac{P(H_0|\boldsymbol{y})}{P(H_1|\boldsymbol{y})} = \frac{\pi_0 f(\boldsymbol{y}|0)}{\pi_1 f(\boldsymbol{y}|1)}.$$

The *Bayes factor* (in favor of H_0)

$$B(\boldsymbol{y}) = \frac{f(\boldsymbol{y}|0)}{f(\boldsymbol{y}|1)} = \frac{P(H_0|\boldsymbol{y})/P(H_1|\boldsymbol{y})}{P(H_0)/P(H_1)} = \frac{posterior\ odds}{prior\ odds}.$$

Let decision $d = 0$ if H_0 is not rejected and $d = 1$ if H_0 is rejected. Let $L(\theta = 0, d = 1) = w_0$ be the loss when a type I error occurs or the loss for rejecting H_0 when H_0 is true. Let $L(\theta = 1, d = 0) = w_1$ be the loss when a type II error occurs or the loss for failing to reject H_0 when H_0 is false. Assume there is no loss (L(0,0) = L(1,1) = 0) for making a correct decision. Then the expected posterior loss for failing to reject H_0 is $E[L(\theta, d = 0)] = 0p(H_0|\boldsymbol{y}) + w_1 p(H_1|\boldsymbol{y})$ and the expected posterior loss for rejecting H_0 is $E[L(\theta, d = 1)] = w_0 p(H_0|\boldsymbol{y}) + 0p(H_1|\boldsymbol{y})$. The ratio of the two expected posterior losses for rejecting and failing to reject H_0 is

$$\frac{w_0 p(H_0|\boldsymbol{y})}{w_1 p(H_1|\boldsymbol{y})} = \frac{w_0 \pi_0 f(\boldsymbol{y}|0)}{w_1 \pi_1 f(\boldsymbol{y}|1)} = \frac{w_0 \pi_0}{w_1 \pi_1} B(\boldsymbol{y}).$$

Want the decision d to minimize the expected loss, and rejecting H_0 is the better decision when the above ratio is less than 1, since rejecting H_0 has smaller expected posterior loss than failing to reject H_0. This argument proves the following theorem.

Theorem 11.3. The **Bayes test** for $H_0 : f(\boldsymbol{y}|\theta) = f(\boldsymbol{y}|0)$ versus $H_1 : f(\boldsymbol{y}|\theta) = f(\boldsymbol{y}|1)$, rejects H_0 if

$$B(\boldsymbol{y}) = \frac{f(\boldsymbol{y}|0)}{f(\boldsymbol{y}|1)} < \frac{w_1 \pi_1}{w_0 \pi_0},$$

and fails to reject H_0, otherwise.

Berger et al. (1997) suggest using 0–1 loss with $w_0 = w_1 = 1$ unit, and $\pi_0 = \pi_1 = 0.5$. Then if $B(\boldsymbol{y}) < 1$, reject H_0 and report the posterior probability that H_0 is true: $P(H_0|\boldsymbol{y}) = \dfrac{B(\boldsymbol{y})}{1 + B(\boldsymbol{y})}$. If $B(\boldsymbol{y}) \geq 1$, fail to reject H_0 and report the posterior probability that H_1 is true: $P(H_1|\boldsymbol{y}) = \dfrac{1}{1 + B(\boldsymbol{y})}$.

11.4 Bayesian Computation

Suppose the data $\boldsymbol{y}$ has been observed, and that the prior $\pi(\boldsymbol{\theta})$ and the likelihood function $f(\boldsymbol{y}|\boldsymbol{\theta})$ have been selected. Then the posterior pdf is

$$p(\boldsymbol{\theta}\,|\,\boldsymbol{y}) = \frac{f(\boldsymbol{y}|\boldsymbol{\theta})\pi(\boldsymbol{\theta})}{\int f(\boldsymbol{y}|\boldsymbol{t})\pi(\boldsymbol{t})d\boldsymbol{t}} = \frac{f(\boldsymbol{y}|\boldsymbol{\theta})\pi(\boldsymbol{\theta})}{m_{\boldsymbol{\eta}}(\boldsymbol{y})},$$

and $\int f(\boldsymbol{y}|\boldsymbol{t})\pi(\boldsymbol{t})d\boldsymbol{t} = m_{\boldsymbol{\eta}}(\boldsymbol{y})$ needs to be computed or estimated. This computation can be difficult if the posterior distribution is not a brand name distribution. Let $\boldsymbol{Z}$ be a random vector with pdf equal to the prior pdf $\pi(\boldsymbol{z})$. Since $\boldsymbol{y}$ is observed, $\boldsymbol{y}$ is treated as a constant. Let the random variable $W = t_{\boldsymbol{y}}(\boldsymbol{Z}) = f(\boldsymbol{y}|\boldsymbol{Z})$. Then $E(W) = E[t_{\boldsymbol{y}}(\boldsymbol{Z})] = E[f(\boldsymbol{y}|\boldsymbol{Z})] = \int f(\boldsymbol{y}|\boldsymbol{z})\pi(\boldsymbol{z})d\boldsymbol{z} = m_{\boldsymbol{\eta}}(\boldsymbol{y})$. Note that this expected value exists for a proper prior pdf since marginal pdfs exist. Replace the integral by a sum for a pmf. Let $m = \max(K,n)$ and generate $\boldsymbol{Z}_1,\ldots,\boldsymbol{Z}_m$ from the distribution with pdf $\pi(\boldsymbol{z})$ and let $W_i = f(\boldsymbol{y}|\boldsymbol{Z}_i)$. Then by the law of large numbers, $\overline{W} = \sum_{i=1}^m W_i/m \xrightarrow{P} E(W) = m_{\boldsymbol{\eta}}(\boldsymbol{y})$ as $m \to \infty$. If $V(W) < \infty$, then by the CLT, $\overline{W}$ is a $\sqrt{n}$ consistent estimator of $m_{\boldsymbol{\eta}}(\boldsymbol{y})$, and taking K to be a large number like $K = 100{,}000$ will often give a good estimate of $m_{\boldsymbol{\eta}}(\boldsymbol{y})$.

Normal approximations for the posterior pdf are sometimes used. Let $\tilde{\boldsymbol{\theta}}$ be the unique posterior mode of the posterior pdf. Let $\boldsymbol{\theta}$ be $1 \times k$ and let the ijth element of the $k \times k$ matrix $\boldsymbol{I}(\tilde{\boldsymbol{\theta}})$ be given by

$$\boldsymbol{I}_{ij}(\tilde{\boldsymbol{\theta}}) = -\left[\frac{\partial^2}{\partial\theta_i\partial\theta_j}\log f(\boldsymbol{y}|\boldsymbol{\theta})\pi(\boldsymbol{\theta})\right]_{\boldsymbol{\theta}=\tilde{\boldsymbol{\theta}}}.$$

Then under regularity conditions, as $n \to \infty$, the posterior pdf $p(\boldsymbol{\theta}\,|\,\boldsymbol{y})$ can be approximated by the multivariate normal $N_k(\tilde{\boldsymbol{\theta}}, \boldsymbol{I}(\tilde{\boldsymbol{\theta}})^{-1})$ pdf. See Carlin and Louis (2009, p. 108).

Computational methods can also be used to approximate the posterior pdf $p(\boldsymbol{\theta}\,|\,\boldsymbol{y})$ where $p(\boldsymbol{\theta}\,|\,\boldsymbol{y}) \propto f(\boldsymbol{y}|\boldsymbol{\theta})\pi(\boldsymbol{\theta}) = L_{\boldsymbol{y}}(\boldsymbol{\theta})\pi(\boldsymbol{\theta})$ where $L_{\boldsymbol{y}}(\boldsymbol{\theta})$ is the likelihood for fixed $\boldsymbol{y}$. Then the likelihood is maximized by the maximum likelihood estimator (MLE) $\hat{\boldsymbol{\theta}}$. Let $M = L_{\boldsymbol{y}}(\hat{\boldsymbol{\theta}}) = f(\boldsymbol{y}|\hat{\boldsymbol{\theta}})$. Then use the following three steps. 1) Generate $\boldsymbol{\theta}_i$ from a distribution with pdf $\pi(\boldsymbol{\theta})$. 2) Generate $U_i \sim U(0,1)$. 3) If

$$U_i \le \frac{L_{\boldsymbol{y}}(\boldsymbol{\theta})\pi(\boldsymbol{\theta})}{M\pi(\boldsymbol{\theta})} = \frac{L_{\boldsymbol{y}}(\boldsymbol{\theta})}{M},$$

then accept $\boldsymbol{\theta}_i$, otherwise repeat steps 1)–3). Then any accepted $\boldsymbol{\theta}_i$ is a (pseudo) random variate from a distribution with pdf

$$\frac{L_{\boldsymbol{y}}(\boldsymbol{\theta})\pi(\boldsymbol{\theta})}{\int L_{\boldsymbol{y}}(\boldsymbol{\theta})\pi(\boldsymbol{\theta})d\boldsymbol{\theta}} = p(\boldsymbol{\theta}\,|\,\boldsymbol{y}).$$

For a fixed $\boldsymbol{\theta}$ in the prior distribution, the probability that $\boldsymbol{\theta}$ is accepted is $L_{\boldsymbol{y}}(\boldsymbol{\theta})/L_{\boldsymbol{y}}(\hat{\boldsymbol{\theta}})$. See Smith and Gelfand (1992). This rejection sampling technique gives a way to generate a pseudo random sample $\boldsymbol{\theta}_1,\ldots,\boldsymbol{\theta}_m$ from a distribution with pdf $p(\boldsymbol{\theta}|\boldsymbol{y})$. This pseudo random sample can be used to estimate the posterior pdf, the posterior mean, the highest density region, etc.

Markov Chain Monte Carlo is another way to generate pseudo random variables. The Gibbs Sampler is one technique. Let $\boldsymbol{x}$ be an $1 \times b$ vector from a distribution with cdf $F(\boldsymbol{x})$, and let $\boldsymbol{x} = (\boldsymbol{w}_1, \boldsymbol{w}_2, \ldots, \boldsymbol{w}_k)$ where the $\boldsymbol{w}_i$ are $1 \times a_i$ vectors with $2 \le k \le b$ and $\sum_{i=1}^k a_i = b$. Assume that the k *full conditional distributions* are known with cdfs $F(\boldsymbol{w}_1|\boldsymbol{w}_2,\ldots,\boldsymbol{w}_k)$, $F(\boldsymbol{w}_2|\boldsymbol{w}_1,\boldsymbol{w}_3,\ldots,\boldsymbol{w}_k)$, $\ldots$, $F(\boldsymbol{w}_j|\boldsymbol{w}_1,\ldots,\boldsymbol{w}_{j-1},\boldsymbol{w}_{j+1},\ldots,\boldsymbol{w}_k)$, $\ldots$, $F(\boldsymbol{w}_k|\boldsymbol{w}_1,\ldots,\boldsymbol{w}_{k-1})$. Pick an arbitrary starting value $\boldsymbol{x}^0 = (\boldsymbol{w}_1^0,\ldots,\boldsymbol{w}_k^0)$ from the support of the unknown distribution of $\boldsymbol{x}$ which may be a mixture distribution where some of the marginal full conditional distributions have pdfs and some have pmfs. Typically the full conditional distribution will have a pdf or pmf, but could be a mixture of continuous and discrete marginal distributions. Generate $\boldsymbol{x}^1 = (\boldsymbol{w}_1^1,\ldots,\boldsymbol{w}_k^1)$ as follows.
Generate $\boldsymbol{w}_1^1$ from the distribution with cdf $F(\boldsymbol{w}_1|\boldsymbol{w}_2^0,\ldots,\boldsymbol{w}_k^0)$.
Generate $\boldsymbol{w}_2^1$ from the distribution with cdf $F(\boldsymbol{w}_2|\boldsymbol{w}_1^1,\boldsymbol{w}_3^0,\ldots,\boldsymbol{w}_k^0)$.
Generate $\boldsymbol{w}_3^1$ from the distribution with cdf $F(\boldsymbol{w}_3|\boldsymbol{w}_1^1,\boldsymbol{w}_2^1,\boldsymbol{w}_4^0,\ldots,\boldsymbol{w}_k^0)$.

$\vdots$

Generate $\boldsymbol{w}_j^1$ from the distribution with cdf $F(\boldsymbol{w}_j|\boldsymbol{w}_1^1,\ldots,\boldsymbol{w}_{j-1}^1,\boldsymbol{w}_{j+1}^0,\ldots,\boldsymbol{w}_k^0)$.

$\vdots$

Generate $\boldsymbol{w}_{k-1}^1$ from the distribution with cdf $F(\boldsymbol{w}_{k-1}|\boldsymbol{w}_1^1,\ldots,\boldsymbol{w}_{k-2}^1,\boldsymbol{w}_k^0)$.
Generate $\boldsymbol{w}_k^1$ from the distribution with cdf $F(\boldsymbol{w}_k|\boldsymbol{w}_1^1,\ldots,\boldsymbol{w}_{k-1}^1)$.

These k steps generate $\boldsymbol{x}^1$ from $\boldsymbol{x}^0$. Continue the iteration to generate $\boldsymbol{x}^i$ from $\boldsymbol{x}^{i-1}$ for $i = 2,\ldots,d$. For large d, $\boldsymbol{x}^d$ is a pseudo random vector from the distribution with cdf $F(\boldsymbol{x})$, in that under regularity conditions, it can be shown that $\boldsymbol{x}^d \xrightarrow{D} \boldsymbol{x}$ as $d \to \infty$, where the random vector $\boldsymbol{x}$ is from the distribution with cdf $F(\boldsymbol{x})$. Note that if $\boldsymbol{w}_i$ is a random vector from the marginal distribution with cdf $F(\boldsymbol{w}_i)$, then $\boldsymbol{w}_i^d$ is a pseudo random vector from that marginal distribution. The Gibbs Sampler could be used to generate pseudo random vectors $\boldsymbol{x}_1^d,\ldots,\boldsymbol{x}_m^d$ for $m = \max(K,n)$ where K is some integer such as $K = 10{,}000$.

Example 11.8. Let $b = 2$ and $a_i = 1$ for $i = 1,2$. Let $w_1 = y$ and $w_2 = \rho$ in Example 11.2. Then interest might be in the marginal distribution with pmf $f(y) = m_{\boldsymbol{\eta}}(y)$. Then $y|\rho \sim \text{binomial}(n,\rho)$ and $\rho|y \sim \text{beta}(\delta_\pi + y, \nu_\pi + n - y)$. To generate (y^0,ρ^0), randomly select an integer from $\{0,1,..,n\}$ and $\rho^0 = U$ where $U \sim U(0,1)$. Then generate y^1 from a binomial(n,ρ^0) random number generator and ρ^1 from a beta$(\delta_\pi + y^1, \nu_\pi + n - y^1)$ random number generator, and repeat the iteration $d = 1{,}000$ steps. Generate $m = 500$ pairs of random variables (y_i^d,ρ_i^d) for $i = 1,\ldots,m$. Suppose the uniform prior with $\delta_\pi = \nu_\pi = 1$ is used. Then $\hat{m}_{\boldsymbol{\eta}}(y) = \frac{1}{m}\sum_{i=1}^m f(y|\rho_i^d)$. Note that $m_{\boldsymbol{\eta}}(y) = E_\rho[f(y|\rho)] = \int f(y|\rho)\pi(\rho)d\rho$, and the ρ_i^d are a pseudo sample from the distribution with pdf $\pi(\rho)$. Of course, for this simple example, $\pi(\rho)$ is known, and computing the exact form of the beta-binomial pmf $m_{\boldsymbol{\eta}}(y)$ is not too hard, but Gibbs Sampling can be used for harder problems. Can fix t and compute $\hat{m}_{\boldsymbol{\eta}}(t) = \frac{1}{m}\sum_{i=1}^m f(t|\rho_i^d)$ for $t = 0,1,\ldots,n$. Casella and George (1992) used $m = 500$

and $d = 10$ for this example with a beta(2,4) prior, and showed that the histogram of $\hat{m}_{\boldsymbol{\eta}}(t)$ was similar to the true histogram (pmf) of $m_{\boldsymbol{\eta}}(t)$. Could also make a histogram of $y_1^d, \ldots, y_m^d$, or equivalently, use $\tilde{m}_{\boldsymbol{\eta}}(t) = \frac{1}{m}\sum_{i=1}^m I(y_i^d = t)$ for $t = 0, 1, \ldots, n$. This estimator tends to be inferior to $\hat{m}_{\boldsymbol{\eta}}(t)$ since $\tilde{m}_{\boldsymbol{\eta}}(t)$ ignores the parametric structure of the model.

11.5 Complements

DeGroot and Schervish (2012) and Berry and Lindgren (1995) give good introductions to frequentist and Bayesian methods, while Carlin and Louis (2009) provides a good treatment of Bayesian methods. Computational methods are covered by Smith and Gelfand (1992), Casella and George (1992), and Robert and Casella (2010). Lavine and Schervish (1999) discuss Bayes factors, and Berger et al. (1997) discuss Bayes testing. Hyndman (1996) discusses highest density regions, and Frey (2013) discusses the coverage of the shorth interval. Lindley (1972) and Morris (1983) show that exponential families have conjugate priors.

Empirical Bayesian models use the data $\boldsymbol{y}$ to estimate the hyperparameters $\boldsymbol{\eta}$. Hierarchical Bayesian models use a prior distribution for the hyperparameters $\boldsymbol{\eta}$. Hence $\boldsymbol{\eta}$ is a random vector. Then there is a likelihood $f(\boldsymbol{y}|\boldsymbol{\theta})$, a prior $\pi(\boldsymbol{\theta}|\boldsymbol{\eta})$, and a hyperprior $\pi(\boldsymbol{\eta})$ that depends on chosen parameters $\boldsymbol{\tau}$. In a more complicated hierarchical model, $\boldsymbol{\tau}$ would be a random vector with a prior. In general, there might be random vectors $\boldsymbol{\eta}_1, \boldsymbol{\eta}_2, \ldots, \boldsymbol{\eta}_k$ where $\boldsymbol{\eta}_k$ has a prior $\pi(\boldsymbol{\eta}_k)$ that depends on constant parameters $\boldsymbol{\eta}_{k+1}$.

11.6 Problems

PROBLEMS WITH AN ASTERISK * ARE ESPECIALLY USEFUL.

Refer to Chap. 10 for the pdf or pmf of the distributions in the problems below.

11.1. Write Bayes' theorem if p, f and π correspond to pmfs.

As problems 11.2–11.8 illustrate, it is often useful to find the exponential family form of the pdf or pmf $f(y|\theta)$ and write $f(y|\theta) \propto$ terms that depend on θ. Terms that do not depend on θ are treated as constants. The MLE and the Bayes estimator are very similar for large n for these problems. Note that $e^{\theta t(y)} = e^{-\theta[-t(y)]}$.

11.2. Assume that r is known, and that conditional on ρ, the random variables $y_1, \ldots, y_n$ are iid negative binomial NB(r, ρ), so that the likelihood $f(\boldsymbol{y}|\rho) \propto \rho^{nr}(1-\rho)^{\sum_{i=1}^n y_i}$. Then the conjugate prior is the beta(δ_π, ν_π) pdf.

a) Find the posterior pdf.

b) For squared error loss, what is the Bayes estimator of ρ?

11.3. Let $\theta = 1/\lambda$, and write the pdf of the power distribution as $f(y|\theta) = \theta y^{-1} e^{\theta \log(y)}$ for $0 < y < 1$. Assume that conditional on θ, the random variables $y_1, \ldots, y_n$ are iid from this power distribution. Then the conjugate prior is the gamma$(\nu, 1/\delta)$ pdf.

a) Find the posterior pdf.

b) For squared error loss, what is the Bayes estimator of θ?

11.4. Assume that conditional on θ, the random variables $y_1, \ldots, y_n$ are iid from an inverse exponential IEXP(θ) distribution. Then the conjugate prior is the gamma$(\nu, 1/\lambda)$ pdf.

a) Find the posterior pdf.

b) For squared error loss, what is the Bayes estimator of θ?

11.5. Assume that conditional on τ, the random variables $y_1, \ldots, y_n$ are iid from a Burr type X BTX(τ) distribution. Then the conjugate prior is the gamma$(\nu, 1/\lambda)$ pdf and

$$f(y|\tau) \propto \tau \exp(\tau \log(1 - e^{-y^2})).$$

a) Find the posterior pdf.

b) For squared error loss, what is the Bayes estimator of τ?

11.6. Assume that conditional on σ, the random variables $y_1, \ldots, y_n$ are iid from a one-side stable OSS(σ) distribution. Then the conjugate prior is the gamma$(\nu, 1/\lambda)$ pdf and

$$f(y|\sigma) \propto \sigma^{1/2} \exp(-\sigma/(2y)).$$

a) Find the posterior pdf.

b) For squared error loss, what is the Bayes estimator of σ?

11.7. Let $\theta = 1/\lambda$. Assume that conditional on θ, the random variables $y_1, \ldots, y_n$ are iid from a truncated extreme value TEV(θ) distribution with

$$f(y|\theta) \propto \theta \exp(-\theta(e^y - 1)).$$

Then the conjugate prior is the gamma$(\nu, 1/\delta)$ pdf.

a) Find the posterior pdf.

b) For squared error loss, what is the Bayes estimator of θ?

11.8. Assume that conditional on ν, the random variables $y_1, \ldots, y_n$ are iid from a Topp–Leone (ν) distribution with

$$f(y|\nu) \propto \nu \exp(-\nu[-\log(2y - y^2)]).$$

Then the conjugate prior is the gamma$(\theta, 1/\lambda)$ pdf.

a) Find the posterior pdf.

b) For squared error loss, what is the Bayes estimator of ν?

Chapter 12
Stuff for Students

To be blunt, many of us are lousy teachers, and our efforts to improve are feeble. So students frequently view statistics as the worst course taken in college.
Hogg (1991)

This chapter tells how to get the book's *R* functions in *sipack* that are useful for Chaps. 9 and 11. Solutions to many of the text problems are also given. These problems and solutions should be considered as additional examples.

12.1 R Statistical Software

R is a statistical software package, and *R* is the free version of *Splus*. A very useful *R* link is (www.r-project.org/#doc).

As of September 2013, the author's personal computer has Version 2.13.1 (July 8, 2011) of *R*.

Downloading the book's R functions *sipack.txt* into *R*:

In Chap. 9, several of the homework problems use *R* functions contained in the book's website (http://lagrange.math.siu.edu/Olive/sipack.txt) under the file name *sipack.txt*. The command

```
source("http://lagrange.math.siu.edu/Olive/
  sipack.txt")
```

can be used to download the *R* functions into *R*. Type *ls()*. About 11 *R* functions from *sipack.txt* should appear.

For Windows, the functions can be saved on a flash drive G, say. Then use the following command.

```
> source("G:/sipack.txt")
```

D.J. Olive, *Statistical Theory and Inference*, DOI 10.1007/978-3-319-04972-4_12,

Alternatively, from the website (http://lagrange.math.siu.edu/Olive/sipack.txt), go to the *Edit* menu and choose *Select All*, then go to the *Edit* menu and choose *Copy*. Next enter *R*, go to the *Edit* menu and choose *Paste*. These commands also enter the *sipack* functions into *R*.

When you finish your *R* session, enter the command *q()*. A window asking "*Save workspace image?*" will appear. Click on *No* if you do not want to save the programs in *R*. (If you do want to save the programs then click on *Yes*.)

This section gives tips on using *R*, but is no replacement for books such as Becker et al. (1988), Chambers (2008), Crawley (2005, 2013), Dalgaard (2002) or Venables and Ripley (2010). Also see MathSoft (1999a,b) and use the website (www.google.com) to search for useful websites. For example enter the search words *R documentation*.

The command *q()* gets you out of *R* or *Splus*.

The commands *help(fn)* and *args(fn)* give information about the function fn, e.g., if fn = rnorm.

Making functions in R is easy.

For example, type the following commands.

```
mysquare <- function(x){
# this function squares x
r <- x^2
r }
```

The second line in the function shows how to put comments into functions.

Modifying your function is easy.

Use the fix command.

fix(mysquare)

This will open an editor such as *Notepad* and allow you to make changes.

In *Splus*, the command *Edit(mysquare)* may also be used to modify the function *mysquare*.

To save data or a function in *R*, when you exit, click on *Yes* when the "*Save worksheet image?*" window appears. When you reenter *R*, type *ls()*. This will show you what is saved. You should rarely need to save anything for this book. In *Splus*, data and functions are automatically saved. To remove unwanted items from the worksheet, e.g., *x*, type *rm(x)*,
pairs(x) makes a scatterplot matrix of the columns of *x*,
hist(y) makes a histogram of *y*,
boxplot(y) makes a boxplot of *y*,
stem(y) makes a stem and leaf plot of y,
scan(), source(), and *sink()* are useful on a *Unix* workstation.
To type a simple list, use *y* < − *c(1,2,3.5)*.
The commands *mean(y), median(y), var(y)* are self explanatory.

The following commands are useful for a scatterplot created by the command *plot(x,y)*.
lines(x,y), lines(lowess(x,y,f=.2))
identify(x,y)
abline(out$coef), abline(0,1)

The usual arithmetic operators are $2+4$, $3-7$, $8*4$, $8/4$, and

`2^{10}.`

The *i*th element of vector y is $y[i]$ while the ij element of matrix x is $x[i,j]$. The second row of x is $x[2,]$ while the 4th column of x is $x[,4]$. The transpose of x is *t(x)*.

The command *apply(x,1,fn)* will compute the row means if fn = mean. The command *apply(x,2,fn)* will compute the column variances if fn = var. The commands *cbind* and *rbind* combine column vectors or row vectors with an existing matrix or vector of the appropriate dimension.

12.2 Hints and Solutions to Selected Problems

1.10. d) See Problem 1.19 with $Y = W$ and $r = 1$.

f) Use the fact that $E(Y^r) = E[(Y^\phi)^{r/\phi}] = E(W^{r/\phi})$ where $W \sim \text{EXP}(\lambda)$. Take $r = 1$.

1.11. d) Find $E(Y^r)$ for $r = 1,2$ using Problem 1.19 with $Y = W$.

f) For $r = 1,2$, find $E(Y^r)$ using the fact that $E(Y^r) = E[(Y^\phi)^{r/\phi}] = E(W^{r/\phi})$ where $W \sim \text{EXP}(\lambda)$.

1.12. a) 200

b) $0.9(10) + 0.1(200) = 29$

1.13. a) $400(1) = 400$

b) $0.9E(Z) + 0.1E(W) = 0.9(10) + 0.1(400) = 49$

1.15. a) $1\frac{A}{A+B} + 0\frac{B}{A+B} = \frac{A}{A+B}$.

b) $\frac{nA}{A+B}$.

1.16. a) $g(x_o)P(X = x_o) = g(x_o)$

b) $E(e^{tX}) = e^{tx_o}$ by a).

c) $m'(t) = x_o e^{tx_o}$, $m''(t) = x_o^2 e^{tx_o}$, $m^{(n)}(t) = x_o^n e^{tx_o}$.

1.17. $m(t) = E(e^{tX}) = e^t P(X = 1) + e^{-t} P(X = -1) = 0.5(e^t + e^{-t})$.

1.18. a) $\sum_{x=0}^n x e^{tx} f(x)$

b) $\sum_{x=0}^{n} x f(x) = E(X)$
c) $\sum_{x=0}^{n} x^2 e^{tx} f(x)$
d) $\sum_{x=0}^{n} x^2 f(x) = E(X^2)$
e) $\sum_{x=0}^{n} x^k e^{tx} f(x)$

1.19. $E(W^r) = E(e^{rX}) = m_X(r) = \exp(r\mu + r^2\sigma^2/2)$ where $m_X(t)$ is the mgf of a $N(\mu, \sigma^2)$ random variable.

1.20. a) $E(X^2) = V(X) + (E(X))^2 = \sigma^2 + \mu^2$.
b) $E(X^3) = 2\sigma^2 E(X) + \mu E(X^2) = 2\sigma^2\mu + \mu(\sigma^2 + \mu^2) = 3\sigma^2\mu + \mu^3$.

1.22. $\frac{1}{\sqrt{2\pi}} \int_{-\infty}^{\infty} \exp\left(-\frac{1}{2}y^2\right) dy = 1$. So $\int_{-\infty}^{\infty} \exp\left(-\frac{1}{2}y^2\right) dy = \sqrt{2\pi}$.

1.23. $\int_{\sigma}^{\infty} f(x|\sigma, \theta) dx = 1$, so

$$\int_{\sigma}^{\infty} \frac{1}{x^{\theta+1}} dx = \frac{1}{\theta \sigma^{\theta}}. \tag{12.1}$$

So

$$EX^r = \int_{\sigma}^{\infty} x^r \theta \sigma^{\theta} \frac{1}{x^{\theta+1}} dx = \theta \sigma^{\theta} \int_{\sigma}^{\infty} \frac{1}{x^{\theta-r+1}} dx = \frac{\theta \sigma^{\theta}}{(\theta - r)\sigma^{\theta-r}}$$

by Eq. (12.1). So

$$EX^r = \frac{\theta \sigma^r}{\theta - r}$$

for $\theta > r$.

1.24.

$$EY^r = \int_0^1 y^r \frac{\Gamma(\delta+\nu)}{\Gamma(\delta)\Gamma(\nu)} y^{\delta-1}(1-y)^{\nu-1} dy =$$

$$\frac{\Gamma(\delta+\nu)}{\Gamma(\delta)\Gamma(\nu)} \frac{\Gamma(\delta+r)\Gamma(\nu)}{\Gamma(\delta+r+\nu)} \int_0^1 \frac{\Gamma(\delta+r+\nu)}{\Gamma(\delta+r)\Gamma(\nu)} y^{\delta+r-1}(1-y)^{\nu-1} dy =$$

$$\frac{\Gamma(\delta+\nu)\Gamma(\delta+r)}{\Gamma(\delta)\Gamma(\delta+r+\nu)}$$

for $r > -\delta$ since $1 = \int_0^1$ beta$(\delta + r, \nu)$ pdf.

1.25. $E(e^{tY}) = \sum_{y=1}^{\infty} e^{ty} \frac{-1}{\log(1-\theta)} \frac{1}{y} \exp[\log(\theta) y]$. But $e^{ty} \exp[\log(\theta)y] = \exp[(\log(\theta) + t)y] = \exp[(\log(\theta) + \log(e^t))y] = \exp[\log(\theta e^t)y]$.
So $E(e^{tY}) = \frac{-1}{\log(1-\theta)} [-\log(1 - \theta e^t)] \sum_{y=1}^{\infty} \frac{-1}{\log(1-\theta e^t)} \frac{1}{y} \exp[\log(\theta e^t)y] = \frac{\log(1-\theta e^t)}{\log(1-\theta)}$ since $1 = \sum$ [logarithmic (θe^t) pmf] if $0 < \theta e^t < 1$ or $0 < e^t < 1/\theta$ or $-\infty < t < -\log(\theta)$.

1.28. a) $EX = 0.9EZ + 0.1EW = 0.9\nu\lambda + 0.1(10) = 0.9(3)(4) + 1 = 11.8$.

b) $EX^2 = 0.9[V(Z) + (E(Z))^2] + 0.1[V(W) + (E(W))^2]$
$= 0.9[\nu\lambda^2 + (\nu\lambda)^2] + 0.1[10 + (10)^2]$
$= 0.9[3(16) + 9(16)] + 0.1(110) = 0.9(192) + 11 = 183.8.$

2.8. a) $F_W(w) = P(W \leq w) = P(Y \leq w - \mu) = F_Y(w - \mu)$. So $f_W(w) = \frac{d}{dw} F_Y(w - \mu) = f_Y(w - \mu)$.

b) $F_W(w) = P(W \leq w) = P(Y \leq w/\sigma) = F_Y(w/\sigma)$. So $f_W(w) = \frac{d}{dw} F_Y(w/\sigma) = f_Y(w/\sigma)\frac{1}{\sigma}$.

c) $F_W(w) = P(W \leq w) = P(\sigma Y \leq w - \mu) = F_Y(\frac{w-\mu}{\sigma})$. So $f_W(w) = \frac{d}{dw} F_Y(\frac{w-\mu}{\sigma}) = f_Y(\frac{w-\mu}{\sigma})\frac{1}{\sigma}$.

2.9. a) See Example 2.14.

2.11. $W = Z^2 \sim \chi^2_1$ where $Z \sim N(0,1)$. So the pdf of W is

$$f(w) = \frac{w^{\frac{1}{2}-1} e^{-\frac{w}{2}}}{2^{\frac{1}{2}} \Gamma(\frac{1}{2})} = \frac{1}{\sqrt{w}\sqrt{2\pi}} e^{-\frac{w}{2}}$$

for $w > 0$.

2.12. $(Y - \mu)/\sigma = |Z| \sim \text{HN}(0,1)$ where $Z \sim N(0,1)$. So $(Y - \mu)^2 = \sigma^2 Z^2 \sim \sigma^2 \chi^2_1 \sim G(0.5, 2\sigma^2)$.

2.16. a) $y = e^{-w} = t^{-1}(w)$, and

$$\left| \frac{dt^{-1}(w)}{dw} \right| = |-e^{-w}| = e^{-w}.$$

Now $P(Y = 0) = 0$ so $0 < Y \leq 1$ implies that $W = -\log(Y) > 0$. Hence

$$f_W(w) = f_Y(t^{-1}(w)) \left| \frac{dt^{-1}(w)}{dw} \right| = \frac{1}{\lambda}(e^{-w})^{\frac{1}{\lambda}-1} e^{-w} = \frac{1}{\lambda} e^{-w/\lambda}$$

for $w > 0$ which is the EXP(λ) pdf.

2.18. a)

$$f(y) = \frac{1}{\lambda} \frac{\phi y^{\phi-1}}{(1 + y^\phi)^{\frac{1}{\lambda}+1}}$$

where y, ϕ, and λ are all positive. Since $Y > 0$, $W = \log(1 + Y^\phi) > \log(1) > 0$ and the support $\mathscr{W} = (0, \infty)$. Now $1 + y^\phi = e^w$, so $y = (e^w - 1)^{1/\phi} = t^{-1}(w)$. Hence

$$\left| \frac{dt^{-1}(w)}{dw} \right| = \frac{1}{\phi}(e^w - 1)^{\frac{1}{\phi}-1} e^w$$

since $w > 0$. Thus

$$f_W(w) = f_Y(t^{-1}(w)) \left| \frac{dt^{-1}(w)}{dw} \right| = \frac{1}{\lambda} \frac{\phi(e^w-1)^{\frac{\phi-1}{\phi}}}{\left(1+(e^w-1)^{\frac{\phi}{\phi}}\right)^{\frac{1}{\lambda}+1}} \frac{1}{\phi}(e^w-1)^{\frac{1}{\phi}-1} e^w$$

$$= \frac{1}{\lambda} \frac{(e^w-1)^{1-\frac{1}{\phi}}(e^w-1)^{\frac{1}{\phi}-1}}{(e^w)^{\frac{1}{\lambda}+1}} e^w$$

$$\frac{1}{\lambda} e^{-w/\lambda}$$

for $w > 0$ which is the EXP(λ) pdf.

2.25. b)

$$f(y) = \frac{1}{\pi\sigma[1+(\frac{y-\mu}{\sigma})^2]}$$

where y and μ are real numbers and $\sigma > 0$. Now $y = \log(w) = t^{-1}(w)$ and $W = e^Y > 0$ so the support $\mathscr{W} = (0, \infty)$. Thus

$$\left| \frac{dt^{-1}(w)}{dw} \right| = \frac{1}{w},$$

and

$$f_W(w) = f_Y(t^{-1}(w)) \left| \frac{dt^{-1}(w)}{dw} \right| = \frac{1}{\pi\sigma} \frac{1}{[1+\left(\frac{\log(w)-\mu}{\sigma}\right)^2]} \frac{1}{w} =$$

$$\frac{1}{\pi\sigma w[1+\left(\frac{\log(w)-\mu}{\sigma}\right)^2]}$$

for $w > 0$ which is the $LC(\mu, \sigma)$ pdf.

2.63. a) $EX = E[E[X|Y]] = E[\beta_o + \beta_1 Y] = \beta_0 + 3\beta_1$.

b) $V(X) = E[V(X|Y)] + V[E(X|Y)] = E(Y^2) + V(\beta_0 + \beta_1 Y) = V(Y) + [E(Y)]^2 + \beta_1^2 V(Y) = 10 + 9 + \beta_1^2 10 = 19 + 10\beta_1^2$.

2.64. a) $X_2 \sim N(100, 6)$.

b)

$$\begin{pmatrix} X_1 \\ X_3 \end{pmatrix} \sim N_2\left(\begin{pmatrix} 49 \\ 17 \end{pmatrix}, \begin{pmatrix} 3 & -1 \\ -1 & 4 \end{pmatrix} \right).$$

c) $X_1 \perp\!\!\!\perp X_4$ and $X_3 \perp\!\!\!\perp X_4$.

d)

$$\rho(X_1, X_2) = \frac{Cov(X_1, X_3)}{\sqrt{\text{VAR}(X_1)\text{VAR}(X_3)}} = \frac{-1}{\sqrt{3}\sqrt{4}} = -0.2887.$$

2.65. a) $Y|X \sim N(49,16)$ since $Y \perp\!\!\!\perp X$. (Or use $E(Y|X) = \mu_Y + \Sigma_{12}\Sigma_{22}^{-1}(X - \mu_x) = 49 + 0(1/25)(X - 100) = 49$ and $\text{VAR}(Y|X) = \Sigma_{11} - \Sigma_{12}\Sigma_{22}^{-1}\Sigma_{21} = 16 - 0(1/25)0 = 16$.)

b) $E(Y|X) = \mu_Y + \Sigma_{12}\Sigma_{22}^{-1}(X - \mu_x) = 49 + 10(1/25)(X - 100) = 9 + 0.4X$.

c) $\text{VAR}(Y|X) = \Sigma_{11} - \Sigma_{12}\Sigma_{22}^{-1}\Sigma_{21} = 16 - 10(1/25)10 = 16 - 4 = 12$.

2.68. Note that the pdf for λ is the EXP(1) pdf, so $\lambda \sim \text{EXP}(1)$.
a) $E(Y) = E[E(Y|\lambda)] = E(\lambda) = 1$.
b) $V(Y) = E[V(Y|\lambda)] + V[E(Y|\lambda)] = E(\lambda) + V(\lambda) = 1 + 1^2 = 2$.

2.71.

y	0	1
$f_{Y_1}(y) = P(Y_1 = y)$	0.76	0.24

So $m(t) = \sum_y e^{ty} f(y) = \sum_y e^{ty} P(Y = y) = e^{t0} 0.76 + e^{t1} 0.24$
$= 0.76 + 0.24e^t$.

2.72. No, $f(x,y) \neq f_X(x) f_Y(y) = \frac{1}{2\pi} \exp[\frac{-1}{2}(x^2 + y^2)]$.

2.73. a) $E(Y) = E[E(Y|P)] = E(kP) = kE(P) = k\frac{\delta}{\delta + \nu} = k4/10 = 0.4k$.

b) $V(Y) = E[V(Y|P)] + V(E(Y|P)] = E[kP(1-P)] + V(kP) =$
$kE(P) - kE(P^2) + k^2V(P) =$

$$k\frac{\delta}{\delta+\nu} - k\left[\frac{\delta\nu}{(\delta+\nu)^2(\delta+\nu+1)} + \left(\frac{\delta}{\delta+\nu}\right)^2\right] + k^2\frac{\delta\nu}{(\delta+\nu)^2(\delta+\nu+1)}$$

$= k0.4 - k[0.021818 + 0.16] + k^2 0.021818 = 0.021818k^2 + 0.21818k$.

2:74. a)

y_2	0	1	2
$f_{Y_2}(y_2)$	0.55	0.16	0.29

b) $f(y_1|2) = f(y_1,2)/f_{Y_2}(2)$ and $f(0,2)/f_{Y_2}(2) = .24/.29$ while
$f(1,2)/f_{Y_2}(2) = .05/.29$

y_1	0	1
$f_{Y_1\|Y_2}(y_1\|y_2 = 2)$	$24/29 \approx 0.8276$	$5/29 \approx 0.1724$

2.77. c) $m_{\boldsymbol{X}}(\boldsymbol{t}) = m_W(1) = \exp(\mu_W + \frac{1}{2}\sigma_W^2) = \exp(\boldsymbol{t}^T\boldsymbol{\mu} + \frac{1}{2}\boldsymbol{t}^T\boldsymbol{\Sigma}\boldsymbol{t})$.

3.1. a) See Sect. 10.3.
b) See Sect. 10.13.
c) See Sect. 10.42.
d) See Example 3.5.

3.2. a) See Sect. 10.1.
b) See Sect. 10.9.
c) See Sect. 10.16.
d) See Sect. 10.35.
e) See Sect. 10.39.

3.3. b) See Sect. 10.19.
c) See Sect. 10.30.
d) See Sect. 10.38.
f) See Sect. 10.43.
g) See Sect. 10.49.
h) See Sect. 10.53.

3.4. a) See Sect. 10.39.
b) See Sect. 10.39.
c) See Sect. 10.16.

3.5. a) See Sect. 10.7.
b) See Sect. 10.12.
c) See Sect. 10.14.
d) See Sect. 10.29.
h) See Sect. 10.41.
i) See Sect. 10.45.
j) See Sect. 10.51.

3.10. Yes, the top version of the pdf multiplied on the left by $I(y > 0)$ is in the form $h(y)c(\nu)\exp[w(\nu)t(y)]$ where $t(Y)$ is given in the problem, $c(\nu) = 1/\nu$ and $w(\nu) = -1/(2\nu^2)$. Hence $\Omega = (-\infty, 0)$.

4.3. See the proof of Theorem 4.5b.

4.4. See Example 4.14.

4.6. The appropriate section in Chap. 10 gives the 1P-REF parameterization. Then the complete minimal sufficient statistic is T_n given by Theorem 3.6.

4.7. The appropriate section in Chap. 10 gives the 1P-REF parameterization. Then the complete minimal sufficient statistic is T_n given by Theorem 3.7.

4.8. The appropriate section in Chap. 10 gives the 2P-REF parameterization. Then the complete minimal sufficient statistic is $\boldsymbol{T} = (T_1(\boldsymbol{Y}), T_2(\boldsymbol{Y}))$ where $T_i(\boldsymbol{Y}) = \sum_{i=1}^n t_i(Y_i)$ by Corollary 4.6.

4.10. b) The 2P-REF parameterization is given in Chap. 10, so $(\sum_{i=1}^n Y_i, \sum_{i=1}^n e^{Y_i})$ is the complete minimal sufficient statistic by Corollary 4.6.

4.26. See Example 4.11.

4.30. Method i): $E_\lambda(\overline{X} - S^2) = \lambda - \lambda = 0$ for all $\lambda > 0$, but $P_\lambda(\overline{X} - S^2 = 0) < 1$ so $\boldsymbol{T} = (\overline{X}, S^2)$ is not complete.

Method ii): The Poisson distribution is a 1P-REF with complete sufficient statistic $\sum X_i$, so $\overline{X}$ is a minimal sufficient statistic. $\boldsymbol{T} = (\overline{X}, S^2)$ is not a function of $\overline{X}$, so $\boldsymbol{T}$ is not minimal sufficient and hence not complete.

4.31.

$$f(x) = \frac{\Gamma(2\theta)}{\Gamma(\theta)\Gamma(\theta)} x^{\theta-1}(1-x)^{\theta-1} = \frac{\Gamma(2\theta)}{\Gamma(\theta)\Gamma(\theta)} \exp[(\theta-1)(\log(x)+\log(1-x))],$$

for $0 < x < 1$, a one-parameter exponential family. Hence $\sum_{i=1}^n (\log(X_i)+\log(1-X_i))$ is a complete minimal sufficient statistic.

4.32. a) and b)

$$f(x) = I_{\{1,2,\ldots\}}(x) \frac{1}{\zeta(\nu)} \exp[-\nu \log(x)]$$

is a one-parameter regular exponential family with $\Omega = (-\infty, -1)$. Hence $\sum_{i=1}^n \log(X_i)$ is a complete minimal sufficient statistic.

c) By the Factorization Theorem, $\boldsymbol{W} = (X_1, \ldots, X_n)$ is sufficient, but $\boldsymbol{W}$ is not minimal since $\boldsymbol{W}$ is not a function of $\sum_{i=1}^n \log(X_i)$.

4.33. $f(x_1,\ldots,x_n) = \prod_{i=1}^n f(x_i) =$

$$\left[\frac{2}{\sqrt{2\pi}}\right]^n \frac{1}{\sigma^n} \exp\left(\frac{-n\mu^2}{2\sigma^2}\right) I[x_{(1)} > \mu] \exp\left(\frac{-1}{2\sigma^2}\sum_{i=1}^n x_i^2 + \frac{\mu}{\sigma^2}\sum_{i=1}^n x_i\right)$$

$$= h(\boldsymbol{x}) g(\boldsymbol{T}(\boldsymbol{x})|\boldsymbol{\theta})$$

where $\boldsymbol{\theta} = (\mu, \sigma)$. Hence $\boldsymbol{T}(\boldsymbol{X}) = (X_{(1)}, \sum_{i=1}^n X_i^2, \sum_{i=1}^n X_i)$ is a sufficient statistic for (μ, σ).

4.34. Following the end of Example 4.4, $X_{(1)} \sim$ EXP(λ/n) with $\lambda = 1$, so $E[X_{(1)}] = 1/n$.

4.35. $F_{X_{(n)}}(x) = [F(x)]^n = x^n$ for $0 < x < 1$. Thus $f_{X_{(n)}}(x) = nx^{n-1}$ for $0 < x < 1$. This pdf is the beta$(\delta = n, \nu = 1)$ pdf. Hence

$$E[X_{(n)}] = \int_0^1 x f_{X_{(n)}}(x)dx = \int_0^1 x n x^{n-1} dx = \int_0^1 n x^n dx = n \left.\frac{x^{n+1}}{n+1}\right|_0^1 = \frac{n}{n+1}.$$

4.36. Now

$$f_X(x) = I(\theta < x < \theta + 1)$$

and

$$\frac{f(\boldsymbol{x})}{f(\boldsymbol{y})} = \frac{I(\theta < x_{(1)} \le x_{(n)} < \theta+1)}{I(\theta < y_{(1)} \le y_{(n)} < \theta+1)}$$

which is constant for all real θ iff $(x_{(1)}, x_{(n)}) = (y_{(1)}, y_{(n)})$. Hence $\boldsymbol{T} = (X_{(1)}, X_{(n)})$ is a minimal sufficient statistic by the LSM theorem. To show that $\boldsymbol{T}$ is not complete, first find $E(\boldsymbol{T})$. Now

$$F_X(t) = \int_\theta^t dx = t - \theta$$

for $\theta < t < \theta + 1$. Hence

$$f_{X_{(n)}}(t) = n[F_X(t)]^{n-1} f_x(t) = n(t-\theta)^{n-1}$$

for $\theta < t < \theta + 1$ and

$$E_\theta(X_{(n)}) = \int t f_{X_{(n)}}(t)dt = \int_\theta^{\theta+1} tn(t-\theta)^{n-1}dt.$$

Use u-substitution with $u = t - \theta, t = u + \theta$ and $dt = du$. Hence $t = \theta$ implies $u = 0$, and $t = \theta + 1$ implies $u = 1$. Thus

$$E_\theta(X_{(n)}) = \int_0^1 n(u+\theta)u^{n-1}du = \int_0^1 nu^n du + \int_0^1 n\theta u^{n-1}du =$$

$$n\frac{u^{n+1}}{n+1}\bigg|_0^1 + \theta\, n\,\frac{u^n}{n}\bigg|_0^1 = \frac{n}{n+1} + \frac{n\theta}{n} = \theta + \frac{n}{n+1}.$$

Now

$$f_{X_{(1)}}(t) = n[1 - F_X(t)]^{n-1} f_x(t) = n(1-t+\theta)^{n-1}$$

for $\theta < t < \theta + 1$ and thus

$$E_\theta(X_{(1)}) = \int_\theta^{\theta+1} tn(1-t+\theta)^{n-1}dt.$$

Use u–substitution with $u = (1 - t + \theta)$ and $t = 1 - u + \theta$ and $du = -dt$. Hence $t = \theta$ implies $u = 1$, and $t = \theta + 1$ implies $u = 0$. Thus

$$E_\theta(X_{(1)}) = -\int_1^0 n(1-u+\theta)u^{n-1}du = n(1+\theta)\int_0^1 u^{n-1}du - n\int_0^1 u^n du =$$

$$n(1+\theta)\frac{u^n}{n}\bigg|_0^1 - n\,\frac{u^{n+1}}{n+1}\bigg|_0^1 = (\theta+1)\frac{n}{n} - \frac{n}{n+1} = \theta + \frac{1}{n+1}.$$

To show that $\boldsymbol{T}$ is not complete try showing $E_\theta(aX_{(1)} + bX_{(n)} + c) = 0$ for some constants a, b and c. Note that $a = -1$, $b = 1$ and $c = -\frac{n-1}{n+1}$ works. Hence

$$E_\theta(-X_{(1)} + X_{(n)} - \frac{n-1}{n+1}) = 0$$

for all real θ but

$$P_\theta(g(\boldsymbol{T}) = 0) = P_\theta(-X_{(1)} + X_{(n)} - \frac{n-1}{n+1} = 0) = 0 < 1$$

for all real θ. Hence $\boldsymbol{T}$ is not complete.

4.37. Note that

$$f(y) = I(y > 0)\; 2y\, e^{-y^2}\; \tau\, \exp[(1-\tau)\left(-\log(1-e^{-y^2})\right)]$$

is a one-parameter exponential family with minimal and complete sufficient statistic $-\sum_{i=1}^n \log(1-e^{-Y_i^2})$.

5.2. The likelihood function $L(\theta) =$

$$\frac{1}{(2\pi)^n} \exp\left(\frac{-1}{2}\left[\sum(x_i - \rho\cos\theta)^2 + \sum(y_i - \rho\sin\theta)^2\right]\right) =$$

$$\frac{1}{(2\pi)^n} \exp\left(\frac{-1}{2}\left[\sum x_i^2 - 2\rho\cos\theta\sum x_i + \rho^2\cos^2\theta + \sum y_i^2 \right.\right.$$

$$\left.\left. -2\rho\sin\theta\sum y_i + \rho^2\sin^2\theta\right]\right)$$

$$= \frac{1}{(2\pi)^n} \exp\left(\frac{-1}{2}\left[\sum x_i^2 + \sum y_i^2 + \rho^2\right]\right) \exp\left(\rho\cos\theta\sum x_i + \rho\sin\theta\sum y_i\right).$$

Hence the log likelihood $\log L(\theta)$

$$= c + \rho\cos\theta\sum x_i + \rho\sin\theta\sum y_i.$$

The derivative with respect to θ is

$$-\rho\sin\theta\sum x_i + \rho\cos\theta\sum y_i.$$

Setting this derivative to zero gives

$$\rho\sum y_i\cos\theta = \rho\sum x_i\sin\theta$$

or

$$\frac{\sum y_i}{\sum x_i} = \tan\theta.$$

Thus

$$\hat{\theta} = \tan^{-1}\left(\frac{\sum y_i}{\sum x_i}\right).$$

Now the boundary points are $\theta = 0$ and $\theta = 2\pi$. Hence $\hat{\theta}_{\text{MLE}}$ equals 0, 2π, or $\hat{\theta}$ depending on which value maximizes the likelihood.

5.6. See Sect. 10.7.

5.7. See Sect. 10.9.

5.8. See Sect. 10.12.

5.9. See Sect. 10.13.

5.10. See Sect. 10.16.

5.11. See Sect. 10.19.

5.12. See Sect. 10.26.

5.13. See Sect. 10.26.

5.14. See Sect. 10.29.

5.15. See Sect. 10.38.

5.16. See Sect. 10.45.

5.17. See Sect. 10.51.

5.18. See Sect. 10.3.

5.19. See Sect. 10.14.

5.20. See Sect. 10.49.

5.23. a) The log likelihood is $\log L(\tau) = -\frac{n}{2}\log(2\pi\tau) - \frac{1}{2\tau}\sum_{i=1}^n (X_i - \mu)^2$. The derivative of the log likelihood is equal to $-\frac{n}{2\tau} + \frac{1}{2\tau^2}\sum_{i=1}^n (X_i - \mu)^2$. Setting the derivative equal to 0 and solving for τ gives the MLE $\hat{\tau} = \frac{\sum_{i=1}^n (X_i-\mu)^2}{n}$. Now the likelihood is only defined for $\tau > 0$. As τ goes to 0 or ∞, $\log L(\tau)$ tends to $-\infty$. Since there is only one critical point, $\hat{\tau}$ is the MLE.

b) By the invariance principle, the MLE is $\sqrt{\frac{\sum_{i=1}^n (X_i-\mu)^2}{n}}$.

5.28. This problem is nearly the same as finding the MLE of σ^2 when the data are iid $N(\mu, \sigma^2)$ when μ is known. See Problem 5.23 and Sect. 10.23. The MLE in a) is $\sum_{i=1}^n (X_i - \mu)^2/n$. For b) use the invariance principle and take the square root of the answer in a).

5.29. See Example 5.5.

5.30.

$$L(\theta) = \frac{1}{\theta\sqrt{2\pi}} e^{-(x-\theta)^2/2\theta^2}$$

$$\ln(L(\theta)) = -\ln(\theta) - \ln(\sqrt{2\pi}) - (x-\theta)^2/2\theta^2$$

$$\frac{d\ln(L(\theta))}{d\theta} = \frac{-1}{\theta} + \frac{x-\theta}{\theta^2} + \frac{(x-\theta)^2}{\theta^3}$$
$$= \frac{x^2}{\theta^3} - \frac{x}{\theta^2} - \frac{1}{\theta} \stackrel{\text{set}}{=} 0$$

by solving for θ,

$$\theta = \frac{x}{2} * (-1 + \sqrt{5}),$$

and

$$\theta = \frac{x}{2} * (-1 - \sqrt{5}).$$

But, $\theta > 0$. Thus, $\hat{\theta} = \frac{x}{2} * (-1 + \sqrt{5})$, when $x > 0$, and $\hat{\theta} = \frac{x}{2} * (-1 - \sqrt{5})$, when $x < 0$.

To check with the second derivative

$$\begin{aligned}\frac{d^2 \ln(L(\theta))}{d\theta^2} &= -\frac{2\theta + x}{\theta^3} + \frac{3(\theta^2 + \theta x - x^2)}{\theta^4} \\ &= \frac{\theta^2 + 2\theta x - 3x^2}{\theta^4}\end{aligned}$$

but the sign of the θ^4 is always positive, thus the sign of the second derivative depends on the sign of the numerator. Substitute $\hat{\theta}$ in the numerator and simplify, you get $\frac{x^2}{2}(-5 \pm \sqrt{5})$, which is always negative. Hence by the invariance principle, the MLE of θ^2 is $\hat{\theta}^2$.

5.31. a) For any $\lambda > 0$, the likelihood function

$$L(\sigma, \lambda) = \sigma^{n/\lambda}\, I[x_{(1)} \geq \sigma]\, \frac{1}{\lambda^n} \exp\left[-\left(1 + \frac{1}{\lambda}\right) \sum_{i=1}^n \log(x_i)\right]$$

is maximized by making σ as large as possible. Hence $\hat{\sigma} = X_{(1)}$.

b)

$$L(\hat{\sigma}, \lambda) = \hat{\sigma}^{n/\lambda}\, I[x_{(1)} \geq \hat{\sigma}]\, \frac{1}{\lambda^n} \exp\left[-\left(1 + \frac{1}{\lambda}\right) \sum_{i=1}^n \log(x_i)\right].$$

Hence $\log L(\hat{\sigma}, \lambda) =$

$$\frac{n}{\lambda} \log(\hat{\sigma}) - n \log(\lambda) - \left(1 + \frac{1}{\lambda}\right) \sum_{i=1}^n \log(x_i).$$

Thus

$$\frac{d}{d\lambda} \log L(\hat{\sigma}, \lambda) = \frac{-n}{\lambda^2} \log(\hat{\sigma}) - \frac{n}{\lambda} + \frac{1}{\lambda^2} \sum_{i=1}^n \log(x_i) \stackrel{\text{set}}{=} 0,$$

or $-n \log(\hat{\sigma}) + \sum_{i=1}^n \log(x_i) = n\lambda$. So

$$\hat{\lambda} = -\log(\hat{\sigma}) + \frac{\sum_{i=1}^n \log(x_i)}{n} = \frac{\sum_{i=1}^n \log(x_i/\hat{\sigma})}{n}.$$

Now

$$\frac{d^2}{d\lambda^2}\log L(\hat{\sigma},\lambda) = \frac{2n}{\lambda^3}\log(\hat{\sigma}) + \frac{n}{\lambda^2} - \frac{2}{\lambda^3}\sum_{i=1}^{n}\log(x_i)\bigg|_{\lambda=\hat{\lambda}}$$

$$= \frac{n}{\hat{\lambda}^2} - \frac{2}{\hat{\lambda}^3}\sum_{i=1}^{n}\log(x_i/\hat{\sigma}) = \frac{-n}{\hat{\lambda}^2} < 0.$$

Hence $(\hat{\sigma},\hat{\lambda})$ is the MLE of (σ,λ).

5.32. a) the likelihood

$$L(\lambda) = \frac{1}{\lambda^n}\exp\left[-\left(1+\frac{1}{\lambda}\right)\sum\log(x_i)\right],$$

and the log likelihood

$$\log(L(\lambda)) = -n\log(\lambda) - \left(1+\frac{1}{\lambda}\right)\sum\log(x_i).$$

Hence

$$\frac{d}{d\lambda}\log(L(\lambda)) = \frac{-n}{\lambda} + \frac{1}{\lambda^2}\sum\log(x_i) \stackrel{\text{set}}{=} 0,$$

or $\sum\log(x_i) = n\lambda$ or

$$\hat{\lambda} = \frac{\sum\log(X_i)}{n}.$$

Notice that

$$\frac{d^2}{d\lambda^2}\log(L(\lambda)) = \frac{n}{\lambda^2} - \frac{2\sum\log(x_i)}{\lambda^3}\bigg|_{\lambda=\hat{\lambda}} =$$

$$\frac{n}{\hat{\lambda}^2} - \frac{2n\hat{\lambda}}{\hat{\lambda}^3} = \frac{-n}{\hat{\lambda}^2} < 0.$$

Hence $\hat{\lambda}$ is the MLE of λ.

b) By invariance, $\hat{\lambda}^8$ is the MLE of λ^8.

5.33. a) The likelihood

$$L(\theta) = c\, e^{-n2\theta}\exp\left[\log(2\theta)\sum x_i\right],$$

and the log likelihood

$$\log(L(\theta)) = d - n2\theta + \log(2\theta)\sum x_i.$$

Hence

$$\frac{d}{d\theta}\log(L(\theta)) = -2n + \frac{2}{2\theta}\sum x_i \stackrel{\text{set}}{=} 0,$$

or $\sum x_i = 2n\theta$, or

$$\hat{\theta} = \overline{X}/2.$$

Notice that

$$\frac{d^2}{d\theta^2}\log(L(\theta)) = \frac{-\sum x_i}{\theta^2} < 0$$

unless $\sum x_i = 0$.

b) $(\hat{\theta})^4 = (\overline{X}/2)^4$ by invariance.

5.34. $L(0|\boldsymbol{x}) = 1$ for $0 < x_i < 1$, and $L(1|\boldsymbol{x}) = \prod_{i=1}^n \frac{1}{2\sqrt{x_i}}$ for $0 < x_i < 1$. Thus the MLE is 0 if $1 \geq \prod_{i=1}^n \frac{1}{2\sqrt{x_i}}$ and the MLE is 1 if $1 < \prod_{i=1}^n \frac{1}{2\sqrt{x_i}}$.

5.35. a) Notice that $\theta > 0$ and

$$f(y) = \frac{1}{\sqrt{2\pi}}\frac{1}{\sqrt{\theta}}\exp\left(\frac{-(y-\theta)^2}{2\theta}\right).$$

Hence the likelihood

$$L(\theta) = c\,\frac{1}{\theta^{n/2}}\exp\left[\frac{-1}{2\theta}\sum(y_i-\theta)^2\right]$$

and the log likelihood

$$\log(L(\theta)) = d - \frac{n}{2}\log(\theta) - \frac{1}{2\theta}\sum(y_i-\theta)^2 =$$

$$d - \frac{n}{2}\log(\theta) - \frac{1}{2}\sum_{i=1}^n\left(\frac{y_i^2}{\theta} - \frac{2y_i\theta}{\theta} + \frac{\theta^2}{\theta}\right)$$

$$= d - \frac{n}{2}\log(\theta) - \frac{1}{2}\frac{\sum_{i=1}^n y_i^2}{\theta} + \sum_{i=1}^n y_i - \frac{1}{2}n\theta.$$

Thus

$$\frac{d}{d\theta}\log(L(\theta)) = \frac{-n}{2}\frac{1}{\theta} + \frac{1}{2}\sum_{i=1}^n y_i^2\frac{1}{\theta^2} - \frac{n}{2} \stackrel{\text{set}}{=} 0,$$

or

$$\frac{-n}{2}\theta^2 - \frac{n}{2}\theta + \frac{1}{2}\sum_{i=1}^n y_i^2 = 0,$$

or

$$n\theta^2 + n\theta - \sum_{i=1}^n y_i^2 = 0. \tag{12.2}$$

Now the quadratic formula states that for $a \neq 0$, the quadratic equation $ay^2 + by + c = 0$ has roots

$$\frac{-b \pm \sqrt{b^2 - 4ac}}{2a}.$$

Applying the quadratic formula to (12.2) gives

$$\theta = \frac{-n \pm \sqrt{n^2 + 4n\sum_{i=1}^n y_i^2}}{2n}.$$

Since $\theta > 0$, a candidate for the MLE is

$$\hat{\theta} = \frac{-n + \sqrt{n^2 + 4n\sum_{i=1}^n Y_i^2}}{2n} = \frac{-1 + \sqrt{1 + 4\frac{1}{n}\sum_{i=1}^n Y_i^2}}{2}.$$

Since $\hat{\theta}$ satisfies (12.2),

$$n\hat{\theta} - \sum_{i=1}^n y_i^2 = -n\hat{\theta}^2. \tag{12.3}$$

Note that

$$\frac{d^2}{d\theta^2}\log(L(\theta)) = \frac{n}{2\theta^2} - \frac{\sum_{i=1}^n y_i^2}{\theta^3} = \frac{1}{2\theta^3}\left[n\theta - 2\sum_{i=1}^n y_i^2\right]\Bigg|_{\theta=\hat{\theta}} =$$

$$\frac{1}{2\hat{\theta}^3}\left[n\hat{\theta} - \sum_{i=1}^n y_i^2 - \sum_{i=1}^n y_i^2\right] = \frac{1}{2\hat{\theta}^3}\left[-n\hat{\theta}^2 - \sum_{i=1}^n y_i^2\right] < 0$$

by (12.3). Since $L(\theta)$ is continuous with a unique root on $\theta > 0$, $\hat{\theta}$ is the MLE.

5.36. a) $L(\theta) = (\theta - x)^2/3$ for $x - 2 \le \theta \le x + 1$. Since $x = 7$, $L(5) = 4/3$, $L(7) = 0$, and $L(8) = 1/3$. So L is maximized at an endpoint and the MLE $\hat{\theta} = 5$.

b) By invariance the MLE is $h(\hat{\theta}) = h(5) = 10 - e^{-25} \approx 10$.

5.37. a) $L(\lambda) = c\frac{1}{\lambda^n}\exp\left(\frac{-1}{2\lambda^2}\sum_{i=1}^n (e^{x_i} - 1)^2\right)$.
Thus

$$\log(L(\lambda)) = d - n\log(\lambda) - \frac{1}{2\lambda^2}\sum_{i=1}^n (e^{x_i} - 1)^2.$$

Hence

$$\frac{d\log(L(\lambda))}{d\lambda} = \frac{-n}{\lambda} + \frac{1}{\lambda^3}\sum(e^{x_i} - 1)^2 \overset{\text{set}}{=} 0,$$

or $n\lambda^2 = \sum(e^{x_i} - 1)^2$, or

$$\hat{\lambda} = \sqrt{\frac{\sum(e^{X_i} - 1)^2}{n}}.$$

Now

$$\frac{d^2\log(L(\lambda))}{d\lambda^2} = \frac{n}{\lambda^2} - \frac{3}{\lambda^4}\sum(e^{x_i} - 1)^2\Big|_{\lambda=\hat{\lambda}}$$

$$= \frac{n}{\hat{\lambda}^2} - \frac{3n}{\hat{\lambda}^4}\hat{\lambda}^2 = \frac{n}{\hat{\lambda}^2}[1 - 3] < 0.$$

So $\hat{\lambda}$ is the MLE.

5.38. a) The likelihood

$$L(\lambda) = \prod f(x_i) = c\left(\prod \frac{1}{x_i}\right) \frac{1}{\lambda^n} \exp\left[\frac{\sum -(\log x_i)^2}{2\lambda^2}\right],$$

and the log likelihood

$$\log(L(\lambda)) = d - \sum \log(x_i) - n\log(\lambda) - \frac{\sum(\log x_i)^2}{2\lambda^2}.$$

Hence

$$\frac{d}{d\lambda}\log(L(\lambda)) = \frac{-n}{\lambda} + \frac{\sum(\log x_i)^2}{\lambda^3} \stackrel{\text{set}}{=} 0,$$

or $\sum(\log x_i)^2 = n\lambda^2$, or

$$\hat{\lambda} = \sqrt{\frac{\sum(\log x_i)^2}{n}}.$$

This solution is unique.

Notice that

$$\frac{d^2}{d\lambda^2}\log(L(\lambda)) = \left.\frac{n}{\lambda^2} - \frac{3\sum(\log x_i)^2}{\lambda^4}\right|_{\lambda=\hat{\lambda}}$$

$$= \frac{n}{\hat{\lambda}^2} - \frac{3n\hat{\lambda}^2}{\hat{\lambda}^4} = \frac{-2n}{\hat{\lambda}^2} < 0.$$

Hence

$$\hat{\lambda} = \sqrt{\frac{\sum(\log X_i)^2}{n}}$$

is the MLE of λ.

b)

$$\hat{\lambda}^2 = \frac{\sum(\log X_i)^2}{n}$$

is the MLE of λ^2 by invariance.

5.39. a) The plot of $L(\theta)$ should be 1 for $x_{(n)} - 1 \le \theta \le x_{(1)}$, and 0 otherwise.
b) $[c,d] = [x_{(n)} - 1,\ x_{(1)}]$.

5.42. See Example 5.6.

5.43. a) $L(\theta) = \dfrac{[\cos(\theta)]^n \exp(\theta \sum x_i)}{\prod 2\cosh(\pi x_i/2)}$. So $\log(L(\theta)) = c + n\log(\cos(\theta)) + \theta\sum x_i$, and

$$\frac{d\log(L(\theta))}{d\theta} = n\frac{1}{\cos(\theta)}[-\sin(\theta)] + \sum x_i \stackrel{\text{set}}{=} 0,$$

or $\tan(\theta) = \overline{x}$, or $\hat{\theta} = \tan^{-1}(\overline{X})$.

Since

$$\frac{d^2 \log(L(\theta))}{d\theta^2} = -n\sec^2(\theta) < 0$$

for $|\theta| < 1/2$, $\hat{\theta}$ is the MLE.

b) The MLE is $\tan(\hat{\theta}) = \tan(\tan^{-1}(\overline{X})) = \overline{X}$ by the invariance principle.

(By properties of the arctan function, $\hat{\theta} = \tan^{-1}(\overline{X})$ iff $\tan(\hat{\theta}) = \overline{X}$ and $-\pi/2 < \hat{\theta} < \pi/2$.)

5.44. a) This is a two-parameter exponential distribution. So see Sect. 10.14 where $\sigma = \lambda$ and $\mu = \theta$.

b)

$$1 - F(x) = \tau(\mu, \sigma) = \exp\left[-\left(\frac{x-\mu}{\sigma}\right)\right].$$

By the invariance principle, the MLE of $\tau(\mu, \sigma) = \tau(\hat{\mu}, \hat{\sigma})$

$$= \exp\left[-\left(\frac{x - X_{(1)}}{\overline{X} - X_{(1)}}\right)\right].$$

5.45. a) Let

$$w = t(y) = \frac{y}{\theta} + \frac{\theta}{y} - 2.$$

Then the likelihood

$$L(\nu) = d\frac{1}{\nu^n}\exp\left(\frac{-1}{2\nu^2}\sum_{i=1}^n w_i\right),$$

and the log likelihood

$$\log(L(\nu)) = c - n\log(\nu) - \frac{1}{2\nu^2}\sum_{i=1}^n w_i.$$

Hence

$$\frac{d}{d\nu}\log(L(\nu)) = \frac{-n}{\nu} + \frac{1}{\nu^3}\sum_{i=1}^n w_i \overset{\text{set}}{=} 0,$$

or

$$\hat{\nu} = \sqrt{\frac{\sum_{i=1}^n w_i}{n}}.$$

This solution is unique and

$$\frac{d^2}{d\nu^2}\log(L(\nu)) = \frac{n}{\nu^2} - \frac{3\sum_{i=1}^n w_i}{\nu^4}\Bigg|_{\nu=\hat{\nu}} = \frac{n}{\hat{\nu}^2} - \frac{3n\hat{\nu}^2}{\hat{\nu}^4} = \frac{-2n}{\hat{\nu}^2} < 0.$$

Thus

$$\hat{\nu} = \sqrt{\frac{\sum_{i=1}^n W_i}{n}}$$

is the MLE of ν if $\hat{\nu} > 0$.

b) $\hat{\nu}^2 = \dfrac{\sum_{i=1}^n W_i}{n}$ by invariance.

5.46. a) The likelihood

$$L(\lambda) = c\frac{1}{\lambda^n}\exp\left[-\frac{1}{\lambda}\sum_{i=1}^n \log(1+y_i^{-\phi})\right],$$

and the log likelihood $\log(L(\lambda)) = d - n\log(\lambda) - \frac{1}{\lambda}\sum_{i=1}^n \log(1+y_i^{-\phi})$. Hence

$$\frac{d}{d\lambda}\log(L(\lambda)) = \frac{-n}{\lambda} + \frac{\sum_{i=1}^n \log(1+y_i^{-\phi})}{\lambda^2} \overset{\text{set}}{=} 0,$$

or $\sum_{i=1}^n \log(1+y_i^{-\phi}) = n\lambda$ or

$$\hat{\lambda} = \frac{\sum_{i=1}^n \log(1+y_i^{-\phi})}{n}.$$

This solution is unique and

$$\frac{d^2}{d\lambda^2}\log(L(\lambda)) = \frac{n}{\lambda^2} - \left.\frac{2\sum_{i=1}^n \log(1+y_i^{-\phi})}{\lambda^3}\right|_{\lambda=\hat{\lambda}} = \frac{n}{\hat{\lambda}^2} - \frac{2n\hat{\lambda}}{\hat{\lambda}^3} = \frac{-n}{\hat{\lambda}^2} < 0.$$

Thus

$$\hat{\lambda} = \frac{\sum_{i=1}^n \log(1+Y_i^{-\phi})}{n}$$

is the MLE of λ if ϕ is known.

b) The MLE is $\hat{\lambda}^2$ by invariance.

5.47. a) The likelihood

$$L(\sigma^2) = c\left(\frac{1}{\sigma^2}\right)^{\frac{n}{2}}\exp\left[\frac{-1}{2\sigma^2}\sum_{i=1}^n \frac{1}{y_i^2}\right],$$

and the log likelihood

$$\log(L(\sigma^2)) = d - \frac{n}{2}\log(\sigma^2) - \frac{1}{2\sigma^2}\sum_{i=1}^n \frac{1}{y_i^2}.$$

Hence

$$\frac{d}{d(\sigma^2)}\log(L(\sigma^2)) = \frac{-n}{2(\sigma^2)} + \frac{1}{2(\sigma^2)^2}\sum_{i=1}^n \frac{1}{y_i^2} \overset{\text{set}}{=} 0,$$

or $\sum_{i=1}^n \frac{1}{y_i^2} = n\sigma^2$ or

$$\hat{\sigma}^2 = \frac{1}{n}\sum_{i=1}^n \frac{1}{y_i^2}.$$

This solution is unique and

$$\frac{d^2}{d(\sigma^2)^2}\log(L(\sigma^2)) =$$

$$\left.\frac{n}{2(\sigma^2)^2} - \frac{\sum_{i=1}^n \frac{1}{y_i^2}}{(\sigma^2)^3}\right|_{\sigma^2=\hat{\sigma}^2} = \frac{n}{2(\hat{\sigma}^2)^2} - \frac{n\hat{\sigma}^2}{(\hat{\sigma}^2)^3}\frac{2}{2} = \frac{-n}{2\hat{\sigma}^4} < 0.$$

Thus

$$\hat{\sigma}^2 = \frac{1}{n}\sum_{i=1}^n \frac{1}{Y_i^2}$$

is the MLE of σ^2.

b) By invariance, $\hat{\sigma} = \sqrt{\hat{\sigma}^2}$.

5.48. Solution. a) The likelihood $L(\theta) = c\ \theta^n \exp\left[-\theta\sum_{i=1}^n \frac{1}{y_i}\right]$, and the log likelihood $\log(L(\theta)) = d + n\log(\theta) - \theta\sum_{i=1}^n \frac{1}{y_i}$. Hence

$$\frac{d}{d\theta}\log(L(\theta)) = \frac{n}{\theta} - \sum_{i=1}^n \frac{1}{y_i} \overset{\text{set}}{=} 0, \text{ or } \hat{\theta} = \frac{n}{\sum_{i=1}^n \frac{1}{y_i}}.$$

Since this solution is unique and $\frac{d^2}{d\theta^2}\log(L(\theta)) = \frac{-n}{\theta^2} < 0$,

$\hat{\theta} = \frac{n}{\sum_{i=1}^n \frac{1}{Y_i}}$ is the MLE of θ.

b) By invariance, the MLE is $1/\hat{\theta} = \frac{\sum_{i=1}^n \frac{1}{Y_i}}{n}$.

5.49. a) The likelihood

$$L(\theta) = c\left(\frac{\theta^2}{1+\theta}\right)^n \exp(-\theta\sum_{i=1}^n y_i),$$

and the log likelihood

$$\log(L(\theta)) = d + n\log\left(\frac{\theta^2}{1+\theta}\right) - \theta\sum_{i=1}^n y_i.$$

Always use properties of logarithms to simplify the log likelihood before taking derivatives. Note that

$$\log(L(\theta)) = d + 2n\log(\theta) - n\log(1+\theta) - \theta\sum_{i=1}^{n} y_i.$$

Hence

$$\frac{d}{d\theta}\log(L(\theta)) = \frac{2n}{\theta} - \frac{n}{(1+\theta)} - \sum_{i=1}^{n} y_i \overset{\text{set}}{=} 0,$$

or

$$\frac{2(1+\theta)-\theta}{\theta(1+\theta)} - \overline{y} = 0 \text{ or } \frac{2+\theta}{\theta(1+\theta)} - \overline{y} = 0$$

or $2+\theta = \overline{y}(\theta+\theta^2)$ or $\overline{y}\theta^2 + \theta(\overline{y}-1) - 2 = 0$. So

$$\hat{\theta} = \frac{-(\overline{Y}-1) + \sqrt{(\overline{Y}-1)^2 + 8\overline{Y}}}{2\overline{Y}}.$$

b) By invariance, the MLE is $1/\hat{\theta}$.

6.7. a) The joint density

$$f(\boldsymbol{x}) = \frac{1}{(2\pi)^{n/2}}\exp\left[-\frac{1}{2}\sum(x_i-\mu)^2\right]$$

$$= \frac{1}{(2\pi)^{n/2}}\exp\left[-\frac{1}{2}\left(\sum x_i^2 - 2\mu\sum x_i + n\mu^2\right)\right]$$

$$= \frac{1}{(2\pi)^{n/2}}\exp\left[-\frac{1}{2}\sum x_i^2\right]\exp\left[n\mu\overline{x} - \frac{n\mu^2}{2}\right].$$

Hence by the factorization theorem $\overline{X}$ is a sufficient statistic for μ.

b) $\overline{X}$ is sufficient by a) and complete since the $N(\mu,1)$ family is a regular one-parameter exponential family.

c) $E(I_{-(\infty,t]}(X_1)|\overline{X}=\overline{x}) = P(X_1 \le t|\overline{X}=\overline{x}) = \Phi(\frac{t-\overline{x}}{\sqrt{1-1/n}})$.

d) By the LSU theorem,

$$\Phi\left(\frac{t-\overline{X}}{\sqrt{1-1/n}}\right)$$

is the UMVUE.

6.14. Note that $\sum X_i \sim G(n,\theta)$. Hence MSE(c) = $\text{Var}_\theta(T_n(c)) + [E_\theta T_n(c) - \theta]^2$ $= c^2\text{Var}_\theta(\sum X_i) + [ncE_\theta X - \theta]^2 = c^2 n\theta^2 + [nc\theta - \theta]^2$.

So

$$\frac{d}{dc}\text{MSE}(c) = 2cn\theta^2 + 2[nc\theta - \theta]n\theta.$$

Set this equation to 0 to get $2n\theta^2[c + nc - 1] = 0$ or $c(n+1) = 1$. So $c = 1/(n+1)$.

The second derivative is $2n\theta^2 + 2n^2\theta^2 > 0$ so the function is convex and the local min is in fact global.

6.17. a) Since this is an exponential family, $\log(f(x|\lambda)) = -\log(\lambda) - x/\lambda$ and

$$\frac{\partial}{\partial\lambda}\log(f(x|\lambda)) = \frac{-1}{\lambda} + \frac{x}{\lambda^2}.$$

Hence

$$\frac{\partial^2}{\partial\lambda^2}\log(f(x|\lambda)) = \frac{1}{\lambda^2} - \frac{2x}{\lambda^3}$$

and

$$I_1(\lambda) = -E\left[\frac{\partial^2}{\partial\lambda^2}\log(f(x|\lambda))\right] = \frac{-1}{\lambda^2} + \frac{2\lambda}{\lambda^3} = \frac{1}{\lambda^2}.$$

b)

$$\text{FCRLB}(\tau(\lambda)) = \frac{[\tau'(\lambda)]^2}{nI_1(\lambda)} = \frac{4\lambda^2}{n/\lambda^2} = 4\lambda^4/n.$$

c) ($T = \sum_{i=1}^n X_i \sim$ Gamma(n, λ) is a complete sufficient statistic. Now $E(T^2) = V(T) + [E(T)]^2 = n\lambda^2 + n^2\lambda^2$. Hence the UMVUE of λ^2 is $T^2/(n+n^2)$.) No, W is a nonlinear function of the complete sufficient statistic T.

6.19.

$$W \equiv S^2(k)/\sigma^2 \sim \chi^2_n/k$$

and

$$\text{MSE}(S^2(k)) = \text{MSE}(W) = \text{VAR}(W) + (E(W) - \sigma^2)^2$$

$$= \frac{\sigma^4}{k^2}2n + \left(\frac{\sigma^2 n}{k} - \sigma^2\right)^2$$

$$= \sigma^4\left[\frac{2n}{k^2} + (\frac{n}{k} - 1)^2\right] = \sigma^4\frac{2n + (n-k)^2}{k^2}.$$

Now the derivative $\frac{d}{dk}\text{MSE}(S^2(k))/\sigma^4 =$

$$\frac{-2}{k^3}[2n + (n-k)^2] + \frac{-2(n-k)}{k^2}.$$

Set this derivative equal to zero. Then

$$2k^2 - 2nk = 4n + 2(n-k)^2 = 4n + 2n^2 - 4nk + 2k^2.$$

Hence

$$2nk = 4n + 2n^2$$

or $k = n + 2$.

Should also argue that $k = n + 2$ is the global minimizer. Certainly need $k > 0$ and the absolute bias will tend to ∞ as $k \to 0$ and the bias tends to σ^2 as $k \to \infty$, so $k = n + 2$ is the unique critical point and is the global minimizer.

6.20. a) Let $W = X^2$. Then $f(w) = f_X(\sqrt{w})\ 1/(2\sqrt{w}) = (1/\theta)\exp(-w/\theta)$ and $W \sim \text{EXP}(\theta)$. Hence $E_\theta(X^2) = E_\theta(W) = \theta$.

b) This is an exponential family and

$$\log(f(x|\theta)) = \log(2x) - \log(\theta) - \frac{1}{\theta}x^2$$

for $x > 0$. Hence

$$\frac{\partial}{\partial \theta} f(x|\theta) = \frac{-1}{\theta} + \frac{1}{\theta^2}x^2$$

and

$$\frac{\partial^2}{\partial \theta^2} f(x|\theta) = \frac{1}{\theta^2} + \frac{-2}{\theta^3}x^2.$$

Hence

$$I_1(\theta) = -E_\theta\left[\frac{1}{\theta^2} + \frac{-2}{\theta^3}x^2\right] = \frac{1}{\theta^2}$$

by a). Now

$$\text{CRLB} = \frac{[\tau'(\theta)]^2}{nI_1(\theta)} = \frac{\theta^2}{n}$$

where $\tau(\theta) = \theta$.

c) This is a regular exponential family so $\sum_{i=1}^n X_i^2$ is a complete sufficient statistic. Since

$$E_\theta\left[\frac{\sum_{i=1}^n X_i^2}{n}\right] = \theta,$$

the UMVUE is $\frac{\sum_{i=1}^n X_i^2}{n}$.

6.21. a) In normal samples, $\overline{X}$ and S are independent, hence

$$\text{Var}_\theta[W(\alpha)] = \alpha^2 \text{Var}_\theta(T_1) + (1-\alpha)^2 \text{Var}_\theta(T_2).$$

b) $W(\alpha)$ is an unbiased estimator of θ. Hence $\text{MSE}[W(\alpha)] \equiv \text{MSE}(\alpha) = \text{Var}_\theta[W(\alpha)]$ which is found in part a).

c) Now

$$\frac{d}{d\alpha}\text{MSE}(\alpha) = 2\alpha \text{Var}_\theta(T_1) - 2(1-\alpha)\text{Var}_\theta(T_2) \stackrel{\text{set}}{=} 0.$$

Hence

$$\hat{\alpha} = \frac{\text{Var}_\theta(T_2)}{\text{Var}_\theta(T_1) + \text{Var}_\theta(T_2)} \approx \frac{\frac{\theta^2}{2n}}{\frac{2\theta^2}{2n} + \frac{\theta^2}{2n}} = 1/3$$

using the approximation and the fact that $\text{Var}(\bar{X}) = \theta^2/n$. Note that the second derivative

$$\frac{d^2}{d\alpha^2}\text{MSE}(\alpha) = 2[\text{Var}_\theta(T_1) + \text{Var}_\theta(T_2)] > 0,$$

so $\alpha = 1/3$ is a local min. The critical value was unique, hence 1/3 is the global min.

6.22. a) $X_1 - X_2 \sim N(0, 2\sigma^2)$. Thus,

$$E(T_1) = \int_0^\infty u \frac{1}{\sqrt{4\pi\sigma^2}} e^{\frac{-u^2}{4\sigma^2}} du$$
$$= \frac{\sigma}{\sqrt{\pi}}.$$

$$E(T_1^2) = \frac{1}{2}\int_0^\infty u^2 \frac{1}{\sqrt{4\pi\sigma^2}} e^{\frac{-u^2}{4\sigma^2}} du$$
$$= \frac{\sigma^2}{2}.$$

$V(T_1) = \sigma^2(\frac{1}{2} - \frac{1}{\pi})$ and

$$\text{MSE}(T_1) = \sigma^2\left[\left(\frac{1}{\sqrt{\pi}} - 1\right)^2 + \frac{1}{2} - \frac{1}{\pi}\right] = \sigma^2\left[\frac{3}{2} - \frac{2}{\sqrt{\pi}}\right].$$

b) $\frac{X_i}{\sigma}$ has a N(0,1) and $\frac{\sum_{i=1}^n X_i^2}{\sigma^2}$ has a chi-square distribution with n degrees of freedom. Thus

$$E\left(\sqrt{\frac{\sum_{i=1}^n X_i^2}{\sigma^2}}\right) = \frac{\sqrt{2}\Gamma(\frac{n+1}{2})}{\Gamma(\frac{n}{2})},$$

and

$$E(T_2) = \frac{\sigma}{\sqrt{n}} \frac{\sqrt{2}\Gamma(\frac{n+1}{2})}{\Gamma(\frac{n}{2})}.$$

Therefore,

$$E\left(\frac{\sqrt{n}}{\sqrt{2}} \frac{\Gamma(\frac{n}{2})}{\Gamma(\frac{n+1}{2})} T_2\right) = \sigma.$$

6.23. This is a regular one-parameter exponential family with complete sufficient statistic $T_n = \sum_{i=1}^n X_i \sim G(n, \lambda)$. Hence $E(T_n) = n\lambda$, $E(T_n^2) = V(T_n) + (E(T_n))^2 = n\lambda^2 + n^2\lambda^2$, and $T_n^2/(n+n^2)$ is the UMVUE of λ^2.

6.24.

$$\frac{1}{X_i} = \frac{W_i}{\sigma} \sim \frac{\chi_1^2}{\sigma}.$$

Hence if

$$T = \sum_{i=1}^n \frac{1}{X_i}, \text{ then } E\left(\frac{T}{n}\right) = \frac{n}{n\sigma},$$

and T/n is the UMVUE since $f(x)$ is an exponential family with complete sufficient statistic $1/X$.

6.25. The pdf of T is

$$g(t) = \frac{2nt^{2n-1}}{\theta^{2n}}$$

for $0 < t < \theta$.
$E(T) = \frac{2n}{2n+1}\theta$ and $E(T^2) = \dfrac{2n}{2n+2}\theta^2$.

$$\text{MSE}(CT) = \left(C\frac{2n}{2n+1}\theta - \theta\right)^2 + C^2\left[\frac{2n}{2n+2}\theta^2 - \left(\frac{2n}{2n+1}\theta\right)^2\right]$$

$$\frac{d\text{MSE}(CT)}{dC} = 2\left[\frac{2Cn\theta}{2n+1} - \theta\right]\left[\frac{2n\theta}{2n+1}\right] + 2C\left[\frac{2n\theta^2}{2n+2} - \frac{4n^2\theta^2}{(2n+1)^2}\right].$$

Solve $\frac{d\text{MSE}(CT)}{dC} \overset{\text{set}}{=} 0$ to get

$$C = 2\frac{n+1}{2n+1}.$$

The MSE is a quadratic in C and the coefficient on C^2 is positive, hence the local min is a global min.

6.26. a) $E(X_i) = 2\theta/3$ and $V(X_i) = \theta^2/18$. So bias of $T = B(T) = Ec\overline{X} - \theta = c\frac{2}{3}\theta - \theta$ and Var(T) =

$$\text{Var}\left(\frac{c\sum X_i}{n}\right) = \frac{c^2}{n^2}\sum Var(X_i) = \frac{c^2}{n^2}\frac{n\theta^2}{18}.$$

So MSE = Var(T) + $[B(T)]^2$ =

$$\frac{c^2\theta^2}{18n} + \left(\frac{2\theta}{3}c - \theta\right)^2.$$

b)

$$\frac{d\text{MSE}(c)}{dc} = \frac{2c\theta^2}{18n} + 2\left(\frac{2\theta}{3}c - \theta\right)\frac{2\theta}{3}.$$

Set this equation equal to 0 and solve, so

$$\frac{\theta^2 2c}{18n} + \frac{4}{3}\theta\left(\frac{2}{3}\theta c - \theta\right) = 0$$

or

$$c\left[\frac{2\theta^2}{18n} + \frac{8}{9}\theta^2\right] = \frac{4}{3}\theta^2$$

or

$$c\left(\frac{1}{9n} + \frac{8}{9}\right)\theta^2 = \frac{4}{3}\theta^2$$

or

$$c\left(\frac{1}{9n} + \frac{8n}{9n}\right) = \frac{4}{3}$$

or

$$c = \frac{9n}{1+8n}\frac{4}{3} = \frac{12n}{1+8n}.$$

This is a global min since the MSE is a quadratic in c^2 with a positive coefficient, or because

$$\frac{d^2\text{MSE}(c)}{dc^2} = \frac{2\theta^2}{18n} + \frac{8\theta^2}{9} > 0.$$

6.27. See Example 6.6.

6.30. See Example 6.3.

6.31. a) $E(T) = cE(Y) = c\alpha\beta = 10c\beta$.
$V(T) = c^2V(\overline{Y}) = c^2\alpha\beta^2/n = 10c^2\beta^2/n$.
$\text{MSE}(T) = V(T) + [B(T)]^2 = 10c^2\beta^2/n + (10c\beta - \beta)^2$.

$$b)\ \frac{d\text{MSE}(c)}{dc} = \frac{2c10\beta^2}{n} + 2(10c\beta - \beta)10\beta \overset{\text{set}}{=} 0$$

or $[20\beta^2/n]\ c + 200\beta^2\ c - 20\beta^2 = 0$
or $c/n + 10c - 1 = 0$ or $c(1/n + 10) = 1$
or

$$c = \frac{1}{\frac{1}{n} + 10} = \frac{n}{10n+1}.$$

This value of c is unique, and

$$\frac{d^2\text{MSE}(c)}{dc^2} = \frac{20\beta^2}{n} + 200\beta^2 > 0,$$

so c is the minimizer.

6.32. a) Since this distribution is a one-parameter regular exponential family, $T_n = -\sum_{i=1}^n \log(2Y_i - Y_i^2)$ is complete.

b) Note that $\log(f(y|v)) = \log(v) + \log(2-2y) + (1-v)[-\log(2y-y^2)]$. Hence

$$\frac{d\ \log(f(y|v))}{dv} = \frac{1}{v} + \log(2y-y^2)$$

and

$$\frac{d^2\ \log(f(y|v))}{dv^2} = \frac{-1}{v^2}.$$

Since this family is a 1P-REF, $I_1(v) = -E\left(\frac{-1}{v^2}\right) = \frac{1}{v^2}$.

c) $\frac{[\tau'(v)]^2}{nI_1(v)} = \frac{v^2}{v^4\ n} = \frac{1}{nv^2}$.

d) $E[T_n^{-1}] = \frac{1}{v^{-1}}\ \frac{\Gamma(-1+n)}{\Gamma(n)} = \frac{v}{n-1}$. So $(n-1)/T_n$ is the UMVUE of v by LSU.

6.33. a) Since $f(y) = \frac{\theta}{2}[\exp[-(\theta+1)\log(1+|y|)]$ is a 1P-REF, $T = \sum_{i=1}^n \log(1+|Y_i|)$ is a complete sufficient statistic.

b) Since this is an exponential family, $\log(f(y|\theta)) = \log(\theta/2) - (\theta+1)\log(1+|y|)$ and

$$\frac{\partial}{\partial\theta}\log(f(y|\theta)) = \frac{1}{\theta} - \log(1+|y|).$$

Hence

$$\frac{\partial^2}{\partial\theta^2}\log(f(y|\theta)) = \frac{-1}{\theta^2}$$

and

$$I_1(\theta) = -E_\theta\left[\frac{\partial^2}{\partial\theta^2}\log(f(Y|\theta))\right] = \frac{1}{\theta^2}.$$

c) The complete sufficient statistic $T \sim G(n, 1/\theta)$. Hence the UMVUE of θ is $(n-1)/T$ since for $r > -n$,

$$E(T^r) = E(T^r) = \left(\frac{1}{\theta}\right)^r \frac{\Gamma(r+n)}{\Gamma(n)}.$$

So

$$E(T^{-1}) = \theta\frac{\Gamma(n-1)}{\Gamma(n)} = \theta/(n-1).$$

6.34. a) Note that $\overline{X}$ is a complete and sufficient statistic for μ and $\overline{X} \sim N(\mu, n^{-1}\sigma^2)$. We know that $E(e^{2\overline{X}})$, the mgf of $\overline{X}$ when $t = 2$, is given by $e^{2\mu+2n^{-1}\sigma^2}$. Thus the UMVUE of $e^{2\mu}$ is $e^{-2n^{-1}\sigma^2}e^{2\overline{X}}$.

b) The CRLB for the variance of unbiased estimator of $g(\mu)$ is given by $4n^{-1}\sigma^2 e^{4\mu}$ whereas

$$V(e^{-2n^{-1}\sigma^2} e^{2\bar{X}}) = e^{-4n^{-1}\sigma^2} E(e^{4\bar{X}}) - e^{4\mu} \tag{12.4}$$
$$= e^{-4n^{-1}\sigma^2} e^{4\mu + \frac{1}{2}16n^{-1}\sigma^2} - e^{4\mu}$$
$$= e^{4\mu}[e^{4n^{-1}\sigma^2} - 1]$$
$$> 4n^{-1}\sigma^2 e^{4\mu}$$

since $e^x > 1 + x$ for all $x > 0$. Hence the CRLB is not attained.

6.36. See Theorem 6.5.

a) $E(W_n) = c\sum_{i=1}^n E(t(Y_i)) = cn\theta E(X)$, and $V(W_n) = c^2 \sum_{i=1}^n V(t(Y_i)) = c^2 n\theta^2 V(X)$. Hence $\text{MSE}(c) \equiv \text{MSE}(W_n) = V(W_n) + [E(W_n) - \theta]^2 = c^2 n\theta^2 V(X) + (cn\theta E(X) - \theta)^2$.

b) Thus

$$\frac{d\text{MSE}(c)}{dc} = 2cn\theta^2 V(X) + 2(cn\theta E(X) - \theta)n\theta E(X) \stackrel{\text{set}}{=} 0,$$

or

$$c(n\theta^2 V(X) + n^2\theta^2[E(X)]^2) = n\theta^2 E(X),$$

or

$$c_M = \frac{E(X)}{V(X) + n[E(X)]^2},$$

which is unique. Now

$$\frac{d^2\text{MSE}(c)}{dc^2} = 2[n\theta^2 V(X) + n^2\theta^2[E(X)]^2] > 0.$$

So $\text{MSE}(c)$ is convex and $c = c_M$ is the minimizer.

c) Let $c_U = \dfrac{1}{nE(X)}$. Then $E[c_U T(\boldsymbol{Y})] = \theta$, hence $c_U T(\boldsymbol{Y})$ is the UMVUE of θ by the Lehmann Scheffé theorem.

6.37. See Example 6.4.

6.38. See Example 6.9.

6.39. a) $E(\overline{X} - \mu)^2 = \text{Var}(\overline{X}) = \frac{\sigma^2}{n}$.

b) From a), $E(\overline{X}^2 - 2\mu\overline{X} + \mu^2) = E(\overline{X}^2) - \mu^2 = \frac{\sigma^2}{n}$, or $E(\overline{X}^2) - \frac{\sigma^2}{n} = \mu^2$, or $E(\overline{X}^2 - \frac{\sigma^2}{n}) = \mu^2$.

Since $\overline{X}$ is a complete and sufficient statistic, and $\overline{X}^2 - \frac{\sigma^2}{n}$ is an unbiased estimator of μ^2 and is a function of $\overline{X}$, the UMVUE of μ^2 is $\overline{X}^2 - \frac{\sigma^2}{n}$ by the Lehmann-Scheffé Theorem.

c) Let $Y = \overline{X} - \mu \sim N(0, \sigma^2)$. Then $E(Y^3) = \int_{-\infty}^{\infty} h(y)dy = 0$, because $h(y)$ is an odd function.

d) $E(\overline{X} - \mu)^3 = E(\overline{X}^3 - 3\mu\overline{X}^2 + 3\mu^2\overline{X} - \mu^3) = E(\overline{X}^3) - 3\mu E(\overline{X}^2) + 3\mu^2 E(\overline{X}) - \mu^3 = E(\overline{X}^3) - 3\mu\left(\frac{\sigma^2}{n} + \mu^2\right) + 3\mu^3 - \mu^3 = E(\overline{X}^3) - 3\mu\frac{\sigma^2}{n} - \mu^3$.

Thus $E(\overline{X}^3) - 3\mu\frac{\sigma^2}{n} - \mu^3 = 0$, so replacing μ with its unbiased estimator $\overline{X}$ in the middle term, we get

$$E\left[\overline{X}^3 - 3\overline{X}\frac{\sigma^2}{n}\right] = \mu^3.$$

Since $\overline{X}$ is a complete and sufficient statistic, and $\overline{X}^3 - 3\overline{X}\frac{\sigma^2}{n}$ is an unbiased estimator of μ^3 and is a function of $\bar{X}$, the UMVUE of μ^3 is $\overline{X}^3 - 3\overline{X}\frac{\sigma^2}{n}$ by the Lehmann-Scheffé Theorem.

6.40. a) The pdf of T is $f(t) = \frac{nt^{n-1}}{\theta^n} I(0 < t < \theta)$. Hence $E(T^k) = \int_0^\theta t^k \frac{nt^{n-1}}{\theta^n} dt = \int_0^\theta \frac{nt^{k+n-1}}{\theta^n} dt = \frac{n\theta^{k+n}}{(k+n)\theta^n} = \frac{n}{k+n}\theta^k.$

b) Thus the UMVUE of θ^k is $\frac{k+n}{n}T^k$.

7.6. For both a) and b), the test is reject Ho iff $\prod_{i=1}^n x_i(1-x_i) > c$ where $P_{\theta=1}[\prod_{i=1}^n x_i(1-x_i) > c] = \alpha$.

7.10. H says $f(x) = e^{-x}$ while K says

$$f(x) = x^{\theta-1}e^{-x}/\Gamma(\theta).$$

The monotone likelihood ratio property holds for $\prod x_i$ since then

$$\frac{f_n(\boldsymbol{x},\theta_2)}{f_n(\boldsymbol{x},\theta_1)} = \frac{(\prod_{i=1}^n x_i)^{\theta_2-1}(\Gamma(\theta_1))^n}{(\prod_{i=1}^n x_i)^{\theta_1-1}(\Gamma(\theta_2))^n} = \left(\frac{\Gamma(\theta_1)}{\Gamma(\theta_2)}\right)^n \left(\prod_{i=1}^n x_i\right)^{\theta_2-\theta_1}$$

which increases as $\prod_{i=1}^n x_i$ increases if $\theta_2 > \theta_1$. Hence the level α UMP test rejects H if

$$\prod_{i=1}^n X_i > c$$

where

$$P_H\left(\prod_{i=1}^n X_i > c\right) = P_H\left(\sum \log(X_i) > \log(c)\right) = \alpha.$$

7.11. See Example 7.8.

7.13. Let $\theta_1 = 4$. By Neyman–Pearson lemma, reject Ho if

$$\frac{f(\boldsymbol{x}|\theta_1)}{f(\boldsymbol{x}|2)} = \left(\frac{\log(\theta_1)}{\theta_1 - 1}\right)^n \theta_1^{\sum x_i} \left(\frac{1}{\log(2)}\right)^n \frac{1}{2^{\sum x_i}} > k$$

iff

$$\left(\frac{\log(\theta_1)}{(\theta_1 - 1)\log(2)}\right)^n \left(\frac{\theta_1}{2}\right)^{\sum x_i} > k$$

iff

$$\left(\frac{\theta_1}{2}\right)^{\sum x_i} > k'$$

iff

$$\sum x_i \log(\theta_1/2) > c'.$$

So reject Ho iff $\sum X_i > c$ where $P_{\theta=2}(\sum X_i > c) = \alpha$.

7.14. a) By NP lemma reject Ho if

$$\frac{f(\boldsymbol{x}|\sigma=2)}{f(\boldsymbol{x}|\sigma=1)} > k'.$$

The LHS =

$$\frac{\frac{1}{2^{3n}} \exp\left[\frac{-1}{8}\sum x_i^2\right]}{\exp\left[\frac{-1}{2}\sum x_i^2\right]}$$

So reject Ho if

$$\frac{1}{2^{3n}} \exp\left[\sum x_i^2 \left(\frac{1}{2} - \frac{1}{8}\right)\right] > k'$$

or if $\sum x_i^2 > k$ where $P_{Ho}(\sum x_i^2 > k) = \alpha$.

b) In the above argument, with any $\sigma_1 > 1$, get

$$\sum x_i^2 \left(\frac{1}{2} - \frac{1}{2\sigma_1^2}\right)$$

and

$$\frac{1}{2} - \frac{1}{2\sigma_1^2} > 0$$

for any $\sigma_1^2 > 1$. Hence the UMP test is the same as in a).

7.15. a) By NP lemma reject Ho if

$$\frac{f(\boldsymbol{x}|\sigma=2)}{f(\boldsymbol{x}|\sigma=1)} > k'.$$

The LHS =

$$\frac{\frac{1}{2^n} \exp\left[\frac{-1}{8}\sum[\log(x_i)]^2\right]}{\exp\left[\frac{-1}{2}\sum[\log(x_i)]^2\right]}$$

So reject Ho if

$$\frac{1}{2^n} \exp\left[\sum[\log(x_i)]^2 \left(\frac{1}{2} - \frac{1}{8}\right)\right] > k'$$

or if $\sum[\log(X_i)]^2 > k$ where $P_{Ho}(\sum[\log(X_i)]^2 > k) = \alpha$.

b) In the above argument, with any $\sigma_1 > 1$, get

$$\sum [\log(x_i)]^2 \left(\frac{1}{2} - \frac{1}{2\sigma_1^2}\right)$$

and

$$\frac{1}{2} - \frac{1}{2\sigma_1^2} > 0$$

for any $\sigma_1^2 > 1$. Hence the UMP test is the same as in a).

7.16. The most powerful test will have the following form.
Reject H_0 iff $\frac{f_1(x)}{f_0(x)} > k$.
But $\frac{f_1(x)}{f_0(x)} = 4x^{-\frac{3}{2}}$ and hence we reject H_0 iff X is small, i.e., reject H_0 is $X < k$ for some constant k. This test must also have the size α, that is we require:
$\alpha = P(X < k)$ when $f(x) = f_0(x)) = \int_0^k \frac{3}{64} x^2 dx = \frac{1}{64} k^3$,
so that $k = 4\alpha^{\frac{1}{3}}$.
For the power, when $k = 4\alpha^{\frac{1}{3}}$
$P[X < k$ when $f(x) = f_1(x)] = \int_0^k \frac{3}{16}\sqrt{x}dx = \sqrt{\alpha}$.
When $\alpha = 0.01$, the power is $= 0.10$.

7.19. See Example 7.5.

7.20. $E[T(X)] = 1/\lambda_1$ and the power $=$ P(test rejects H_0) $= P_{\lambda_1}(T(X) < \log(100/95)) = F_{\lambda_1}(\log(100/95))$
$= 1 - \exp(-\lambda_1 \log(100/95)) = 1 - (95/100)^{\lambda_1}$.
a) Power = $1 - \exp(-\log(100/95)) = 1 - \exp(\log(95/100)) = 0.05$.
b) Power = $1 - (95/100)^{50} = 0.923055$.
c) Let T_0 be the observed value of $T(X)$. Then p-value = $P(W \le T_0)$ where $W \sim$ exponential(1) since under H_0, $T(X) \sim$ exponential(1). So p-value = $1 - \exp(-T_0)$.

7.21. Note that

$$f(x) = I(x > 0)\ 2x\, e^{-x^2}\ \tau\ \exp[(\tau - 1)(\log(1 - e^{-x^2}))]$$

is a one-parameter exponential family and $w(\tau) = \tau - 1$ is an increasing function of τ. Thus the UMP test rejects H_0 if $T(\boldsymbol{x}) = \sum_{i=1}^n \log(1 - e^{-x_i^2}) > k$ where $\alpha = P_{\tau=2}(T(\boldsymbol{X}) > k)$.
Or use NP lemma.
a) Reject Ho if

$$\frac{f(\boldsymbol{x}|\tau = 4)}{f(\boldsymbol{x}|\tau = 1)} > k.$$

The LHS =

$$\frac{4^n}{2^n} \frac{\prod_{i=1}^n (1 - e^{-x_i^2})^{4-1}}{\prod_{i=1}^n (1 - e^{-x_i^2})} = 2^n \prod_{i=1}^n (1 - e^{-x_i^2})^2.$$

So reject Ho if

$$\prod_{i=1}^{n}(1-e^{-x_i^2})^2 > k'$$

or

$$\prod_{i=1}^{n}(1-e^{-x_i^2}) > c$$

or

$$\sum_{i=1}^{n}\log(1-e^{-x_i^2}) > d$$

where

$$\alpha = P_{\tau=2}\left(\prod_{i=1}^{n}(1-e^{-x_i^2}) > c\right).$$

b) Replace $4-1$ by $\tau_1 - 1$ where $\tau_1 > 2$. Then reject H_0 if

$$\prod_{i=1}^{n}(1-e^{-x_i^2})^{\tau_1-2} > k'$$

which gives the same test as in a).

7.22. By exponential family theory, the UMP test rejects H_0 if $T(\boldsymbol{x}) = -\sum_{i=1}^{n}\frac{1}{x_i} > k$ where $P_{\theta=1}(T(\boldsymbol{X}) > k) = \alpha$.
Alternatively, use the Neyman–Pearson lemma:
a) reject Ho if

$$\frac{f(\boldsymbol{x}|\theta=2)}{f(\boldsymbol{x}|\theta=1)} > k'.$$

The LHS =

$$\frac{2^n \exp\left(-2\sum \frac{1}{x_i}\right)}{\exp\left(-\sum \frac{1}{x_i}\right)}.$$

So reject Ho if

$$2^n \exp\left[(-2+1)\sum \frac{1}{x_i}\right] > k'$$

or if $-\sum \frac{1}{x_i} > k$ where $P_1(-\sum \frac{1}{x_i} > k) = \alpha$.
b) In the above argument, reject H_0 if

$$2^n \exp\left[(-\theta_1+1)\sum \frac{1}{x_i}\right] > k'$$

or if $-\sum \frac{1}{x_i} > k$ where $P_1(-\sum \frac{1}{x_i} > k) = \alpha$ for any $\theta_1 > 1$. Hence the UMP test is the same as in a).

7.23. a) We reject H_0 iff $\frac{f_1(x)}{f_0(x)} > k$. Thus we reject H_0 iff $\frac{2x}{2(1-x)} > k$. That is $\frac{1-x}{x} < k_1$, that is $\frac{1}{x} < k_2$, that is $x > k_3$. Now $0.1 = P(X > k_3)$ when $f(x) = f_0(x)$, so $k_3 = 1 - \sqrt{0.1}$.

7.24. a) Let $k = [2\pi\sigma_1\sigma_2(1-\rho^2)^{1/2}]$. Then the likelihood $L(\boldsymbol{\theta}) =$

$$\frac{1}{k^n}\exp\left(\frac{-1}{2(1-\rho^2)}\sum_{i=1}^{n}\left[\left(\frac{x_i-\mu_1}{\sigma_1}\right)^2 - 2\rho\left(\frac{x_i-\mu_1}{\sigma_1}\right)\left(\frac{y_i-\mu_2}{\sigma_2}\right)+\left(\frac{y_i-\mu_2}{\sigma_2}\right)^2\right]\right).$$

Hence

$$L(\hat{\boldsymbol{\theta}}) = \frac{1}{[2\pi\hat{\sigma}_1\hat{\sigma}_2(1-\hat{\rho}^2)^{1/2}]^n}\exp\left(\frac{-1}{2(1-\hat{\rho}^2)}[T_1 - 2\hat{\rho}T_2 + T_3]\right)$$

$$= \frac{1}{[2\pi\hat{\sigma}_1\hat{\sigma}_2(1-\hat{\rho}^2)^{1/2}]^n}\exp(-n)$$

and

$$L(\hat{\boldsymbol{\theta}}_0) = \frac{1}{[2\pi\hat{\sigma}_1\hat{\sigma}_2]^n}\exp\left(\frac{-1}{2}[T_1 + T_3]\right)$$

$$= \frac{1}{[2\pi\hat{\sigma}_1\hat{\sigma}_2]^n}\exp(-n).$$

Thus $\lambda(\boldsymbol{x},\boldsymbol{y}) =$

$$\frac{L(\hat{\boldsymbol{\theta}}_0)}{L(\hat{\boldsymbol{\theta}})} = (1-\hat{\rho}^2)^{n/2}.$$

So reject H_0 if $\lambda(\boldsymbol{x},\boldsymbol{y}) \le c$ where $\alpha = \sup_{\boldsymbol{\theta}\in\Theta_o} P(\lambda(\boldsymbol{X},\boldsymbol{Y}) \le c)$. Here Θ_o is the set of $\boldsymbol{\theta} = (\mu_1,\mu_2,\sigma_1,\sigma_2,\rho)$ such that the μ_i are real, $\sigma_i > 0$ and $\rho = 0$, i.e., such that X_i and Y_i are independent.

b) Since the unrestricted MLE has one more free parameter than the restricted MLE, $-2\log(\lambda(\boldsymbol{X},\boldsymbol{Y})) \approx \chi^2_1$, and the approximate LRT rejects H_0 if $-2\log\lambda(\boldsymbol{x},\boldsymbol{y}) > \chi^2_{1,1-\alpha}$ where $P(\chi^2_1 > \chi^2_{1,1-\alpha}) = \alpha$.

7.25. Parts a), b), e), f), g), i), j), k), n), o), and q) are similar with power = 0.9. Parts c), d), h), and m) are similar with power = 0.99.

l) See Example 7.7.

m) See Problem 7.27.

p) The power = 0.9. See Problem 7.28.

q) See Problem 7.26.

7.26. b) This family is a regular one-parameter exponential family where $w(\lambda) = -1/\lambda$ is increasing. Hence the level α UMP test rejects H_0 when $\sum_{i=1}^n (e^{y_i} - 1) > k$ where $\alpha = P_2(\sum_{i=1}^n (e^{Y_i} - 1) > k) = P_2(T(\boldsymbol{Y}) > k)$.

c) Since $T(\boldsymbol{Y}) \sim \frac{\lambda}{2}\chi^2_{2n}$, $\frac{2T(\boldsymbol{Y})}{\lambda} \sim \chi^2_{2n}$. Hence

$$\alpha = 0.05 = P_2(T(\boldsymbol{Y}) > k) = P(\chi^2_{40} > \chi^2_{40,1-\alpha}),$$

and $k = \chi^2_{40,1-\alpha} = 55.758$. Hence the power

$$\beta(\lambda) = P_\lambda(T(\boldsymbol{Y}) > 55.758) = P\left(\frac{2T(\boldsymbol{Y})}{\lambda} > \frac{2(55.758)}{\lambda}\right) = P\left(\chi^2_{40} > \frac{2(55.758)}{\lambda}\right)$$

$$= P\left(\chi^2_{40} > \frac{2(55.758)}{3.8386}\right) = P(\chi^2_{40} > 29.051) = 1 - 0.1 = 0.9.$$

7.27. b) This family is a regular one-parameter exponential family where $w(\sigma^2) = -1/(2\sigma^2)$ is increasing. Hence the level α UMP test rejects H_0 when $\sum_{i=1}^n y_i^2 > k$ where $\alpha = P_1(\sum_{i=1}^n Y_i^2) > k) = P_1(T(\boldsymbol{Y}) > k)$.

c) Since $T(\boldsymbol{Y}) \sim \sigma^2 \chi^2_n$, $\dfrac{T(\boldsymbol{Y})}{\sigma^2} \sim \chi^2_n$. Hence

$$\alpha = 0.05 = P_1(T(\boldsymbol{Y}) > k) = P(\chi^2_{20} > \chi^2_{20,1-\alpha}),$$

and $k = \chi^2_{20,1-\alpha} = 31.410$. Hence the power

$$\beta(\sigma) = P_\sigma(T(\boldsymbol{Y}) > 31.41) = P\left(\frac{T(\boldsymbol{Y})}{\sigma^2} > \frac{31.41}{\sigma^2}\right) = P\left(\chi^2_{20} > \frac{31.41}{3.8027}\right)$$

$$= P(\chi^2_{20} > 8.260) = 1 - 0.01 = 0.99.$$

7.28. a) Let $X = Y^2/\sigma^2 = t(Y)$. Then $Y = \sigma\sqrt{X} = t^{-1}(X)$. Hence

$$\frac{dt^{-1}(x)}{dx} = \frac{\sigma}{2}\frac{1}{\sqrt{x}}$$

and the pdf of X is

$$g(x) = f_Y(t^{-1}(x))\left|\frac{dt^{-1}(x)}{dx}\right| = \frac{\sigma\sqrt{x}}{\sigma^2}\exp\left[\frac{-1}{2}\left(\frac{\sigma\sqrt{x}}{\sigma}\right)^2\right]\frac{\sigma}{2\sqrt{x}} = \frac{1}{2}\exp(-x/2)$$

for $x > 0$, which is the χ^2_2 pdf.

b) This family is a regular one-parameter exponential family where $w(\sigma) = -1/(2\sigma^2)$ is increasing. Hence the level α UMP test rejects H_0 when $\sum_{i=1}^n y_i^2 > k$ where $\alpha = P_1(\sum_{i=1}^n Y_i^2 > k) = P_1(T(\boldsymbol{Y}) > k)$.

c) Since $T(\boldsymbol{Y}) \sim \sigma^2 \chi^2_{2n}$, $\dfrac{T(\boldsymbol{Y})}{\sigma^2} \sim \chi^2_{2n}$. Hence

$$\alpha = 0.05 = P_1(T(\boldsymbol{Y}) > k) = P(\chi^2_{40} > \chi^2_{40,1-\alpha}),$$

and $k = \chi^2_{40,1-\alpha} = 55.758$. Hence the power

$$\beta(\sigma) = P_\sigma(T(\boldsymbol{Y}) > 55.758) = P\left(\frac{T(\boldsymbol{Y})}{\sigma^2} > \frac{55.758}{\sigma^2}\right) = P\left(\chi^2_{40} > \frac{55.758}{\sigma^2}\right)$$

$$= P\left(\chi^2_{40} > \frac{55.758}{1.9193}\right) = P(\chi^2_{40} > 29.051) = 1 - 0.1 = 0.9.$$

7.29. See Example 7.11.

7.30. a) Suppose the observed values of N_i is n_i. Then
$f(\boldsymbol{x}|\theta) = d_{\boldsymbol{x}}\ p_1^{n_1} p_2^{n_2} p_3^{n_3} = d_{\boldsymbol{x}}\ (\theta^2)^{n_1}(2\theta(1-\theta))^{n_2}((1-\theta)^2)^{n_3} =$
$2^{n_2}\theta^{2n_1+n_2}(1-\theta)^{n_2+2n_3}$.
Since $n_2 + 2n_3 = 2n - (2n_1 + n_2)$, the above is

$$= d_{\boldsymbol{x}}\ 2^{n_2}\theta^{2n_1+n_2}(1-\theta)^{2n-(2n_1+n_2)} = d_{\boldsymbol{x}}\ 2^{n_2}\left(\frac{\theta}{1-\theta}\right)^{2n_1+n_2}(1-\theta)^{2n}.$$

Then $$\frac{f(\boldsymbol{x}|\theta_1)}{f(\boldsymbol{x}|\theta_0)} = \frac{\left(\frac{\theta_1}{1-\theta_1}\right)^{2n_1+n_2}(1-\theta_1)^{2n}}{\left(\frac{\theta_0}{1-\theta_0}\right)^{2n_1+n_2}(1-\theta_0)^{2n}} = \left(\frac{\theta_1}{\theta_0}\ \frac{1-\theta_0}{1-\theta_1}\right)^{2n_1+n_2}\left(\frac{1-\theta_1}{1-\theta_0}\right)^{2n},$$

which is an increasing function of $2n_1 + n_2$ as $0 < \theta_0 < \theta_1 < 1$ (since $\frac{\theta_1}{\theta_0} > 1$ and $\frac{1-\theta_0}{1-\theta_1} > 1$ and n is a constant).

(b) Using the NP Lemma, the most powerful test is given by $\frac{f(\boldsymbol{x}|\theta_1)}{f(\boldsymbol{x}|\theta_0)} > k$, for some k. Thus, the test that rejects H_0 if and only if $2N_1 + N_2 \geq c$, for some c, is a most powerful test.

7.31. See Example 7.12.

8.1. c) The histograms should become more like a normal distribution as n increases from 1 to 200. In particular, when $n = 1$ the histogram should be right skewed while for $n = 200$ the histogram should be nearly symmetric. Also the scale on the horizontal axis should decrease as n increases.

d) Now $\overline{Y} \sim N(0, 1/n)$. Hence the histograms should all be roughly symmetric, but the scale on the horizontal axis should be from about $-3/\sqrt{n}$ to $3/\sqrt{n}$.

8.3. a) $E(X) = \frac{3\theta}{\theta+1}$, thus
$\sqrt{n}(\overline{X} - E(X)) \xrightarrow{D} N(0, V(X))$, where
$V(X) = \frac{9\theta}{(\theta+2)(\theta+1)^2}$. Let $g(y) = \frac{y}{3-y}$, thus $g'(y) = \frac{3}{(3-y)^2}$. Using the delta method,
$\sqrt{n}(T_n - \theta) \xrightarrow{D} N(0, \frac{\theta(\theta+1)^2}{\theta+2})$.

b) It is asymptotically efficient if $\sqrt{n}(T_n - \theta) \xrightarrow{D} N(0, \nu(\theta))$, where

$$\nu(\theta) = \frac{\frac{d}{d\theta}(\theta)}{-E(\frac{d^2}{d\theta^2}lnf(x|\theta))}.$$

But, $E((\frac{d^2}{d\theta^2}lnf(x|\theta)) = \frac{1}{\theta^2}$. Thus $\nu(\theta) = \theta^2 \neq \frac{\theta(\theta+1)^2}{\theta+2}$.

c) $\overline{X} \to \frac{3\theta}{\theta+1}$ in probability. Thus $T_n \to \theta$ in probability.

8.5. See Example 8.8.

8.7. a) See Example 8.7.

8.13. a) $Y_n \stackrel{D}{=} \sum_{i=1}^n X_i$ where the X_i are iid χ_1^2. Hence $E(X_i) = 1$ and $\text{Var}(X_i) = 2$. Thus by the CLT,

$$\sqrt{n}\left(\frac{Y_n}{n} - 1\right) \stackrel{D}{=} \sqrt{n}\left(\frac{\sum_{i=1}^n X_i}{n} - 1\right) \stackrel{D}{\to} N(0,2).$$

b) Let $g(\theta) = \theta^3$. Then $g'(\theta) = 3\theta^2$, $g'(1) = 3$, and by the delta method,

$$\sqrt{n}\left[\left(\frac{Y_n}{n}\right)^3 - 1\right] \stackrel{D}{\to} N(0, 2(g'(1))^2) = N(0,18).$$

8.23. See the proof of Theorem 6.3.

8.27. a) See Example 8.1b.

b) See Example 8.3.

c) See Example 8.14.

8.28. a) By the CLT, $\sqrt{n}(\overline{X} - \lambda)/\sqrt{\lambda} \stackrel{D}{\to} N(0,1)$. Hence $\sqrt{n}(\overline{X} - \lambda) \stackrel{D}{\to} N(0,\lambda)$.

b) Let $g(\lambda) = \lambda^3$ so that $g'(\lambda) = 3\lambda^2$ then $\sqrt{n}[(\overline{X})^3 - (\lambda)^3] \stackrel{D}{\to} N(0, \lambda[g'(\lambda)]^2) = N(0, 9\lambda^5)$.

8.29. a) $\overline{X}$ is a complete sufficient statistic. Also, we have $\dfrac{(n-1)S^2}{\sigma^2}$ has a chi square distribution with $df = n-1$, thus since σ^2 is known the distribution of S^2 does not depend on μ, so S^2 is ancillary. Thus, by Basu's Theorem $\overline{X}$ and S^2 are independent.

b) by CLT (n is large) $\sqrt{n}(\overline{X} - \mu)$ has approximately normal distribution with mean 0 and variance σ^2. Let $g(x) = x^3$, thus, $g'(x) = 3x^2$. Using delta method $\sqrt{n}(g(\overline{X}) - g(\mu))$ goes in distribution to $N(0, \sigma^2(g'(\mu))^2)$ or $\sqrt{n}(\overline{X}^3 - \mu^3)$ goes in distribution to $N(0, \sigma^2(3\mu^2)^2)$.

8.30. a) According to the standard theorem, $\sqrt{n}(\hat{\theta}_n - \theta) \to N(0,3)$.

b) $E(Y) = \theta, \text{Var}(Y) = \frac{\pi^2}{3}$, according to CLT we have $\sqrt{n}(\overline{Y}_n - \theta) \to N(0, \frac{\pi^2}{3})$.

c) $\text{MED}(Y) = \theta$, then $\sqrt{n}(\text{MED}(n) - \theta) \to N(0, \frac{1}{4f^2(\text{MED}(Y))})$ and $f(\text{MED}(Y))$ $= \dfrac{\exp(-(\theta-\theta))}{[1+\exp(-(\theta-\theta))]^2} = \frac{1}{4}$. Thus $\sqrt{n}(\text{MED}(n) - \theta) \to N(0, \frac{1}{4\frac{1}{16}}) \to$ $\sqrt{n}(\text{MED}(n) - \theta) \to N(0,4)$.

d) All three estimators are consistent, but $3 < \frac{\pi^2}{3} < 4$, therefore the estimator $\hat{\theta}_n$ is the best, and the estimator $\text{MED}(n)$ is the worst.

8.32. a) $F_n(y) = 0.5 + 0.5y/n$ for $-n < y < n$, so $F(y) \equiv 0.5$.

b) No, since $F(y)$ is not a cdf.

8.33. a) $F_n(y) = y/n$ for $0 < y < n$, so $F(y) \equiv 0$.

b) No, since $F(y)$ is not a cdf.

8.34. a)

$$\sqrt{n}\left(\overline{Y} - \frac{1-\rho}{\rho}\right) \xrightarrow{D} N\left(0, \frac{1-\rho}{\rho^2}\right)$$

by the CLT.

c) The method of moments estimator of ρ is $\hat{\rho} = \frac{\overline{Y}}{1+\overline{Y}}$.

d) Let $g(\theta) = 1 + \theta$ so $g'(\theta) = 1$. Then by the delta method,

$$\sqrt{n}\left(g(\overline{Y}) - g\left(\frac{1-\rho}{\rho}\right)\right) \xrightarrow{D} N\left(0, \frac{1-\rho}{\rho^2}1^2\right)$$

or

$$\sqrt{n}\left((1+\overline{Y}) - \frac{1}{\rho}\right) \xrightarrow{D} N\left(0, \frac{1-\rho}{\rho^2}\right).$$

This result could also be found with algebra since $1 + \overline{Y} - \frac{1}{\rho} = \overline{Y} + 1 - \frac{1}{\rho} = \overline{Y} + \frac{\rho-1}{\rho} = \overline{Y} - \frac{1-\rho}{\rho}$.

e) $\overline{Y}$ is the method of moments estimator of $E(Y) = (1-\rho)/\rho$, so $1+\overline{Y}$ is the method of moments estimator of $1 + E(Y) = 1/\rho$.

8.35. a) $\sqrt{n}(\bar{X} - \mu)$ is approximately $N(0, \sigma^2)$. Define $g(x) = \frac{1}{x}$, $g'(x) = \frac{-1}{x^2}$. Using delta method, $\sqrt{n}(\frac{1}{\overline{X}} - \frac{1}{\mu})$ is approximately $N(0, \frac{\sigma^2}{\mu^4})$. Thus $1/\overline{X}$ is approximately $N(\frac{1}{\mu}, \frac{\sigma^2}{n\mu^4})$, provided $\mu \neq 0$.

b) Using part a)

$\frac{1}{\overline{X}}$ is asymptotically efficient for $\frac{1}{\mu}$ if

$$\frac{\sigma^2}{\mu^4} = \left[\frac{(\tau'(\mu))^2}{E_\mu\left(\frac{\partial}{\partial\mu}\ln f(X/\mu)\right)^2}\right]$$

$$\tau(\mu) = \frac{1}{\mu}$$

$$\tau'(\mu) = \frac{-1}{\mu^2}$$

$$\ln f(x|\mu) = \frac{-1}{2}\ln 2\pi\sigma^2 - \frac{(x-\mu)^2}{2\sigma^2}$$

$$E\left[\frac{\partial}{\partial\mu}\ln f(X/\mu)\right]^2 = \frac{E(X-\mu)^2}{\sigma^4} = \frac{1}{\sigma^2}$$

Thus

$$\frac{(\tau'(\mu))^2}{E_\mu\left[\frac{\partial}{\partial\mu}\ln f(X/\mu)\right]^2} = \frac{\sigma^2}{\mu^4}.$$

8.36. a) $E(Y^k) = 2\theta^k/(k+2)$ so $E(Y) = 2\theta/3, E(Y^2) = \theta^2/2$ and $V(Y) = \theta^2/18$. So $\sqrt{n}\left(\overline{Y} - \frac{2\theta}{3}\right) \xrightarrow{D} N\left(0, \frac{\theta^2}{18}\right)$ by the CLT.

b) Let $g(\tau) = \log(\tau)$ so $[g'(\tau)]^2 = 1/\tau^2$ where $\tau = 2\theta/3$. Then by the delta method,

$\sqrt{n}\left(\log(\overline{Y}) - \log\left(\frac{2\theta}{3}\right)\right) \xrightarrow{D} N\left(0, \frac{1}{8}\right).$

c) $\hat{\theta}^k = \frac{k+2}{2n}\sum Y_i^k$.

8.37. a) $\sqrt{n}\left(\overline{Y} - \frac{r(1-\rho)}{\rho}\right) \xrightarrow{D} N\left(0, \frac{r(1-\rho)}{\rho^2}\right)$ by the CLT.

b) Let $\theta = r(1-\rho)/\rho$. Then

$$g(\theta) = \frac{r}{r + \frac{r(1-\rho)}{\rho}} = \frac{r\rho}{r\rho + r(1-\rho)} = \rho = c.$$

Now

$$g'(\theta) = \frac{-r}{(r+\theta)^2} = \frac{-r}{(r + \frac{r(1-\rho)}{\rho})^2} = \frac{-r\rho^2}{r^2}.$$

So

$$[g'(\theta)]^2 = \frac{r^2\rho^4}{r^4} = \frac{\rho^4}{r^2}.$$

Hence by the delta method

$$\sqrt{n}\,(\,g(\overline{Y}) - \rho\,) \xrightarrow{D} N\left(0, \frac{r(1-\rho)}{\rho^2}\frac{\rho^4}{r^2}\right) = N\left(0, \frac{\rho^2(1-\rho)}{r}\right).$$

c) $\overline{Y} \overset{\text{set}}{=} r(1-\rho)/\rho$ or $\rho\overline{Y} = r - r\rho$ or $\rho\overline{Y} + r\rho = r$ or $\hat{\rho} = r/(r+\overline{Y})$.

8.38. a) By the CLT,

$$\sqrt{n}\left(\overline{X} - \frac{\theta}{2}\right) \xrightarrow{D} N\left(0, \frac{\theta^2}{12}\right).$$

b) Let $g(y) = y^2$. Then $g'(y) = 2y$ and by the delta method,

$$\sqrt{n}\left(\overline{X}^2 - \left(\frac{\theta}{2}\right)^2\right) = \sqrt{n}\left(\overline{X}^2 - \frac{\theta^2}{4}\right) = \sqrt{n}\left(g(\overline{X}) - g\left(\frac{\theta}{2}\right)\right) \xrightarrow{D}$$

$$N\left(0, \frac{\theta^2}{12}\left[g'\left(\frac{\theta}{2}\right)\right]^2\right) = N\left(0, \frac{\theta^2}{12}\frac{4\theta^2}{4}\right) = N\left(0, \frac{\theta^4}{12}\right).$$

9.1. a) $\sum_{i=1}^n X_i^b$ is minimal sufficient for a.

b) It can be shown that $\frac{X^b}{a}$ has an exponential distribution with mean 1. Thus, $\frac{2\sum_{i=1}^n X_i^b}{a}$ is distributed χ^2_{2n}. Let $\chi^2_{2n,\alpha/2}$ be the upper $100(\frac{1}{2}\alpha)\%$ point of the chi-square distribution with $2n$ degrees of freedom. Thus, we can write

$$1-\alpha = P\left(\chi^2_{2n,1-\alpha/2} < \frac{2\sum_{i=1}^n X_i^b}{a} < \chi^2_{2n,\alpha/2}\right)$$

which translates into

$$\left(\frac{2\sum_{i=1}^n X_i^b}{\chi^2_{2n,\alpha/2}}, \frac{2\sum_{i=1}^n X_i^b}{\chi^2_{2n,1-\alpha/2}}\right)$$

as a two sided $(1-\alpha)$ confidence interval for a. For $\alpha = 0.05$ and $n = 20$, we have $\chi^2_{2n,\alpha/2} = 34.1696$ and $\chi^2_{2n,1-\alpha/2} = 9.59083$. Thus the confidence interval for a is

$$\left(\frac{\sum_{i=1}^n X_i^b}{17.0848}, \frac{\sum_{i=1}^n X_i^b}{4.795415}\right).$$

9.4c). Tables are from simulated data but should be similar to the table below.

```
n    p   ccov   acov
50  .01 .4236 .9914 ACT CI better
100 .01 .6704 .9406 ACT CI better
150 .01 .8278 .9720 ACT CI better
200 .01 .9294 .9098 the CIs are about the same
250 .01 .8160 .8160 the CIs are about the same
300 .01 .9158 .9228 the CIs are about the same
350 .01 .9702 .8312 classical is better
400 .01 .9486 .6692 classical is better
450 .01 .9250 .4080 classical is better
```

9.11. The simulated coverages should be close to the values below. The pooled t CI has coverage that is too small.

```
pcov  mpcov wcov
0.847 0.942 0.945
```

9.12. a) Let $W_i \sim U(0,1)$ for $i = 1,\ldots,n$ and let $T_n = Y/\theta$. Then

$$P(\frac{Y}{\theta} \le t) = P(\max(W_1,\ldots,W_n) \le t) =$$

P(all $W_i \leq t) = [F_{W_i}(t)]^n = t^n$ for $0 < t < 1$. So the pdf of T_n is

$$f_{T_n}(t) = \frac{d}{dt} t^n = n t^{n-1}$$

for $0 < t < 1$.

b) Yes, the distribution of $T_n = Y/\theta$ does not depend on θ by a).

c) See Example 9.21.

12.3 Tables

Tabled values are F(0.95,k,d) where $P(F < F(0.95,k,d)) = 0.95$.
00 stands for ∞. Entries produced with the `qf(.95,k,d)` command in R. The numerator degrees of freedom are k while the denominator degrees of freedom are d.

k	1	2	3	4	5	6	7	8	9	00
d										
1	161	200	216	225	230	234	237	239	241	254
2	18.5	19.0	19.2	19.3	19.3	19.3	19.4	19.4	19.4	19.5
3	10.1	9.55	9.28	9.12	9.01	8.94	8.89	8.85	8.81	8.53
4	7.71	6.94	6.59	6.39	6.26	6.16	6.09	6.04	6.00	5.63
5	6.61	5.79	5.41	5.19	5.05	4.95	4.88	4.82	4.77	4.37
6	5.99	5.14	4.76	4.53	4.39	4.28	4.21	4.15	4.10	3.67
7	5.59	4.74	4.35	4.12	3.97	3.87	3.79	3.73	3.68	3.23
8	5.32	4.46	4.07	3.84	3.69	3.58	3.50	3.44	3.39	2.93
9	5.12	4.26	3.86	3.63	3.48	3.37	3.29	3.23	3.18	2.71
10	4.96	4.10	3.71	3.48	3.33	3.22	3.14	3.07	3.02	2.54
11	4.84	3.98	3.59	3.36	3.20	3.09	3.01	2.95	2.90	2.41
12	4.75	3.89	3.49	3.26	3.11	3.00	2.91	2.85	2.80	2.30
13	4.67	3.81	3.41	3.18	3.03	2.92	2.83	2.77	2.71	2.21
14	4.60	3.74	3.34	3.11	2.96	2.85	2.76	2.70	2.65	2.13
15	4.54	3.68	3.29	3.06	2.90	2.79	2.71	2.64	2.59	2.07
16	4.49	3.63	3.24	3.01	2.85	2.74	2.66	2.59	2.54	2.01
17	4.45	3.59	3.20	2.96	2.81	2.70	2.61	2.55	2.49	1.96
18	4.41	3.55	3.16	2.93	2.77	2.66	2.58	2.51	2.46	1.92
19	4.38	3.52	3.13	2.90	2.74	2.63	2.54	2.48	2.42	1.88
20	4.35	3.49	3.10	2.87	2.71	2.60	2.51	2.45	2.39	1.84
25	4.24	3.39	2.99	2.76	2.60	2.49	2.40	2.34	2.28	1.71
30	4.17	3.32	2.92	2.69	2.53	2.42	2.33	2.27	2.21	1.62
00	3.84	3.00	2.61	2.37	2.21	2.10	2.01	1.94	1.88	1.00

Tabled values are $t_{\alpha,d}$ where $P(t < t_{\alpha,d}) = \alpha$ where t has a t distribution with d degrees of freedom. If $d > 29$ use the $N(0,1)$ cutoffs $d = Z = \infty$.

					alpha					pvalue
d	0.005	0.01	0.025	0.05	0.5	0.95	0.975	0.99	0.995	left tail
1	-63.66	-31.82	-12.71	-6.314	0	6.314	12.71	31.82	63.66	
2	-9.925	-6.965	-4.303	-2.920	0	2.920	4.303	6.965	9.925	
3	-5.841	-4.541	-3.182	-2.353	0	2.353	3.182	4.541	5.841	
4	-4.604	-3.747	-2.776	-2.132	0	2.132	2.776	3.747	4.604	
5	-4.032	-3.365	-2.571	-2.015	0	2.015	2.571	3.365	4.032	
6	-3.707	-3.143	-2.447	-1.943	0	1.943	2.447	3.143	3.707	
7	-3.499	-2.998	-2.365	-1.895	0	1.895	2.365	2.998	3.499	
8	-3.355	-2.896	-2.306	-1.860	0	1.860	2.306	2.896	3.355	
9	-3.250	-2.821	-2.262	-1.833	0	1.833	2.262	2.821	3.250	
10	-3.169	-2.764	-2.228	-1.812	0	1.812	2.228	2.764	3.169	
11	-3.106	-2.718	-2.201	-1.796	0	1.796	2.201	2.718	3.106	
12	-3.055	-2.681	-2.179	-1.782	0	1.782	2.179	2.681	3.055	
13	-3.012	-2.650	-2.160	-1.771	0	1.771	2.160	2.650	3.012	
14	-2.977	-2.624	-2.145	-1.761	0	1.761	2.145	2.624	2.977	
15	-2.947	-2.602	-2.131	-1.753	0	1.753	2.131	2.602	2.947	
16	-2.921	-2.583	-2.120	-1.746	0	1.746	2.120	2.583	2.921	
17	-2.898	-2.567	-2.110	-1.740	0	1.740	2.110	2.567	2.898	
18	-2.878	-2.552	-2.101	-1.734	0	1.734	2.101	2.552	2.878	
19	-2.861	-2.539	-2.093	-1.729	0	1.729	2.093	2.539	2.861	
20	-2.845	-2.528	-2.086	-1.725	0	1.725	2.086	2.528	2.845	
21	-2.831	-2.518	-2.080	-1.721	0	1.721	2.080	2.518	2.831	
22	-2.819	-2.508	-2.074	-1.717	0	1.717	2.074	2.508	2.819	
23	-2.807	-2.500	-2.069	-1.714	0	1.714	2.069	2.500	2.807	
24	-2.797	-2.492	-2.064	-1.711	0	1.711	2.064	2.492	2.797	
25	-2.787	-2.485	-2.060	-1.708	0	1.708	2.060	2.485	2.787	
26	-2.779	-2.479	-2.056	-1.706	0	1.706	2.056	2.479	2.779	
27	-2.771	-2.473	-2.052	-1.703	0	1.703	2.052	2.473	2.771	
28	-2.763	-2.467	-2.048	-1.701	0	1.701	2.048	2.467	2.763	
29	-2.756	-2.462	-2.045	-1.699	0	1.699	2.045	2.462	2.756	
Z	-2.576	-2.326	-1.960	-1.645	0	1.645	1.960	2.326	2.576	
CI						90%	95%		99%	
	0.995	0.99	0.975	0.95	0.5	0.05	0.025	0.01	0.005	right tail
	0.01	0.02	0.05	0.10	1	0.10	0.05	0.02	0.01	two tail

References

Abuhassan, H. (2007), *Some Transformed Distributions*, Ph.D. Thesis, Southern Illinois University, see (http://lagrange.math.siu.edu/Olive/shassan.pdf).

Abuhassan, H., and Olive, D.J. (2008), "Inference for the Pareto, Half Normal and Related Distributions," unpublished manuscript, (http://lagrange.math.siu.edu/Olive/pppar.pdf).

Adell, J.A., and Jodrá, P. (2005), "Sharp Estimates for the Median of the $\Gamma(n+1,1)$ Distribution," *Statistics & Probability Letters*, 71, 185–191.

Agresti, A., and Caffo, B. (2000), "Simple and Effective Confidence Intervals for Proportions and Difference of Proportions Result by Adding Two Successes and Two Failures," *The American Statistician*, 54, 280–288.

Agresti, A., and Coull, B.A. (1998), "Approximate is Better than Exact for Interval Estimation of Binomial Parameters," *The American Statistician*, 52, 119–126.

Al-Mutairi, D.K., Ghitany, M.E., and Kundu, D. (2013), "Inference on Stress-Strength Reliability from Lindley Distributions," *Communications in Statistics: Theory and Methods*, 42, 1443–1463.

Anderson, T.W. (1984), *An Introduction to Multivariate Statistical Analysis,* 2nd ed., Wiley, New York, NY.

Apostol, T.M. (1957), *Mathematical Analysis: a Modern Approach to Advanced Calculus,* Addison-Wesley, Reading, MA.

Arnold, S.F. (1990), *Mathematical Statistics*, Prentice Hall, Upper Saddle River, NJ.

Ash, C. (1993), *The Probability Tutoring Book: an Intuitive Course for Engineers and Scientists (and Everyone Else!)*, IEEE Press, Piscataway, NJ.

Ash, R.B. (1972), *Real Analysis and Probability,* Academic Press, San Diego, CA.

Ash, R.B. (2011), *Statistical Inference: a Concise Course*, Dover, Mineola, NY.

Ash, R.B. (2013), *Lectures on Statistics*, online at (http://www.math.uiuc.edu/~r-ash/).

Ash, R.B., and Doleans-Dade, C.A. (1999), *Probability and Measure Theory,* 2nd ed., Academic Press, San Diego, CA.

Azzalini, A. (1996), *Statistical Inference Based on Likelihood*, Chapman & Hall/CRC, Boca Raton, FL.

D.J. Olive, *Statistical Theory and Inference*, DOI 10.1007/978-3-319-04972-4,

Bahadur, R.R. (1958), "Examples of Inconsistency of Maximum Likelihood Estimators," *Sankhyā,* 20, 207–210.

Bain, L.J. (1978), *Statistical Analysis of Reliability and Life-Testing Models,* Marcel Dekker, New York, NY.

Bain, L.J., and Engelhardt, M. (1992), *Introduction to Probability and Mathematical Statistics*, Duxbury Press, Boston, MA.

Barker, L. (2002), "A Comparison of Nine Confidence Intervals for a Poisson Parameter When the Expected Number of Events ≤ 5," *The American Statistician*, 56, 85–89.

Barndorff-Nielsen, O. (1978), *Information and Exponential Families in Statistical Theory*, Wiley, New York, NY.

Barndorff-Nielsen, O. (1982), "Exponential Families," in *Encyclopedia of Statistical Sciences,* Vol. 2, eds. Kotz, S. and Johnson, N.L., Wiley, New York, NY, 587–596.

Bartle, R.G. (1964), *The Elements of Real Analysis*, Wiley, New York, NY.

Basu, D. (1959), "The Family of Ancillary Statistics," *Sankhyā, A*, 21, 247–256.

Becker, R.A., Chambers, J.M., and Wilks, A.R. (1988), *The New S Language: a Programming Environment for Data Analysis and Graphics,* Wadsworth and Brooks/Cole, Pacific Grove, CA.

Berger, J.O., Boukai, B., and Wang, Y. (1997), "Unified Frequentist and Bayesian Testing of a Precise Hypothesis," *Statistical Science*, 12, 133–160.

Berk, R. (1967), "Review 1922 of 'Invariance of Maximum Likelihood Estimators' by Peter W. Zehna," *Mathematical Reviews,* 33, 342–343.

Berk, R.H. (1972), "Consistency and Asymptotic Normality of MLE's for Exponential Models," *The Annals of Mathematical Statistics*, 43, 193–204.

Berry, D.A., and Lindgren, B.W. (1995), *Statistics, Theory and Methods,* 2nd ed., Duxbury Press, Belmont, CA.

Bertsekas, D.P. (1999), *Nonlinear Programming*, 2nd ed., Athena Scientific, Nashua, NH.

Besbeas, P., and Morgan, B.J.T. (2004), "Efficient and Robust Estimation for the One-Sided Stable Distribution of Index $1/2$," *Statistics & Probability Letters,* 66, 251–257.

Bickel, P.J., and Doksum, K.A. (2007), *Mathematical Statistics: Basic Ideas and Selected Topics,* Vol. 1., 2nd ed., Updated Printing, Pearson Prentice Hall, Upper Saddle River, NJ.

Bierens, H.J. (2004), *Introduction to the Mathematical and Statistical Foundations of Econometrics,* Cambridge University Press, Cambridge, UK.

Billingsley, P. (1995), *Probability and Measure,* 3rd ed., Wiley, New York, NY.

Birkes, D. (1990), "Generalized Likelihood Ratio Test and Uniformly Most Powerful Tests," *The American Statistician,* 44, 163–166.

Bolstad, W.M. (2004, 2007), *Introduction to Bayesian Statistics*, 1st and 2nd ed., Wiley, Hoboken, NJ.

Boos, D.D., and Hughes-Oliver, J.M. (1998), "Applications of Basu's Theorem," *The American Statistician*, 52, 218–221.

Bowman, K.O., and Shenton, L.R. (1988), *Properties of Estimators for the Gamma Distribution,* Marcel Dekker, New York, NY.

Broffitt, J.D. (1986), "Zero Correlation, Independence, and Normality," *The American Statistician,* 40, 276–277.

Brown, L.D. (1986), *Fundamentals of Statistical Exponential Families with Applications in Statistical Decision Theory*, Institute of Mathematical Statistics Lecture Notes – Monograph Series, IMS, Haywood, CA.

Brown, L.D., Cai, T.T., and DasGupta, A. (2001), "Interval Estimation for a Binomial Proportion," (with discussion), *Statistical Science*, 16, 101–133.

Brown, L.D., Cai, T.T., and DasGupta, A. (2002), "Confidence Intervals for a Binomial Proportion and Asymptotic Expansions," *The Annals of Statistics*, 30, 150–201.

Brown, L.D., Cai, T.T., and DasGupta, A. (2003), "Interval Estimation in Exponential Families," *Statistica Sinica*, 13, 19–49.

Brownstein, N., and Pensky, M. (2008), "Application of Transformations in Parametric Inference," *Journal of Statistical Education*, 16 (online).

Buckland, S.T. (1984), "Monte Carlo Confidence Intervals," *Biometrics*, 40, 811–817.

Bühler, W.J., and Sehr, J. (1987), "Some Remarks on Exponential Families," *The American Statistician,* 41, 279–280.

Buxton, L.H.D. (1920), "The Anthropology of Cyprus," *The Journal of the Royal Anthropological Institute of Great Britain and Ireland,* 50, 183–235.

Byrne, J., and Kabaila, P. (2005), "Comparison of Poisson Confidence Intervals," *Communications in Statistics: Theory and Methods*, 34, 545–556.

Cambanis, S., Huang, S., and Simons, G. (1981), "On the Theory of Elliptically Contoured Distributions," *Journal of Multivariate Analysis,* 11, 368–385.

Carlin, B.P., and Louis, T.A. (2009), *Bayesian Methods for Data Analysis*, 3rd ed., Chapman & Hall/CRC Press, Boca Raton, FL.

Casella, G., and Berger, R.L. (2002), *Statistical Inference,* 2nd ed., Duxbury, Belmont, CA.

Casella, G., and George, E.I. (1992), "Explaining the Gibbs Sampler," *The American Statistician*, 46, 167–174.

Castillo, E. (1988), *Extreme Value Theory in Engineering,* Academic Press, Boston, MA.

Chambers, J.M. (2008), *Software for Data Analysis: Programming With R*, Springer, New York, NY.

Chen, J., and Rubin, H. (1986), "Bounds for the Difference Between Median and Mean of Gamma and Poisson Distributions," *Statistics & Probability Letters,* 4, 281–283.

Chernoff, H. (1956), "Large-Sample Theory: Parametric Case," *The Annals of Mathematical Statistics*, 27, 1–22.

Chmielewski, M.A. (1981), "Elliptically Symmetric Distributions: a Review and Bibliography," *International Statistical Review,* 49, 67–74.

Cohen, A.C., and Whitten, B.J. (1988), *Parameter Estimation in Reliability and Life Span Models,* Marcel Dekker, New York, NY.

Consonni, G., and Veronese, P. (1992), "Conjugate Priors for Exponential Families Having Quadratic Variance Functions," *Journal of the American Statistical Association*, 87, 1123–1127.

Cook, R.D. (1998), *Regression Graphics: Ideas for Studying Regression Through Graphics,* Wiley, New York, NY.

Cooke, D., Craven, A.H., and Clarke, G.M. (1982), *Basic Statistical Computing,* Edward Arnold Publishers, London, UK.

Cox, C. (1984), "An Elementary Introduction to Maximum Likelihood Estimations for Multinomial Models: Birch's Theorem and the Delta Method," *The American Statistician*, 38, 283–287.

Cox, D.R., and Hinkley, D.V. (1974), *Theoretical Statistics,* Chapman and Hall, London, UK.

Cramér, H. (1946), *Mathematical Methods of Statistics,* Princeton University Press, Princeton, NJ.

Crawley, M.J. (2005), *Statistics: an Introduction Using R,* Wiley, Hoboken, NJ.

Crawley, M.J. (2013), *The R Book,* 2nd ed., Wiley, Hoboken, NJ.

Croux, C., Dehon, C., Rousseeuw, P.J., and Van Aelst, S. (2001), "Robust Estimation of the Conditional Median Function at Elliptical Models," *Statistics & Probability Letters,* 51, 361–368.

Dalgaard, P. (2002), *Introductory Statistics with R*, Springer, New York, NY.

DasGupta, A. (2008), *Asymptotic Theory of Statistics and Probability*, Springer, New York, NY.

Datta, G.S. (2005), "An Alternative Derivation of the Distributions of the Maximum Likelihood Estimators of the Parameters in an Inverse Gaussian Distribution," *Biometrika*, 92, 975–977.

Datta, G.S., and Sarker, S.K. (2008), "A General Proof of Some Known Results of Independence Between Two Statistics," *The American Statistician*, 62, 141–143.

David, H.A. (1981), *Order Statistics*, Wiley, New York, NY.

David, H.A. (1995), "First (?) Occurrences of Common Terms in Mathematical Statistics," *The American Statistician,* 49, 121–133.

David, H.A. (2006–7), "First (?) Occurrences of Common Terms in Statistics and Probability," Publications and Preprint Series, Iowa State University, (www.stat.iastate.edu/preprint/hadavid.html).

Davidson, J. (1994), *Stochastic Limit Theory*, Oxford University Press, Oxford, UK.

deCani, J.S., and Stine, R.A. (1986), "A Note on Deriving the Information Matrix for a Logistic Distribution," *The American Statistician,* 40, 220–222.

DeGroot, M.H., and Schervish, M.J. (2012), *Probability and Statistics,* 4th ed., Pearson Education, Boston, MA.

Dekking, F.M., Kraaikamp, C., Lopuhaä, H.P., and Meester, L.E. (2005), *A Modern Introduction to Probability and Statistics Understanding Why and How*, Springer, London, UK.

Dudley, R.M. (2002), *Real Analysis and Probability*, Cambridge University Press, Cambridge, UK.

Durrett, R. (1995), *Probability, Theory and Examples*, 2nd ed., Duxbury Press, Belmont, CA.

Eaton, M.L. (1986), "A Characterization of Spherical Distributions," *Journal of Multivariate Analysis,* 20, 272–276.

Efron, B., and Tibshirani, R.J. (1993), *An Introduction to the Bootstrap*, Chapman & Hall/CRC, New York, NY.

Fang, K.T., and Anderson, T.W. (eds.) (1990), *Statistical Inference in Elliptically Contoured and Related Distributions,* Allerton Press, New York, NY.

Fang, K.T., Kotz, S., and Ng, K.W. (1990), *Symmetric Multivariate and Related Distributions,* Chapman & Hall, New York, NY.

Feller, W. (1957), *An Introduction to Probability Theory and Its Applications,* Vol. I, 2nd ed., Wiley, New York, NY.

Feller, W. (1971), *An Introduction to Probability Theory and Its Applications,* Vol. II, 2nd ed., Wiley, New York, NY.

Ferguson, T.S. (1967), *Mathematical Statistics: a Decision Theoretic Approach,* Academic Press, New York, NY.

Ferguson, T.S. (1996), *A Course in Large Sample Theory*, Chapman & Hall, New York, NY.

Fisher, R.A. (1922), "On the Mathematical Foundations of Theoretical Statistics," *Philosophical Transactions of the Royal Statistical Society A*, 222, 309–368.

Forbes, C., Evans, M., Hastings, N., and Peacock, B. (2011), *Statistical Distributions,* 4th ed., Wiley, Hoboken, NJ.

Frey, J. (2013), "Data-Driven Nonparametric Prediction Intervals," *Journal of Statistical Planning and Inference*, 143, 1039–1048.

Gabel, R.A., and Roberts, R.A. (1980), *Signals and Linear Systems*, Wiley, New York, NY.

Garwood, F. (1936), "Fiducial Limits for the Poisson Distribution," *Bio-metrika*, 28, 437–442.

Gathwaite, P.H., Jolliffe, I.T., and Jones, B. (2002), *Statistical Inference*, 2nd ed., Oxford University Press, Oxford, UK.

Gaughan, E.D. (2009), *Introduction to Analysis*, 5th ed., American Mathematical Society, Providence, RI.

Greenwood, J.A., and Durand, D. (1960), "Aids for Fitting the Gamma Distribution by Maximum Likelihood," *Technometrics,* 2, 55–56.

Grosh, D. (1989), *A Primer of Reliability Theory*, Wiley, New York, NY.

Grübel, R. (1988), "The Length of the Shorth," *The Annals of Statistics,* 16, 619–628.

Guenther, W.C. (1969), "Shortest Confidence Intervals," *The American Statistician,* 23, 22–25.

Guenther, W.C. (1978), "Some Easily Found Minimum Variance Unbiased Estimators," *The American Statistician,* 32, 29–33.

Gupta, A.K., and Varga, T. (1993), *Elliptically Contoured Models in Statistics,* Kluwar Academic Publishers, Dordrecht, The Netherlands.

Hahn, G.J., and Meeker, M.Q. (1991), *Statistical Intervals: a Guide for Practitioners*, Wiley, Hoboken, NJ.

Halmos, P.R., and Savage, L.J. (1949), "Applications of the Radon-Nikodym Theorem to the Theory of Sufficient Statistics," *The Annals of Mathematical Statistics,* 20, 225–241.

Hamza, K. (1995), "The Smallest Uniform Upper Bound on the Distance Between the Mean and the Median of the Binomial and Poisson Distributions," *Statistics & Probability Letters,* 23, 21–25.

Hanley, J.A., Julien, M., Moodie, E.E.M. (2008), "*t* Distribution Centennial: Student's *z*, *t*, and *s*: What if Gosset had R?" *The American Statistician*, 62, 64–69.

Headrick, T.C., Pant, M.D., and Sheng, Y. (2010), "On Simulating Univariate and Multivariate Burr Type III and Type XII Distributions," *Applied Mathematical Sciences*, 4, 2207–2240.

Hoel, P.G., Port, S.C., and Stone, C.J. (1971), *Introduction to Probability Theory,* Houghton Mifflin, Boston, MA.

Hogg, R.V. (1991), "Statistical Education: Improvements are Badly Needed," *The American Statistician*, 45, 342–343.

Hogg, R.V., McKean, J.W., and Craig, A.T. (2012), *Introduction to Mathematical Statistics,* 7th ed., Pearson Education, Boston, MA.

Hogg, R.V., and Tanis, E.A. (2005), *Probability and Statistical Inference,* 7th ed., Prentice Hall, Englewood Cliffs, NJ.

Hudson, H.M. (1978), "A Natural Identity for Exponential Families with Applications in Multiparameter Estimation," *The Annals of Statistics*, 6, 473–484,

Hyndman, R.J. (1996), "Computing and Graphing Highest Density Regions," *The American Statistician,* 50, 120–126.

Jiang, J. (2010), *Large Sample Techniques for Statistics*, Springer, New York, NY.

Jöckel, K.H. (1986), "Finite Sample Properties and Asymptotic Efficiency of Monte Carlo Tests," *The Annals of Statistics*, 14, 336–347.

Johanson, S. (1979), *Introduction to the Theory of Regular Exponential Families,* Institute of Mathematical Statistics, University of Copenhagen, Copenhagen, Denmark.

Johnson, M.E. (1987), *Multivariate Statistical Simulation,* Wiley, New York, NY.

Johnson, N.L., and Kotz, S. (1970ab), *Distributions in Statistics: Continuous Univariate Distributions,* Vol. 1–2, Houghton Mifflin Company, Boston, MA.

Johnson, N.L., and Kotz, S. (1972), *Distributions in Statistics: Continuous Multivariate Distributions,* Wiley, New York, NY.

Johnson, N.L., Kotz, S., and Kemp, A.K. (1992), *Distributions in Statistics: Univariate Discrete Distributions,* 2nd ed., Wiley, New York, NY.

Johnson, R.A., Ladella, J., and Liu, S.T. (1979), "Differential Relations, in the Original Parameters, Which Determine the First Two Moments of the Multi-parameter Exponential Family," *The Annals of Statistics*, 7, 232–235.

Johnson, R.A., and Wichern, D.W. (1988), *Applied Multivariate Statistical Analysis,* 2nd ed., Prentice Hall, Englewood Cliffs, NJ.

Joshi, V.M. (1976), "On the Attainment of the Cramér-Rao Lower Bound," *The Annals of Statistics,* 4, 998–1002.

Kalbfleisch, J.D., and Prentice, R.L. (1980), *The Statistical Analysis of Failure Time Data*, Wiley, New York, NY.

Karakostas, K.X. (1985), "On Minimum Variance Estimators," *The American Statistician,* 39, 303–305.

Keener, R.W. (2010), *Theoretical Statistics: Topics for a Core Course*, Springer, New York, NY.

Kelker, D. (1970), "Distribution Theory of Spherical Distributions and a Location Scale Parameter Generalization," *Sankhyā, A*, 32, 419–430.

Kennedy, W.J., and Gentle, J.E. (1980), *Statistical Computing,* Marcel Dekker, New York, NY.

Kiefer, J. (1961), "On Large Deviations of the Empiric D. F. of a Vector of Chance Variables and a Law of Iterated Logarithm," *Pacific Journal of Mathematics*, 11, 649–660.

Knight, K. (2000), *Mathematical Statistics*, Chapman & Hall/CRC, Boca Raton, FL.

Koehn, U., and Thomas, D.L. (1975), "On Statistics Independent of a Sufficient Statistic: Basu's Lemma," *The American Statistician*, 29, 40–42.

Kotz, S., and Johnson, N.L. (editors) (1982ab), *Encyclopedia of Statistical Sciences,* Vol. 1–2, Wiley, New York, NY.

Kotz, S., and Johnson, N.L. (editors) (1983ab), *Encyclopedia of Statistical Sciences,* Vol. 3–4, Wiley, New York, NY.

Kotz, S., and Johnson, N.L. (editors) (1985ab), *Encyclopedia of Statistical Sciences,* Vol. 5–6, Wiley, New York, NY.

Kotz, S., and Johnson, N.L. (editors) (1986), *Encyclopedia of Statistical Sciences,* Vol. 7, Wiley, New York, NY.

Kotz, S., and Johnson, N.L. (editors) (1988ab), *Encyclopedia of Statistical Sciences,* Vol. 8–9, Wiley, New York, NY.

Kotz, S., and van Dorp, J.R. (2004), *Beyond Beta: Other Continuous Families of Distributions with Bounded Support and Applications*, World Scientific, Singapore.

Kowalski, C.J. (1973), "Non-Normal Bivariate Distributions with Normal Marginals," *The American Statistician,* 27, 103–106.

Krishnamoorthy, K., and Mathew, T. (2009), *Statistical Tolerance Regions: Theory, Applications and Computation*, Wiley, Hoboken, NJ.

Lancaster, H.O. (1959), "Zero Correlation and Independence," *Australian Journal of Statistics,* 21, 53–56.

Larsen, R.J., and Marx, M.L. (2011), *Introduction to Mathematical Statistics and Its Applications*, 5th ed., Prentice Hall, Upper Saddle River, NJ.

Lavine, M., and Schervish, M.J. (1999), "Bayes Factors: What They Are and What They Are Not," *The American Statistician*, 53, 119–122.

Leemis, L.M., and McQueston, J.T. (2008), "Univariate Distribution Relationships," *The American Statistician*, 62, 45–53.

Lehmann, E.L. (1980), "Efficient Likelihood Estimators," *The American Statistician*, 34, 233–235.

Lehmann, E.L. (1983), *Theory of Point Estimation,* Wiley, New York, NY.

Lehmann, E.L. (1986), *Testing Statistical Hypotheses,* 2nd ed., Wiley, New York, NY.

Lehmann, E.L. (1999), *Elements of Large–Sample Theory*, Springer, New York, NY.

Lehmann, E.L., and Casella, G. (1998), *Theory of Point Estimation,* 2nd ed., Springer, New York, NY.

Lehmann, E.L., and Romano, J.P. (2005), *Testing Statistical Hypotheses,* 3rd ed., Springer, New York, NY.

Lehmann, E.L., and Scheffé, H. (1950), "Completeness, Similar Regions, and Unbiased Estimation," *Sankhyā,* 10, 305–340.

Levy, M.S. (1985), "A Note on Nonunique MLE's and Sufficient Statistics", *The American Statistician,* 39, 66.

Liero, H., and Zwanzig, S. (2012), *Introduction to the Theory of Statistical Inference*, CRC Press, Boca Raton, FL.

Lindgren, B.W. (1993), *Statistical Theory,* 4th ed., Chapman & Hall/CRC, Boca Raton, FL.

Lindley, D.V. (1972), *Bayesian Statistics: a Review*, SIAM, Philadelphia, PA.

Lindsey, J.K. (1996), *Parametric Statistical Inference*, Oxford University Press, Oxford, UK.

Lindsey, J.K. (2004), *Introduction to Applied Statistics: a Modelling Approach*, 2nd ed., Oxford University Press, Oxford, UK.

Mahmoud, M.A.W., Sultan, K.S., and Amer, S.M. (2003), "Order Statistics for Inverse Weibull Distributions and Characterizations," *Metron*, LXI, 389–402.

Mann, H. B., and Wald, A. (1943), "On Stochastic Limit and Order Relationships," *The Annals of Mathematical Statistics*, 14, 217–226.

Mann, N.R., Schafer, R.E., and Singpurwalla, N.D. (1974), *Methods for Statistical Analysis of Reliability and Life Data*, Wiley, New York, NY.

Marden, J.I. (2012), *Mathematical Statistics, Old School*, course notes from (http://istics.net/pdfs/mathstat.pdf).

Mardia, K.V., Kent, J.T., and Bibby, J.M. (1979), *Multivariate Analysis,* Academic Press, London, UK.

Marsden, J.E., and Hoffman, M.J. (1993), *Elementary Classical Analysis*, 2nd ed., W.H. Freeman, New York, NY.

Marshall, A.W., and Olkin, I. (2007), *Life Distributions*, Springer, New York, NY.

Massart, P. (1990), "The Tight Constant in the Dvoretzky-Kiefer-Wolfo-Witcz Inequality," *The Annals of Probability*, 3, 1269–1283.

MathSoft (1999a), *S-Plus 2000 User's Guide,* Data Analysis Products Division, MathSoft, Seattle, WA.

MathSoft (1999b), *S-Plus 2000 Guide to Statistics,* Vol. 2, Data Analysis Products Division, MathSoft, Seattle, WA.

McCulloch, R.E. (1988), "Information and the Likelihood Function in Exponential Families," *The American Statistician*, 42, 73–75.

Meeker, W.Q., and Escobar, L.A. (1998), *Statistical Methods for Reliability Data,* Wiley, New York, NY.

Melnick, E.L., and Tenebien, A. (1982), "Misspecifications of the Normal Distribution," *The American Statistician,* 36, 372–373.

Mood, A.M., Graybill, F.A., and Boes, D.C. (1974), *Introduction to the Theory of Statistics,* 3rd ed., McGraw-Hill, New York, NY.

Moore, D.S. (1971), "Maximum Likelihood and Sufficient Statistics," *The American Mathematical Monthly,* 78, 50–52.

Moore, D.S. (2007), *The Basic Practice of Statistics,* 4th ed., W.H. Freeman, New York, NY.

Morris, C.N. (1982), "Natural Exponential Families with Quadratic Variance Functions," *The Annals of Statistics*, 10, 65–80.

Morris, C.N. (1983), "Natural Exponential Families with Quadratic Variance Functions: Statistical Theory," *The Annals of Statistics*, 11, 515–529.

Muirhead, R.J. (1982), *Aspects of Multivariate Statistical Theory,* Wiley, New York, NY.

Mukhopadhyay, N. (2000), *Probability and Statistical Inference,* Marcel Dekker, New York, NY.

Mukhopadhyay, N. (2006), *Introductory Statistical Inference,* Chapman & Hall/CRC, Boca Raton, FL.

Olive, D.J. (2004), "Does the MLE Maximize the Likelihood?" Unpublished Document, see (http://lagrange.math.siu.edu/Olive/infer.htm).

Olive, D.J. (2007), "A Simple Limit Theorem for Exponential Families," Unpublished Document, see (http://lagrange.math.siu.edu/Olive/infer.htm).

Olive, D.J. (2008), "Using Exponential Families in an Inference Course," Unpublished Document, see (http://lagrange.math.siu.edu/Olive/infer.htm).

Olive, D.J. (2008b), *Applied Robust Statistics,* Unpublished Online Text, see (http://lagrange.math.siu.edu/Olive/ol-bookp.htm).

Olive, D.J. (2013), "Asymptotically Optimal Regression Prediction Intervals and Prediction Regions for Multivariate Data," *International Journal of Statistics and Probability*, 2, 90–100.

O'Reilly, F., and Rueda, R. (2007), "Fiducial Inferences for the Truncated Exponential Distribution," *Communications in Statistics: Theory and Methods*, 36, 2207–2212.

Pal, N., and Berry, J.C. (1992), "On Invariance and Maximum Likelihood Estimation," *The American Statistician,* 46, 209–212.

Panjer, H.H. (1969), "On the Decomposition of Moments by Conditional Moments," *The American Statistician,* 23, 170–171.

Parzen, E. (1960), *Modern Probability Theory and Its Applications*, Wiley, New York, NY.

Patel, J.K., Kapadia C.H., and Owen, D.B. (1976), *Handbook of Statistical Distributions,* Marcel Dekker, New York, NY.

Pawitan, Y. (2001), *In All Likelihood: Statistical Modelling and Inference Using Likelihood*, Oxford University Press, Oxford, UK.

Peressini, A.L., Sullivan, F.E., and Uhl, J.J. (1988), *The Mathematics of Nonlinear Programming*, Springer, New York, NY.

Perlman, M.D. (1972), "Maximum Likelihood–an Introduction," *Proceedings of the Sixth Berkeley Symposium on Mathematical Statistics and Probability,* 1, 263–281.

Pewsey, A. (2002), "Large-Sample Inference for the Half-Normal Distribution," *Communications in Statistics: Theory and Methods,* 31, 1045–1054.

Pfanzagl, J. (1968), "A Characterization of the One Parameter Exponential Family by the Existence of Uniformly Most Powerful Tests," *Sankhyā, A,* 30, 147–156.

Pfanzagl, J. (1993), "Sequences of Optimal Unbiased Estimators Need Not be Asymptotically Optimal," *Scandinavian Journal of Statistics,* 20, 73–76.

Pires, A.M., and Amado, C. (2008), "Interval Estimators for a Binomial Proportion: Comparison of Twenty Methods," *REVSTAT-Statistical Journal*, 6, 165–197.

Polansky, A.M. (2011), *Introduction to Statistical Limit Theory*, CRC Press, Boca Raton, FL.

Poor, H.V. (1994), *An Introduction to Signal Detection and Estimation,* 2nd ed., Springer, New York, NY.

Portnoy, S. (1977), "Asymptotic Efficiency of Minimum Variance Unbiased Estimators," *The Annals of Statistics,* 5, 522–529.

Pourahmadi, M. (1995), "Ratio of Successive Probabilities, Moments and Convergence of (Negative) Binomial to Poisson Distribution," Unpublished Manuscript.

Pratt, J.W. (1959), "On a General Concept of 'in Probability'," *The Annals of Mathematical Statistics,* 30, 549–558.

Pratt, J.W. (1968), "A Normal Approximation for Binomial, F, Beta, and Other Common, Related Tail Probabilities, II," *Journal of the American Statistical Association,* 63, 1457–1483.

Press, S.J. (2005), *Applied Multivariate Analysis: Using Bayesian and Frequentist Methods of Inference,* 2nd ed., Dover, New York, NY.

Rahman, M.S., and Gupta, R.P. (1993), "Family of Transformed Chi-Square Distributions," *Communications in Statistics: Theory and Methods,* 22, 135–146.

Rao, C.R. (1965, 1973), *Linear Statistical Inference and Its Applications,* 1st and 2nd ed., Wiley, New York, NY.

Resnick, S. (1999), *A Probability Path*, Birkhäuser, Boston, MA.

Rice, J. (1988), *Mathematical Statistics and Data Analysis,* Wadsworth, Belmont, CA.

Rice, J. (2006), *Mathematical Statistics and Data Analysis,* 3rd ed., Dux-bury, Belmont, CA.

Robert, C.P., and Casella, G. (2010), *Monte Carlo Statistical Methods*, Springer, New York, NY.

Robinson, J. (1988), "Discussion of 'Theoretical Comparison of Bootstrap Confidence Intervals' by P. Hall," *The Annals of Statistics*, 16, 962–965.

Rohatgi, V.K. (1976), *An Introduction to Probability Theory and Mathematical Statistics,* Wiley, New York, NY.

Rohatgi, V.K. (1984), *Statistical Inference,* Wiley, New York, NY.

Rohatgi, V.K., and Ehsanes Saleh, A.K.M.D. (2001), *An Introduction to Probability and Statistics,* 2nd ed., Wiley, NY.

Romano, J.P., and Siegel, A.F. (1986), *Counterexamples in Probability and Statistics,* Wadsworth, Belmont, CA.

Rosenlicht, M. (1985), *Introduction to Analysis*, Dover, New York, NY.

Ross, K.A. (1980), *Elementary Analysis: The Theory of Calculus*, Springer, New York, NY.

Ross, S. (2009), *A First Course in Probability*, 8th ed., Prentice Hall, Upper Saddle River, NJ.

Roussas, G. (1997), *A Course in Mathematical Statistics*, 2nd ed., Academic Press, San Diego, CA.

Rousseeuw, P.J., and Croux, C. (1993), "Alternatives to the Median Absolute Deviation," *Journal of the American Statistical Association,* 88, 1273–1283.

Rudin, W. (1964), *Principles of Mathematical Analysis,* 2nd ed., McGraw Hill, New York, NY.

Sampson, A., and Spencer, B. (1976), "Sufficiency, Minimal Sufficiency, and the Lack Thereof," *The American Statistician,* 30, 34–35.

Sankaran, P.G., and Gupta, R.D. (2005), "A General Class of Distributions: Properties and Applications," *Communications in Statistics: Theory and Methods*, 34, 2089–2096.

Savage, L.J. (1976), "On Rereading R.A. Fisher," *The Annals of Statistics,* 4, 441–500.

Schervish, M.J. (1995), *Theory of Statistics,* Springer, New York, NY.

Schwarz, C.J., and Samanta, M. (1991), "An Inductive Proof of the Sampling Distributions for the MLE's of the Parameters in an Inverse Gaussian Distribution," *The American Statistician,* 45, 223–225.

Scott, W.F. (2007), "On the Asymptotic Distribution of the Likelihood Ratio Statistic," *Communications in Statistics: Theory and Methods*, 36, 273–281.

Searle, S.R. (1982), *Matrix Algebra Useful for Statistics*, Wiley, New York, NY.

Seber, G.A.F., and Lee, A.J. (2003), *Linear Regression Analysis,* 2nd ed., Wiley, New York, NY.

Sen, P.K., and Singer, J.M. (1993), *Large Sample Methods in Statistics: an Introduction with Applications,* Chapman & Hall, New York, NY.

Sen, P.K., Singer, J.M., and Pedrosa De Lima, A.C. (2010), *From Finite Sample to Asymptotic Methods in Statistics,* Cambridge University Press, New York, NY.

Serfling, R.J. (1980), *Approximation Theorems of Mathematical Statistics,* Wiley, New York, NY.

Severini, T.A. (2005), *Elements of Distribution Theory*, Cambridge University Press, New York, NY.

Shao, J. (1989), "The Efficiency and Consistency of Approximations to the Jackknife Variance Estimators," *Journal of the American Statistical Association*, 84, 114–119.

Shao, J. (2003), *Mathematical Statistics,* 2nd ed., Springer, New York, NY.

Silvey, S.D. (1970), *Statistical Inference*, Penguin Books, Baltimore, MD.

Smith, A.F.M, and Gelfand, A.E. (1992), "Bayesian Statistics Without Tears: a Sampling-Resampling Perspective," *The American Statistician,* 46, 84–88.

Solomen, D.L. (1975), "A Note on the Non-Equivalence of the Neyman Pearson and Generalized Likelihood Ratio Tests for Testing a Simple Null Hypothesis Versus a Simple Alternative Hypothesis," *The American Statistician,* 29, 101–102.

Spanos, A. (1999), *Probability Theory and Statistical Inference: Econometric Modeling with Observational Data,* Cambridge University Press, Cambridge, UK.

Spiegel, M.R. (1975), *Probability and Statistics*, Shaum's Outline Series, McGraw-Hill, New York, NY.

Staudte, R.G., and Sheather, S.J. (1990), *Robust Estimation and Testing,* Wiley, New York, NY.

Stein, C. (1981), "Estimation of the Mean of a Multivariate Normal Distribution," *The Annals of Statistics*, 9, 1135–1151.

Stigler, S.M. (1984), "Kruskal's Proof of the Joint Distribution of $\overline{X}$ and s^2," *The American Statistician,* 38, 134–135.

Stigler, S.M. (2007), "The Epic Journey of Maximum Likelihood," *Statistical Science*, 22, 598–620.

Stigler, S.M. (2008), "Karl Pearson's Theoretical Errors and the Advances They Inspired," *Statistical Science*, 23, 261–271.

Sundaram, R.K. (1996), *A First Course in Optimization Theory,* Cambridge University Press, Cambridge, UK.

Swift, M.B. (2009), "Comparison of Confidence Intervals for a Poisson Mean–Further Considerations," *Communications in Statistics: Theory and Methods,* 38, 748–759.

Tucker, A. (1984), *Applied Combinatorics,* 2nd ed., Wiley, New York, NY.

van der Vaart, A.W. (1998), *Asymptotic Statistics*, Cambridge University Press, Cambridge, UK.

Vardeman, S.B. (1992), "What About Other Intervals?," *The American Statistician,* 46, 193–197.

Venables, W.N., and Ripley, B.D. (2010), *Modern Applied Statistics with S*, 4th ed., Springer, New York, NY.

Wackerly, D.D., Mendenhall, W., and Scheaffer, R.L. (2008), *Mathematical Statistics with Applications,* 7th ed., Thomson Brooks/Cole, Belmont, CA.

Wade, W.R. (2000), *Introduction to Analysis*, 2nd ed., Prentice Hall, Upper Saddle River, NJ.

Wald, A. (1949), "Note on the Consistency of the Maximum Likelihood Estimate," *The Annals of Mathematical Statistics*, 20, 595–601.

Walpole, R.E., Myers, R.H., Myers, S.L., and Ye, K. (2006), *Probability & Statistics for Engineers & Scientists,* 8th ed., Prentice Hall, Upper Saddle River, NJ.

Wasserman, L. (2004), *All of Statistics: a Concise Course in Statistical Inference*, Springer, New York, NY.

Welch, B.L. (1937), "The Significance of the Difference Between Two Means When the Population Variances are Unequal," *Biometrika*, 29, 350–362.

Welsh, A.H. (1996), *Aspects of Statistical Inference*, Wiley, New York, NY.

White, H. (1984), *Asymptotic Theory for Econometricians,* Academic Press, San Diego, CA.

Wijsman, R.A. (1973), "On the Attainment of the Cramér-Rao Lower Bound, *The Annals of Statistics,* 1, 538–542.

Yuen, K.K. (1974), "The Two-Sample Trimmed t for Unequal Population Variances," *Biometrika,* 61, 165–170.

Zabell, S.L. (2008), "On Student's 1908 Article 'The Probable Error of a Mean'," *Journal of the American Statistical Association*, 103, 1–7.

Zacks, S. (1971), *Theory of Statistical Inference,* Wiley, New York, NY.

Zehna, P.W. (1966), "Invariance of Maximum Likelihood Estimators," *The Annals of Mathematical Statistics,* 37, 744.

Zehna, P.W. (1991), "On Proving that $\overline{X}$ and S^2 are Independent," *The American Statistician,* 45, 121–122.

Index

D.J. Olive, *Statistical Theory and Inference*, DOI 10.1007/978-3-319-04972-4,